MW00835335

SURFACE ELECTROMYOGRAPHY

IEEE Press
445 Hoes Lane
Piscataway, NJ 08854

IEEE Press Editorial Board
Tariq Samad, *Editor in Chief*

George W. Arnold	Ziaoou Li	Ray Perez
Giancarlo Fortino	Vladimir Lumelsky	Linda Shafer
Dmitry Goldgof	Pui-In Mak	Zidong Wang
Ekram Hossain	Jeffrey Nanzer	MengChu Zhou

Kenneth Moore, *Director of IEEE Book and Information Services (BIS)*

Technical Reviewers

Philip A. Parker, *University of New Brunswick*
Dejan Popović, *University of Belgrade*
Cathi Disselhorst-Klug, *RWTH Aachen University*

SURFACE ELECTROMYOGRAPHY

Physiology, Engineering, and Applications

Edited by

ROBERTO MERLETTI
DARIO FARINA

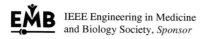

IEEE Engineering in Medicine
and Biology Society, *Sponsor*

IEEE Press Series in Biomedical Engineering
Metin Akay, *Series Editor*

IEEE PRESS

WILEY

Copyright © 2016 by The Institute of Electrical and Electronics Engineers, Inc.

Published by John Wiley & Sons, Inc., Hoboken, New Jersey. All rights reserved
Published simultaneously in Canada

No part of this publication may be reproduced, stored in a retrieval system, or transmitted in any form or by any means, electronic, mechanical, photocopying, recording, scanning, or otherwise, except as permitted under Section 107 or 108 of the 1976 United States Copyright Act, without either the prior written permission of the Publisher, or authorization through payment of the appropriate per-copy fee to the Copyright Clearance Center, Inc., 222 Rosewood Drive, Danvers, MA 01923, (978) 750-8400, fax (978) 750-4470, or on the web at www.copyright.com. Requests to the Publisher for permission should be addressed to the Permissions Department, John Wiley & Sons, Inc., 111 River Street, Hoboken, NJ 07030, (201) 748-6011, fax (201) 748-6008, or online at http://www.wiley.com/go/permission.

Limit of Liability/Disclaimer of Warranty: While the publisher and author have used their best efforts in preparing this book, they make no representations or warranties with respect to the accuracy or completeness of the contents of this book and specifically disclaim any implied warranties of merchantability or fitness for a particular purpose. No warranty may be created or extended by sales representatives or written sales materials. The advice and strategies contained herein may not be suitable for your situation. You should consult with a professional where appropriate. Neither the publisher nor author shall be liable for any loss of profit or any other commercial damages, including but not limited to special, incidental, consequential, or other damages.

For general information on our other products and services or for technical support, please contact our Customer Care Department within the United States at (800) 762-2974, outside the United States at (317) 572-3993 or fax (317) 572-4002.

Wiley also publishes its books in a variety of electronic formats. Some content that appears in print may not be available in electronic formats. For more information about Wiley products, visit our web site at www.wiley.com.

Library of Congress Cataloging-in-Publication Data is available.

ISBN: 978-1-118-98702-5

Printed in the United States of America

10 9 8 7 6 5 4 3 2 1

CONTENTS

v

INTRODUCTION

In 2004, the book *Electromyography: Physiology, Engineering and Noninvasive Applications*, edited by R. Merletti and P. Parker, was published by IEEE Press and Wiley-Interscience. After more than a decade from that publication, the techniques and the equipment adopted in the study of muscles and muscle signals, by means of surface electrodes, underwent major advances. New tools are available for the detection, processing, and interpretation of surface electromyographic (sEMG) signals, new experience and knowledge have been acquired in the field, and new applications are now possible. These advances are related to electrode arrays and "EMG Imaging" techniques, signal amplifiers, signal transmission, EMG decomposition, as well as to many applications of these methodologies.

For many reasons, this work is not a second edition of the 2004 publication but rather a completely new book. First, it focuses only on surface EMG and not on invasive methods. Second, although it still provides the basic background, it emphasizes the new developments on grid recordings and EMG imaging in several applications. In this perspective, some topics discussed in the previous book have been eliminated while new chapters have been added.

The technical progresses in signal sensing, conditioning, processing, and interpretation techniques, however, have not always been exploited in the applied fields. Clinical applications of new methodologies are still lagging, mostly because of insufficient activities in technology transfer and in education/training efforts. The gap between researchers and practitioners widened in the last decade because of the acceleration of research and the inertia of educational and clinical institutions. This issue requires attention by research-supporting agencies at the European and national levels, especially when economical restrictions limit the spread of innovations.

OUTLINE OF THE BOOK

The areas concerning basic physiological and biophysical issues overlap with those of the book published in 2004, although new knowledge and new points of view are presented, especially in Chapters 1 and 3. Recent advanced approaches for signal detection are illustrated in Chapter 3, which also deals with the issue of electrode–skin interface (impedance and noise), while signal processing approaches for single channel EMG are described in Chapter 4.

One of the relevant new developments of the last decade is the technology of two-dimensional EMG (2D-EMG) or EMG imaging, based on electrode grids. A similar technology was developed earlier for EEG to facilitate its interpretation. 2D-EMG (or high-density EMG, HDEMG) provides a wealth of anatomical and physiological data concerning the muscle(s) below the electrode grid, including information about the innervation zone, the recruitment, de-recruitment, discharge rate, and conduction velocity of the detected motor units.

A wide spectrum of new applications in rehabilitation and movement sciences is opened up by this technology, described in Chapter 5, including ergonomics, occupational medicine, posture analysis, obstetrics, and new forms of biofeedback and rehabilitation training, described in the last 11 chapters.

A long-lasting theory in motor control is based on the modular organization of spinal neuronal networks, which can be identified by the analysis of sEMG. Muscular activation patterns, associated with a large number of movements, appear to be based on a limited number of fundamental patterns (basis functions) whose linear combinations produce such movements which are therefore defined by the weights of the linearly combined basis functions. Chapter 6 illustrates this concept and its applications.

The signals obtained from geometrically different viewpoints of the signal sources in the muscle comprise the inputs to unscrambling algorithms designed to identify and separate the contributions of the individual sources. This process is referred to as decomposition of the sEMG and is described in Chapter 7. It provides a window not only on the muscle but also on the control mechanisms and driving signals provided by the spinal cord networks to the muscle(s).

Mathematical modeling of sEMG is an important research and teaching tool for acquiring and transferring knowledge and for answering questions such as "what if . . .?" that cannot be answered by experiments. Testing new signal processing algorithms and defining their performance and limitations is another important application of the models described in Chapter 8.

Surface EMG data experimentally recorded from the major superficial muscle groups have been successfully used as a direct input drive to musculoskeletal models of human limbs. These models were demonstrated to be an effective way to predict muscle dynamics and joint moments in both healthy and pathological subjects and are described in Chapter 9.

Chapter 10 deals with myoelectric manifestations of muscle fatigue, which is one of the earliest fields of application of surface EMG and also the most treacherous. The statement made more than 20 years ago by Professor Carlo J. De Luca (Wartenweiler Memorial Lecture, International Society for Biomechanics, 1993)—"To its detriment, electromyography is too easy to use and consequently too easy to abuse."—is still very much true in this field, despite the advances made in the last decade.

Muscles can be activated voluntarily or by electrical stimulation. In the second case the discharge frequency and the number of motor units are respectively controlled by the frequency and the amplitude of the stimulation pulses, thereby reducing a number of confounding factors and allowing external control of these parameters. Chapter 11 describes the electrical stimulation technique and the muscle features that can be investigated with it.

The chapters that follow deal with clinical applications of the techniques described in the previous 11 chapters. The list is certainly not comprehensive. Chapter 12 deals with neurophysiological investigations, with particular focus on reflex studies, while Chapter 13 addresses the applications in ergonomics and occupational medicine whose social and economic relevance are substantial.

Chapter 14 addresses applications in obstetrics and proctology and the issue of prevention of iatrogenic lesions due to episiotomy, a surgical intervention performed too frequently during child delivery and potentially increasing the likelihood of later fecal incontinence.

The issue of posture analysis is addressed in Chapter 15, which deals with monitoring the activity of the triceps surae during quiet standing. The muscles involved are pinnate, and the information provided by sEMG is different from that provided by muscles with fibers parallel to the skin.

Movement and, in particular, gait analysis is one of the fields with current clinical applications. This topic is addressed in Chapter 16 and is also a treacherous one because, in dynamic situations, the movement of the muscle under the skin causes sEMG alterations which reflect geometrical changes too often wrongly attributed to neurophysiological factors.

Physical therapy is the main field of sEMG application where the technique is used to plan and monitor treatment and assess its effectiveness. Timing, amplitude, and distribution of muscle activities, as well as monitoring of fatigue and of changes due to rehabilitation treatments, are the issues presented in Chapter 17. Chapter 18 expands the issue of sEMG biofeedback, which is gaining interest because of the recent sEMG imaging techniques.

Exercise physiology and sports is another important field of sEMG application. The issues of co-activation, muscle timing, and characterization of exercise are analyzed in Chapter 19 with focus on muscle coordination. Chapter 20 describes the use of surface EMG in man–machine interfacing for rehabilitation technologies. Examples of these applications include active prostheses and orthoses.

Other applications of more limited current clinical relevance, such as in space medicine, yoga relaxation studies, and other fields, are not discussed in this book.

OPEN TECHNICAL AND SCIENTIFIC ISSUES

Despite recent progress, sEMG technology is still developing and relatively far from being perfected. As the number of electrodes increases, the cable connection between the electrodes, the amplifiers, and the computer becomes more problematic because of the high rate of information transfer that is required. The availability of gloves or sleeves with up to a few hundred electrodes covering a limb is very near in the future. The development of thin multi-lead connections between electrodes and amplifiers, with reduced movement artifacts, is a challenge being addressed. A possible solution is the incorporation of battery-powered amplifiers and wireless transmitters into such sleeves.

At the moment, wearable (pocket-size) devices with thin and wide-band fiber-optic connections (up to 50 m long) to a PC are commercially available, but a wireless connection is obviously preferred. In this case the bandwidth limitation is a bottleneck constraining the number of channels. A number of solutions are being considered in research laboratories. Since the signals to be transmitted are highly correlated, compression is a possible solution to reduce the bit rate. This approach has been investigated by researchers who demonstrated that lossless compression can reduce the bit rate by ~60% whereas lossy compression can reduce it by >90% while maintaining a signal-to-noise ratio greater than 20 dB. An alternative approach is the use of a data logger with a removable memory card and wireless transmission of a fraction of the data (for example, 0.1 s every second) to allow the operator to check for signal quality during acquisition.

High-density arrays imply small electrodes whose electrode–skin impedance and noise increase as the contact surface decreases. Skin treatments and paste-less electrode technologies are being investigated to reduce both impedance and noise and increase wearability of electrode arrays. Noninvasive detection of sEMG from deep muscles is still an open problem.

The estimation of force produced by individual muscles and the sharing of the global load among agonists and antagonists muscles, as well as the issue of co-contraction, are still unsolved problems. This issue is still biasing the study of sEMG–force relationship because sEMG is measured from one or few muscles whereas force (or torque at a joint) is produced by many more muscles whose electrical activity, at this time, cannot be entirely detected by noninvasive techniques.

Estimating crosstalk among muscles, and compensating for it, is not a satisfactorily solved problem, although high-density sEMG is a promising tool to address it. The relatively poor repeatability of the surface EMG measures is also an unresolved issue, although also for this problem high-density EMG may be a good approach to the solution.

The absolute values of EMG amplitude often need to be normalized because of confounding factors that should be compensated for. However, despite the many approaches proposed for sEMG normalization, no consensus exists on the most appropriate normalization technique.

Most of the scientific investigations are still limited to isometric conditions that are, however, far from the natural functioning of the muscles in daily-life activities. Extensions of the advanced methods developed for EMG analysis to fully dynamic tasks is challenging and still at a preliminary stage of investigation.

Although methods for decomposing the sEMG into the constituent single motor unit activities have progressed substantially in the last decade, these approaches still have limitations with respect to the number of conditions and muscles that can be analyzed.

EDUCATION, TRAINING, AND STANDARDIZATION IN THE FIELD OF EMG IMAGING

Despite the availability of free teaching material on the web (www.lisin.polito.it, www.seniam.org, among others) and textbooks on the topic, very few Schools of Movement Sciences, Physical Therapy, or Rehabilitation Medicine include either traditional or advanced EMG technology in the training curricula of rehabilitation, sport, and occupational medicine. This fact results in limited user awareness of the potentialities of some of the new EMG tools available from research laboratories. This is a general problem which hinders clinical applications of sEMG, but its relevance is higher for multichannel sEMG because recently developed techniques are removing many of the problems that were pointed out, in the past, as limiting clinical applications of the technique. Clinical application of sEMG is much more limited by lack of dissemination than by technical limitations.

The issue of EMG best practice was addressed by the European Project "Surface Electromyography for Non-Invasive Assessment of Muscles (SENIAM, www .seniam.org)" whose recommendations (2000) are becoming outdated and do not include multichannel sEMG. A strong need is felt for upgrading recommendations to encompass recent advances. Activities in this direction were proposed in a special section of the 2014 Congress of the International Society for Electrophysiology and Kinesiology (ISEK).

FINAL REMARKS

The contributors to this book include only a very small number of senior members of the community of sEMG researchers. We chose them with the aim of merging as smoothly as possible, in the same book, physiology, engineering, and some important applications by providing suggestions and recommendations to the authors. We believe that all contributors did an excellent job in producing a harmonious result that can be appreciated by readers of different backgrounds. Any errors or failings of this work are certainly not attributable to the contributors but are strictly the responsibility of the editors.

ACKNOWLEDGMENTS

The Editors had the privilege of coordinating an excellent team of contributors and are greatly indebted to them for their efforts and results. They have devoted, without compensation, a considerable portion of their time to this endeavor.

The Editors owe a great debt to the reviewers who dedicated time and patience in reading this book. Their comments and criticisms have been of great help, not only for detecting a number of inconsistencies and drawbacks but also for placing the material in a better framework. Mrs. Antonietta Stango and Mr. Domenico Signorile greatly contributed to the preparation of the material for this book.

The book reports and disseminates knowledge that was in large part acquired within the following European and National Projects:

- "Surface Electromyography for Non-Invasive Assessment of Muscles (SENIAM)"
- "Prevention of Neuromuscular Disorders in the Use of Computer Input Devices (PROCID)"
- "Neuromuscular Assessment of the Elderly Worker (NEW)"
- "Decomposition of Multichannel Surface Electromyograms (DEMUSE)"
- "Cybernetic Manufacturing Systems (CyberManS)"
- "On Asymmetry in Sphincters (OASIS)"
- "Technologies for Anal Sphincter Analysis and Incontinence (TASI)"

- "Decoding the Neural Code of Human Movements for a New Generation of Man–Machine Interfaces (DEMOVE)"
- "A Novel Concept for Support to Diagnosis and Remote Management of Tremor (NeuroTREMOR)"

PROFESSOR ROBERTO MERLETTI

Laboratory for Engineering of the
Neuromuscular System, Politecnico di Torino,
Torino, Italy

PROFESSOR DARIO FARINA

Department of Neurorehabilitation Engineering,
University Medical Center, Göttingen, Germany

CONTRIBUTORS

B. Afsharipour
Laboratory for Engineering of the Neuromuscular System, Politecnico di Torino, Torino, Italy

U. Barone
Laboratory for Engineering of the Neuromuscular System, Politecnico di Torino, Torino, Italy

S. Baudry
Laboratory of Applied Biology and Neurophysiology, ULB Neuroscience Institute (UNI), Université Libre de Bruxelles (ULB), Brussels, Belgium

T. F. Besier
Department of Engineering Science and Auckland Bioengineering Institute, University of Auckland, Auckland, New Zealand

G. Boccia
Motor Science Research Center, SUISM University of Turin, Turin, Italy; and CeRiSM Research Center "Sport, Mountain, and Health," Rovereto (TN), Italy

A. Botter
Laboratory for Engineering of the Neuromuscular System (LISiN) Politecnico di Torino, Torino, Italy

I. CAMPANINI
LAM—Motion Analysis Laboratory, Department of Rehabilitation, AUSL of Reggio Emilia, Correggio, Italy

E. A. CLANCY
Electrical & Computer Engineering Department; Biomedical Engineering Department, Worcester Polytechnic Institute, Worcester, Massachusetts

A. D'AVELLA
*Laboratory of Neuromotor Physiology, Santa Lucia Foundation, Rome, Italy
Department of Biomedical and Dental Sciences and Morphofunctional Imaging, University of Messina, Messina, Italy*

J. DIDERIKSEN
Department of Neurorehabilitation Engineering, Universitätsmedizin Göttingen, Georg-August-Universität, Göttingen, Germany

JACQUES DUCHATEAU
Laboratory of Applied Biology and Neurophysiology, ULB Neuroscience Institute (UNI), Université Libre de Bruxelles (ULB), Brussels, Belgium

ROGER M. ENOKA
Department of Integrative Physiology, University of Colorado, Boulder, Colorado

D. FALLA
*Institute for Neurorehabilitation Systems, Bernstein Center for Computational Neuroscience, University Medical Center Göttingen, Georg-August University, Göttingen, Germany
Pain Clinic, Center for Anesthesiology, Emergency and Intensive Care Medicine, University*

D. FARINA
Department of Neurorehabilitation Engineering, Bernstein Focus Neurotechnology Göttingen, Bernstein Center for Computational Neuroscience, University Medical Center Göttingen, Georg-August University, Göttingen, Germany

J. W. FERNANDEZ
Department of Engineering Science and Auckland Bioengineering Institute, University of Auckland, Auckland, New Zealand

A. GALLINA
*Laboratory for Engineering of the Neuromuscular System, Politecnico di Torino, Italy
University of British Columbia—Rehabilitation Science, Vancouver, BC, Canada*

M. GAZZONI
Laboratory for Engineering of the Neuromuscular System, Politecnico di Torino, Torino, Italy

A. HOLOBAR
Faculty of Electrical Engineering and Computer Science, University of Maribor, Maribor, Slovenia

Y. P. IVANENKO
Laboratory of Neuromotor Physiology, Santa Lucia Foundation, Rome, Italy

F. LACQUANITI
Laboratory of Neuromotor Physiology, Santa Lucia Foundation, Rome, Italy
Center of Space Biomedicine, University of Rome "Tor Vergata", Rome, Italy
Department of Systems Medicine, University of Rome "Tor Vergata", Rome, Italy

D. G. LLOYD
Centre for Musculoskeletal Research, Griffith Health Institute, Griffith University, Gold Coast, QLD, Australia

I. D. LORAM
Cognitive Motor Function Research Group, School of Healthcare Science, Manchester Metropolitan University, United Kingdom

M. M. LOWERY
School of Electrical & Electronic Engineering, University College, Dublin, Ireland

R. MERLETTI
Laboratory for Engineering of the Neuromuscular System (LISiN), Politecnico di Torino, Torino, Italy

A. MERLO
LAM—Motion Analysis Laboratory, Department of Rehabilitation, AUSL of Reggio Emilia, Correggio, Italy

M. A. MINETTO
Division of Endocrinology, Diabetology and Metabolism, Department of Medical Sciences, University of Turin, Turin, Italy

T. MORITANI
Professor of Applied Physiology, Graduate School of Human and Environmental Studies, Kyoto University, Sakyo-ku, Kyoto, Japan

F. NEGRO
Department of Neurorehabilitation Engineering, Bernstein Center for Computational Neuroscience, University Medical Center Goettingen, Goettingen, Germany

A. RAINOLDI
Professor of Physiology, Department of Medical Sciences, Motor Science Research Center, SUISM University of Turin, Turin, Italy

M. SARTORI
Department of Neurorehabilitation Engineering, Bernstein Focus Neurotechnology Göttingen, Bernstein Center for Computational Neuroscience, University Medical Center Göttingen, Georg-August University, Göttingen, Germany

D. F. STEGEMAN
Department of Neurology/Clinical Neurophysiology, Radboud University Nijmegen Medical Centre and Donders Institute for Brain, Cognition and Behaviour, The Netherlands

T. M. VIEIRA
Laboratory for Engineering of the Neuromuscular System, Politecnico di Torino, Torino, Italy, and Escola de Educação Física e Desportos, Universidade Federal do Rio de Janeiro, Rio de Janeiro, Brasil

D. ZAZULA
Faculty of Electrical Engineering and Computer Science, University of Maribor, Maribor, Slovenia

1

PHYSIOLOGY OF MUSCLE ACTIVATION AND FORCE GENERATION

R. M. Enoka[1] and J. Duchateau[2]

[1]*Department of Integrative Physiology, University of Colorado, Boulder, Colorado*
[2]*Laboratory of Applied Biology and Neurophysiology, ULB Neuroscience Institute (UNI), Université Libre de Bruxelles (ULB), Brussels, Belgium*

1.1 INTRODUCTION

To extract information about the control of movement by the nervous system from electromyographic (EMG) signals, it is necessary to understand the processes underlying both the generation of the activation signal and the torques exerted by the involved muscles. As a foundation for the subsequent chapters in this book, the goal of this chapter is to describe the physiology of muscle activation and force generation. We discuss the anatomy of the final common pathway from the nervous system to muscle, the electrical properties of motor neurons and muscle fibers, the contractile properties of muscle fibers and motor units, the concept of motor unit types, and the control of muscle force by modulating the recruitment and rate coding of motor unit activity.

1.2 ANATOMY OF A MOTOR UNIT

The basic functional unit of the neuromuscular system is the motor unit. It comprises a motor neuron, including its dendrites and axon, and the muscle fibers innervated by the axon [28]. The motor neuron is located in the ventral horn of the spinal cord or

Surface Electromyography: Physiology, Engineering, and Applications, First Edition.
Edited by Roberto Merletti and Dario Farina.
© 2016 by The Institute of Electrical and Electronics Engineers, Inc. Published 2016 by John Wiley & Sons, Inc.

brain stem where it receives sensory and descending inputs from other parts of the nervous system. The axon of each motor neuron exits the spinal cord through the ventral root, or through a cranial nerve in the brain stem, and projects in a peripheral nerve to its target muscle and the muscle fibers it innervates. Because the generation of an action potential by a motor neuron typically results in the generation of action potentials in all of the muscle fibers belonging to the motor unit, EMG recordings of muscle fiber action potentials provide information about the activation of motor neurons in the spinal cord or brain stem.

1.2.1   Motor Nucleus

The population of motor neurons that innervate a single muscle is known as a motor nucleus or motor neuron pool [51]. The number of motor neurons in a motor nucleus ranges from a few tens to several hundred [40,58] (Table 1.1). The motor neuron pool for each muscle typically extends longitudinally for a few segments of the spinal cord (Fig. 1.1), and at each segmental level the pools for proximal muscles tend to be more ventral and lateral than those for distal muscles and the pools for anterior muscles are more lateral than those for posterior muscles [59]. Nonetheless, the extensive dendritic projections of motor neurons intermingle across motor neuron pools.

1.2.2   Muscle Fibers

The muscle fibers innervated by a single axon are known as the muscle unit (Fig. 1.1), the size of which varies across each motor unit pool. The motor units first recruited

TABLE 1.1   Motor Neuron Locations and Numbers for Selected Forelimb Muscles

Muscle	Spinal Location	Number
Biceps brachii	C5–C7	1051
Triceps brachii	C6–T1	1271
Flexor carpi radialis	C7–C8	235
Extensor carpi radialis	C5–C7	890
Flexor carpi ulnaris	C7–T1	314
Extensor carpi ulnaris	C7–T1	216
Extensor pollicis longus	C8–T1	14
Abductor pollicis longus	C8–T1	126
Flexor digitorum superficialis	C8–T1	306
Extensor digitorum communis	C8–T1	273
Flexor digitorum profundus	C8–T1	475
Extensor digiti secundi proprius	C8–T1	87
Abductor pollicis brevis and flexor pollicis brevis	C8–T1	115
Adductor pollicis	C8–T1	370
First dorsal interosseus	C8–T1	172
Lateral lumbricalis	C8–T1	57

Data are from Jenny and Inukai [58] and are listed as pairs of antagonistic muscles.

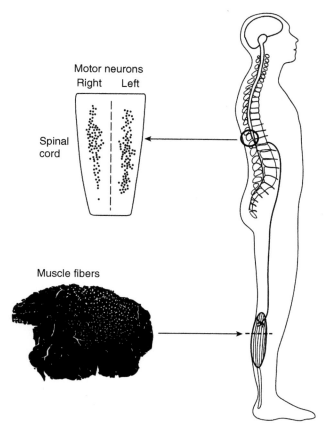

FIGURE 1.1 Muscle force is controlled by a population of motor units (motor unit pool) located in the spinal cord with each motor unit innervating a number of muscle fibers (muscle unit). The muscle fibers belonging to a single muscle unit are indicated by the white dots in the cross-sectional view of the muscle. A typical motor unit pool spans several spinal segments, and muscle units are usually limited to discrete parts of the muscle. Modified from Enoka [30] with permission.

during a voluntary contraction innervate fewer muscle fibers and hence have smaller muscle units than those that are recruited later in the contraction. Most motor units in a muscle have small muscle units and only a few have large muscle units [76,102,107] (Fig. 1.2A). Based on the association between muscle unit size and maximal motor unit force, Enoka and Fuglevand [32] estimated the innervation numbers (muscle unit size) for the 120 motor units in a human hand muscle (first dorsal interosseus) ranged from 21 to 1770 (Fig. 1.2B). Similar relations likely exist for most muscles [51]. Due to the exponential distribution of innervation number across a motor unit pool, it is necessary to distinguish between the number of motor unit action potentials discharged from the spinal cord and the number of muscle fiber action potentials recorded in the muscle with EMG electrodes. This distinction is indicated with the

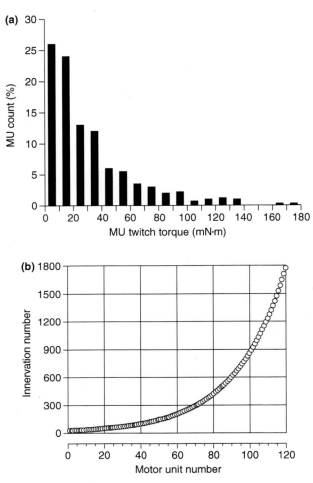

FIGURE 1.2   Variation in muscle unit size across the motor unit pool. (**a**) Distribution of motor unit (MU) twitch torques for 528 motor units in the tibialis anterior muscle of 10 subjects [107]. (**b**) Estimated distribution of innervation numbers across the 120 motor units comprising the first dorsal interosseus muscle [32].

term "neural drive" to denote the motor unit action potentials and "muscle activation" to indicate the muscle fiber action potentials [26,31,36].

The fibers in each muscle unit are located in a subvolume of the muscle and intermingle with the fibers of other muscle units (Fig. 1.1). The spatial distribution of the fibers belonging to a muscle unit is referred to as the motor unit territory. Counts of muscle unit fibers indicate that motor unit territories can occupy from 10% to 70% of the cross-sectional area of a muscle and that the density of muscle unit fibers ranges from 3 to 20 per 100 muscle fibers [51]. Moreover, the fibers of a single muscle unit often do not extend from one end of the muscle to the other, but instead terminate within a muscle fascicle [48,108]. As a consequence of muscle unit anatomy, the forces generated by individual muscle fibers must be transmitted through various layers of connective tissues before reaching the skeleton and contributing to the movement. Such interactions attenuate the unique contribution of individual fibers to the net muscle force during a movement and thereby reduce the influence of differences in contractile properties among muscle fibers.

1.3 MOTOR NEURON

The motor unit is classically considered to be the final common pathway in that sensory and descending inputs converge onto a single neuron that discharges an activation signal to the muscle fibers it innervates [28]. The motor neuron has extensive dendritic branches that receive up to 50,000 synaptic contacts with each contact capable of eliciting inward or outward currents across the membrane and thereby generate an excitatory or inhibitory postsynaptic potential. The inputs are integrated and will generate an action potential in the trigger zone (axon hillock) when the change in membrane potential exceeds voltage threshold (Fig. 1.3). Motor neurons have four main types of receptors and ion channels that produce the responses to the synaptic inputs [50]:

1. *Leak Channels.* These primarily pass an outward K current and are largely responsible for establishing the resting membrane potential, which is approximately −70 mV in motor neurons.
2. *Voltage-Gated Channels.* These receptors are activated by a change in the membrane potential, such as activation of Na, K, and Ca channels by depolarization of the membrane. Na currents are the key elements in the generation of action potentials, and Ca-activated K channels are responsible for the afterhyperpolarization phase of the action potential.
3. *Ionotropic Synaptic Channels.* These are ligand-gated receptors that bind neurotransmitters and pass currents that produce excitatory or inhibitory postsynaptic potentials. Excitatory currents that depolarize the membrane potential are mainly produced by glutamate-gated receptors, whereas inhibitory currents that hyperpolarize the membrane typically involve either glycine- or GABA-gated receptors.
4. *Neuromodulatory Receptors.* Once a neurotransmitter binds to these receptors, they activate intracellular second messenger pathways that can modulate the function of leak, voltage-gated, and ionotropic channels. Neuromodulatory receptors, therefore, control motor neuron excitability by modulating its responsiveness to ionotropic input. Two neurotransmitters with potent neuromodulatory effects on motor neuron excitability are serotonin and noradrenaline.

The change in motor neuron membrane potential in response to the synaptic inputs it receives depends on the electrical interaction among its ion channels. These interactions can be characterized with Ohm's law: $V = I/g$, where V = potential difference across a

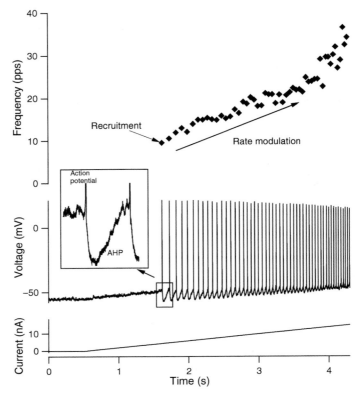

FIGURE 1.3   Relation between the current received by a motor neuron, the rate at which it discharges action potentials, and the force exerted by the muscle unit. (**Bottom trace**) The current injected into a motor neuron with a microelectrode. (**Middle trace**) The change in membrane potential (voltage) of the motor neuron in response to the progressive increase in injected current. When the change in membrane potential exceeds voltage threshold, the motor neuron is activated (recruitment threshold) and begins discharging action potentials. The inset shows the membrane potential trajectory between action potentials, which have been truncated to emphasize the afterhyperpolarization (AHP) phase. (**Upper trace**) Plot of the instantaneous discharge rate (pps = pulses per second) in response to the increase in current and the corresponding increase in the force produced by the muscle unit. Modified from Heckman and Enoka [50] with permission.

patch of the cell membrane, I = membrane current density (current per unit area), and g = input conductance (inverse of resistance) per unit area. The change in motor neuron membrane potential in response to a synaptic current varies with input conductance, which is largely determined by the number and size of its dendrites. Small motor neurons have the least extensive network of dendrites (low input conductance) and therefore experience the greatest change in membrane potential in response to synaptic current [50]. With similar values for voltage threshold among all motor neurons, the change in membrane potential in the trigger zone (axon hillock) will exceed the voltage threshold with the least amount of synaptic current in small motor neurons, and therefore these neurons will be recruited first with progressively increasing synaptic input. Moreover, the rate at which action potentials are discharged by a motor neuron (rate coding) after it has been recruited (recruitment threshold) increases in direct proportion to the current it receives (Fig. 1.3).

The dendrites account for 95% of the surface area of a motor neuron and ~95% of the synaptic contacts occur on the dendrites. Ionotropic inputs to the motor neuron arise from the cerebral cortex, brain stem, and peripheral sensory receptors, but are transmitted via interneurons to the motor neurons. The key descending pathways that contribute to the control of movement include the corticospinal, rubrospinal, vestibulospinal, and reticulospinal tracts. The distribution of synaptic input within the dendrites of motor neurons is known for only a few systems [93]. Although excitatory input is generally greatest in the largest motor neurons and least in the smallest, recruitment order still proceeds in the order from smallest to largest motor neurons when combined with the intrinsic properties of motor neurons. As an exception to this pattern of input distribution, however, the excitation arising from the length detector in muscle—the muscle spindle—is greatest in the smallest motor neurons. In contrast, the inhibitory inputs studied to date often seem to generate approximately equal currents in all motor neurons.

Transmission of the postsynaptic potentials elicited by the synaptic currents to the trigger zone was initially assumed to occur passively by electrotonic conduction. More recent evidence, however, indicates that the postsynaptic potentials in the dendrites of motor neurons are augmented by the modulation of voltage-sensitive channels [51]. For example, motor neurons contain ~1000 to 1500 synapses with receptors that bind serotonin or noradrenaline and are capable of amplifying and prolonging synaptic inputs with persistent inward Ca and Na currents. Due to the profound influence of the persistent inward currents on the gain of relation between synaptic input and discharge rate [51], the neuromodulatory input is considered critical in defining the excitability of the motor neuron pool, including recruitment threshold [50]. Neuromodulatory input, for example, can amplify ionotropic input by as much as fivefold [8,49,55], which can saturate the capacity of the motor neuron to respond to further increases in synaptic input [56,68]. The function of these monosynaptic projections from the brain stem is likely to enhance motor neuron excitability during motor activity (serotonin) and conditions that require modulation of physiological arousal (noradrenaline).

1.4 MUSCLE UNIT

The muscle unit comprises the muscle fibers innervated by a single motor neuron and corresponds to the peripheral element of the motor unit. In a healthy person, the discharge of an action potential by a motor neuron invariably results in the activation of the muscle fibers it innervates. The force produced by a muscle unit largely depends on its innervation number, which varies exponentially within the motor unit pool (Fig. 1.2B), and average innervation numbers differ across muscles (Table 1.1). The ability to grade force precisely during weak contractions likely depends on the size and number of small muscle units in the involved muscles. In contrast, the largest muscle units may only be engaged during rapid or powerful contractions.

1.4.1   Muscle Fiber Action Potentials

As nerve–muscle synapses typically provide secure transmission between a motor neuron and its muscle fibers, axonal action potentials invariably generate end-plate potentials that exceed voltage threshold and produce muscle fiber action potentials that engage the contractile proteins. Despite both the electrical signal (EMG) and the contractile activity (muscle fiber force) originating from the convolution of the neural drive to the muscle, the summation of the two signals diverges due to differences in the shape and sensitivity of each basic element. Because each action potential comprises positive and negative phases, the summation of multiple action potentials is algebraic. In contrast, the twitch, which is the force response of muscle to a single action potential, has only a positive phase and the summation of multiple twitches is only positive. Moreover, the shapes of the action potential and twitch are affected differently by changes in the physiological state of the neuromuscular system.

A muscle fiber is electrically similar to a large-diameter, unmyelinated axon and requires high transmembrane currents to propagate the action potential along the sarcolemma. The transmembrane current density associated with the action potential is proportional to the second derivative in the spatial domain of the potential recorded in extracellular space [97]. The currents underlying the propagation of action potentials along multiple muscle fibers sum to generate extracellular field potentials that can be readily detected with appropriate electrodes. The shapes of the recorded potentials depend on the properties and location of the electrodes and on the anatomy and physiology of the muscle fibers and associated tissues [37].

Each muscle fiber action potential begins at the nerve–muscle synapse and propagates in both directions toward the ends of the muscle fiber [37,48,78]. Although the speed at which the action potential propagates along the muscle fiber, which is referred to as its conduction velocity, depends on the diameter of muscle fiber [9,81], it is modulated by such factors as changes in muscle length [104], skin temperature [34], and extracellular concentrations of metabolites [46]. Nonetheless, there is a statistically significant, albeit moderate, association between the contractile properties and conduction velocity of motor units in tibialis anterior [2].

Although the presence of a muscle fiber action potential provides an index of muscle activation, the actual interaction of the contractile proteins to produce the force depends on the controlled release and reuptake of Ca^{2+} from the sarcoplasmic reticulum to the sarcoplasm and the level of phosphorylation of myosin light chains. When Ca^{2+} kinetics are compromised, such as during some types of fatiguing contractions [92], the association between the number of muscle fiber action potentials and muscle force is disrupted. As an example of the magnitude of the dissociation between muscle activation (number of muscle fiber action potentials) and force due to dysfunction of excitation-contraction coupling and other impairments, Dideriksen et al. [25] compared the simulated relations between EMG amplitude and

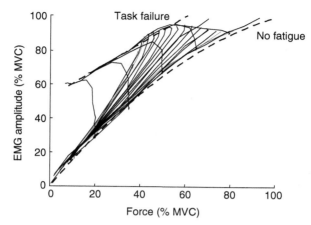

FIGURE 1.4 The relation between the amplitude of the surface EMG signal and muscle force during three simulated protocols that involved fatiguing contractions. The thin lines denote the simulated relations for the different fatigue protocols, the details of which are described in Dideriksen et al. [25]. The observed associations between EMG amplitude and force were bounded by the absence of fatigue (*lower dashed line*) and the relation when the simulated contractions were continued beyond task failure for sustained submaximal contractions (*upper dashed line*). The results indicate that EMG amplitude was not uniquely related to muscle force during the simulated fatiguing contractions. Modified from Dideriksen et al. [25] with permission.

muscle force after three fatigue protocols. The associations between EMG amplitude and muscle force across the three protocols were bounded by the absence of fatigue and the adjustments observed when the simulated contractions were sustained longer than task failure (Fig. 1.4). Fatiguing contractions, therefore, resulted in the same muscle force being associated with EMG amplitudes that differed by up to 25% of the MVC value.

Conversely, the peak force achieved during a muscle twitch in response to a single electrical stimulus is transiently increased immediately after a maximal voluntary contraction [99]. The increase in twitch force, referred to as post-activation potentiation, is obtained without any change in the size of the compound muscle action potential [3]. Therefore, as with the adjustments during fatiguing contractions, the force exerted by a muscle is not directly related to the amplitude of the activation signal (number of muscle fiber action potentials) when the muscle is in a state of post-activation potentiation [4].

1.4.2 Muscle Unit Force

The maximal force capacity of a muscle unit depends on the average cross-sectional area of the muscle fibers (μm^2), the specific force of the fibers ($mN/\mu m^2$), and the innervation number. Of these three factors, the most significant is the number of fibers in the muscle unit [10,63,103]. Therefore, the weakest muscle units have the lowest innervation numbers, whereas the strongest muscle units comprise the greatest number of muscle fibers. Consequently, adaptations in motor unit force can be associated with changes in innervation number. For example, the greater tetanic force of the weakest motor units in the medial gastrocnemius muscle of old rats was associated with an increase in innervation number and a decrease in the number of strong motor units compared with middle-aged rats [60,62]. Nonetheless, both the cross-sectional area and specific force of muscle fibers [13] can change in response to interventions that modulate physical activity level and thereby contribute to adaptations in the peak tetanic force of motor units [60,62,91].

Many activities of daily living, however, are limited by the capacity of motor units to produce power rather than tetanic force. Because power is the product of force and velocity, muscle unit function is also modulated by differences in maximal shortening velocity, which varies with the dominant myosin heavy chain (MHC) isoform in the muscle fibers that comprise the muscle unit. Adult human muscle fibers can express three types of MHC isoforms (types 1, 2A, and 2X) that can also be combined in two types of hybrid fibers (types 1–2A and 2AX) [13]. Type 1 fibers have the slowest unloaded shortening velocity and type 2X fibers the fastest, mainly due to differences in the time that ADP is bound to myosin during the power stroke of the crossbridge cycle [14]. Due to differences in tetanic force and unloaded shortening velocity, the power production capacity of muscle fibers is least in type 1 fibers and greatest in type 2X fibers, although there is considerable overlap between the different type 2 fibers (Fig. 1.5).

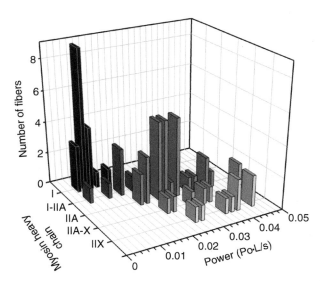

FIGURE 1.5 Maximal power production for 67 skinned fibers from the vastus lateralis muscle of humans. The fibers were classified on the basis of the three myosin heavy chain isoform (I, IIA, and IIX) or combinations of isoforms (I–IIA, IIA–IIX). Despite differences in the average values for the different types of fibers, there was considerable overlap in the distributions across fiber types. Data from Bottinelli et al. [11].

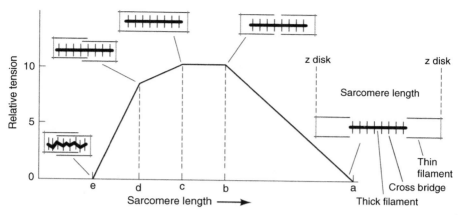

FIGURE 1.6 Length–tension relation for a sarcomere. The force (tension) that the contractile proteins in a sarcomere can exert varies with the amount of overlap between the thick and thin filaments. The graph shows the force generated at five different sarcomere lengths with varying amounts of overlap. The force is least at the shortest and longest sarcomere lengths, and greatest at intermediate lengths. The force exerted by a muscle fiber exhibits a similar relation with length. The force–length relation of the sarcomere and muscle fiber is characterized as comprising an ascending limb, a plateau, and a descending limb. Modified from Enoka and Pearson [33].

Because most muscle actions rarely involve maximal contractions, the forces contributed by individual muscle units to a movement depends on the rate at which the motor neurons discharge action potentials and the length and rate of change in length of the muscle fibers [70]. For a given discharge rate, the force developed by a muscle unit will depend on the length and rate of change in length of its muscle fibers. The variation in muscle unit force as a function of length is attributable to the overlapping structure of thick and thin filaments in the sarcomere and the availability of cross-bridge attachments. Fewer attachment sites are available at both long and short muscle fiber lengths and force peaks at an intermediate length when the number of cross-bridges is maximal (Fig. 1.6). However, the length at which peak muscle unit force is achieved becomes longer with a decrease in activation rate. The relation is similar for all muscle units because this property depends on sarcomere dynamics alone [42]. Based on measurements of sarcomere length, muscle units usually, but not always, function on the ascending and plateau regions of the length–force relation [70]. However, the length–force relation of individual sarcomeres, and therefore muscle units, is modulated by the contractile history of the sarcomere [54].

Maximal muscle unit force also varies as a function of the rate of change in muscle length, decreasing as shortening speed increases and increasing as the activated muscle unit is lengthened (Fig. 1.7). The decrease in maximal force with an increase in shortening velocity is attributable to crossbridges missing some available binding sites and a reduction in the average work performed per crossbridge cycle. The increase in force produced by a muscle fiber during a lengthening contraction is caused by an increase in the amount of Ca^{2+} released by the sarcoplasmic reticulum in

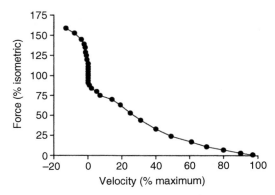

FIGURE 1.7   The force–velocity relation of muscle fibers. The maximal force that a muscle fiber can exert varies with the rate of change and the direction of its change in length. Maximal force declines nonlinearly with an increase in the velocity of shortening contractions (right-hand side of the graph), increases when an external load lengthens the activated muscle fiber (left-hand side of the graph), and has an intermediate value the muscle fiber performs an isometric contraction and does not change length (zero velocity). The maximal rate of shortening (V_{max}) occurs when the muscle fiber acted against a zero load. Data from Edman [29].

response of muscle fiber action potentials, stretching of less completely activated sarcomeres, more rapid cycling of crossbridges, and more work performed by each crossbridge [18,71]. The shape of the force–velocity relation differs among muscle units based on differences in contraction speed; the decline in shortening velocity occurs more rapidly in slower contracting muscle units [89]. Due to differences in contraction speed, slower contracting muscle units produce maximal power at lower contraction velocities compared with faster contracting muscle units [90].

The contribution of a muscle unit to the force exerted by a muscle on the skeleton is also modulated by the lateral transmission of its force throughout associated connective tissue structures [101]. In the longest muscle in the human body, for example, Harris et al. [48] found that only approximately one-third of the motor unit action potentials were recorded at both the proximal and distal electrodes over the sartorius muscle, which indicated that only the muscle unit fibers of these motor units extended the length of the muscle. The muscle fibers of the other motor units either began or ended within a fascicle and therefore the force generated by these muscle fibers was transmitted laterally via various connective tissues. Similarly, muscle unit territories in medial gastrocnemius occupy approximately 10% of the length of the muscle [108]. The lateral transmission of muscle unit force due to the staggered arrangement of muscle fibers reduces differences in the contractile properties of neighboring fibers, such as those attributed to differences in the myosin heavy chain isoforms.

1.4.3   Motor Unit Types

Because the contractile properties of the motor units in a population vary over a fivefold range for twitch contraction time and a 100-fold range for peak tetanic

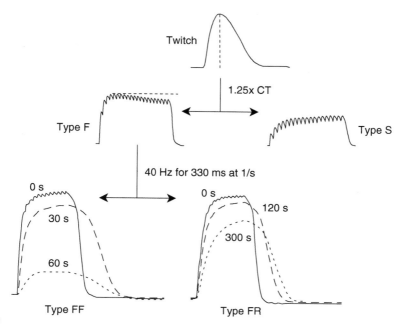

FIGURE 1.8 Burke protocol for classifying motor units in the cat medial gastrocnemius muscle as either type S, FR, or FF [12]. The protocol comprised three measurements: (1) time to peak force (contraction time, CT) for a twitch; (2) profile of a submaximal tetanus (stimulation rate = 1.25 × CT) to determine whether or not it exhibited sag; and (3) fatigability in response trains of electrical stimuli. Those motor units that exhibited sag were classified as type F units and then were characterized as being fatigue-sensitive (type FF) or fatigue-resistant (type FR) based on the decline in tetanic force during the fatigue test. Reprinted from Enoka [31] with permission.

force [51], the function of the slowest and weakest motor units is often contrasted with that of the fastest and strongest motor units. Such comparisons typically involve classifying motor units into discrete categories based on the contractile properties of the muscle units. One scheme identifies three types of motor units based on the profile of a submaximal tetanus and a measure of fatigability [12]. The first test in classifying a motor unit involves stimulating the motor axon at a rate of 1.25× the twitch contraction time of the motor unit for 0.5–2.0 s and determining whether the force increases progressively or it sags (Fig. 1.8). In this scheme, fast units (type F) exhibit sag and slow units (type S) do not. The second test subjects the type F units to repeated trains of electric stimuli and quantifies the decline in force during the submaximal tetani. Those units that experience the greatest decrease in force are classified as fatigue-sensitive (type FF), whereas lesser reductions in force result in the unit being classified as fatigue-resistant (type FR) (Fig. 1.8).

Contrary to common usage, the Burke scheme does not distinguish fast- and slow-twitch motor units, but rather fast and slow motor units based on the shape of a submaximal tetanus. Although many investigators do interpret muscle biopsy data in terms of fast and slow muscle fibers, such measurements are based on the biochemical (e.g., myosin ATPase) and molecular (e.g., myosin heavy chain isoforms) properties of the muscle fibers and not the physiological measurement of motor unit contraction speed. Indeed, twitch contraction times in both animals [12] and humans [107] are distributed continuously from slow to fast and do not comprise discrete categories. Nonetheless, even discrimination of fast and slow motor units with the sag response, which is caused by a transient reduction in the duration of the contractile state, is not a reliable measure [15,95]. Not surprisingly, it is not possible to distinguish motor unit types in human muscles based on the approach proposed by Burke and colleagues [12]. Due to this limitation, human motor units are most appropriately compared based on the force at which the motor unit begins discharging action potentials repetitively during a slow increase in muscle force during an isometric

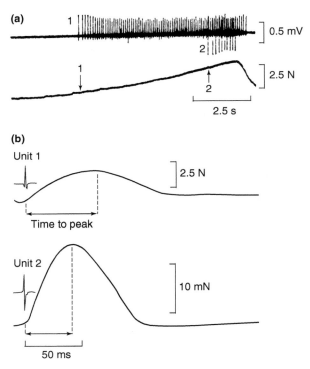

FIGURE 1.9   Association between recruitment threshold and contractile properties for two motor units in a hand muscle (first dorsal interosseus). (a) The subject gradually increased the abduction force exerted by the index finger during an isometric contraction, and the action potentials discharged by two motor units were recorded with an intramuscular electrode. The force at which each motor unit began discharging action potentials is referred to as it recruitment threshold. (b) The spike-triggered average forces for two motor units (1 and 2). The time to peak force was 65 ms for Unit 1 and 39 ms for Unit 2, and the peak forces were 2.9 and 14.7 mN, respectively. Adapted from Desmedt and Godaux [21].

contraction (Fig. 1.9). This property is referred to as recruitment threshold and indicates the relative location of a motor unit in a population [77].

The other unfortunate aspect of Burke's scheme is the misinterpretation of the proposed index of fatigability. The fatigue test used by Burke et al. [12] comprised trains of electric stimuli applied to a motor axon at a frequency that minimizes failure of muscle fiber activation and thereby stresses the physiological processes responsible for force generation—that is, those processes distal to the muscle fiber action potential. In the performance of voluntary actions, however, adjustments in motor unit activity (recruitment threshold, discharge rate, discharge variability) depend more on the relative duration of activity rather than a classification of being either fatigue-sensitive or fatigue-resistant [35]. Therefore, in addition to human motor units not being distinguishable as either fast or slow, the fatigability of motor units cannot be characterized solely on the basis of the contractile properties of the muscle units. Rather, each functional property of motor units is distributed along a continuum and should be considered in such a framework [31,66].

1.5 RECRUITMENT AND RATE CODING

The magnitude and direction of the force exerted by a muscle depends on the number of motor units that are activated (recruitment) and the rates at which they discharge action potentials (rate coding) [28]. Over most of the operating range of a muscle, both processes occur concurrently, although recruitment dominates at low forces and rate coding is more significant at high forces and during fast contractions.

1.5.1 Orderly Recruitment

Early on in the study of motor unit function, it was noted that motor units tend to be recruited in a relatively fixed order during voluntary contractions [28]. The orderly recruitment of motor units provided the foundation for the size principle in which small motor neurons with weak muscle units and relatively long contraction times were recruited before progressively larger motor neurons with stronger muscle units and faster contraction times [53] (Fig. 1.9). The orderly recruitment of motor units is largely due to differences in motor neuron size (soma and dendrites) as indicated by the smallest motor neuron requiring the least current to reach voltage threshold [50]. In human motor units, the size principle is manifested as a positive relation between twitch force and recruitment threshold (Fig. 1.10).

The upper limit of motor unit recruitment varies across muscles in the range of 50% to 90% MVC force [51]. Due to the exponential distribution of recruitment thresholds across a motor unit pool, most motor units are recruited at relatively low forces and progressively fewer are recruited at moderate-to-high forces [44,107]. Therefore, high forces largely depend on the capacity of motor neurons to achieve high discharge rates, which may be limited by the modulation of motor neuron excitability by neuromodulatory input.

Due to recruitment order being defined by an intrinsic motor neuron property, it is relatively insensitive to changes in the source of the synaptic input [22,23,27].

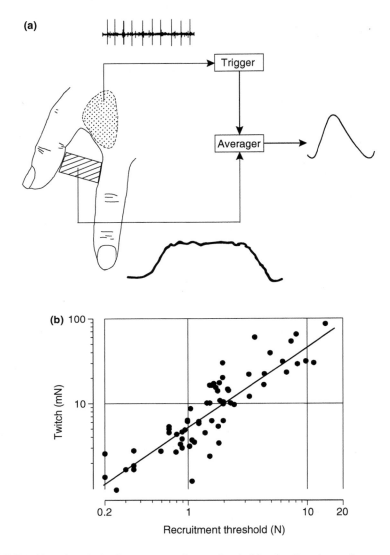

FIGURE 1.10 The relation between recruitment threshold and spike-triggered average force. (a) The experimental arrangement to estimate the contractile properties of motor units during voluntary contractions with a hand muscle (first dorsal interosseus). The isolated action potential of a motor unit was used to trigger an averaging device that accumulated a running average from the net force exerted by the muscle. Each time an action potential was detected, a 100- to 200-ms epoch of the force signal was added to an accumulating average until several hundred events had been detected and the force contributed by the single motor unit to the net force could be identified. Adapted from McComas [75]. (b) Relation between spike-triggered average force and recruitment threshold ($r^2 = 0.582$) for motor units in first dorsal interosseus. Data from Milner-Brown et al. [77].

Nonetheless, there are at least two conditions in which recruitment order can change. One condition in which there can be some variability in recruitment order is when there is change in the action to which the muscle contributes, such as a flexion or abduction force [24]. Similarly, recruitment order seems to be more variable in muscles that have extensive attachments onto the skeleton, and the direction of the net force vector can change [109]. The second condition involves the influence of afferent feedback on recruitment order. Manipulation of cutaneous feedback, for example, can have a substantial influence of the recruitment order of motor units in a hand muscle [61], but this may be related to the critical role of cutaneous feedback in the manipulation of objects by the hand. In general, therefore, recruitment order is typically consistent during voluntary actions, although there can be some variability among motor units with similar recruitment thresholds [51].

In contrast, the activation order of motor units is random when the muscle is activated artificially with brief pulses (0.1–1.0 ms) of electrical stimulation. When the electrodes through which the current is applied are placed over the motor point of a muscle (the region of greatest excitability), the stimulus evokes action potentials in motor axons and motor unit action potentials are elicited in response to the release of neurotransmitter at the nerve–muscle synapse [57]. Therefore, the motor units activated by a specific current depends on a combination of factors that includes motor axon diameter and the location of the axon and its branches relative to the stimulating electrodes. As a consequence, electrical stimulation activates motor units in a random order and not in order of increasing motor unit size as occurs during voluntary contractions [41]. In contrast, longer pulses (e.g., 1 ms) of electrical stimulation delivered through electrodes placed over the muscle nerve more selectively elicit action potentials in Ia fibers originating from muscle spindles and recruitment order is more similar to that observed during voluntary contractions [6,17].

1.5.2 Rate Coding

As defined by the force–frequency relation (Fig. 1.11), the force contributed by a motor unit to an action depends on the rate at which the motor neuron discharges action potentials. The discharge rate achieved by a motor neuron is a consequence of (a) the quality and quantity of ionotropic and neuromodulatory input it receives and (b) time-dependent processes that modulate its responsiveness to the input. The modulation of discharge rate, referred to as rate coding, varies substantially across motor tasks.

Rate coding is typically quantified during submaximal isometric contractions with slow changes in the applied force. Such actions are accomplished by concurrent changes in recruitment and rate coding, with earlier recruited motor units achieving greater discharge rates [88,110]. As the target force approaches the maximal MVC value, however, later recruited motor units discharge action potentials at greater rates than lower threshold motor units [79,85]. The combination of recruitment and rate coding that increases force during a submaximal isometric contraction differs from that controlling the decrease in force. Although discharge rate declines to lower values

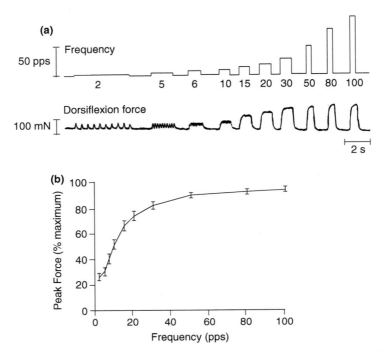

FIGURE 1.11 Relation between stimulus frequency and the force exerted by single motor units located in muscles that extend the toes and flex the ankle. (**a**) Motor units activated by intraneural stimulation (*upper trace*) evoked a dorsiflexion force (*lower trace*). (**b**) Relation between stimulus frequency and normalized force (mean ± SE) for 13 motor units. Modified from Macefield et al. [73] with permission.

during the decrease in force, the force at which a motor unit is derecruited relative to its recruitment threshold is variable [51].

The peak discharge rates observed during slow changes in the force exerted during isometric contractions typically reaches 30 to 50 pps (action potentials per second), which is less than the plateau of the force–frequency relation recorded electrical stimulation of single motor axons (Fig. 1.11). In contrast, rapid increases in force during an isometric contraction involve discharge rates of 60–120 pps [23,105]. Moreover, several weeks of training with rapid contractions substantially increased the average discharge rate of motor units in tibialis anterior from ~75 to 105 pps and the rate of torque development by 82% during ballistic contractions [106]. The capacity of motor neurons to discharge action potentials at rates on the plateau of the force–frequency relation during rapid contractions suggests that rate coding is limited during slow changes in force during isometric contractions.

When an action involves an angular displacement about a joint, rate coding in the involved motor units differs with the direction of the change in length of the active muscles. Due to the greater force capacity of muscle during lengthening contractions [1], less motor unit activity is required to lower an inertial load than to lift it, and discharge rates are typically lower [27]. Shortening contractions involve greater modulation of discharge rate and the recruitment of additional motor units that are subsequently derecruited during the lengthening contraction [87]. Similar relations are observed whether the action involves lifting an inertial load, resisting a torque motor, or exerting a force against an elastic load [51]. In contrast to the parallel changes in discharge rate for motor units recruited during brief isometric and anisometric contractions, adjustments in discharge rate can differ among motor units during submaximal isometric contractions sustained at forces less than the upper limit of motor unit recruitment. Motor units recruited from the onset of such a contraction exhibit a decrease in discharge rate, whereas newly recruited motor units experience first an increase in discharge rate and then a subsequent decline [16]. The concurrent decline in discharge rate of active motor units and recruitment of new motor units indicates that the progressive increase in neural drive to the motor unit pool was superimposed on adjustments in intrinsic motor neuron properties and changes in afferent feedback [16,45]. Moreover, the adjustments in rate coding vary with the compliance of the load against which the limb acts, being greater for high-compliance loads [80,98]. Despite the increase in descending drive to spinal motor neurons during a sustained submaximal contraction, some motor units can stop discharging action potentials and resume activity later in the contraction [5,86,96]. Such adjustments presumably involve changes in the balance between ionotropic input and neuromodulation of motor neuron properties rather than a deliberate recruitment strategy [51].

1.5.3 Discharge Rate Patterns

The discharge characteristics of motor units are often used to infer details about the synaptic input and the responsiveness of motor neurons. The discharge times of motor units, for example, can provide information about the relative amount of common input received by motor neurons, the time course of the afterhyperpolarization phase of the action potential, and the relative state of excitability of the motor neurons [38].

Time- and frequency-domain procedures can be used to estimate the amount of common input from the degree of correlation in the discharge times of action potentials by two or more motor neurons [19,20,38,74]. The underlying assumption with this approach is that the strength of the correlation in the sets of discharge times is proportional to the relative amount of common input received by the motor neurons [65]. By varying the bandwidth of the filter used to characterize the discharge times, the results of different approaches have been interpreted to indicate distinct features of the underlying physiological processes that produce the correlated activity. One measure, for example, represents the discharge times for pairs of motor units as rectangular 1-ms pulses and constructs a cross-correlation histogram to denote the amount of correlated activity as short-term motor unit synchronization. A peak in the histogram indicates a greater-than-chance number of coincidental discharge times by the pair of motor units due to either (a) branched common input to the motor neurons or (b) presynaptic synchronization in efferent axons from the motor cortex [100]. The strength of short-term synchronization is relatively weak, but varies across some conditions [51]. Another measure of correlation filters the discharge times with a Hanning window (e.g., 150 or

400 ms) and quantifies the peak in the cross-correlation function at zero lag as a measure of common drive [20]. Common drive is purported to provide an index of descending input to a motor nucleus, but likely originates somewhere other than the motor cortex [39] and is not shared by motor units in co-activated agonist and antagonist muscles [69]. The amount of common drive is associated with the fluctuations in force during low-force steady contractions ([84] and Fig. 1.12).

The capacity of such indexes to quantify common synaptic input has at least two limitations. First, due to the nonlinear association between the amount of correlation in

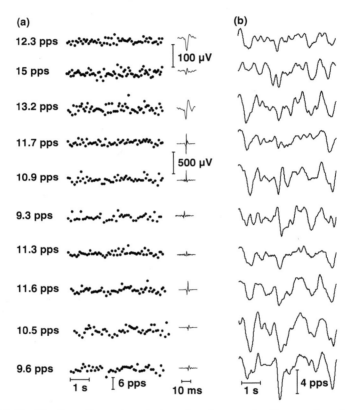

FIGURE 1.12 The force fluctuations during a low-force contraction are strongly associated with the first principal component of motor unit discharge rates in a hand muscle. (a) Instantaneous discharge rates (pulses per second; pps) for 10 motor units in the abductor digiti minimi muscle during an isometric contraction at 5% MVC. The 100-μV calibration refers to the first three motor units, which were detected in surface recordings, and the 500-μV calibration applies to the bottom seven motor units, which were discriminated from intramuscular recordings. (b) Smoothed discharge rates. (c) First four principal components extracted from the set of 10 detrended discharge rates and the variance in the signal accounted for by each component. (d) Detrended force (*gray line*) for a contraction at 7.5% MVC force of the abductor digiti minimi muscle and first common component (FCC) of the smoothed motor unit discharge rate (*black line*). Modified from Negro et al. [84] with permission.

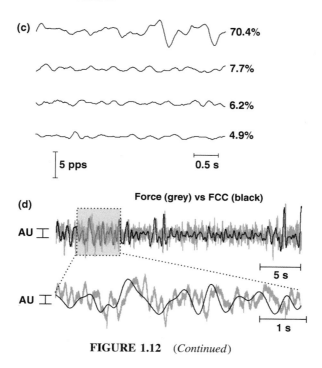

FIGURE 1.12 (*Continued*)

the input and output signals, the strength of the correlation derived from the discharge times varies with discharge rate up to a limit that corresponds to the number of samples required to reconstruct the oscillations in the input current [82]. Although the frequency content of the input signal is unknown, the sampling available in the output signal is limited by the discharge rates of the motor units used in an analysis. One solution is to increase the number of motor units included in the output signal by accumulating multiple trains of action potentials into a composite spike train [82]. Second, the strength of a time-domain index varies with the filter bandwidth applied to the discharge times and indexes derived with different filters produce correlation estimates that are not related. However, frequency-domain coherence measures are not influenced by the properties of the filter and the information contained in a time-domain index can be extracted from the appropriate bandwidth of the coherence function [82]. These limitations can be minimized by performing a correlation analysis on the unfiltered discharge times of multiple motor units to produce a coherence function that approximates the power spectrum of the common synaptic input [83].

Although the rate at which motor neurons discharge action potentials is proportional to the synaptic input (Fig. 1.3), the time between successive action potentials (interspike interval) is more variable when the membrane potential of a motor neuron is close to its voltage threshold. The coefficient of variation for interspike interval is greatest (~35%) when the force exerted by a muscle during a brief contraction is slightly above the recruitment threshold for a motor unit, and then it decreases to

~10% as the force increases [79]. The variability in interspike intervals at low discharge rates can be used to estimate the time course of the afterhyperpolarization (AHP) trajectory, albeit with some limitations [94]. Nonetheless, the time constant of the AHP trajectory in humans is related to twitch contraction time and minimal discharge rate [72].

Rate coding can also include brief interspike intervals (≤5 ms), which are referred to as double discharges or doublets. Double discharges likely involve the generation of a second action potential when a motor neuron is in a state of increased depolarization (delayed repolarization of membrane potential) due to either (a) the antidromic invasion of the dendrites by the outgoing action potential or (b) the development of a plateau potential [51]. Double discharges occur during tasks that involve rapid changes in muscle force [21], but not always [67]. Because training-induced changes in the rate of torque development were accompanied by an increased incidence of double discharges [106], double discharges likely augment the rate of torque development, at least during rapid submaximal contractions.

Some aspects of rate coding have been used to infer the activation of persistent inward currents (PICs) by neuromodulatory input to motor neurons [52]. One approach, known as the paired motor unit technique, involves recording the discharge of two motor units during gradual increases and decreases in muscle force and using the difference in rate coding to estimate the relative contribution of a PIC to the discharge of a motor unit [47]. The results suggest that ~40% of the rate coding exhibited by motor units during low-force isometric contractions is provided by a PIC. Although it is technically challenging to provide convincing evidence of a significant role for PICs in the rate coding of human motor units [43], the sustained discharge of motor units after the muscle has either been vibrated or exposed to wide-pulse, high-frequency stimulation is attributed to PIC activity [17,64]. Moreover, PICs are presumed to contribute to the dysfunction observed in spastic-paretic muscle [51].

1.6 SUMMARY

As the final common pathway from the nervous system to muscle, the motor unit transmits the activation signal that engages the contractile proteins to produce the muscle force needed for movement. Due to the exponential distribution of innervation number across the motor units that comprise a muscle, the number of motor unit action potentials emerging from the spinal cord is not directly related to the number of muscle fiber action potentials that can be detected with EMG electrodes. Nonetheless, progressive increases in the amplitude of the activation signal is attributable to the activation of motor units in order of increasing size and concurrent modulation of the rates at which they discharge action potentials. Moreover, the recruitment order generally results in motor units being activated in an order that progresses from slowest to fastest, weakest to strongest, and least to most powerful, but not as discrete types of motor units based on twitch contraction time and fatigability. The combinations of motor unit recruitment and rate coding that contribute to the actions performed by a muscle vary across its operating range, which requires caution when interpreting details about the activation signal from surface EMG recordings. Although a prescribed relation between EMG activity and muscle force is often assumed in many applications, this assumption is tenuous, as is discussed in subsequent chapters.

REFERENCES

1. Altenburg, T. M., C. J. de Ruiter, P. W. Verdijk, W. van Mechelen, and A. de Haan, "Vastus lateralis surface and single motor unit electromyography during shortening, lengthening and isometric contractions corrected for mode-dependent differences in force-generating capacity," *Acta Physiol. (Oxf.)* **196**, 315–328 (2009).

2. Andreassen, S., and L. Arendt-Nielsen, "Muscle fibre conduction velocity in motor units of the human anterior tibial muscle: a new size principle parameter," *J. Physiol.* **391**, 561–571 (1987).

3. Baudry, S., and J. Duchateau, "Post-activation potentiation in human muscle is not related to the type of maximal conditioning contraction," *Muscle Nerve* **30**, 328–336 (2004).

4. Baudry, S., and J. Duchateau, "Postactivation potentiation in a human muscle: effect on the rate of torque development of tetanic and voluntary isometric contractions," *J. Appl. Physiol.* **102**, 1394–1401 (2007).

5. Bawa, P., and C. Murnaghan, "Motor unit rotation in a variety of human muscles," *J. Neurophysiol.* **102**, 2265–2272 (2009).

6. Bergquist, A. J., M. J. Wiest, and D. F. Collins, "Motor unit recruitment when neuromuscular electrical stimulation is applied over a nerve trunk compared with a muscle belly: quadriceps femoris," *J. Appl. Physiol.* **113**, 78–89 (2012).

7. Bigland-Ritchie, B., A. J. Fuglevand, and C. K. Thomas, "Contractile properties of human motor units: is man a cat?," *Neuroscientist* **4**, 240–249 (1998).

8. Binder, M. D., "Integration of synaptic and intrinsic dendritic currents in cat spinal motoneurons," *Brain Res. Brain Res. Rev.* **40**, 1–8 (2002).

9. Blijham, P. J., H. J. ter Laak, H. J. Schelhaas, B. G. M. van Engelen, D. F. Stegeman, and M. J. Zwarts, "Relation between muscle fiber conduction velocity and fiber size in neuromuscular disorders," *J. Appl. Physiol.* **100**, 1837–141 (2006).

10. Bodine, S. C., R. R. Roy, E. Eldred, and V. R. Edgerton, "Maximal force as a function of anatomical features of motor units in cat tibialis anterior," *J. Neurophysiol.* **57**, 1730–1745 (1987).

11. Bottinelli, R., M. Canepari, M. A. Pellegrino, and C. Reggiani, "Force–velocity properties of human skeletal muscle fibres: myosin heavy chain isoform and temperature dependence," *J. Physiol.* **495**, 573–586 (1996).

12. Burke, R. E., D. N. Levine, P. Tsairis, F. E. Zajac III, "Physiological types and histochemical profiles in motor units of the cat gastrocnemius," *J. Physiol.* **234**, 723–748 (1973).

13. Canepari, M., M. A. Pellegrino, G. D'Antona, and R. Bottinelli, "Skeletal muscle fibre diversity and the underlying mechanisms," *Acta Physiol.* **199**, 465–476 (2010).

14. Capitano, M., M. Canepari, P. Cacciafesta, V. Lombardi, R. Cicchi, M. Maffei, F. S. Pavone, and R. Bottinelli, "Two independent mechanical events in the interaction cycle of skeletal muscle myosin with actin," *Proc. Natl. Acad. Sci. USA* **103**, 87–91 (2006).

15. Carp, J. S., P. A. Herchenroder, X. Y. Chen, and J. R. Wolpaw, "Sag during unfused tetanic contractions in rat triceps surae motor units," *J. Neurophysiol.* **81**, 2647–2661 (1999).

16. Carpentier, A., J. Duchateau, and K. Hainaut, "Motor unit behavior and contractile changes during fatigue in first dorsal interosseus," *J. Physiol.* **534**, 903–912 (2001).

17. Collins, D. F., D. Burke, and S. C. Gandevia, "Sustained contractions produced by plateau-like behaviour in human motoneurones," *J. Physiol.* **538**, 289–301 (2002).

18. Colombini, B., M. Nocella, G. Benelli, G. Cecchi, and M. A. Bagni, "Crossbridge properties during force enhancement by slow stretching in single intact frog muscle fibres," *J. Physiol.* **585**, 607–615, (2007).

19. Datta, A. K., and J. A. Stephens, "Synchronization of motor unit activity during voluntary contraction in man," *J. Physiol.* **422**, 397–419 (1990).

20. DeLuca, C. J., R. S. LeFever, M. P. McCue, and A. P. Xenakis "Control scheme governing concurrently active human motor units during voluntary contractions," *J. Physiol.* **329**, 129–142 (1982).

21. Desmedt, J. E., and E. Godaux, "Fast motor units are not preferentially activated in rapid voluntary contractions in man," *Nature* **267**, 717–719 (1977).

22. Desmedt, J. E., and E. Godaux, "Ballistic contractions in fast or slow human muscles: discharge patterns of single motor units," *J. Physiol.* **285**, 185–196 (1978).

23. Desmedt, J. E., and E. Godaux, "Voluntary motor commands in human ballistic movements," *Ann. Neurol.* **5**, 415–421 (1979).

24. Desmedt, J. E., and E. Godaux, "Spinal motoneuron recruitment in man: rank deordering with direction but not with speed of voluntary movement," *Science* **214**, 933–936 (1981).

25. Dideriksen, J. L., D. Farina, and R. M. Enoka, "Influence of fatigue on the simulated relation between the amplitude of the surface electromyogram and muscle force," *Philos. Trans. R. Soc. B* **368**, 2765–2781 (2010)

26. Dideriksen, J. L., R. M. Enoka, and D. Farina, "Neuromuscular adjustments that constrain submaximal EMG amplitude at task failure of sustained isometric contractions," *J. Appl. Physiol.* **111**, 485–494 (2011).

27. Duchateau, J., and R. M. Enoka "Neural control of shortening and lengthening contractions: influence of task constraints," *J. Physiol.* **586**, 5853–5864 (2008).

28. Duchateau, J., and R. M. Enoka, "Human motor unit recordings: origins and insight into the integrated motor system," *Brain Res.* **1409**, 42–61 (2011).

29. Edman, K. A. P., "Double-hyperbolic force-velocity relation in frog muscle fibres," *J. Physiol.* **404**, 301–321 (1988).

30. Enoka, R. M., *Neuromechanics of Human Movement*, 5th ed., Human Kinetics, Champaign, IL, 2015.

31. Enoka, R. M., "Muscle fatigue—from motor units to clinical symptoms," *J. Biomech.* **45**, 427–433 (2012).

32. Enoka, R. M., and A. J. Fuglevand, "Motor unit physiology: some unresolved issues," *Muscle Nerve* **24**, 4–17 (2001).

33. Enoka, R. M., and K. G. Pearson, "The Motor Unit and Muscle Action," in E. R. Kandel, J. H. Schwartz, T. M. Jessell, S. A. Siegelbaum and A. J. Hudspeth, *Principles of Neural Science*, 5th ed. McGraw Hill, New York, pp. 768–788, 2013.

34. Farina, D., L. Arendt-Nielsen, and T. Graven-Nielsen, "Effect of temperature on spike-triggered average torque and electrophysiological properties of low-threshold motor units," *J. Appl. Physiol.* **99**, 197–203 (2005).

35. Farina, D., A. Holobar, M. Gazzoni, D. Zazula, R. Merletti, and R. M. Enoka, "Adjustments differ among low-threshold motor units during intermittent, isometric contractions," *J. Neurophysiol.* **101**, 350–359 (2009).

36. Farina, D., A. Holobar, R. Merletti, and R. M. Enoka, "Decoding the neural drive to muscles from the surface electromyogram," *Clin. Neurophysiol.* **121**, 1616–1623 (2010).

37. Farina, D., and R. Merletti, "Estimation of average muscle fiber conduction velocity from two-dimensional surface EMG recordings," *J. Neurosci. Meth* **134**, 199–208 (2004).

38. Farina, D., R. Merletti, and R. M. Enoka, "The extraction of neural strategies from the surface EMG," *J. Appl. Physiol.* **96**, 1486–1495 (2004).

39. Farmer, S. F., F. D. Bremner, D. M. Halliday, J. R. Rosenberg, and J. A. Stephens, "The frequency content of common synaptic inputs to motoneurones studies during voluntary isometric contraction in man," *J. Physiol.* **470**, 127–155 (1993).

40. Feinstein, B., B. Lindegård, E. Nyman, and G. Wohlfart, "Morphologic studies of motor units in normal human muscles," *Acta Anat.* **23**, 127–142 (1955).

41. Feiereisen, P., J. Duchateau, and K. Hainaut, "Motor unit recruitment order during voluntary and electrically induced contractions in the tibialis anterior," *Exp. Brain Res.* **114**, 117–123 (1997).

42. Filippi, G. M., and D. Troiani, "Relations among motor unit types, generated forces and muscle length in single motor units of anaesthetized cat peroneus longus muscle," *Exp. Brain Res.* **101**, 406–414 (1994).

43. Fuglevand, A. J., A. P. Dutoit, R. K. Johns, and D. A. Keen, "Evaluation of plateau potential-mediated 'warm up' in human motor units," *J. Physiol.* **571**, 683–693 (2006).

44. Fuglevand, A. J., D. A. Winter, and A. E. Patla, "Models of recruitment and rate coding organization in motor-unit pools," *J. Neurophysiol.* **70**, 2470–2488 (1993).

45. Garland, S. J., R. M. Enoka, L. P. Serrano, and G. A. Robinson, "Behavior of motor units in human biceps brachii during a submaximal fatiguing contraction," *J. Appl. Physiol.* **76**, 2411–2419 (1994).

46. Gazzoni, M,. F. Camelia, and D. Farina, "Conduction velocity of quiescent muscle fibers decreases during sustained contraction," *J. Neurophysiol.* **94**, 387–394 (2005).

47. Gorassini, M,. J. F. Yang, M. Siu, and D. J. Bennett, "Intrinsic activation of human motoneurons: possible contribution to motor unit excitation," *J. Neurophysiol.* **87**, 1850–1858 (2002).

48. Harris A. J., M. J. Duxon, J. E. Butler, P. W. Hodges, J. L. Taylor, and S. C. Gandevia, "Muscle fiber and motor unit behavior in the longest human skeletal muscle," *J. Neurosci.* **25**, 8528–8533 (2005).

49. Heckman, C. J., R. H. Lee, and R. M. Brownstone, "Hyperexcitable dendrites in motoneurons and their neuromodulatory control during motor behavior," *Trends Neurosci.* **26**, 688–695 (2003).

50. Heckman, C. J., and R. M. Enoka, "Physiology of the motor neuron and the motor unit," in: A. Eisen, ed., *Clinical Neurophysiology of Motor Neuron Diseases: Handbook of Clinical Neurophysiology*, vol. **4**. Elsevier B. V., Amsterdan, pp. 119–147, 2004.

51. Heckman, C. J., and R. M. Enoka, "Motor unit," *Comp. Physiol.* **2**, 2629–2682 (2012).

52. Heckman, C. J., C. Mottram, K. Quinlan, R. Theiss, and J. Schuster, "Motoneuron excitability: the importance of neuromodulatory inputs," *Clin. Neurophysiol.* **120**, 2040–2054 (2009).

53. Henneman, E., "Relation between size of neurons and their susceptibility to discharge," *Science* **126**, 1345–1347 (1957).

54. Herzog, W., V. Joumaa, and T. R. Leonard, "The force–length relationship of mechanically isolated sarcomeres," *Adv. Exp. Med. Biol.* **682**, 141–161 (2010).

55. Hultborn, H., "Plateau potentials and their role in regulating motoneuronal firing," *Adv. Exp. Med. Biol.* **508**, 213–218 (2002).

56. Hultborn, H., M. E. Denton, J. Wienecke, and J. B. Nielsen, "Variable amplification of synaptic input to cat spinal motoneurones by dendritic persistent inward current," *J. Physiol.* **552**, 945–952 (2003).

57. Hultman, E., H. Sjöholm, I. Jäderhaolm-E.k., and J. Krynicki, "Evaluation of methods for electrical stimulation of human skeletal muscle in situ," *Pflügers Arch.* **398**, 139–141 (1983).

58. Jenny, A. B., and J. Inukai, "Principles of motor organization of the monkey cervical spinal cord," *J. Neurosci.* **3**, 567–575 (1983).

59. Jessell, T. M., G. Sürmeli, and J. S. Kelly, "Motor neurons and the sense of place," *Neuron* **72**, 419–424 (2011).

60. Kadhiresan, V. A., C. A. Hassett, and J. A. Faulkner, "Properties of single motor units in medial gastrocnemius muscles of adult and old rats," *J. Physiol.* **493**, 543–552 (1996).

61. Kanda, K. and J. E. Desmedt, "Cutaneous facilitation of large motor units and motor control of human fingers in precision grip," *Adv. Neurol.* **39**, 253–261 (1983).

62. Kanda, K., and K. Hashizume, "Changes in properties of the medial gastrocnemius motor units in aging rats," *J. Neurophysiol.* **61**, 737–746 (1989).

63. Kanda, K., and K. Hashizume, "Factors causing difference in force output among motor units in the rat medial gastrocnemius muscle," *J. Physiol.* **448**, 677–695 (1992).

64. Kiehn, O., and T. Eken, "Prolonged firing in motor units: evidence of plateau potentials in human motoneurons?," *J. Neurophysiol.* **78**, 3061–3068 (1997).

65. Kirkwood, P. A., "On the use and interpretation of cross-correlation measurements in the mammalian central nervous system," *J. Neurosci. Meth.* **1**, 107–132 (1979).

66. Kluger, B. M., L. B. Krupp, and R. M. Enoka, "Fatigue and fatigability in neurologic illnesses: proposal for a unified taxonomy," *Neurology* **80**, 409–416 (2013).

67. Kudina, L. P., and N. L. Alexeeva, "Repetitive doublets of human motoneurones: analysis of interspike intervals and recruitment pattern," *Electroencephalogr. Clin. Neurophysiol.* **85**, 243–247 (1992).

68. Lee, R. H., and C. J. Heckman, "Adjustable amplification of synaptic input in the dendrites of spinal motoneurons in vivo," *J. Neurosci.* **20**, 6734–6740 (2000).

69. Lévénez, M. C., S. J. Garland, M. Klass, and J. Duchateau, "Cortical and spinal modulation of antagonist coactivation during a submaximal fatiguing contraction in humans," *J. Neurophysiol.* **99**, 554–562 (2008).

70. Lieber, R. L., and S. R. Ward, "Skeletal muscle design to meet functional demands," *Philos. Trans. R. Soc. B* **366**, 1466–1476 (2011).

71. Lombardi, V., and G. Piazzesi, "The contractile response during lengthening of stimulated frog muscle fibres," *J. Physiol.* **431**, 141–171 (1990).

72. MacDonnell, C. W., T. D. Ivanova, and S. J. Garland, "Afterhyperpolarization time-course and minimal discharge rate in low threshold motor units in humans," *Exp. Brain Res.* **189**, 23–33 (2008).

73. Macefield, V. G., A. J. Fuglevand, J. N. Howell, and B. Bigland-Ritchie, "Contractile properties of single motor units in human toe extensors assessed by intraneural motor axon stimulation," *J. Neurophysiol.* **75**, 2509–2519 (1996).

74. Marsden, J. F., S. F. Farmer, D. M. Halliday, J. R. Rosenberg, and P. Brown "The unilateral and bilateral control of motor unit pairs in the first dorsal interosseus and paraspinal muscles in man," *J. Physiol.* **521**, 553–564 (1999).

75. McComas, A. J., *Neuromuscular Function and Disorders*, Butterworths, Boston, 1977.

76. Milner-Brown, H. S., R. B. Stein, and R. Yemm, "The contractile properties of human motor units during voluntary isometric contractions," *J. Physiol.* **228**, 285–306 (1973a).

77. Milner-Brown, H. S., R. B. Stein, and R. Yemm, "The orderly recruitment of human motor units during voluntary isometric contractions," *J. Physiol.* **230**, 359–370 (1973b).

78. Mito, K., and K. Sakamoto, "On the evaluation of muscle fiber conduction velocity considering waveform properties of an electromyogram in M. biceps brachii during voluntary isometric contraction," *Electromyogr. Clin. Neurophysiol.* **42**, 137–149 (2002).

79. Mortiz, C. T., B. K. Barry, M. A. Pascoe, and R. M. Enoka "Discharge rate variability influences the variation in force fluctuations across the working range of a hand muscle," *J. Neurophysiol.* **93**, 2449–2459 (2005).

80. Mottram, C. J., J. M. Jakobi, J. G. Semmler, and R. M. Enoka, "Motor-unit activity differs with load type during a fatiguing contraction," *J. Neurophysiol.* **93**, 1381–1392 (2005).

81. Nandedkar, S. D., and E. Stålberg, "Simulation of single muscle fibre action potentials," *Med. Biol. Eng. Comput.* **21**, 158–165 (1983).

82. Negro, F., and D. Farina "Factors influencing the estimates of correlation between motor unit activities in humans," *PLoS One*, **7** (9):e44894 (2012).

83. Negro, F., and D. Farina, "Linear transmission of cortical oscillations to the neural drive to muscles is mediated by common projections to populations of motoneurons in humans," *J. Physiol.* **589**, 929–637 (2011).

84. Negro, F., A. Holobar, and D. Farina, "Fluctuations in isometric muscle force can be described by one linear projection of low-frequency components of motor unit discharge rates," *J. Physiol.* **587**, 5925–5938 (2009).

85. Oya, T., S. Riek, and A. G. Cresswell, "Recruitment and rate coding organisation for soleus motor units across entire range of voluntary isometric plantar flexions," *J. Physiol.* **587**, 4737–4748 (2009).

86. Pascoe, M. A., M. R. Holmes, and R. M. Enoka, "Discharge characteristics of biceps brachii motor units at recruitment when older adults sustained an isometric contraction," *J. Neurophysiol.* **105**, 571–581 (2011).

87. Pasquet, B., A. Carpentier, and J. Duchateau, "Specific modulation of motor unit discharge for a similar change in fascicle length during shortening and lengthening contractions in humans," *J. Physiol.* **577**, 753–765 (2006).

88. Person, R. S., and L. P. Kudina, "Discharge frequency and discharge pattern of human motor units during voluntary contraction of muscle," *Electroencephalogr. Clin. Neurophysiol.* **32**, 471–483 (1972).

89. Petit, J., M. Chua, and C. C. Hunt, "Maximum shortening speed of motor units of various types in cat lumbrical muscles," *J. Neurophysiol.* **69**, 442–448 (1993).

90. Petit, J., M. A. Giroux-Metges, and M. Gioux, "Power developed by motor units of the peroneus tertius muscle of the cat," *J. Neurophysiol.* **90**, 3095–3104 (2003).

91. Pierotti, D. J., R. R. Roy, S. C. Bodine-Fowler, J. A. Hodgson, and V. R. Edergton, "Mechanical and morphological properties of chronically inactive cat tibialis anterior motor units," *J. Physiol.* **444**, 175–192 (1991).

92. Place, N., T. Yamada, J. D. Bruton and H. Westerblad, "Muscle fatigue: from observations in humans to underlying mechanisms studied in intact single muscle fibres," *Eur. J. Appl. Physiol.* **110**, 1–15 (2010).

93. Powers, R. K., and M. D. Binder, "Input–output functions of mammalian motoneurons," *Rev. Physiol. Biochem. Pharmacol.* **143**, 137–263 (2001).

94. Powers, R. K., K. S. Türker, and M. D. Binder, "What can be learned about motoneurone properties from studying firing patterns?," *Adv. Exp. Med. Biol.* **50**, 199–205 (2002).

95. Reinking, R. M., J. A. Stephens, and D. G. Stuart, "The motor units of cat medial gastrocnemius: problem of their categorisation on the basis of mechanical properties," *Exp. Brain Res.* **23**, 301–313 (1975).

96. Riley, Z. A., A. H. Maerz, J. C. Litsey, and R. M. Enoka, "Motor unit recruitment in human biceps brachii during sustained voluntary contractions," *J. Physiol.* **586**, 2183–2193 (2008).

97. Rosenfalck, P., "Intra and extracellular fields of active nerve and muscle fibers. A physic–mathematical analysis of different models," *Acta Physiol. Scand.* **321**, 1–49 (1969).

98. Rudroff, T., K. Jordan, J. A. Enoka, S. D. Matthews, S. Baudry, and R. M. Enoka, "Discharge of biceps brachii motor units is modulated by load compliance and forearm posture," *Exp. Brain Res.* **202**, 111–120 (2010).

99. Sale, D. G., "Postactivation potentiation: role in human performance," *Exerc. Sport Sci. Rev.* **30**, 138–143 (2002).

100. Semmler, J. G., "Motor unit synchronization and neuromuscular performance," *Exerc. Sport Sci. Rev.* **30**, 8–14 (2002).

101. Sheard, P. W., "Tension delivery from short fibers in long muscles," *Exerc. Sport Sci. Rev.* **28**, 51–56 (2000).

102. Thomas, C. K., R. S. Johansson, G. Westling, and B. Bigland-Ritchie, "Twitch properties of human thenar motor units measured in response to intraneural motor-axon stimulation," *J. Neurophysiol.* **64**, 1339–1346 (1990).

103. Tötösy de Zepetnek, J. E., H. V. Zung, S. Erdebil, and T. Gordon, "Innervation number is an important determinant of force in normal and reinnervated rat tibialis anterior muscles," *J. Neurophysiol.* **67**, 1385–1403 (1992).

104. Trontelj, J. V., "Muscle fiber conduction velocity changes with length," *Muscle Nerve* **16**, 506–512 (1993).

105. Van Cutsem, M., and J. Duchateau, "Preceding muscle activity influences motor unit discharge and rate of torque development during ballistic contractions in humans," *J. Physiol.* **562**, 635–644 (2005).

106. Van Cutsem, M., J. Duchateau, and K. Hainaut, "Neural adaptations mediate increase in muscle contraction speed and change in motor unit behaviour after dynamic training," *J. Physiol.* **513**, 295–305 (1998).

107. Van Cutsem, M., P. Feiereisen, J. Duchateau, and K. Hainaut, "Mechanical properties and behavior of motor units in the tibialis anterior during voluntary contractions," *Can. J. Appl. Physiol.* **22**, 585–597 (1997).

108. Vieira, T. M., I. D. Loram, S. Muceli, R. Merletti, and D. Farina, "Postural activation of the human medial gastrocnemius muscle: are the muscle units spatially located?," *J. Physiol.* **589**, 431–443 (2011).

109. Westad, C., R. H. Westgaard, and C. J. DeLuca, "Motor unit recruitment and derecruitment induced by brief increase in contraction amplitude of the human trapezius muscle," *J. Physiol.* **552**, 645–656 (2003).

110. Westgaard, R. H., and C. J. DeLuca, "Motor control of low-threshold motor units in the human trapezius muscle," *J. Neurophysiol.* **85**, 1777–1781 (2001).

2

BIOPHYSICS OF THE GENERATION OF EMG SIGNALS

D. Farina,[1] D. F. Stegeman,[2] and R. Merletti[3]

[1]Department of Neurorehabilitation Engineering, Bernstein Focus Neurotechnology Göttingen, University Medical Center Göttingen, Georg-August University, Göttingen, Germany
[2]Department of Neurology/Clinical Neurophysiology, Radboud University Nijmegen Medical Centre and Donders Institute for Brain, Cognition and Behaviour, The Netherlands
[3]Laboratory for Engineering of the Neuromuscular System, Department of Electronics, Politecnico di Torino, Torino, Italy

2.1 INTRODUCTION

The electromyographic (EMG) signal is a representation of the electric potential field generated by the depolarization of the outer muscle fiber membrane (the sarcolemma). Its detection involves the use of intramuscular or surface electrodes which are placed at a certain distance from the sources. The tissue separating the sources and the recording electrodes acts as a so-called volume conductor. The volume conductor properties largely determine the features of the detected signals, for instance the frequency content.

In this chapter, basic concepts of generation and detection of EMG signals will be described. Specific emphasis is devoted to the generation of muscle fiber action potentials at the fiber end plates, their propagation along the sarcolemma, and their extinction at the tendons. The topics of crosstalk between nearby muscles and selectivity of the recording systems are also addressed. The chapter terminates with a discussion on the relationships between muscle force and the surface EMG, a topic further addressed in Chapter 10.

Surface Electromyography: Physiology, Engineering, and Applications, First Edition.
Edited by Roberto Merletti and Dario Farina.
© 2016 by The Institute of Electrical and Electronics Engineers, Inc. Published 2016 by John Wiley & Sons, Inc.

The reading of parts of this chapter may require familiarity with some of the concepts presented in Chapter 5, on filtering and sampling. Moreover, the issues discussed in this chapter provide a basis for the EMG modeling approaches presented in Chapter 8.

2.2   EMG SIGNAL GENERATION

The EMG signal is generated by the electrical activity of the muscle fibers active during a contraction. The signal sources are the depolarizing and repolarizing zones of the muscle fibers. The sources of the signal are separated from the recording electrodes by biological tissues which act as spatial and temporal low-pass filters on the potential distribution [3]. For intramuscular recordings, the effect of the tissues between electrodes and muscle fibers is relatively small due to the closeness of the recording electrodes to the sources. On the contrary, for surface recordings the volume conductor constitutes an important deforming effect on the signal. This volume-conductor-based effect is principally spatial. This concept will be better elucidated in later sections.

2.2.1   Signal Source

If a micropipette electrode is inserted intracellularly into a muscle fiber, a membrane resting potential of 70–90 mV, negative inside the cell with respect to the extracellular environment, is measured [38]. The maintenance of this potential is mediated by the sodium–potassium pump (NaK ATP-ase) working against the concentration gradients of ions flowing through the membrane [19]. When the action potential generated by a motor neuron reaches the neuromuscular junction, it causes the emission of acetylcholine in the gap between the nerve terminal and the muscle fiber membrane. Acetylcholine excites the fiber membrane, so that a potential gradient is locally generated in the fiber. An inward current density (*depolarization zone*) corresponds to this potential gradient.

The depolarization zone propagates along the muscle fibers from the neuro-muscular junctions to the tendons' endings. This propagating intracellular action potential (IAP) causes an ionic transmembrane current profile also propagating along the sarcolemma. For muscle fibers, the length of the depolarization and repolarization zones is on the order of millimeters. Therefore, a muscle fiber can be considered as a very thin tube in which current is only flowing axially, which is the so-called *line source model* [37]. The model depicted in Fig. 2.1a represents in its center a portion Δz of fiber membrane in the assumption of line source condition. In the following derivations on the fiber membrane currents and voltages, the effect of the volume conductor is neglected for simplicity. The decrease in potential per unit length is equal to the product of the resistance per unit length and the current flowing through the resistance. For the extracellular and intracellular path, this results in:

$$\frac{\partial \phi_e}{\partial z} = -I_e r_e, \qquad \frac{\partial \phi_i}{\partial z} = -I_i r_i \qquad (2.1)$$

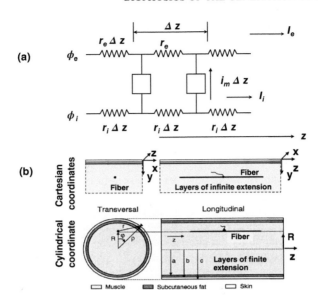

FIGURE 2.1 **(a)** Linear core–conductor model representing a portion of the fiber membrane. For graphical representation the structure is shown as a repetitive network of finite length Δz, but in fact $\Delta z \rightarrow 0$ and the analysis is based on the continuum. The open box is a symbol representing the equivalent circuit of the membrane which depends on the membrane state, that is, a passive structure during the resting period and a circuit with time-dependent components during the active phase, as described in Chapter 1. **(b)** Representation of muscle fiber position in Cartesian and cylindrical coordinate systems.

with r_e and r_i the resistance per unit length (Ω/cm) of the extracellular and intracellular path, respectively. The above-mentioned volume conduction effects are neglected here for clarity. Moreover, the conservation of current requires that the axial rate of decrease in the intracellular longitudinal current be equal to the transmembrane current per unit length:

$$\frac{\partial I_i}{\partial z} = -i_m \tag{2.2}$$

The extracellular longitudinal current may decrease with increasing z either because of a decrement of current that crosses the membrane (transmembrane current) or a loss that is carried outside by indwelling electrodes:

$$\frac{\partial I_e}{\partial z} = i_m + i_p \tag{2.3}$$

where i_p is the current flowing through the electrodes. The transmembrane voltage is given by

$$V_m = \phi_i - \phi_e \tag{2.4}$$

thus, its first derivative is obtained as [substituting Eq. (2.1)]:

$$\frac{\partial V_m}{\partial z} = \frac{\partial \phi_i}{\partial z} - \frac{\partial \phi_e}{\partial z} = -r_i I_i + r_e I_e \tag{2.5}$$

and deriving a second time [substituting Eqs. (2.2) and (2.3)]:

$$\frac{\partial^2 V_m}{\partial z^2} = r_i i_m + r_e(i_m + i_p) = (r_e + r_i)i_m + r_e i_p \tag{2.6}$$

which shows that the second derivative of the transmembrane potential is almost proportional to the transmembrane current in the hypothesis of the line source model. Figure 2.1b shows the two coordinate systems (rectangular and cylindrical) adopted in the literature to study the field generated by the depolarized area of a muscle fiber. Figure 2.2 depicts the surface potential generated by a motor unit.

The velocity at which the action potential propagates is referred to as conduction velocity (CV) or as propagation velocity. It depends on the fiber diameter [51]. The observed relation between fiber diameter and CV remains valid for affected muscles in neuromuscular diseases [8].

The IAP shape can be approximated by simple functions, such as the analytical expression provided by Rosenfalck [57] or a triangular approximation [46]. In general, the IAP can be characterized by a depolarization phase, a repolarization

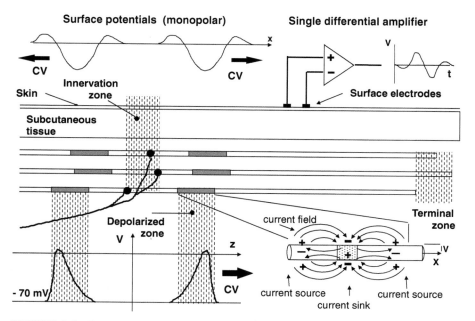

FIGURE 2.2 Representation of a motor unit (MU) and of a motor unit action potential (MUAP). Zoomed view of the source.

phase, and a hyperpolarizing long afterpotential. The IAP shape may change due to the conditions of the muscle. In particular, during fatigue, few stages of IAP alteration can be distinguished [35]. In the beginning of fatigue, the IAP spike width in space increases mainly because of the slowing of the repolarization phase. In this phase, the rate of increase of the IAP remains practically unchanged while the amplitude slightly decreases. The amplitude of the afterpotential increases. In the following stages, the absolute value of the resting potential, spike amplitude, and rate of the IAP rise decrease together with further slowing of the IAP falling phase and increasing of the afterpotential amplitude.

2.2.2   Generation and Extinction of the Intracellular Action Potential

The generation and extinction of the IAP have been modeled using different mathematical approaches using the same basic assumptions [12,15,32–34,47]. For this purpose, it is assumed that the integral of the transmembrane current over the entire muscle fiber length is zero at all times. On this basis, Gootzen et al. [31,32] replaced the current source at the end plate and at the tendons with an equivalent source proportional to the first derivative of the IAP. The same approach has been used in a model in which the volume conductor is simulated numerically by a finite element approach [45].

Dimitrov and Dimitrova [13] started their description by considering the first derivative of the IAP and assumed its progressive appearance at the end plate and disappearance at the tendons. This second approach is computationally and conceptually attractive and has been also used by a number of researchers who applied different computation techniques [24,47]. Approximating the IAP with a triangular function, its first derivative is a function which assumes two constant values of opposite signs. The current density source is in this case approximated by a current tripole, sometimes divided conceptually in a leading and a trailing dipole part [18]. The life cycle of the IAP is schematically presented in Fig. 2.3. Assuming the progressive disappearance of the first IAP derivative at the tendon, when the tripole reaches the tendon, the first pole stops and the other two poles get closer to the first; when the second pole coincides with the first, the leading dipole has disappeared and eventually also the trailing one disappears at the tendon. A similar mechanism in opposite direction occurs at the end plate. From the basic understanding of the extinction of an IAP, independent of the way in which it is described, it can be concluded that the end-of-fiber potential has the same wave shape of the IAP [65]. As a consequence of the above, EMG signals at different positions along the fiber length are not simply delayed versions of each other. On the basis of the concepts outlined by Dimitrov and Dimitrova [13], Farina and Merletti [24] proposed a general description of the current density source traveling at velocity v along the fiber with an origin and an end point:

$$i(z,t) = \frac{d}{dz}\left[\psi(z - z_i - vt)p_{L_1}(z - z_i - L_1/2) - \psi(-z + z_i - vt)p_{L2}(z - z_i + L_2/2)\right]$$

$$(2.7)$$

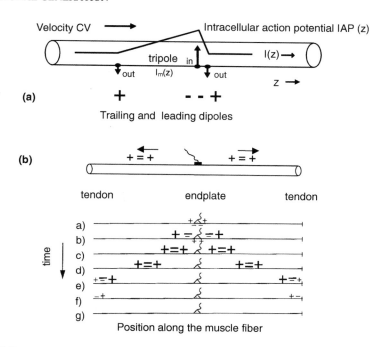

FIGURE 2.3 (a) Stylized presentation of an intracellular action potential as a function of position along a muscle fiber during steady action potential propagation. Indications of propagation velocity CV, the short depolarization, and the (longer tail) repolarization phase are included, as well as the tripole transmembrane ionic current I_m and its ultimate simplification in terms of a leading $(- +)$ and trailing $(+ -)$ dipole pair. (b) The spatiotemporal development ("life cycle") of the transmembrane current along a muscle fiber. The transmembrane ionic current starts as two dipoles at the end plate after excitation by a motoneuron action potential (a). After full development of these first (leading) dipoles, a second (trailing) pair emerges (b). The balanced, double pair then propagates as two tripoles in opposite directions $(+ = +)$ (c, d). On arrival at the tendon, the leading dipoles decline in strength and disappears, leaving the trailing dipole (e). Subsequently, also the trailing dipoles decline (f) and disappear (g). Reproduced from Roeleveld et al. [55] with permission.

where $i(z,t)$ is the current density source, $\psi(z) = dV_m(z)/dz$ ($V_m(z)$ being the IAP), $p_L(z)$ a function which takes value *1* for $-L/2 \le z \le L/2$ and 0 otherwise, z_i the position of the end plate, and L_1 and L_2 the length of the fiber from the end plate to the right and to the left tendon respectively.

Equation (2.7) is general and does not assume any approximation of the current density source. Special cases, such as the tripole approximation, can be described with Eq. (2.7). From (2.7) we again note that the sources of EMG signals are not plane waves traveling at constant velocity from minus to plus infinity.

Specifically, when the potential stops at the tendon junction, it generates a signal component which is not propagating and may have different shapes at different locations along the muscle. These components are referred to as end-of-fiber signal or end-of-fiber effect. Their properties will be discussed in the next sections.

2.2.3 Volume Conductor

The generation of the intracellular action potential determines an electric field in the surrounding space. The potential generated by a motor unit (MU) can thus be detected also in locations relatively far from the source. The biological tissues separating the sources and the detecting electrodes are referred to as *volume conductor*, and their characteristics strongly influence the detected signal.

Under the static hypothesis, in a volume conductor, the current density, the electric field, and the potential satisfy the following relations [54]:

$$\nabla \cdot J = I, \qquad J = \bar{\bar{\sigma}} E, \qquad E = -\nabla \varphi \tag{2.8}$$

where J is the current density in the volume conductor ($A \cdot m^{-2}$), I the current density of the source ($A \cdot m^{-3}$), E the electric field ($V \cdot m^{-1}$), and φ the potential (V).

From Eq. (2.8), the Poisson's equation is obtained:

$$-\frac{\partial}{\partial x}\left(\sigma_x \frac{\partial \varphi}{\partial x}\right) - \frac{\partial}{\partial y}\left(\sigma_y \frac{\partial \varphi}{\partial y}\right) - \frac{\partial}{\partial z}\left(\sigma_z \frac{\partial \varphi}{\partial z}\right) = I \tag{2.9}$$

where σ_x, σ_y, and σ_z are the conductivities of the medium in the three spatial directions. Equation (2.9) is the general relation (in Cartesian coordinates) between the potential and the current density in a nonhomogeneous and anisotropic medium. If the medium is homogeneous, the conductivities do not depend on the site and the following equation is obtained:

$$-\sigma_x \frac{\partial^2 \varphi}{\partial x^2} - \sigma_y \frac{\partial^2 \varphi}{\partial y^2} - \sigma_z \frac{\partial^2 \varphi}{\partial z^2} = I \tag{2.10}$$

Equation (2.10) can be also written in cylindrical coordinates (ρ, z) (Figure 2.1b), resulting in:

$$\frac{\partial^2 \varphi}{\partial \rho^2} + \frac{1}{\rho}\frac{\partial \varphi}{\partial \rho} + \frac{1}{\rho^2}\frac{\partial^2 \varphi}{\partial \phi^2} + \frac{\partial^2 \varphi}{\partial z^2} = -\frac{I}{\sigma} \tag{2.11}$$

The solution of Eq. (2.10) or (2.11) provides the potential in any point in space when the characteristics of the source (I) and of the medium (σ) are known. This solution can be obtained only if the boundary conditions can be described analytically in simple coordinate systems. This problem has been solved for different degrees of simplification. The simplest assumption for the solution of the Poisson's equation is to deal with a homogeneous, isotropic, infinite volume conductor. In this case, assuming a source distributed along a line in z coordinate direction, the potential distribution in the volume conductor is given by the following relationship:

$$\varphi(r, z) = \frac{1}{2\sigma} \int_{-\infty}^{+\infty} \frac{I(s)}{\sqrt{r^2 + (z - s)^2}} \, ds \tag{2.12}$$

where $I(s)$ is the current density source and σ is the conductivity of the medium. For a semi-infinite medium with different conductivities in the longitudinal and radial directions (such as is the case for muscle tissue), we get

$$\varphi(r, z) = \frac{1}{\sigma_r} \int_{-\infty}^{+\infty} \frac{I(s)}{\sqrt{r^2 \dfrac{\sigma_z}{\sigma_r} + (z - s)^2}} \, ds \tag{2.13}$$

where σ_z and σ_r are the longitudinal (read: fiber direction) and radial (read: across fibers) conductivities, respectively. The method of images may be applied to compute the surface potential distribution in the case of a semi-space of conductive medium (tissue) and a semi-space of insulation material (air). It is a basic observation from electrostatics that in that case the surface potential doubles with respect to the case of an infinite medium [47] [refer also to Stegeman, D. F., J. P., de Weerd, and E. G. Eijkman, A volume conductor study of compound action potentials of nerves in situ: the forward problem, *Biol. Cybern.* 33(2), 97–111 (1979).]

More complex descriptions of the volume conductor have been proposed and include a nonhomogeneous medium comprised of layers of different conductivities (see also Chapter 8). In the case of layered geometries, Eqs. (2.10) and (2.11) can be solved independently in the different layers. The final solution is then obtained by imposing the boundary conditions at the surfaces between the layers. Boundary conditions are the continuity of the current in the direction perpendicular to the boundary surface and continuity of the potential over the boundary. A nondivergent potential in all points of the volume conductor, except for the locations of the sources, is the final condition for deriving a unique solution.

For Cartesian coordinates and layered medium (infinite layers parallel to the xz plane), the boundary conditions are:

$$\sigma_{i+1} \frac{\partial \varphi_{i+1}(x, y, z)}{\partial y}\bigg|_{y=h_i} = \sigma_i \frac{\partial \varphi_i(x, y, z)}{\partial y}\bigg|_{y=h_i}$$

$$\frac{\partial \varphi_{i+1}(x, y, z)}{\partial x}\bigg|_{y=h_i} = \frac{\partial \varphi_i(x, y, z)}{\partial x}\bigg|_{y=h_i} \quad ; \quad \frac{\partial \varphi_{i+1}(x, y, z)}{\partial z}\bigg|_{y=h_i} = \frac{\partial \varphi_i(x, y, z)}{\partial z}\bigg|_{y=h_i} \tag{2.14}$$

for the interfaces (here $y = h_i$) between adjacent layers. Similar expressions can be derived in the case of a cylindrical coordinate system [9].

A cylindrical description of the volume conductor is more realistic since it accounts for the finiteness of the volume conductor. A three-layer model (muscle, fat, and skin) has been developed by Blok et al. [9] as an elaboration of the two-layer model described by Gootzen et al. [31,32]. The general solution of Poisson's equation for each of the three concentric cylindrical layers of this configuration for the potential detected over the skin layer reads as follows:

$$\Phi_3(\rho, \phi, k) = \frac{d}{2\sigma_{3_r}} K_0\left(r\sqrt{\frac{\sigma_{3_z}}{\sigma_{3_r}}}|k|\right) G(k) + \sum_{n=-\infty}^{\infty} e^{-in\phi}$$
$$\times \left[E_n(k)I_n\left(\rho\sqrt{\frac{\sigma_{3_z}}{\sigma_{3_r}}}k\right) + F_n(k)K_n\left(\rho\sqrt{\frac{\sigma_{3_z}}{\sigma_{3_r}}}|k|\right)\right]$$

(2.15)

In Eq. (2.15), ρ and ϕ are the cylindrical coordinates, r is the distance of the fiber axis from the detection point, k is the spatial angular frequency in the axial direction, and $G(k)$ is the Fourier transform of the electric current source function to the spatial frequency domain. The conductivity of the skin layer in the radial direction is represented by the parameter σ_{3_r}, and that in the axial direction is denoted by σ_{3_z}. All three layers were allowed to be anisotropic in the cylindrical coordinates r and z. The functions K_n and I_n are modified Bessel functions of order n, of the first and second kind, respectively, and $E_n(k)$ and $F_n(k)$ are unknowns which—for each n and k—have to be determined from the boundary conditions. Together with expressions similar to Eq. (2.15), for the two other layers (muscle, subcutaneous fat), five of such unknowns have to be determined by the use of five boundary conditions. Finally, d is the diameter of the fiber. Because the K_n and I_n Bessel functions tend toward very large or very small values for increasing values of n and for small or large values of k, the solution system becomes ill-conditioned and its solution inaccurate. This problem can be mitigated as described by Gootzen et al. [31].

It has to be noted that the finiteness of the volume conductor has peculiar consequences especially for the appearance of the end-of-fiber potentials since they then do belong to the category of so-called *far-field* potentials [18,32,62], characterized by a nondecaying nature in a finite volume conductor [62]. Examples of monopolarly recorded single muscle fiber action potentials, showing the increasing and differential influence of volume conduction on the propagating and nonpropagating components with increasing observation distance, are depicted in Fig. 2.4.

More advanced descriptions of the volume conductor may involve tissue inhomogeneities [59] or the presence of bones [45] and, in case of needle EMG, the presence of the recording needle itself [63].

2.2.4　EMG Detection, Electrode Montages, and Electrode Size

EMG can be detected by intramuscular electrodes or by electrodes mounted on the skin surface. The insertion of electrodes directly into the muscle allows the detection of electric potentials very close to the source, and thus the influence of the volume conductor on the current sources at the fiber membranes is minimal. For this reason the action potentials of the different MUs are reasonably separated in time and can be identified relatively easy at low/medium force levels.

When surface electrodes are applied, the distance between the source and the detection point is significant and the positioning of the electrodes becomes a relevant issue. To remove the common mode components caused by technical interference (such as from

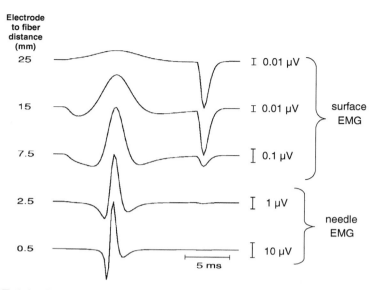

FIGURE 2.4 Calculated single-fiber action potentials from the same muscle fiber. The observation distance to the muscle fiber decreases from 25 mm (**upper trace**, monopolar skin recording from a single fiber in a deep MU) to 0.5 mm (**lowest trace**, monopolar needle electrode recording near the fiber). Note the large amplitude and waveform differences. Note the differences in vertical scaling. The propagation velocity is 4 m/s, half fiber length is 60 mm, and the electrode is positioned at 20 mm from the end-plate zone. Positive deflection downward. Adapted from Blok et al. [9] with permission.

power line) and to partially compensate for the spatial low-pass filtering effect of the tissue separating sources and electrodes (also known in the EEG domain as "blurring"), the surface signals are usually detected as a linear combination of the signals recorded at more than one electrode [21]. Such operation (leading to different electrode "montages") can be viewed as a spatial (mostly high-pass) filtering of the surface EMG signal as it can also be recorded "monopolarly" with a far-away reference electrode (Chapter 5) [17]. The simplest form is the differential detection, the "classic" bipolar montage, using two (+, −) electrodes at a short distance (1–3 cm) in the muscle fiber direction.

The interpretation of the effect of the electrode configuration as a spatial filtering operation and the relation between the time and space domains (Chapter 5) led in the past to the observation that, by properly selecting the weights of the different electrodes, it is possible to introduce zeros of the transfer function of the filter. If they are within the spatial bandwidth of the EMG signal and assuming pure propagation of the IAP along the muscle fibers, these zeros are reflected in the (time-related) frequency spectrum of the EMG and are referred to as spectral dips. The presence of spectral dips has been theoretically shown by Lindstrom and Magnusson [43] for one-dimensional differential detection systems. Spectral dips were found experimentally for one-dimensional differentially detected signals by Lindstrom et al. [44] and later by a number of investigators. More in general, considering both one- and two-dimensional systems with point electrodes or with electrodes with physical dimensions, Farina and

Merletti [24] demonstrated that a sufficient (but not necessary) condition for having a dip in the EMG spectrum at the spatial frequency f_{z0} is:

$$H_{ele}(f_x, f_{z0}) = 0 \qquad \forall f_x \qquad (2.16)$$

where f_x and f_z are the spatial frequencies in the direction transversal and parallel to the muscle fibers, respectively, and $H_{ele}(f_x, f_z)$ is the transfer function of the detection system (including the spatial filter due to the linear combination of signals and to the physical dimensions of electrodes).

The physical size of an electrode also influences the EMG signal. When the electrode–skin impedance between the electrode material and the skin surface is uniformly distributed and when it is low compared to the input impedance of the amplifier, but high compared to the impedances within the tissue, the potential measured by an electrode approximately equals the average of the potential distribution over the skin under it [68]. As a consequence, the influence of the electrode size can also be described as a spatial low-pass filter whereby the electrode dimensions define the filter shaping (Fig. 2.5). As in the case of electrode montages, also the influence of electrode size is largely dependent on structural elements of the EMG

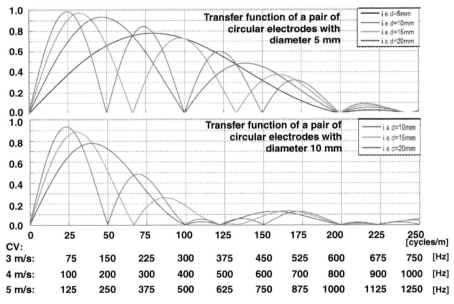

FIGURE 2.5 Transfer function (along the direction of the fibers) for a differential derivation with two circular electrodes of varying interelectrode distance and diameter of 5 mm (**upper plot**) and 10 mm (**lower plot**). The *x*-axis reports the spatial frequency (in cycles/m) as well as the equivalent temporal frequency for three values of conduction velocity. The first dip in the transfer function is inversely related to the interelectrode distance. See also Fig. 19.12 showing the effect of two transfer functions on the sEMG signal.

sources, such as the direction of the muscle fibers with respect to the electrode length or width and the conduction velocity of the muscle fibers.

2.3   ANATOMICAL, PHYSICAL, AND DETECTION SYSTEM PARAMETERS INFLUENCING EMG FEATURES

EMG signal features depend on a number of anatomical, physical, and detection system parameters. The most important of these factors are (1) the thickness of the subcutaneous tissue layer (only for surface recordings), (2) the depth of the sources within the muscle (for surface recordings) and the distance from the source to the electrodes (for intramuscular recordings), (3) the inclination of the detection system with respect to the muscle fiber orientation (mainly for surface recordings), (4) the length of the fibers (mainly for surface recordings), (5) the location of the electrodes over the muscle (or within the muscle in case of intramuscular recordings), (6) the spatial filter (electrode montage) used for signal detection, including the inter-electrode distance, (7) the electrode size and shape (for surface recordings), and (8) crosstalk among nearby muscles (for surface recordings). The influence of all these factors on EMG features has been measured, simulated, and discussed extensively in the specialized literature [14,15,22,25–27,36,42,56,58,64,66].

With increasing thickness of the subcutaneous tissue layer, the signal amplitude decreases [55] and the signal bandwidth is reduced [spatial (and, as a consequence of the propagation, also temporal) low-pass filter] [28]. The effect of the inclination of the muscle fibers with respect to the detection system is difficult to predict since it depends on the spatial filter used, electrode size and shape, and many other factors as well (refer to Farina et al. [22]. The length of the fibers determines the amplitude of the end-of-fiber components. Shorter fibers have relatively large end-of-fiber potentials, with respect to propagating signals. The effect of electrode location, in particular with respect to the tendon and the innervation zones, has been discussed [36,58] and it is well established that for bipolar derivations the best estimates of amplitude, spectral variables, and CV are obtained with electrode locations between the innervation zone and the tendon. For the most common bipolar electrode montage, the innervation zone and tendon regions generate a bias towards high frequencies for the spectral features of the EMG and towards low amplitude values. The spatial filter used (e.g., interelectrode distance in a bipolar montage) for signal detection determines the attenuation of the end-of-fiber components and performs a high-pass temporal filtering of the propagating components. As a consequence, the amplitude is reduced and the frequency content is increased when highly selective spatial filters are used [21]. Interelectrode distance changes the spatial transfer function of the detection system and thus also influences the signal properties (Chapter 5). Finally, electrode size has the effect of decreasing amplitude and frequency content [25] and influences CV estimates when the detection system is not perfectly aligned with respect to the fiber orientation. Because of the above effects, methods based on the interference EMG analysis for extracting neural information (E.G., [5,7,30,53,60]) may be critical.

2.4   CROSSTALK

Crosstalk refers to the signal that is detected over a certain muscle, but is generated by another, mostly nearby muscle [72]. The phenomenon is present exclusively in surface recordings when the distance of the detection points from the sources is of the same order of magnitude for the sources in different muscles. Crosstalk is due to the volume conduction properties in combination with the source properties, and it is one of the most important sources of error in interpreting surface EMG signals. The problem is particularly relevant when the timing of activation of different muscles is of importance, such as in gait analysis (Chapter 16). In the past there have been many attempts to investigate and reduce problems caused by crosstalk.

Morrenhof and Abbink [50] used the cross-correlation coefficient between signals as an indicator of crosstalk, assuming minor shape changes of the signals generated by the same source and detected in different locations over the skin. This assumption was based on a simple model of surface EMG signal generation that did not take into account the generation and extinction of the IAP at the end plate and tendon. Their approach for the verification of the presence of crosstalk was also based on the joint recording of intramuscular and surface EMG signals. The same method was followed by Perry et al. [52], who proposed crosstalk indices based on the ratio between the amplitudes of the surface and intramuscular recordings.

DeLuca and Merletti [11] investigated crosstalk by electrical stimulation of a single muscle and detection from nearby muscles. They provided reference results of crosstalk magnitude for the muscles of the leg and proved the theoretically higher selectivity of the so-called double differential with respect to the bipolar recording (at least for propagating signal components). A similar technique was more recently applied for crosstalk quantification by other investigators with more complex spatial filtering schemes [69].

2.4.1   Crosstalk and Detection System Selectivity

The selectivity of a detection system for surface EMG recording can be defined as the volume of muscle from which the system records signals that are above noise level. Roeleveld et al. [55] did a comprehensive experimental study of the contribution of the potentials of single MUs of the biceps brachii muscle to the surface EMG. A cross-sectional impression of the influence of bipolar electrode montage versus a monopolar recording for a superficial and a deep MU respectively is presented in Fig. 2.6. Intramuscularly detected MU potentials (MUAPs) served as triggers for an averaging process aimed at extracting the surface potentials in different locations over the muscle. The difference between the monopolar and the bipolar recording system is obvious as is the influence of MU depth. It should also be noted from Fig. 2.6 that the potentials detected bipolarly decrease to very small amplitudes within 20–25 mm from the source.

The previous findings might suggest that the potential generated by a source decays rather fast in space, and therefore crosstalk should be a limited problem. However, from Fig. 2.6b, it can be observed that for deep MUs there hardly is a spatial

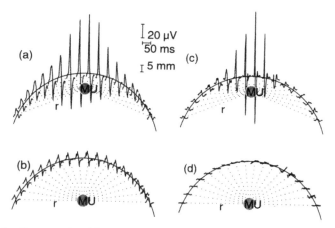

FIGURE 2.6 Cross-sectional impression, over the surface of the biceps brachii, of the action potentials of a superficial (**a, c**) and a deep (**b, d**) MU. Monopolar (far away reference) (**a, b**) and bipolar (**c, d**) detection with a transversal electrode array. These data on individual MUs were experimentally obtained in a study in which so-called intramuscular scanning EMG was combined with surface EMG recordings. The interelectrode distance was 6 mm. Reprinted from Roeleveld et al. [56] with permission.

gradient between different detection locations. Although the contribution of the deep MU in the bipolar montage (Fig. 2.6d) seems negligible, there is a very small difference between the electrodes right above the MU and those lying further aside.

Farina et al. [26] applied the technique proposed by DeLuca and Merletti [11] (selective muscle stimulation) to the extensor leg muscles and recorded signals by eight contact linear electrode arrays (see also Chapter 5). Their results, in agreement with references 20 and 71, are summarized by the representative signals in Fig. 2.7. In this case the signals detected far from the source are generated when the action potentials of the active muscle extinguish at the tendon region and generate non-propagating components. From Fig. 2.7, different considerations can be drawn: (1) Crosstalk is mainly due to the so-called far-field signals [62,65] generated by the extinction of the potentials at the tendons, (2) the shape of crosstalk signals is different from that of signals detected over the active muscle, (3) as a consequence of point 2, the cross-correlation coefficient is not indicative of the amount of crosstalk, and (4) the bandwidth of crosstalk signals may be even larger than that of signals from the active muscle, and thus crosstalk reduction cannot be achieved by temporal high-pass filtering of the surface EMG signals [16].

It is emphasized that spatial and temporal frequency characteristics are simply related by the conduction velocity v only for components which are traveling at the velocity v without relevant shape changes along the fiber (Chapter 5). In the latter case the temporal frequency f_t is related by a scaling factor to the spatial (in the direction of propagation) frequency f_z (see also Chapter 5). As a consequence, low (high) spatial frequencies result in low (high) temporal frequencies. For nonpropagating

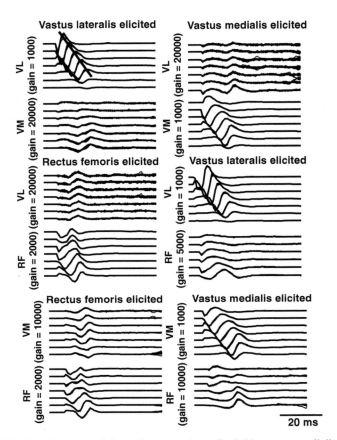

FIGURE 2.7 Signals recorded from the vastus lateralis (VL), vastus medialis (VM), and rectus femoris (RF) muscles in six conditions of selective electrical stimulation (2-Hz stimulation frequency) of one muscle and recording from a muscle pair (one stimulated and one nonstimulated). In each case, the responses to 20 stimuli are shown. For this subject, the distances between the arrays (center to center) are 77 mm for the pair VL-RF, 43 mm for the pair VM-RF, and 39 mm for the pair VL-VM. Note the different gains used for the stimulated and nonstimulated muscles. In this case, the ratio between average ARV values (along the array) is, for the six muscle pairs, 2.8% (VL-VM), 2.5% (VM-VL), 6.5% (RF-VL), 6.6% (VL-RF), 15.6% (RF-VM), and 7.0% (VM-RF). The signals detected from a pair of muscles are normalized with respect to a common factor (apart from the different gain) and so are comparable in amplitude, whereas the signals detected from different muscle pairs are normalized with respect to different factors. Reprinted from Farina et al. [26] with permission.

components this does not occur, as illustrated in Fig. 2.7. Indeed, if we consider the crosstalk signals (nontraveling components which arise over the nonstimulated muscle), we may observe that their temporal frequencies are often higher than those of the signals detected from the stimulated muscles. However, considering the rate of variation of the signal in the spatial direction (that is, along the array), we note that the signal is almost constant (being nontraveling), thus that it is spatially almost "DC" but changes rapidly in time. High-pass filtering in the temporal domain may therefore enhance, rather than reduce, the crosstalk components, which have high temporal and low spatial frequency content. For the same reason, it is not possible to predict the amount of crosstalk from models that do not take into consideration the end-of-fiber components; indeed, the rate of decay of the propagating part of the signal is considerably faster than that of the nontraveling potentials [18,62], so that the latter are mostly responsible for crosstalk. Clinical considerations concerning the relevance of crosstalk are mentioned in Chapter 16.

The theoretical transfer functions, in the spatial domain, of spatial filters used for EMG detection are also based on the assumptions of signals propagating with the same shape along the direction of the filter (Chapter 5). When discussing the frequency content of the crosstalk signals, as discussed above, the correspondence between time and space is not valid. Thus, the theoretical high-pass transfer function of the spatial filters cannot in general be applied to signals detected far from the source. Considering the nontraveling components as signals in the time domain which may be different when detected at different locations along the muscle fibers, the application of a spatial filter with electrodes with weights a_i provide the following signal:

$$V(t) = \sum_{i=0}^{M} a_i R_i(t) \tag{2.17}$$

with M being the number of electrodes. Since the (monopolar) signals $R_i(t)$ depends on a number of anatomical and physical factors, there is no possibility to predict their shape. Equation (2.17) provides the general expression of a nontraveling signal as detected by a spatial filter comprised of point electrodes. It is not possible to write Eq. (2.17) as a linear filtering operation and, thus, to derive the transfer function in space domain for components which are not traveling along the fiber direction. In particular, the effect of specific spatial filter transfer function on the signals $R_i(t)$ cannot be predicted, as indicated in Fig. 2.8 [21]. Figure 2.9 shows the decrease of signal amplitude with increasing distance from the source for different detection systems. The propagating and the nonpropagating components of the signal both decrease with the distance, depending on the spatial filters. Faster decrease of the propagating signal component does not imply also faster decrease of the nonpropagating signal components. Selectivity with respect to these components should be addressed separately.

2.5   EMG AMPLITUDE AND FORCE

Force production in a muscle is regulated by two main mechanisms (see Chapter 1), namely, the recruitment of additional MUs and the increase in firing rate of the already active MUs [39,67]. These two mechanisms are present in different proportions in different muscles. As for the developed force, also the amplitude of the surface EMG

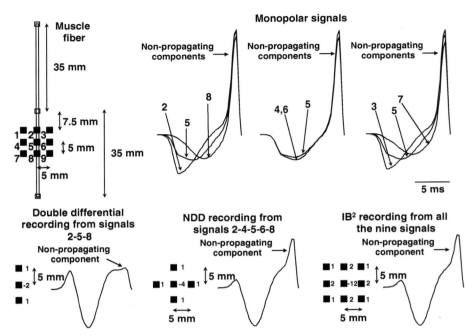

FIGURE 2.8   Simulated examples of monopolar signals detected with square (1 mm × 1 mm) electrodes in the longitudinal, transversal, and diagonal direction. The waveform filtered by the double differential (*DD*), Laplacian (*NDD*), and inverse binomial filter of the second-order (*IB²*) filters ([17], Chapter 7) are also shown. The signals are generated by the same fiber with equal semilengths of 35 mm, at a depth of 7 mm within the muscle. Thickness of the fat and skin layer is 4.5 mm and 1 mm, respectively. Reprinted from Farina et al. [22] with permission.

signal depends on both the number of active MUs and their firing rates [10]. Since both EMG amplitude (estimated in one of the ways presented in Chapters 4 and 10) and force increase as a consequence of the same mechanisms, it is expected that muscle force may be estimated from surface EMG analysis [61].

The possibility of estimating muscle force from the EMG signal is very attractive since it allows the assessment of the contributions of single muscles to the total force exerted by a muscle group. This is the main reason why EMG is and will probably always be the method of choice for force estimation in kinesiological studies. The problem has been addressed experimentally by many researchers in the past (e.g., Milner-Brown and Stein [48]). In some muscles, such as those controlling the fingers, the relationship between force and EMG amplitude was found to be linear [6,73] while in others the relation is closer to a parabolic shape [41,73]. Differences in the percentage of recruitment and rate coding have been considered the most likely explanation for these different relations. It should be noted, however, that even a linear relation between EMG and force cannot be explained when starting straightforwardly from the neural drive to a muscle. Indeed both the force and the EMG amplitude are in most circumstances nonlinearly related to the neural drive. For

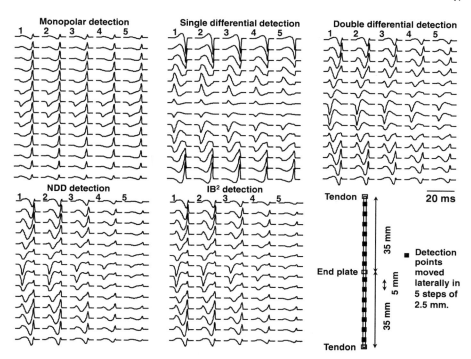

FIGURE 2.9 Simulated single fiber potentials detected with linear arrays with 14 longitudinal detection points separated by 5 mm. Monopolar, single differential (*SD*), double differential (*DD*), Laplacian (*NDD*), and inverse binomial filter of the second-order (*IB²*) signals ([17], Chapter 7) are shown. Signals detected at a lateral distance from the fiber in the horizontal plane from 0 mm to 10 mm in steps of 2.5 mm are shown (labeled by numbers from 1 to 5). All signals are generated by the same fiber with equal semilengths of 35 mm, at a depth of 7 mm within the muscle. Thickness of the fat and skin layer is 4.5 mm and 1 mm, respectively. Reprinted from Farina et al. [22] with permission.

instance, doubling the number of recruited MUs or doubling the firing rate of an already active population of MUs will, under the simplest signal theoretical assumptions, lead to an EMG increase by a factor of $\sqrt{2}$. A doubling in MU firing rate also does lead to less than a doubling of the force of that MU. Apparently, both nonlinearities in the relation between neural drive and EMG, on the one hand, and drive and force, on the other hand, seem to balance each other, leading to a often close-to-linear relation between EMG and force [61,70].

Simulation studies are relevant in quantifying these relations. Fuglevand et al. [29] performed a simulation study, which included a number of recruitment modalities. The relationship between EMG amplitude (average rectified value) and force varied, depending on the recruitment strategy and the uniformity among the peak firing rates of the different MUs (Fig. 2.10).

When discussing the issue of the relation between EMG amplitude and force, a number of other factors should be taken into account. First, the surface EMG

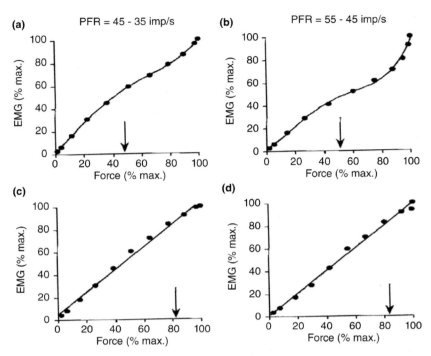

FIGURE 2.10 EMG amplitude–force relationship in simulated signals for narrow (**a, b**) and broad (**c, d**) recruitment range conditions. Peak firing rates were inversely related to the motor unit recruitment threshold and ranged from 35 pps to 45 pps (**a, c**) and from 45 pps to 55 pps (**b, d**). Arrows on force axes show the force at which last unit was recruited for each condition. When recruitment operated over a narrow range (**a, b**), the EMG–force relation was nonlinear. When recruitment operated over a broad range (**c, d**), the relationship between EMG and force was linear. Reprinted from Fuglevand et al. [29] with permission.

amplitude depends strongly on the electrode location. For locations in which EMG amplitude is very sensitive to small electrode displacements, it is expected that the relation between EMG and force may be poorer than in other locations. Jensen et al. [36] elegantly addressed the issue of electrode location in relation to the EMG–force characteristic for the upper trapezius muscle. These authors showed that, while an almost linear relationship between EMG and force could be observed for electrode locations far from the innervation zone, the relationship was far from linear when considering the electrode location over the muscle belly.

Considering an "optimal" electrode placement with minimal sensitivity of EMG amplitude to electrode placement, the relation between force and EMG may depend on the subcutaneous fat layer thickness, the inclination of the fibers with respect to the detection system, the distribution of conduction velocities of the active MUs, the interelectrode distance, the spatial filter applied for EMG recording, the presence of crosstalk, and the degree of synchronization of the active motor units. Moreover, there is a basic factor of variability among different subjects, namely, the location of the MUs within the muscle. The same control strategy may generate signals with different amplitude trends due to the different locations of the MUs within the muscle [23]. All these factors make it impossible to consider any specific EMG-force relation valid in general. This relation should be identified on a subject-by-subject and muscle-by-muscle basis and is likely to be poorly repeatable in different experimental sessions even on muscles of the same subject. The experimental construction of such a relation is difficult, if not impossible, when selective activation of a specific muscle of a group is problematic and other deep muscles (agonists or antagonists) may contribute to force, but not to EMG, as is often the case (see Chapter 10).

Finally, it has to be considered that a relation between force and amplitude should be adapted to the muscle condition like muscle length (joint angle), muscle temperature, and muscle fatigue.

2.6   CONCLUSION/SUMMARY

The intracellular action potentials (IAPs) have a shape which depends on the muscle status and which changes with fatigue. The IAPs are the sources of EMG signals and propagate without shape changes along the muscle fibers. At the end plate and tendon the IAP originates and extinguishes respectively, so that the total current density over the entire muscle fiber length is zero at all times. The generation and extinction of the IAPs produce the so-called end-plate and end-of-fiber effects, which are simultaneously detected over the entire muscle length (nonpropagating) and define the distance beyond which the EMG signal can no longer be detected. This distance is related to the selectivity of the detection, although care should be taken when discussing selectivity with respect to different (propagating and nonpropagating) signal sources. The rate of decrease of signal amplitude with observation distance for the propagating and the nonpropagating components (the two main signal components) of the EMG signal may be very different. These considerations are of importance when the issue of crosstalk is discussed.

Considering all the factors related to the volume conductor and the signal sources that influence the characteristics of the EMG signal, a reliable relation between EMG amplitude and force needs a subject- and condition-specific calibration.

REFERENCES

1. Andreassen, S., and L. Arendt-Nielsen, "Muscle fiber conduction velocity in motor units of the human anterior tibial muscle: a new size principle parameter," *J. Physiol.* **391**, 561–571 (1987).

2. Arendt-Nielsen, L., and K. R. Mills, "The relationship between mean power frequency of the EMG spectrum and muscle fibre conduction velocity," *Electroencephalogr. Clin. Neurophysiol.* **60**, 130–134 (1985).

3. Basmajian, J. V., and C. J. DeLuca, *Muscle Alive*, Williams & Wilkins, Baltimore, 1985.

4. Bernardi, M., F. Felici, M. Marchetti, F. Montellanico, M. F. Piacentini, and M. Solomonow, "Force generation performance and motor unit recruitment strategy in muscles of contralateral limbs," *J. Electromyogr. Kinesiol.* **9**, 121–130 (1999).

5. Bernardi, M., M. Solomonow, G. Nguyen, A. Smith, and R. Baratta, "Motor unit recruitment strategies changes with skill acquisition," *Eur. J. Appl. Physiol.* **74**, 52–59 (1996).

6. Bigland, B., and O. C. J. Lippold, "The relation between force, velocity and integrated electrical activity in human muscles," *J. Physiol.* **123**, 214–224 (1954).

7. Bilodeau, M., A. B. Arsenault, D. Gravel, and D. Bourbonnais, "EMG power spectra of elbow extensors during ramp and step isometric contractions," *Eur. J. Appl. Physiol.* **63**, 24–28 (1991).

8. Blijham, P. J., B. G. M. Van Engelen, and M. J. Zwarts, "Correlation between muscle fiber conduction velocity and fiber diameter in vivo," *Clin. Neurophysiol.* **113**, 39 (2002).

9. Blok, J. H., D. F. Stegeman, and A. van Oosterom, "Three-layer volume conductor model and software package for applications in surface electromyography," *Ann. Biomed. Eng.* **30**, 566–577 (2002).

10. DeLuca, C. J., "Physiology and mathematics of myoelectric signals," *IEEE Trans. Biomed. Eng.* **26**, 313–325 (1979).

11. DeLuca, C. J., and R. Merletti, "Surface myoelectric signal cross-talk among muscles of the leg," *Electroencephalogr. Clin. Neurophysiol.* **69**, 568–575 (1988).

12. Dimitrov, G. V., "Changes in the extracellular potentials produced by unmyelinated nerve fibre resulting from alterations in the propagation velocity or the duration of the action potential," *Electromyogr. Clin. Neurophysiol.* **27**, 243–249 (1987).

13. Dimitrov, G. V., and N. A. Dimitrova, "Precise and fast calculation of the motor unit potentials detected by a point and rectangular plate electrode," *Med. Eng. Phys.* **20**, 374–381 (1998).

14. Dimitrova, N. A., A. G. Dimitrov, and G. V. Dimitrov, "Calculation of extracellular potentials produced by an inclined muscle fiber at a rectangular plate electrode," *Med. Eng. Phys.* **21**, 583–588 (1999).

15. Dimitrova, N. A., G. V. Dimitrov, and V. N. Chihman, "Effect of electrode dimensions on motor unit potentials," *Med. Eng. Phys.* **21**, 479–486 (1999).

16. Dimitrova, N. A., G. V. Dimitrov, and O. A. Nikitin, "Neither high-pass filtering nor mathematical differentiation of the EMG signals can considerably reduce cross-talk," *J. Electromyogr. Kinesiol.* **12**, 235–246 (2002).

17. Disselhorst-Klug, C., J. Silny, and G. Rau, "Improvement of spatial resolution in surface-EMG: a theoretical and experimental comparison of different spatial filters," *IEEE Trans. Biomed. Eng.* **44**, 567–574 (1997).

18. Dumitru, D., "Physiologic basis of potentials recorded in electromyography," *Muscle Nerve* **23**, 1667–1685 (2000).

19. Dumitru. D., *Electrodiagnostic Medicine*, Hanley & Belfus, Philadelphia, 1995.

20. Dumitru, D., and J. C. King, "Far-field potentials in muscle," *Muscle & Nerve* **14**, 981–989 (1991).

21. Farina, D., L. Arendt-Nielsen, R. Merletti, B. Indino, and T. Graven-Nielsen, "Selectivity of spatial filters for surface EMG detection from the tibialis anterior muscle," *IEEE Trans. Biomed. Eng.* **50**, 354–364 (2003).

22. Farina, D., C. Cescon, and R. Merletti, "Influence of anatomical, physical and detection system parameters on surface EMG," *Biol. Cybern.* **86**, 445–456 (2002).

23. Farina, D., M. Fosci, and R. Merletti, "Motor unit recruitment strategies investigated by surface EMG variables," *J. Appl. Physiol.* **92**, 235–247 (2002).

24. Farina, D., and R. Merletti, "A novel approach for precise simulation of the EMG signal detected by surface electrodes," *IEEE Trans. Biomed. Eng.* **48**, 637–645 (2001).

25. Farina, D., and R. Merletti, "Effect of electrode shape on spectral features of surface detected motor unit action potentials," *Acta Physiol. Pharmacol. Bulg.* **26**, 63–66 (2001).

26. Farina, D., R. Merletti, B. Indino, M. Nazzaro, and M. Pozzo, "Cross-talk between knee extensor muscles. Experimental and model results," *Muscle Nerve* **26**, 681–695 (2002).

27. Farina, D., R. Merletti, M. Nazzaro, and I. Caruso, "Effect of joint angle on surface EMG variables for the muscles of the leg and thigh," *IEEE Eng. Med. Biol. Mag.* **20**, 62–71 (2001).

28. Farina, D., and A. Rainoldi, "Compensation of the effect of sub-cutaneous tissue layers on surface EMG: a simulation study," *Med. Eng. Phys.* **21**, 487–496 (1999).

29. Fuglevand, A. J., D. A. Winter, and A. E. Patla, "Models of recruitment and rate coding organization in motor unit pools," *J. Neurophysiol.* **70**, 2470–2488 (1993).

30. Gerdle, B., N. E. Eriksson, and L. Brundin, "The behaviour of the mean power frequency of the surface electromyogram in biceps brachii with increasing force and during fatigue. With special regard to the electrode distance," *Electromyogr. Clin. Neurophysiol.* **30**, 483–489 (1990).

31. Gootzen, T. H. J. M., D. F. Stegeman, and A. Heringa, "On numerical problems in analytical calculations of extracellular fields in bounded cylindrical volume conductors," *J. Appl. Phys.* **66**, 4504–4508 (1989).

32. Gootzen, T. H., D. F. Stegeman, and A. Van Oosterom, "Finite limb dimensions and finite muscle length in a model for the generation of electromyographic signals," *Electroencephalogr. Clin. Neurophysiol.* **81**, 152–162 (1991).

33. Griep, P., F. Gielen, K. Boon, L. Hoogstraten, C. Pool, and W. Wallinga de Jonge, "Calculation and registration of the same motor unit action potential," *Electroencephalogr. Clin. Neurophysiol.* **53**, 388–404 (1982).

34. Gydikov, A., L. Gerilovski, N. Radicheva, and N. Troyanova, "Influence of the muscle fiber end geometry on the extracellular potentials," *Biol. Cybern.* **54**, 1–8 (1986).

35. Hanson, J., and A. Persson, "Changes in the action potential and contraction of isolated frog muscle after repetitive stimulation," *Acta Physiol. Scand.* **81**, 340–348 (1971).

36. Jensen, C., O. Vasseljen, and R. H. Westgaard, "The influence of electrode position on bipolar surface electromyogram recordings of the upper trapezius muscle," *Eur. J. Appl. Physiol.* **67**, 266–273 (1993).

37. Johannsen, G., "Line source models for active fibers," *Biol. Cybern.* **54**, 151–158 (1986).

38. Katz, B., "The electrical properties of the muscle fibre membrane," *Proc. R. Soc. Br. (B)*, **135**, 506–534 (1948).

39. Kukulka, C. G., and H. P. Clamann, "Comparison of the recruitment and discharge properties of motor units in human brachial biceps and adductor pollicis during isometric contractions," *Brain Res.* **219**, 45–55 (1981).

40. Lago, P., and N. B. Jones, "Effect of motor unit firing time statistics on EMG spectra," *Med. Biol. Eng. Comput.* **15**, 648–655 (1977).

41. Lawrence, J. H., and C. J. De Luca, "Myoelectric signal versus force relationship in different human muscles," *J. Appl. Physiol.* **54**, 1653–1659 (1983).

42. Li, W., and K. Sakamoto, "The influence of location of electrode on muscle fiber conduction velocity and EMG power spectrum during voluntary isometric contractions measured with surface array electrodes," *Appl. Human Sci.* **15**, 25–32 (1996).

43. Lindstrom, L., and R. Magnusson, "Interpretation of myoelectric power spectra: a model and its applications," *Proc. IEEE* **65**, 653–662 (1977).

44. Lindstrom, L., R. Magnusson, and I. Petersen, "Muscular fatigue and action potential conduction velocity changes studied with frequency analysis of EMG signals," *Electromyography* **10**, 341–356 (1970).

45. Lowery, M. M., N. S. Stoykov, A. Taflove, and. T. Kuiken, "A multiple-layer finite-element model of the surface EMG signal," *IEEE Trans. Biomed. Eng.* **49**, 446–454 (2002).

46. Merletti, R., L. Lo Conte, E. Avignone, and P. Guglielminotti, "Modelling of surface EMG signals. Part I: model and implementation," *IEEE Trans. Biomed. Eng.* **46**, 810–820 (1999).

47. Merletti, R., M. Knaflitz, and C. J. De Luca, "Myoelectric manifestations of fatigue in voluntary and electrically elicited contractions," *J. Appl. Physiol.* **69**, 1810–1820 (1990).

48. Milner-Brown, H. S., and R. B. Stein, "The relation between the surface electromyogram and muscular force," *J. Physiol.* **246**, 549–569 (1975).

49. Moritani, T., and M. Muro, "Motor unit activity and surface electromyogram power spectrum during increasing force of contraction," *Eur. J. Appl. Physiol.* **56**, 260–265 (1987).

50. Morrenhof, J. W., and H. J. Abbink, "Cross-correlation and cross-talk in surface electromyography," *Electromyogr. Clin. Neurophysiol.* **25**, 73–79 (1985).

51. Nandedkar, S. D., D. B. Sanders, and E. V. Stålberg, "Simulation techniques in electromyography," *IEEE Trans. Biomed. Eng.* **32**, 775–785 (1985).

52. Perry, J., C. Schmidt Easterday, and D. J. Antonelli, "Surface versus intramuscular electrodes for electromyography of superficial and deep muscles," *Phys. Ther.* **61**, 7–15 (1981).

53. Petrofsky, J. S., and A. R. Lind, "Frequency analysis of the surface EMG during sustained isometric contractions," *Eur. J. Appl. Physiol.* **43**, 173–182 (1980).

54. Plonsey, R., "Action potential sources and their volume conductor fields," *IEEE Trans. Biomed. Eng.* **56**, 601–611 (1977).

55. Roeleveld, K., J. H. Blok, D. F. Stegeman, and A. van Oosterom, "Volume conduction models for surface EMG: confrontation with measurements," *J Electromyogr Kinesiol.* **7** (4),221–232 (1997).

56. Roeleveld, K., D. F. Stegeman, H. M. Vingerhoets, and A. Van Oosterom, "The motor unit potential distribution over the skin surface and its use in estimating the motor unit location," *Acta Physiol. Scand.* **161**, 465–472 (1997).

57. Rosenfalck, P., "Intra and extracellular fields of active nerve and muscle fibers. A physicomathematical analysis of different models," *Acta Physiol. Scand.* **321**, 1–49 (1969).

58. Roy, S. H., C. J. DeLuca, and J. Schneider, "Effects of electrode location on myoelectric conduction velocity and median frequency estimates," *J. Appl. Physiol.* **61**, 1510–1517 (1986).

59. Schneider, J., J. Silny, and G. Rau, "Influence of tissue inhomogeneities on noninvasive muscle fiber conduction velocity measurements investigated by physical and numerical modelling," *IEEE Trans. Biomed. Eng.* **38**, 851–860 (1991).

60. Solomonow, M., C. Baten, J. Smith, R. Baratta, H. Hermens, R. D'Ambrosia, and H. Shoji, "Electromyogram power spectra frequencies associated with motor unit recruitment strategies," *J. Appl. Physiol.* **68**, 1177–1185 (1990).

61. Staudenmann, D., K. Roeleveld, D. F. Stegeman, and J. H. van Dieën, "Methodological aspects of SEMG recordings for force estimation—a tutorial and review," *J. Electromyogr. Kinesiol.* **20**, 375–87 (2010).

62. Stegeman, D. F., D. Dumitru, J. C. King, and K. Roeleveld, "Near- and far-fields: source characteristics and the conducting medium in neurophysiology," *J. Clin. Neurophysiol.* **14**, 429–442 (1997).

63. Stegeman, D. F., T. H. Gootzen, M. M. Theeuwen, and H. J. Vingerhoets, "Intramuscular potential changes caused by the presence of the recording EMG needle electrode," *Electroencephalogr. Clin. Neurophysiol.* **93**, 81–90 (1994).

64. Stegeman, D. F., and W. H. J. P. Linssen, "Muscle-fiber action-potential changes and surface emg - a simulation study," *J. Electromyogr. Kines.* **2**, 130–140 (1992).

65. Stegeman, D. F., A. Van Oosterom, and E. J. Colon, "Far-field evoked potential components induced by a propagating generator: computational evidence," *Electroencephalogr. Clin. Neurophysiol.* **67**, 176–187 (1987).

66. Stulen, F. B., and C. J. DeLuca, "Frequency parameters of the myoelectric signals as a measure of muscle conduction velocity," *IEEE Trans. Biomed. Eng.* **28**, 515–523 (1981).

67. Van Bolhuis, B. M., W. P. Medendorp, and C. C. Gielen, "Motor unit firing behavior in human arm flexor muscles during sinusoidal isometric contractions and movements," *Exp. Brain Res.* **117**, 120–130 (1997).

68. van Dijk, J. P., M. M. Lowery, B. G. Lapatki, and D. F. Stegeman, "Evidence of potential averaging over the finite surface of a bioelectric surface electrode," *Ann. Biomed. Eng.* **37**, 1141–1151 (2009).

69. Van Vugt, J. P., and J. G. van Dijk, "A convenient method to reduce crosstalk in surface EMG," *Clin. Neurophysiol.*, **112**, 583–592 (2001).

70. Viitasalo, J. T., and P. V. Komi, "Interrelationships of EMG signal characteristics at different levels of muscle tension during fatigue," *Electromyogr. Clin. Neurophysiol.* **18**, 167–178 (1978).

71. Wee, A. S., and R. A. Ashley, "Volume-conducted or "far-field" compound action potentials originating from the intrinsic-hand muscles," *Electromyogr. Clin. Neurophysiol.* **30**, 325–333 (1990).

72. Winter, D. A., A. J. Fuglevand, and S. E. Archer, "Cross-talk in surface electromyography: Theoretical and practical estimates," *Journ. Electromyogr. Kinesiol.* **4**, 15–26 (1994).

73. Woods, J. J., and B. Bigland-Ritchie, "Linear and non-linear surface EMG–force relationships in human muscle," *Am. J. Phys. Med.* **62**, 287–299 (1983).

3

DETECTION AND CONDITIONING OF SURFACE EMG SIGNALS

R. Merletti, A. Botter, and U. Barone

Laboratory for Engineering of the Neuromuscular System, Politecnico di Torino, Torino, Italy

3.1   INTRODUCTION

Although it is well known that the electromyographic (EMG) signal and its features are strongly influenced by interelectrode distance, electrode size, and placement, the traditional modality for detection of surface EMG (sEMG) signals is still mostly based on a pair of individual electrodes placed on the skin in the region above the muscle. A differential amplifier provides the signal to be displayed and interpreted by the user or by a computer software. The signal detected in this way strongly depends on the location, the interelectrode distance (IED), and the size of the electrode pair and may have very different amplitude and spectral characteristics, depending on the position of the electrode pair along the fiber direction. See Chapter 5 and references 3, 24, 25, 49, and 53.

More advanced techniques are now widely used in research laboratories and are being adopted in clinical settings. Such techniques are based on multichannel detection by means of one- (1-D) or two-dimensional (2-D) electrode arrays [7,19,22,26,45,73]. Figure 3.1 shows some examples of recently developed 2-D arrays. These arrays provide either (a) a spatially sampled *image* or *map* of the analog instantaneous monopolar surface potential distribution or a spatially filtered version of it (e.g., single differential) or (b) a spatially sampled *image* or *map* of EMG features such as amplitude or spectral indicators (e.g., mean or median frequency)

Surface Electromyography: Physiology, Engineering, and Applications, First Edition.
Edited by Roberto Merletti and Dario Farina.
© 2016 by The Institute of Electrical and Electronics Engineers, Inc. Published 2016 by John Wiley & Sons, Inc.

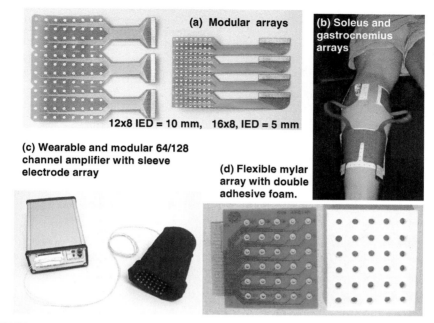

FIGURE 3.1   Examples of electrode arrays. (**a**) Flexible electrode array made by separable modules of 32 contacts each. (**b**) Example of arrays placed on soleus and gastrocnemius muscles. (**c**) Stretchable and flexible sleeve electrode array connected to a portable amplifier system. (**d**) 5 × 6 electrode flexible mylar array with double adhesive foam whose cavities are filled with conductive gel.

estimated over a given time interval (epoch). The extraction and the interpretation of information from such images is discussed in Chapter 5.

The proper acquisition of EMG signals from an electrode array implies the solution of a number of problems. The main ones are listed below and will be discussed in this chapter.

The Single Electrode–Skin Interface, Its Impedance and Noise.   The electrode–skin contact may be *dry* (no gel or conductive paste) or *wet* (gelled), that is, mediated by conductive gel or paste. This contact is intrinsically noisy, with a widely variable impedance depending on the electrode metal and size as well as the gel and skin condition. Table 3.1 provides the order of magnitude of impedance and noise of "small" (few square millimeters) and "large" (hundreds of square millimeters) gelled EMG electrodes.

Sensitivity to Power Line Interference.   Parasitic capacitances with the power line and with ground form voltage dividers and may generate interference voltages much larger than the EMG signal on the skin. The detection system should reject this

TABLE 3.1   Range of Impedance and Noise of Gelled Electrodes of Different Size

Electrode Size	Range of Impedance of an Electrode Pair	Range of Impedance Unbalance	Noise of an Electrode Pair
Small ($5\,mm^2$)	Large (856–2804 kΩ)	Large (30–343 kΩ)	Large (1.62–13.00 μV_{RMS})
Large ($380\,mm^2$)	Small (78–147 kΩ)	Small (1.9–14.1 kΩ)	Small (1.16–2.32 μV_{RMS})

Small electrodes: Four-electrode array (Spes Medica, Battipaglia, Italy), 5×1 mm with10-mm IED.
Large electrodes: Circular electrodes (Kendall Arbo H27PG/F), diameter: 22 mm, with 30-mm IED.
No treatment was applied to the skin. Impedances were measured at 50 Hz and noise was measured in the bandwidth 10–1000 Hz. Both impedance and noise may decrease following skin treatment [55]. IED = interelectrode distance.

interference as much as possible in order to reduce (a) the risk of saturating the amplifier and (b) the need for subsequent processing to reduce the interference off-line.

Balancing the Contact Impedances.   Obtaining electrode–skin contact impedances that are similar and much smaller than the common mode amplifier input impedance is important in order to reduce the effect of common mode voltages, mostly due to parasitic couplings with the power line that are converted into differential mode by the electrode impedance unbalance.

The Transfer Function of the Electrode System.   Whether the detection is monopolar or differential, the finite electrode area implies some degree of spatial low-pass filtering (smoothing) which may be negligible in the case of the small (few mm^2) electrodes of high-density arrays (see Chapter 2).

Electrodes do *sample* the analog instantaneous surface potential distribution in space, and the effect of this sampling operation must be considered carefully. Differential detection implies a form of spatial filtering of the signal whose consequences must be acknowledged and taken into consideration. The distance of detectable sources, generating surface signals separable from the noise level, depends on the electrode montage and on the electrode surface and interelectrode distance [24,25].

3.2   THE ELECTRODE–SKIN INTERFACE AND THE FRONT-END AMPLIFIER STAGE

3.2.1   Electrode–Skin Impedance

Charge is carried by electrons in metals and by ions in electrolytes, such as gels or biological tissues. The continuous bidirectional exchange of carriers at the boundary, even for zero net current, generates noise. Charge layers are created (Helmholtz layer) at the metal–skin (or metal–gel and gel–skin) interface [6]. They contribute to the imaginary (capacitive) component of the impedance. The impedance measured between two electrodes separated by a layer of gel is much lower than the impedance between the same gelled electrodes applied on the skin. For a gelled electrode, most of the impedance is contributed by the gel–skin interface and not by the metal–gel interface, suggesting that further research must be done on the issue of skin treatment and gel composition.

Any single electrode impedance can be described by a series model (impedance $Z_s(\omega) = V(\omega)/I(\omega)$) $Z_s = R_s + jX_s$ or by a parallel model (admittance $Y_p(\omega) = I(\omega)/V(\omega)$) $Y_p = G_p + jB_p$ as described in Fig. 3.2a. To be noted, real and imaginary components do not necessarily represent physical resistors and capacitors and may both be functions of frequency [1]. The single electrode–skin model provided in Fig 3.2b, although very primitive and linear, is more realistic than the one presented in Fig. 3.2a and it includes the half-cell voltage and the generator of the interface noise. The electrode "impedances" (Z_{e1}, Z_{e2}, ... Z_{eN} and Z_{ref} in Fig. 3.3) represent the single contact equivalent Z_e depicted in Fig. 3.2b. More complex models, involving a constant phase element [1], will not be discussed in this book.

Figure 3.3a depicts a front-end stage that can be electronically switched between the monopolar modes (switches in position 2) and the differential modes (switches in position 1). In the classic monopolar configuration, voltages are measured with respect to a reference electrode (Ref) which corresponds to the central point of the front-end power supply (switch S closed). Alternatively, the monopolar montage may use a Ref electrode separated from the isolated ground (switch S open). In this case the detection is actually differential with respect to a common Ref electrode. The

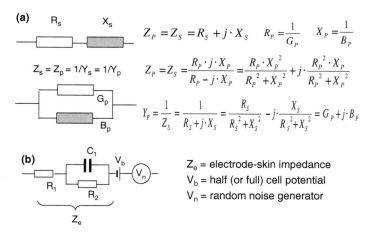

FIGURE 3.2 (a) Series and parallel model of electrical impedance and admittance of an electrode contact. The relations between the real and imaginary parts of the series and parallel models are indicated. Both real and imaginary parts of Z_S and Y_P are functions of frequency and do not represent classic physical resistors and capacitors. (b) A simple physical model of an electrode contact (or of a pair of electrodes) is indicated with the inclusion of the half- (or full-) cell potential (V_b) and of the equivalent noise generator (V_n).

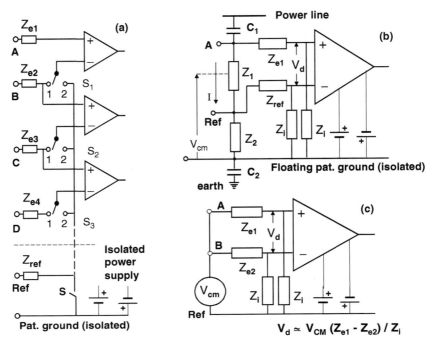

FIGURE 3.3 (a) Switchable monopolar/differential front-end amplifier reading signals with respect to a Ref point which may or may not (Switch S) coincide with the patient ground (center point of the power supply). The electronic switches S_1 . . . .S_N are controlled together by a one bit command. (b) Monopolar floating front-end amplifier. Example: Assuming $C_1 = 1$ pF, $C_2 = 10$ pF, $Z_1 = Z_2 = 10$ kΩ and the power line at $220\,V_{rms}$, $50\,Hz$, we obtain $I = 70\,nA$, $V_{cm} = 1$ mV and $V_{Z1} = 0.7$ mV which is differential and in the range of the EMG amplitude values. As a further example, assuming $Z_i = 300\,M\Omega$ and $Z_{e1} - Z_{ref} = 0.5\,M\Omega$, an additional power line contribution of $1.7\,\mu V$ to V_d is generated. For the same input voltage, this interference is about double and triple at the second and third harmonic of the power line. The effect of additional parasitic capacitances is discussed in Section 5.5 of Merletti and Parker [46]. Other techniques for reducing the common mode voltage, such as the driven right leg (DRL), are described in [67,68] and in Fig. 5.6 of Merletti and Parker [46]. (c) partition of the common mode voltage V_{cm} into the differential voltage V_d. The equation holds for $Z_{e1} \ll Z_i$ and $Z_{e2} \ll Z_i$. If the two input impedance values are different, Z_i may be taken as their average.

reference electrode (theoretically located at a distance from the source, where the potential is assumed to be zero) is placed in a region of no or very limited EMG activity.

In some cases the reference electrode is "virtual" and obtained by averaging (in space, for every sample of time) the voltages of all the electrodes. This solution, referred to as Virtual Reference (VR), often used in ECG and EEG, removes the common mode voltage due to the power line but introduces an often undesirable spatial filtering effect [44].

Figure 3.3b depicts a monopolar detection where the signal of each electrode (A, B, etc.) is detected with respect to a common reference electrode Ref. This monopolar montage is still differential in nature (since each individual electrode voltage is measured with respect to a common reference electrode) but different from the truly differential configuration of Fig. 3.3c. This configuration (Fig. 3.3b) takes advantage of the common mode rejection ratio (CMRR) of each amplifier and is less sensitive to the common mode voltages (e.g., power line interference) with respect to the case of Ref coinciding with the central point of the power supply (Fig. 3.3a with S closed).

Figure 3.3c depicts a traditional differential montage where the EMG signal is detected between two adjacent electrodes, A and B. The sensitivity to the common mode power line voltage is not only due to the finite CMRR but also a result of the unbalance between the individual electrode impedances $Z_{e1} - Z_{e2}$, as described below.

With reference to Fig. 3.3c, the difference ΔZ between Z_{e1} and Z_{e2} unbalances the voltage dividers $Z_i/(Z_{e1} + Z_i)$ and $Z_i/(Z_{e2} + Z_i)$, where Z_i is the input impedance of the system (assumed to be the same for both inputs, Fig. 3.3c), converting a fraction of V_{cm} (common mode voltage mostly due to power-line interference) into V_d which is amplified with the full differential gain. If, as is usually the case, Z_{e1} and Z_{e2} are both much smaller than Z_i, then

$$V_d \approx V_{cm}(Z_{e1} - Z_{e2})/Z_i \qquad (3.1)$$

The output voltage ($V_{cm\text{-}out}$) of the front-end amplifier due to the common mode voltage is therefore due to two additive terms, one due to A_c and one differential (due to ΔZ).

$$V_{cm\text{-}out} = A_c V_{cm} + A_d V_{cm}\Delta Z/Z_i \qquad \text{with } \Delta Z = Z_{e1} - Z_{e2} \qquad (3.2)$$

where A_c and A_d are the common mode and the differential gains of the instrumentation amplifier.

The input impedance of the amplifier, Z_i, is the parallel combination of a resistance (1–100 GΩ) and a capacitance (1–10 pF). The amplifier input impedance at the power-line frequency (50–60 Hz) is dominated by the capacitance effect, is in the range of 0.3–1.0 GΩ, and drops respectively to about one-half and one-third at the second and third harmonic of the power line frequency, enhancing their effect, as indicated in Eqs. (3.1) and (3.2). The example given in Fig. 3.3b indicates that such interference is present, although at a lesser degree, even if the front-end amplifier is floating (isolated from ground). See also Fig. 5.7 and 5.8 of Merletti and Parker [46].

It is therefore important to know and monitor the electrode–skin impedances and their mismatch, before (possibly during) recording, to assess the signal quality. One way to achieve this goal is to apply an artificial common mode voltage and read the resulting signals at the outputs of the differential amplifiers as proposed for ECG detection by Degen and Jäckel [16]. For each differential channel, this output voltage is proportional to ΔZ.

The considerations given above indicate that if differential signals are obtained online or offline (by numerically computing the difference between monopolar signals) from an electrode array, it is important to reduce each *single* electrode–skin impedance and make it similar to the others. To do so, such *single contact* impedance must be measured when procedures to reduce it are investigated. The electrode–skin impedance is not necessarily linear and should be measured using current values close to the real ones, which are in the order of pA. Since this is not easy with currently available instrumentation, values in the range of tens of nA are used.

The impedance between two electrodes is the sum of two electrode–skin impedances plus the interposed tissue impedance. For the sake of simplicity let us consider a discrete model. The impedance of a single electrode–skin contact could be measured using a second much larger electrode whose contact impedance is negligible with respect to the first. Alternatively the method proposed by Grimnes could be adopted [30]. This method uses three electrodes: A current generator is applied between two of them while the third (whose current is forced to be zero by a feedback system) is acting as a tissue voltage detector with no voltage drop on the contact impedance. When an array of electrodes is used, another possible approach is an indirect measure based on the models described in Fig. 3.4.

Figure 3.4a shows how the problem of indirect measure of *single* electrode–skin impedances may be addressed and solved using multiple measurements between the electrode pairs of a four-electrode array and a model of discrete impedance arrangement [55]. The model includes four discrete contact impedances and three discrete tissue impedances. The tissue impedances (Z_T) between contacts A and B and between contacts C and D are in series with Z_A and Z_D, respectively, and are incorporated into $Z'_A = Z_A + Z_T$ and $Z'_D = Z_D + Z_T$ from which they cannot be separated, unless the model is iterated in space by shifting it by one electrode along a longer array.

The matrix A of the resulting system of six linear equations in five unknowns [Eq. (3.3)] has rank 5 (one equation is a linear combination of other equations). The system can be solved either by (a) finding and eliminating the dependent equation and then inverting the resulting square matrix or (b) using the Moore–Penrose pseudo-inversion of matrix A. Impedances Z_B, Z_C, and Z_T can then be obtained (as well as Z_A and Z_D, if the tissue impedance Z_T is assumed to be the same under the three contact pairs $A–B$, $B–C$, and $C–D$). The electrode–skin contact impedances depend on the electrode surface, the subject, and the skin treatment (see Section 3.2.2) and are, in general, one order of magnitude greater than Z_T. If an approximated estimate of the contact impedance is acceptable, Z_T can be neglected with respect to the other impedances and the system reduces to the one described in Fig. 3.4b, where matrix B is square and the system is invertible. All the Z values are, of course, complex.

$$
\begin{bmatrix} Z_1 \\ Z_2 \\ Z_3 \\ Z_4 \\ Z_5 \\ Z_6 \end{bmatrix} = A \begin{bmatrix} Z'_A \\ Z_B \\ Z_C \\ Z'_D \\ Z_T \end{bmatrix}, \quad \text{where} \quad A = \begin{bmatrix} 1 & 1 & 0 & 0 & 0 \\ 0 & 1 & 1 & 0 & 1 \\ 0 & 0 & 1 & 1 & 0 \\ 1 & 0 & 1 & 0 & 1 \\ 0 & 1 & 0 & 1 & 1 \\ 1 & 0 & 0 & 1 & 1 \end{bmatrix} \tag{3.3}
$$

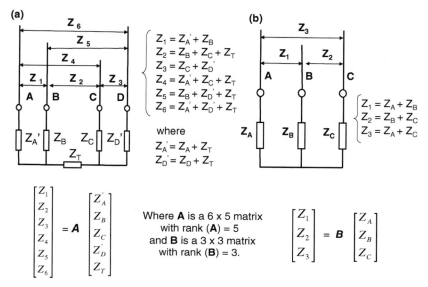

FIGURE 3.4   (a) Model of electrode–gel–skin impedance. A four-electrode array provides six differential impedances between six electrode pairs generating a system of six linear equations in five unknowns, where matrix $\boldsymbol{A}$ is 5×6 and has rank 5. Its Moore–Penrose pseudoinverse provides estimates of the individual impedances (3.3) and (3.4). Under the assumption that the three tissue impedances Z_T are identical, and that the model is valid at each frequency in the EMG bandwidth, Z_A and Z_D can be estimated in the frequency range of interest. If the three Z_T values may not be assumed to be identical, only the central one can be estimated. A longer array would provide additional equations. (b) Experimental measurements indicate that, as a first approximation, Z_T can often be neglected with respect to the contact impedances Z_A, Z_B, Z_C, Z_D and the system reduces to three independent linear equations in three unknowns (3.5) and (3.6). With skin washed with soap and water Z_T is less than 10% of the estimates of Z_A, Z_B, Z_C, Z_D but may be more if the contact impedance is lowered by rubbing the skin with abrasive paste [55]. Other issues concerning the electrode–gel–skin impedance are discussed in [43].

$$
\begin{bmatrix} Z'_A \\ Z_B \\ Z_C \\ Z'_D \\ Z_T \end{bmatrix} = \boldsymbol{A}^{\#}\begin{bmatrix} Z_1 \\ Z_2 \\ Z_3 \\ Z_4 \\ Z_5 \\ Z_6 \end{bmatrix}, \qquad \text{where} \quad \boldsymbol{A}^{\#} = 0.25 \cdot \begin{bmatrix} 2 & -1 & 0 & 1 & -1 & 1 \\ 2 & 1 & 0 & -1 & 1 & -1 \\ 0 & 1 & 2 & 1 & -1 & -1 \\ 0 & -1 & 2 & -1 & 1 & 1 \\ -2 & 1 & -2 & 1 & 1 & 1 \end{bmatrix}
$$

(3.4)

where $\boldsymbol{A}^{\#}$ is the Moore–Penrose pseudoinverse of $\boldsymbol{A}$.

$$
\begin{bmatrix} Z_1 \\ Z_2 \\ Z_3 \end{bmatrix} = \boldsymbol{B}\begin{bmatrix} Z_A \\ Z_B \\ Z_C \end{bmatrix}, \qquad \text{where} \quad \boldsymbol{B} = \begin{bmatrix} 1 & 1 & 0 \\ 0 & 1 & 1 \\ 1 & 0 & 1 \end{bmatrix}
$$

(3.5)

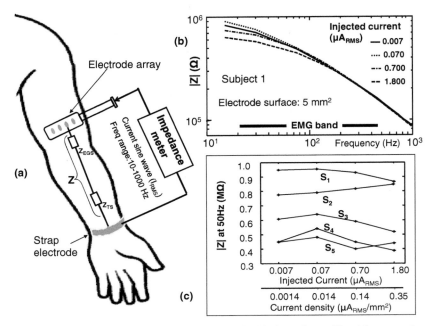

FIGURE 3.5 (a) Measurement of single electrode–gel–skin impedance (Z_{EGS}) between the electrode of interest and a strap with much smaller impedance (less than 5% of that of the single electrode). A four- electrode array (each electrode is 1×5 mm) is placed on the biceps brachii and a wet strap is applied to the wrist. Skin is washed with soap and water and dried with a cloth. (b) Magnitude of Z versus frequency and versus injected current for one subject. (c) Magnitude of Z at 50 Hz for five subjects (S_1 to S_5), showing large intersubject variability.

$$
\begin{bmatrix} Z_A \\ Z_B \\ Z_C \end{bmatrix} = \boldsymbol{B}^{-1} \begin{bmatrix} Z_1 \\ Z_2 \\ Z_3 \end{bmatrix}, \qquad \text{where} \quad \boldsymbol{B}^{-1} = 0.5 \cdot \begin{bmatrix} 1 & -1 & 1 \\ 1 & 1 & -1 \\ -1 & 1 & 1 \end{bmatrix} \tag{3.6}
$$

where $\boldsymbol{B}^{-1}$ is the inverse of $\boldsymbol{B}$.

The relative difference between single impedances measured indirectly through the model and directly (with respect to a large electrode, as in Fig. 3.5a) is $-0.54\% \pm 8.66\%$ (mean $\pm$ SD) (data from Piervirgili et al. [55]). Figure 3.5 shows how the single contact impedance changes with frequency and current applied to the electrode.

Figure 3.6 shows that the electrode impedance decreases with increasing electrode size. This desirable effect has a negative counterpart due to the low-pass filtering effect introduced in space by the nonnegligible electrode size (see Section 3.2.4 and Chapter 2).

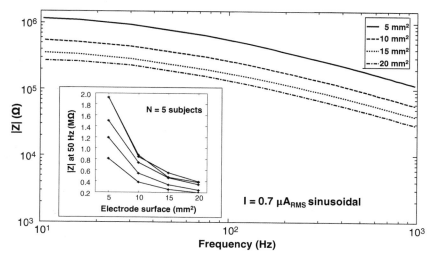

FIGURE 3.6   Effect of electrode surface on the electrode–skin impedance (after washing with soap and water). The inset shows the electrode–skin impedance versus electrode area in five subjects.

The two methods described above (Figs. 3.4 and 3.5) provide the following estimates of ΔZ at 50 Hz (skin washed with soap, 1×5-mm gelled electrodes): median = 88.7 kΩ, interquartile range = 48.7–157 kΩ.

This means that, *for each volt of common mode voltage*, the differential voltage V_d (Fig. 3.3b, c) across the inputs of an instrumentation amplifier having $Z_i = 300$ MΩ at 50 Hz would have, in different subjects, a median of 295 μV and an interquartile range of 162–523 μV. These values cover the entire range of EMG expected amplitude. The situation is worse if the power-line voltage has harmonics, as is often the case. Since the common mode voltage may reach a few volts, the power line interference can be expected to exceed the EMG amplitude, occasionally even by an order of magnitude. It is therefore obvious that it is very important to reduce the value of ΔZ and choose an instrumentation amplifier with very high input impedance, that is, very low input parasitic capacitance.

3.2.2   Effects of Skin Treatment on Impedance and Noise

Impedance.   As described in Section 3.2.1, the electrode–gel–skin impedance (Z_{EGS}) strongly affects the quality of the detected signal. High and nonuniform values of Z_{EGS} in an electrode array imply high power-line interferences in either the monopolar or single differential EMG signals. Since Z_{EGS} is mostly due to the gel–skin contact, different skin treatments have been proposed.

Commonly used skin preparations are: (1) no treatment, (2) rubbing the skin with ethyl alcohol or other solvents to remove oily substances, (3) rubbing the skin with abrasive conductive paste and then cleaning it with a wet cloth, (4) stripping with

TABLE 3.2   Single EGS Contact Impedances and ΔZ (merging $t = 0$ min and $t = 30$ min) Estimated on the Relaxed Biceps Brachii Muscle (No Hair) of 10 Subjects Using Eq. (3.4), at 50 Hz, $I = 200$ nA$_{pp}$, Four Values per Subject. Reproduced with permission from [55].

Skin Treatment	Mean $\pm$ SD	Median (2nd and 3rd Interquartiles)	Range
Magnitude of Single Contact Impedance $\lvert Z_{EGS} \rvert$ (kΩ)			
1. No treatment	845 ± 327	860 (648–1062)	287–1465
2. Rubbing with ethyl alcohol	668 ± 468	655 (266–1054)	23.8–1776
3. Rubbing with abrasive paste	21.8 ± 17.0	15.80 (10.05–29.45)	4.9–78.2
4. Stripping with adhesive tape	725 ± 400	677 (425–1039)	127–1739
5. Washing with soap (30 s) and rinsing	396 ± 351	264 (160–536)	6.5–1612
Impedance difference $\Delta Z = \lvert Z_{e1} - Z_{e2} \rvert$ (kΩ)			
1. No treatment	150 ± 86	163 (70.7–208)	30.2–343
2. Rubbing with alcohol	276 ± 340	161 (55.1–308)	34.3–1317
3. Rubbing with abrasive paste	12.2 ± 12.8	9.35 (2.35–17.05)	0.3–51.5
4. Stripping with adhesive tape	278 ± 226	206 (125–335)	17.90–981
5. Washing with soap (30 s) and rinsing	130 ± 148	88.7 (48.7–157)	4.9–662

Stripping was applied five times with office tape. Gel composition: 1% KCl, 3% HEC (hydroxyethyl cellulose), 1% propylene glycol, 95% preserved water. Electrode area = 5 mm^2, IED = 10 mm. Differences between mean and median indicate asymmetric distribution of the data. Modified with permission from [55].

adhesive tape (apply and remove a strip of adhesive tape several times, to reduce the stratum corneum), and (5) washing the skin with soap and rinsing it with tap water. The effects of these treatments have been investigated by Piervirgili et al. [55] by measuring Z_{EGS} at 50 Hz on the biceps brachii of 10 male subjects using the algorithms described in Fig. 3.4a and the four-contact electrode array depicted in Fig. 3.5a (5×1 mm with interelectrode distance of 10 mm). Measurements were performed with sinusoidal current ($I = 200$ nA$_{pp}$ at 50 Hz) at $t = 0$ min (application time) and $t = 30$ min to study time-related changes. Since the Z_{EGS} values at $t = 0$ min and $t = 30$ min showed small, although statistically significant, decrements with time, they have been merged in Table 3.2. The central impedances Z_B and Z_C (Fig. 3.4a) were obtained and used to estimate ΔZ.

Table 3.2 shows some results in conflict with common beliefs. There is no significant difference between no treatment, rubbing with alcohol or stripping five times with commercial office-type adhesive tape. Rubbing with ethyl alcohol, or other solvents, removes oily substances but leaves the skin very dry and does not increase its conductance. Washing with soap (and rinsing) lowers the impedances and makes them somewhat more uniform. Rubbing with abrasive conductive paste and rinsing is the most effective treatment for lowering Z_{EGS} and ΔZ and therefore for reducing the power line interference (as well as motion artifacts).

Figure 3.6 shows that the contact impedance is reduced when the electrode surface area is increased. Larger electrodes are therefore preferable when a single electrode pair is used but small (often pin-like) electrodes must be used in high-density electrode arrays. Large electrodes imply low-pass spatial filtering, as discussed in Section 3.2.4, in Chapter 2, and in Chapter 5.

Noise. The noise that is present at the output of an instrumentation amplifier (IA) amplifier is due to the intrinsic voltage and current noise referred to the input of the amplifier and to the noise generated at the electrode–skin interface, according to Eq. (3.7), which assumes uncorrelated noise sources:

$$V_{tot}^2 = V_{EGS}^2 + V_{IA}^2 + (Z_{EGS}I_{IA})^2 \qquad (3.7)$$

where V_{tot}^2 is the total noise variance, V_{EGS}^2 is the noise variance of the two electrode–gel–skin systems included in the circuit, V_{IA}^2 is the variance of the instrumentation amplifier noise referred to the input, and $(Z_{EGS} I_{IA})^2$ is the variance of the voltage noise due to the IA input noise current (I_{IA}) flowing through the impedance present between the electrode pair. The amplifier voltage and current noise variances are computed from power densities (V^2/Hz) provided by the amplifier data sheet multiplied by the EMG amplifier bandwidth (500 Hz). All values are referred to the input.

The V_{EGS}^2 contribution may be obtained from Eq. (3.7) by subtracting the other two terms from V_{tot}^2 (measured) and includes the Johnson noise term $4kTBR$, where k is the Boltzman constant, T is the absolute temperature, B is the bandwidth of the amplifier, and R is the real part of Z_{EGS}. The fact that V_{EGS}^2 is not correlated with R [55] indicates that sources other than the Johnson noise generated by R are the dominant component of V_{EGS}^2. Figure 3.7a shows that the noise RMS value (V_{EGS}) between electrode pairs is inversely related to the electrode surface, as expected, due to the averaging effect of the equipotential electrode area. The noise of pin-like contacts often used in high-density electrode arrays is comparable with low-level EMG amplitudes (Table 3.3).

It is tempting to estimate the variance of the noise generated by single gelled electrodes in the same way as the individual impedances are estimated (Fig. 3.4b) by replacing the impedances with voltage noise generators. This model leads to estimates of the variances of the individual sources that are negative, suggesting that this simple approach is inadequate presumably because the noises are not uncorrelated. This issue deserves further investigation.

The variance of the voltage noise generated by an electrode pair depends on the nature of the contacts and therefore on the skin treatment. Table 3.3 shows the RMS value of the V_{EGS} noise obtained from Eq. (3.7) for different skin treatments.

Two gelled Ag–AgCl electrodes (1×5 mm) 10 mm apart generate a noise ranging from about 0.63 μV_{RMS} (treatment number 3) to about 11.8 μV_{RMS} (treatment number 2). The most effective treatment for reducing noise, among those listed in Table 3.3, is rubbing with abrasive conductive paste (treatment number 3). Rubbing with alcohol or stripping does not provide improvement with respect to no treatment.

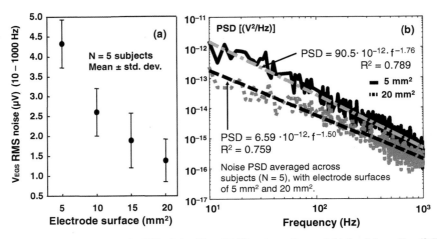

FIGURE 3.7 (a) Means (± SD) of the V_{EGS} RMS noise voltages [obtained from Eq. (3.7)] generated in five subjects by electrode pairs of 5 mm^2, 10 mm^2, 15 mm^2, and 20 mm^2 surface. Average noise levels drop by a factor of 3 as the electrode surface increases by a factor of 4. (b) Spectral density of the noise generated by a pair of electrodes with 5-mm^2 and 20-mm^2 surface. In the 10-Hz to 1000-Hz bandwidth the power spectral density is linear in log–log scale and described by the $P = k \cdot f^{-a}$ model, where k and a depend on the electrode surface. Spectral lines at the frequency of power line and its harmonics have been removed with spectral interpolation. No skin treatment.

TABLE 3.3 Noise RMS Values Between Electrode Pairs for Different Skin Treatments: Measurements with Four-Contact Array Providing Six Electrode Pairs. Reproduced with permission from [55].

Skin Treatment	RMS (mean $\pm$ SD, μV)	RMS Median (2nd and 3rd Interquartile, μV)	RMS Range (μV)
1. No treatment	4.96 ± 2.75	4.14 (2.74–6.49)	1.62–13.0
2. Rubbing with ethyl alcohol	4.32 ± 2.08	3.86 (2.35–5.14)	1.51–11.8
3. Rubbing with abrasive paste	1.67 ± 0.85	1.48 (1.07–1.96)	0.63–6.48
4. Stripping with adhesive tape	4.62 ± 1.48	4.44 (3.64–5.22)	2.17–9.50
5. Washing with soap (30s) and rinsing	3.26 ± 1.23	3.16 (2.41–3.84)	1.24–9.42

Frequency band = 10–1000 Hz. Ten subjects, six measurements/subject, at $t = 0$ min and $t = 30$ min from application (data are merged for $N = 10$ subjects × 6 electrode pairs × 2 measurement times = 120). Differences between mean and median values indicate asymmetric distribution.

3.2.3 Capacitive Electrodes

Conventional electrodes, either wet or dry, behave like transducers converting ionic current (in tissue and gel) into flow of electrons in the metal. These electrical sensors require a careful skin preparation to reduce the impedance and noise associated to this interface (Tables 3.2 and 3.3 and Figs. 3.5, 3.6, and 3.7).

A different approach to the detection of biopotential is to interface the skin surface capacitively with insulating electrodes that do not have a galvanic contact with the body. Capacitive electrodes are made of a conductive plate covered by a layer of dielectric material. As compared to conventional electrodes, they detect electrical displacement currents instead of real charge currents flowing through the electrode–skin interface. With such an electrode, the series resistance (R_E, Fig. 3.8a) of the electrode–skin model is replaced by a coupling capacitance C_E (Fig. 3.8b) ranging from of few picofarads to hundreds of picofarads, depending on the electrode geometry [12].

Since the signal quality does not rely on a good resistive contact, these electrodes require neither skin preparation nor the use of conductive gels. This is a significant advantage for long-term monitoring of the subject (gel might dry or cause skin irritations) and for all the applications where medical staff is unavailable or no subject preparation is easy or possible (e.g., home health monitoring, telemedicine, space applications, etc.) [56,58].

Interfacing the skin surface with capacitive electrodes requires a careful design of the front-end electronics to reduce the occurrence of artifacts in the detected signal. Indeed, the capacitive value associated with the electrode depends on the distance between the metal surface and the skin. This distance might change because of movements inducing temporary variation of C_E and consequently low-frequency artifacts caused by the resulting fluctuations of C_E voltage. In addition, electrode capacitance is very small, resulting in high electrode impedance. Capacity values may range from a few nanofarads, in the case of large aluminum electrodes (with diameter of a few centimeters) covered by a thin layer of oxide and applied directly on the skin surface, to a few picofarads in the case of electrodes positioned over clothing [61,64]. Such capacitances require front-end amplifiers with ultra-high input impedance in order to avoid signal attenuation and ensure an appropriate low-frequency response. In addition, insulating electrodes do not allow the amplifier's bias currents to flow through the body as occurs with conventional electrodes. If a DC path is not provided, bias currents charge the electrode capacitance C_E driving the amplifier into saturation.

The design of front-end amplifiers for capacitive electrodes implies achieving low noise levels and high input impedances providing, at the same time, a DC path for bias currents and a fast recovery from possible saturations due to movement artifacts. Different approaches have been proposed in literature, and they have recently been described in a comprehensive review by Spinelli, Haberman, and co-workers [61,62].

A number of methods have been proposed to provide the DC path required by the bias currents without compromising the high input impedance of the front end amplifier. All the proposed networks behave like a high value resistance R_B (a few teraohms) connecting C_E directly to the reference of the amplifier power supply (Fig. 3.8). The maximum admissible value of R_B is limited by the ratio V_{cc}/I_b (where V_{cc} is the power supply voltage and I_b is the bias current) to avoid saturation of the front-end amplifier. It is worth noting that R_B has an effect on the noise performance of the amplifier (thermal noise and amplifier current noise contributions). Moreover, high values of R_B increase the time constant of the amplifier that, on one hand, do not degrade the low-frequency components of the signal, but on the other hand call for the design of fast recovery networks to recover from possible saturations due to artifacts [62].

In commercially available amplifiers, the input impedance has resistive and capacitive components usually in the range of hundreds of gigaohms and few picofarads, respectively. When capacitive electrodes are used, C_E (the electrode capacitance) becomes comparable with C_1 (Fig. 3.3b) and with C_{IN} (Fig. 3.8), resulting in signal attenuation and in poor rejection of external common mode signals (e.g., power line interference) [50]. Input capacitance can be significantly reduced with the neutralization technique which is based on the injection, through a positive feedback, of a current equal to the current flowing in C_{IN}, thus neutralizing the effect of this capacitance. This technique requires fine adjustments of the feedback gain in order to obtain the exact amount of current needed to compensate that flowing in C_{IN} while avoiding instabilities. Moreover, a neutralization circuit could increase the overall noise level of the first amplification stage (see Spinelli and Haberman [61] for details). A possible complementary approach to increase the input impedance is to use the "power supply bootstrapping" [38,57] that minimizes the current flowing through the input impedance by driving, through feedback, the voltage on the terminal of the input impedance that is usually connected to ground or to a reference.

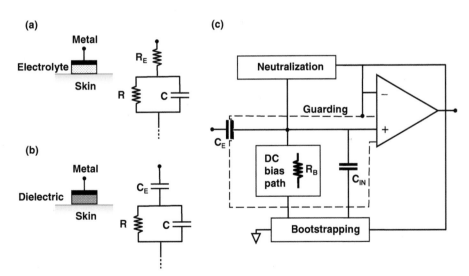

FIGURE 3.8 (a) Gelled electrode and its approximate equivalent circuit. (b) Capacitive electrode and its approximate equivalent circuit. (c) Analog front-end circuit for a capacitive electrode. Circuits that provide closing of bias currents, neutralization of input capacitance, guarding, and bootstrapping (to increase the input impedance) are indicated. See Chi et al. [12] for further information.

The aforementioned methods provide extremely low values of C_{IN} (tens of femtofarads) required for high-quality recordings [13,58]. However, such tight constraints on the amplifier properties require us to address the issue of stray capacitances and resistive paths that could reduce the required high input impedance. For this reason the input of these amplifiers is generally surrounded by a guard screen which is driven at the same potential of the electrode (the output of the input voltage follower). This technique, referred to as "guarding," reduces the current flowing through stray capacitances whose effect is therefore reduced.

Recent advancement in this field provided high-quality recordings through insulating layers and cloths and even in noncontact configurations [32,33,40,57,64]. For certain applications, a limit of this technique is related to the technical difficulties in miniaturizing the electrodes: Currently an electrode area on the order of few square centimeters is needed to obtain acceptable capacitive values and to embody the front-end electronics. This factor limits considerably the number of electrodes that can be used in high-density EMG, where a proper spatial sampling requires closely spaced detection points (see Chapter 2, Chapter 5, and Section 3.2.4). Although attempts to use capacitive electrodes in multichannel configuration can be found in the literature [13,29,33], the current applications are mostly limited to bipolar recordings for ECG monitoring, assistive technology, and gaming applications.

3.2.4   The Transfer Function of the Electrode System

Reading a biosignal from the skin implies detecting the potential difference between one or more points P_i and a "reference" location far from the signal source. This is referred to as monopolar detection from P_i. The voltage difference between two points P_i and P_j is referred to as bipolar or "single differential" detection or "single differential montage." Each EMG signal, either monopolar or single differential, is referred to as an "EMG channel."

A real electrode has a finite conductive surface S which forces the points below the electrode to be equipotential. An electrode is therefore a low-pass filter in space. As a first approximation, let us assume that the electrode voltage is the average of the voltages that would be present, under the electrode surface, without the electrode in place (see Chapter 2 and Chapter 8); that is, each electrode element of unit area contributes with weight $1/S$ and the impulse response of the electrode is therefore $h(x, y) = 1/S$ for x and y within the electrode area, while $h(x, y) = 0$ elsewhere. This is an approximation since the conductive electrode surface somewhat modifies the potential distribution in its surroundings and the exact detected voltage is not exactly the average of the voltage distribution on the surface S with the electrode absent (see Chapter 2 and Chapter 8).

The transfer function of this filter is the Fourier transform of its impulse response. For a rectangular electrode of size $a \times b$ the impulse response is $h(x, y) = 1/ab$ (for $-a/2 < x < a/2$, $-b/2 < y < b/2$, and $h(x, y) = 0$ elsewhere) and the transfer function is:

$$H(f_x, f_y) = \sin(\pi a f_x)/(\pi a f_x) \cdot \sin(\pi b f_y)/(\pi b f_y) \tag{3.8}$$

where f_x and f_y are the spatial frequencies in the x and y directions, expressed in cycles/meter. This function shows a first zero for spatial frequencies $f_x = 1/a$ cycles/m and $f_y = 1/b$ cycles/m. For a square electrode of 5 mm × 5 mm the zero would be for $f_x = f_y = 200$ cycles/m and the 3-dB point at about 89 cycles/m.

For a sinusoidal signal propagating under the electrode at velocity v in the x direction it is:

$$v = \lambda/T = f_t/f_s \qquad (3.9)$$

where λ is the wavelength (meters) of the sinusoid in space, T is the period (seconds), $f_t = 1/T$ (cycles/second or Hertz), and $f_s = 1/\lambda$ (cycles/meter). For $v = 4$ m/s the first zero of the filter is at 800 Hz and the 3-dB point is at about 356 Hz. It is clear that electrodes larger than 5 × 5 mm would begin to affect the EMG spectrum in space and time. For a more detailed analysis see Chapter 2 (Fig. 2.5) and Merletti and Parker [46].

For a circular electrode the transfer function is $H(f_x, f_y) = 2J_1(kr)/kr$, where J_1 is the Bessel function of first kind and order 1, r is the radius of the circle and $k = 2\pi(f_x^2 + f_y^2)^{1/2}$. Circular electrodes with radius greater than 5–6 mm would begin to affect the EMG spectrum in space and time.

Each point-like electrode provides a sample in space of the instantaneous (time variable) potential distribution on the skin surface. A linear array of electrodes provides a sequence of spatial samples along the array, just like sampling a signal in time provides a sequence of samples along time. Figure 3.9 recalls some basic concepts of sampling theory in either space or time, the concept of Fourier transform of sampled signal, and the concept of aliasing (see also Chapter 5).

The traditional EMG detection in kinesiology is based on one electrode pair. The EMG detection is performed using a differential montage with center-to-center interelectrode distance (IED) e in the direction of the fibers (longitudinal single differential). The effect of IED and electrode surface on the sEMG features is quite evident when the signal is collected, for teaching purposes, using two electrode sets, as depicted in Fig. 9.12.

Multiple pairs are used, in the direction of the fibers, to estimate the conduction velocity (CV) of the motor unit action potentials (MUAP), as described in Chapter 5.

If the electrodes are not point-like, sampling along x and y implies the averaging effect described above, resulting in low-pass filtering.

Figure 3.10 illustrates the concept of monodimensional (along a column or a row of the array) and bidimensional sampling in space, associated to sampling in time.

Consider now the differential detection of a potential distribution in space which is moving at a velocity v and consider one harmonic at a time. Consider the harmonic of wavelength λ. If $e = n\lambda$ with $n = 1, 2, \ldots$ the detected voltage will be zero, regardless of the propagation velocity v, because $1, 2, \ldots$ periods in space (wavelengths) will fit exactly within e and the two electrodes will detect two identical sinewaves whose difference is zero. On the other hand, if $e = n\lambda \pm \lambda/2$ (with $n = 0, 1, 2, \ldots$) the detected voltage in time will be a sinusoid with the maximal possible

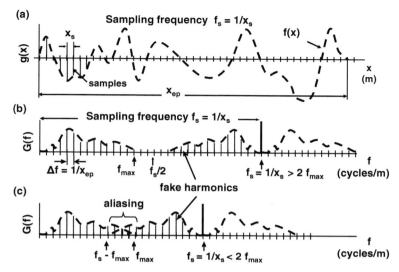

FIGURE 3.9 (a) Example of signal in space $f(x)$ sampled along x with frequency $f_s = 1/x_s$ (samples/meter) with $f_s > 2f_{max}$. (b) Amplitude spectrum of the signal given in (a). Since the signal is sampled, the spectrum is periodic. (c) Amplitude spectrum in the case $f_s < 2f_{max}$. Aliasing is present and spectral cycles partially overlap. The information content of the signal is altered. The same happens in the case of a signal sampled in time.

peak value since the two electrodes will detect the difference between two equal and opposite sinewaves. In intermediate situations the output of the system will be a sinusoid whose amplitude is a function of f_s. As a consequence, the differential detection system is a spatial filter whose output depends on the (spatial) frequency of the input. Recalling the basic relationships between v, time, and spatial frequencies f_t and f_s [Eq. (3.9)], the impulse response and the transfer function in time and in space will be given by the equations reported in Fig. 3.11a. The above considerations and the indications of Fig. 3.11a hold for propagating potentials (not for MUAP generation and extinction effects which contribute nonpropagating components). The magnitude of the transfer function of the differential system is a sinusoid, that is, for $e \ll \lambda$ the system behaves like a differentiator, it shows a maximum for $e = \lambda/2$, it approximates an integrator as λ approaches e, and it presents its first zero or "dip" for $e = \lambda$, indicating an overall band-pass filter behavior (Figs. 3.11b and 3.11c). The location of the dip in the spatial frequency domain is $f_{s\text{-}dip} = 1/e$ (cycles/meter) while that in the temporal frequency domain is $f_{t\text{-}dip} = v/e$ (cycles/second or Hertz) [41]. Two examples of values of the dip frequency are given in Figs. 3.11b and 3.11c.

It is evident from Fig. 3.11a that the Fourier transform of the single differential signal will result from the multiplication of the Fourier transform of the monopolar signal by the transfer function of the filter (in either the spatial or temporal frequency domain). For low conduction velocity values ($v < 4$ m/s) and/or large interelectrode distances ($e > 10$ mm) the first dip of the electrode transfer function will seriously modify the waveform and also seriously modify the amplitude and power spectra of

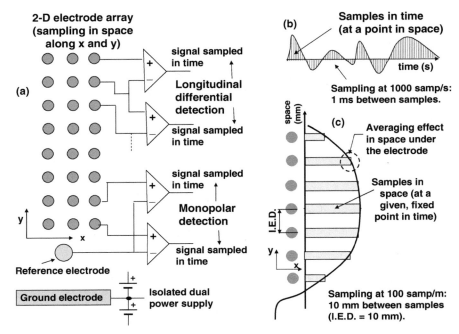

FIGURE 3.10   (**a**) 2-D sampling in space followed by sampling in time of each monopolar or differential channel. (**b**) Sampling in time of one EMG channel. (**c**) Sampling in space of one column of a 2-D electrode array. The samples in space are taken simultaneously. See Chapter 2.

the single differential signal. Examples of "dip" frequencies are reported in Fig. 3.11d. This spatial filtering may or may not be important, depending on the purpose for which the EMG is detected (decomposition, biofeedback, movement analysis). In experimental situations, the dips are "smoothed out" by the spread of the CV values of different motor units which are in the range of 3–5 m/s, corresponding to dips in the range of temporal frequencies of 300–500 Hz for $e = 10$ mm. For this reason the "dips" are not always visible in the amplitude or power spectra of the EMG signal.

Another relevant electrode arrangement, with a different filter transfer function, is the concentric ring system [23]. Further considerations on other spatial filters are given in Chapter 2, Chapter 5, and references 19,24,25.

3.3   STATE OF THE ART ON EMG SIGNAL CONDITIONING AND INTERFACING SOLUTIONS

3.3.1   General Device Features and Specifications

This section is aimed to the technical expert/designer. The features and the design criteria of a single channel EMG amplifier have been extensively described in the literature and summarized in Merletti and Parker [46]. Recent high-density detection systems for "EMG Imaging" introduce additional requirements to manage the issue of

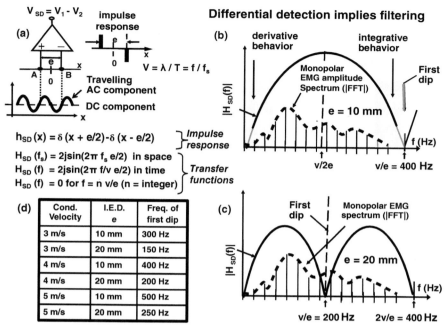

FIGURE 3.11 (a) impulse response and (spatial and temporal) transfer function of a differential detection system. (b) Transfer function of a differential detection system applied to the Fourier transform of the monopolar EMG signal (interelectrode distance $e = 10$ mm). (c) Same as in (b) with $e = 20$ mm. (d) Table of values of the first dip frequencies of the transfer function for three values of conduction velocity and two values of interelectrode distance. See Chapter 2.

"bad channels" or "outliers" [42,44,47] that may jeopardize the extraction of image features.

During EMG measurement sessions the properties of the electrode–skin interfaces are different for different electrodes and change in time due to environmental, chemical, and mechanical factors. For this reason a careful qualitative signal inspection should be performed before, during, and after EMG recording sessions to check the number of bad channels and avoid the recording/processing of poor or meaningless signals. To some degree, this process can be made automatic. Under certain conditions, signals from "bad channels" can be identified and replaced with signals reconstructed using neighboring channels [47]. The most common disturbing events and unpredictable interfering signals are:

1. Motion artifacts caused by variation, or momentary loss, of electrode–skin contact.
2. Movements of cables with consequent variation of parasitic capacitances between charged loose wires.

3. Power-line interference coupling, often associated with contact impedance fluctuations.
4. Fluctuations of electrode polarization and of charge distribution on the skin and on the connecting cables.
5. Change of noise levels due to contact quality fluctuations.

To some degree, the impact of these events on the signal depends on the features of the analog front end (AFE) of the biopotential amplifier.

Several articles and technical notes have been published about the amplifier front end (AFE) whose input is characterized by low amplitude (order of microvolts), frequency bandwidth below 1 kHz, high DC offset (up to ±0.5 VDC), and low SNR (<5 dB). General design criteria have been proposed [28,34,65,70] for system-on-chip and for wearable electronic sensors. A brief summary of amplifier topologies and configurations is given in Table 3.4. SENIAM Recommendations [36], although

TABLE 3.4   Classification of Analog Front-End (AFE) Circuits and Literature References

Feature	Method or Technique		Reference
Supply system	Single power supply (e.g., 0–5 V or 0–3.3 V)		21,48,59
	Dual power supply (e.g., ±5 V or ±2.5 V)		
DC offset suppression technique	AC coupling using second-stage active filtering		52,60
	Input high-pass passive filtering		
	Feed-forward CMOS-based DC removal input circuit		9,54
	Active common mode feedback (CMFB)	Bootstrapped feedback	20,27
		Capacitive extraction with buffered feedback	15
		Continuous-time Integrator	52
		Continuous-time Integrator with DAC-based control	15
		Switched capacitor integrator Floating gate	11
		Differential-to-differential feedback	52
		Digitally assisted DC offset trimming and Input Impedance boosting loop	70
Input stage	Op-amp-based active buffer		48,59
	Transistor-based active buffer (e.g., P-MOS)		17
	Fully differential operational transconductance amplifier (OTA)		34,35,66
	Chopper modulator		8,14
Amplifier circuit topology	Traditional difference amplifier or triple op-amp (instrumentation amplifier)		2,10
	Chopper-stabilized instrumentation amplifier		18,31,69
	Current mode instrumentation amplifier (CMIA)		39

somewhat obsolete, provide specifications and guidelines about EMG signal detection, conditioning, and digitizing rules.

The desirable features of a multichannel AFE for EMG signals are (Fig. 3.12):

1. Efficient removal of the DC due to electrode polarization effect.
2. Very flat differential voltage gain within EMG bandwidth.
3. Low gain mismatch among channels (<0.5%), and good linearity within full voltage dynamics.
4. Transfer function with low group delay within EMG frequency band.
5. Very high common mode rejection within EMG bandwidth (CMRR > 100 dB).
6. High power supply rejection (PSRR > 80 dB).
7. Very high input impedance ($|Z_I| > 100\,M\Omega$) within the EMG frequency band.
8. Small referred-to-input total noise floor level with respect to electrode–skin interface noise (e.g., $<1\,\mu V_{RMS}$ within EMG bandwidth [20–500 Hz]).
9. High accuracy (<0.05%), very low noise ($<3\,\mu V_{PP}$) voltage reference for A/D conversion.
10. Programmable sampling frequency (>1 kS/s/ch) and simultaneous digitizing process of EMG signals. Multiplexers should follow a sample-and-hold circuit (SH) but are often used without it. In this case the delay between samples must be known and taken into account for specific applications (e.g., muscle fiber conduction velocity estimation).
11. High-performance optical isolating interfaces are required to guarantee patient safety (IEC-60601) for systems powered by the power line.

Secondary issues concern low-power and low-voltage operating conditions (for battery-based systems), small-sized integrated circuit, high-quality electrode gel/paste.

A variety of circuits have been developed and are commercially available. Although a detailed description of their operating principles exceeds the purpose of this book, Section 3.3.2 reports a collection of current devices and technologies suitable for multichannel EMG acquisition for EMG imaging.

3.3.2 High-Performance Instrumentation Amplifiers and Acquisition Systems on the Market

Instrumentation amplifier configuration (INA) is commonly used for differential signal measurements. The monolithic instrumentation amplifiers available on the market are grouped in the following categories:

1. Adjustable or fixed gain, traditional triple op-amp configuration:
   a. Digitally selectable gain (internally laser-trimmed).
   b. Externally configurable gain with high-precision resistors.

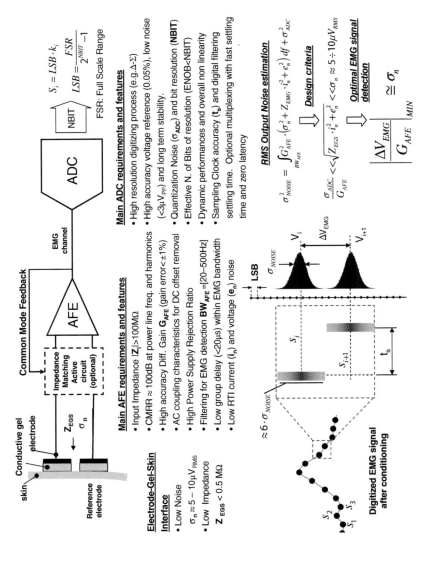

FIGURE 3.12 Summary of one-channel EMG detection chain. Three main blocks represent the recording chain: detection, amplification/conditioning, A/D conversion. Each section is described by a list of the main features, options, and requirements to consider in order to optimize the performances of the whole recording process. Specifically, device selections, circuit topologies (AFE), and high-resolution digitizing process (ADC) are required to implement a high-performance EMG detection system. Further details are provided in [4].

2. Zero drift flying capacitor input stage (patented solutions available from Texas Instruments, Analog Devices, and Linear Technology). Chopper stabilization circuits with internal clock source and adjustable gain by means of external high-precision resistor. An internal low-pass anti-aliasing filter reduces the output noise by external capacitor setting.

3. INA amplifier with additional integrated general-purpose op-amp for signal conditioning implementation (e.g., anti-aliasing filter implementation, AC-coupling circuit, reference voltage circuit).

The main characteristics of commercially available INA circuits are summarized in Table 3.5. Chopper-stabilized input stages, auto-zeroing control blocks, digital potentiometers, and switching capacitors are examples of available advanced options. The chopper technique and the auto-zero loop are continuously self-calibrating circuits which remove the offset voltage drift at the INA inputs and cancel the low-frequency noise (e.g., $1/f$ noise). Patented solutions implement also custom analog blocks able to filter and/or attenuate undesired sources.

The gain accuracy of the instrumentation amplifier (range from $\pm 0.01\%$ to $\pm 0.5\%$) is very important when monopolar signals are acquired and later differentiated in space by software. Any variation between the gains of the amplifiers will be reflected not only in the amplitude of differential signals but primarily in the differential common mode voltage caused by slightly different gains of the monopolar amplifiers. This accuracy depends on both the tolerance of the external resistor (external gain setting option) or the tolerance of the internal resistor network. An external low-pass filter is always required at the output to remove undesired, high-frequency components (e.g., from 10 kHz to 350 kHz according to the chopping frequency) caused by internal circuits. INA333 from Texas Instruments, LT1789-1, LT1167 from Linear Technology, AD8227 and AD8231 from Analog Devices, and MAX419x from MAXIM are monolithic INAs with improved three op-amp topology. Specifically, INA333 is a very low power ($I_{supply} < 50\,\mu A$), chopper-based INA with gain programmable by external resistor. The very high input impedance ($100\,G\Omega \| 3\,pF$) and excellent CMRR (100 dB) are good reasons to select this amplifier for EMG detection in portable applications. LT1789-1 provides high CMRR (>110 dB), good protection of the inputs from static discharges (10 kV), micro-power consumption, and high input impedance ($1.6\,G\Omega \| 1.6\,pF$). AD8227 is a low-cost amplifier which provides an excellent CMRR (110 dB) with low power consumption (less than $350\,\mu A$) but with low input impedance ($400\,k\Omega \| 2\,pF$) compared with that of the electrode–skin interface (Fig. 3.5 and Fig. 3.6).

LT1167 guarantees very low bias current (<350 pA). Both high input impedance ($1000\,G\Omega \| 1.6\,pF$) and CMRR (>140 dB with a differential gain of 1000 V/V) are provided. The power consumption (>1.3 mA) is a drawback for multichannel portable acquisition systems.

AD8231 is a digitally programmable gain amplifier with optional buffer amplifier. Moreover, the digital I/O interface allows the chip selection and power-down features that are useful for battery-powered devices.

TABLE 3.5 Comparison of Selected INA Devices Available on the Market

Model, Manufacturer	CMRR [dB]	PSRR [dB]	Voltage Noise Density RTI [nV/$\sqrt{Hz}$]	Current Noise Density RTI [fA/$\sqrt{Hz}$]	Slew Rate [V/μs]	Full Power BW [kHz]	Max Gain [V/V $\pm e\%$]	Input impedance [GΩ $\|$ pF]	Power [V_{CC}/I_{CC}]
AD8231, Analog Devices	>95	>110	60 $f<10$ kHz	20 $f<2$ kHz	1.1	10	128/$\pm$0.8%	10 $\|$ 5	5 V/4 mA
AD8227, Analog Devices	>100	>60	100 @ 1 Hz 67 @ 500 Hz	250 @ 1 Hz 100 @ 500 Hz $f_{1/f}{}^a=10$ Hz	0.6	>2	1000/$\pm$0.15%	0.8 $\|$ 2	5 V/0.31 mA
AD8235, Analog Devices	>60	>85	400 @ 1 Hz 80 @ 500 Hz $f_{1/f}=20$ Hz	15 (BB^b)	0.009	>0.8	100/$\pm$0.2%	440 $\|$ 1.6	5 V/0.32 mA
AD8553, Analog Devices	>100	>100	1000 (BB)	N A.^{c}	0.005	N.A.	1000/$\pm$0.4%	0.05 $\|$ 1	5 V/1.3 mA
LT1789-1, Linear Technology	>110	>95	550 @ 1 Hz 350 @500 Hz $f_{1/f}=10$ Hz	800 @ 1 Hz 50 @ 500 Hz $f_{1/f}=1$ kHz	0.023	>1	100/$\pm$0.55%	1.6 $\|$ 1.6	5 V/0.07 mA
LT1167A, Linear Technology	>90	>90	300 @ 1 Hz 70 @ 500 Hz $f_{1/f}=10$ Hz	400 @ 1 Hz 60 @ 500 Hz $f_{1/f}=40$ Hz	1.3	>10	1000/$\pm$0.4%	1000 $\|$ 1.6	5 V/0.95 mA
LT6800, Linear Technology	>110	>100	175 (BB) $f<500$ Hz	N.A.	0.2	200	100/$\pm$0.1%	N.A.	5 V/1.3 mA
MAX4194?, MAXIM	>85	>80	800 @ 1 Hz 70 @ 500 Hz $f_{1/f}=5$ Hz	N.A.	0.06	>150	100/$\pm$0.5%	1 $\|$ 4	5 V/0.095 mA

MAX4208, MAXIM	>140	>100	140 @ 10 Hz	N.A.	0.08	0.8	5000/ ± 0.25%	2 ∥ N.A.	5 V/2.3 mA
MAX4461, MAXIM	>80	>10	3000 @ 1 Hz 10 @ 100 kHz $f_{1/f} = 10\,\text{kHz}$	N.A.	0.25	2.5	100/ ± 0.35%	2 ∥ N.A.	5 V/2.8 mA
INA326, Texas Instruments	>60	>70	2000 @ 1 Hz 800 @ 500 Hz $f_{1/f} = 40\,\text{Hz}$ 50 (BB)	200 @ 1 Hz 150 @ 500 Hz $f_{1/f} = 20\,\text{Hz}$ 100 (BB)	Filter-limited	1	$10^{3}/ \pm 0.2\%$	10 ∥ 2	5 V/3.4 mA
INA333, Texas Instruments	>90	>60			0.05	>0.35	1000/ ± 0.5%	100 ∥ 3	5 V/0.05 mA

[a] 1/f noise corner frequency.
[b] Broadband noise shaping response, 1/f noise nulling technique available.
[c] N.A., not available.

MAX419x provides a good gain accuracy (±0.01%) with fixed values (1, 10, 100 V/V). Its shutdown option is useful for power-saving operations in portable solutions. The common mode input impedance (1 GΩ||4 pF) is less than optimal for EMG detection. The implementation of on-chip laser-trimmed resistor network is avoided with current mode, auto-zero instrumentation amplifier topology. This circuit topology works as a differential voltage-to-current converter using a preamplifier stage implemented with sampling capacitors and/or transconductance amplifier (OTA) to store the differential voltage and convert it into a current. An auxiliary offset-nulling amplifier automatically removes the offset drift and low-frequency noise. The device shows a very high CMRR (range from 110 dB to 160 dB), but its input stage has a typical input impedance 10 times lower than that of an equivalent three op-amp topology.

AD8553 from Analog Devices, LT6800 from Linear Technology, and MAX4208 and MAX4461 from MAXIM are examples of monolithic current mode INA. Specifically, the LT6800 implements a 3-kHz sampling capacitor input stage plus a zero drift inverting amplifier to obtain an high CMRR (>110 dB) differential voltage detector.

The INA326 from Texas Instruments implements an unconventional three op-amp topology with a current mode block (90-kHz clocked charge pumps) to overcome the problem of mismatches of resistors which limits the CMRR. The circuit architecture, based on current mirrors, provides a CMRR greater than 110 dB (with a gain of 1000 V/V). The common mode input impedance has a high input capacitance (10 GΩ||14 pF), and the high power consumption (2.4 mA) limits its applications in multichannel portable equipment. In addition, a 1-kHz cutoff low-pass filter is required at the output of the amplifier for the presence of spurious frequency components. These undesired effects reduce the suitability of this device for High-Density EMG detection.

Matching the entire set of requirements for multichannel EMG conditioning is not an easy task. The main INA characteristics, extracted from the datasheets, are reported in Table 3.5 for comparison and to highlight the need for compromises.

For example, very high CMRR devices, such as MAX4208 (140 dB) and LT6800 (110 dB), show low common mode input impedances and higher power consumption than devices with lower CMRR (e.g., MAX4194 or INA333). On the contrary, low-power INA such as INA333 amplifier (50 μA) provide relatively low CMRR (>90 dB) but very high input impedance. The patented circuits implemented on these devices (e.g., auto-zeroing techniques in MAX4208, chopper technique in INA333, current mode signal conversion, and processing in the INA326) aim to reduce input offset error, drift over time and temperature, and the effect of $1/f$ noise. These features are, in general, desirable but not really critical for EMG due to the high-pass filter usually set at 10–20 Hz.

The tuning of the circuit performances and the selection of the right INA to use should be based on requirements such as conditioning of monopolar or differential signals, number of channels and power requirements, acceptable cost, etc., as well as on more specific technical issues (e.g., gain bandwidth product, input bias currents, input impedance, CMRR, PSRR, referred-to-input current and voltage noise densities, gain accuracy, etc.).

TABLE 3.6 Comparison Table Defining Qualitative Device Profiles and Trends ($\downarrow$ = low; $\uparrow$ = high; $\updownarrow$ = intermediate) of Important Parameters Used to Classify High-Performance, OPA-Based, Monolithic Instrumentation Amplifier Devices (INA)

INA Profile	Parameters					
	Power Consumption	CMRR	Gain Bandwidth Product	SNR[a]	Z_{in}	Models
Low power		$\downarrow$	$\downarrow$	$\downarrow$	$\uparrow$	INA333, LT1789-1, MAX4194
High CMRR	$\uparrow$		$\downarrow$	$\updownarrow$	$\downarrow$	MAX4208, LT6800, AD8553
Low noise	$\uparrow$	$\updownarrow$	$\downarrow$		$\uparrow$	AD8231, AD8227, AD8235
General purpose	$\updownarrow$	$\updownarrow$	$\updownarrow$	$\updownarrow$	$\updownarrow$	MAX4461, INA326, LT1167A

[a]The signal-to-noise ratio (SNR) is the ratio between the signal amplitude and the noise (RMS values).

For example, INA devices with chopper input stage and auto-zeroing circuits (e.g., INA333) compensating the input offset voltage and reducing the low-frequency noise (below 100 Hz) is an alternative to high-pass filtering. The high-pass filtering at 10 Hz is also a solution for reducing low-frequency fluctuations, $1/f$ noise, and DC offset, as well as the slow components of movement artifacts.

It must be emphasized that high CMRR by itself is not a guarantee of high power-line rejection since this is also a function of Z_{in} (Fig. 3.3c). For example, an INA with very high CMRR, such as MAX4208 (CMRR = 140 dB), has an unknown input capacitance and a possibly low input impedance at 50–60 Hz which might degrade power-line rejection. The design rule reported in Eq. (3.10) is useful to evaluate the relevance of the voltage divider effect (see Section 3.2.1) and compare it with the magnitude of the not rejected common mode voltage due to a finite CMRR. The rule links the CMRR value to the input impedance Z_i and to the expected maximum electrode–skin impedance mismatch.

$$V_{cm} \cdot A_d \cdot \frac{\Delta Z_e}{Z_i} \ll \frac{V_{cm}}{\text{CMRR}} \ll A_d \cdot V_{\text{EMG min}} \Rightarrow \begin{cases} Z_i \gg A_d \cdot \text{CMRR} \cdot \Delta Z_e \\ \text{CMRR} \gg V_{cm}/(A_d \cdot V_{\text{EMG min}}) \\ \Delta Z_e = Z_{e1} - Z_{e2} \end{cases}$$

$$(3.10)$$

In the case of the MAX4208, and considering an electrode–skin impedance mismatch of 100 kΩ at power-line frequency (50 Hz/60 Hz), when a unity differential gain is set, the input impedance should be significantly greater than 1 TΩ ($Z_i \gg 1 \text{ V/V} \cdot 10^7$.

$100 \, k\Omega = 1 \, T\Omega$) at 50–60 Hz to avoid degrading the common mode rejection capability level of 140 dB. For EMG signals this makes the excellent CMRR not so important. Similarly, expensive very low input noise amplifiers do not improve the system and are not justified unless the noise of the electrode skin interface is equally low, which is usually not the case (Table 3.3 and Fig. 3.7).

The number of EMG amplifiers equals the number of EMG channels. When a high number of channels is required, the cost of the acquisition system becomes an important factor. Analog multiplexing techniques, based on sample-and-hold (SH) circuits and analog switches, could be considered to reduce the power consumption, the cost, and the size of the hardware. Unfortunately, multiplexing of low-level signals in the EMG amplitude range (tens to hundreds of microvolts) introduces unacceptable artifacts and measurement errors due to tolerances, charge injection into parasitic capacitances, and couplings due to fabrication technology. Specifically, the process of "sample-and-hold" of the EMG signal by storing an electric charge in a capacitor is potentially affected by parasitic components. In particular, the analog switching devices are digitally driven and internally affected by parasitic capacitances which cause undesired charge injection during switching transients and channel-to-channel crosstalk. The amplitude of the voltage spikes produced during switching limits the voltage amplitude detectable with acceptable artifacts. For this reason an EMG amplifier should be interposed between each electrode and the analog multiplexer to increase the ratio between the EMG signal and the undesired transients generated before the A/D conversion (ADC). The tolerances of the passive components, the nonidentical gains and transfer functions of the preamplifiers, and the charge losses of the SH stage affect the accuracy of the stored charge, thereby reducing the overall accuracy of the EMG recording chain. For this reason, when a multiplexing device is used, it is preferable to place it after the amplifier chain, before the ADC input. The current technology does not allow multiplexing the original low-level signals which must therefore be amplified with one amplifier per channel.

Analog multiplexing, after signal amplification, reduces the number of ADC converters. Fast ADC devices, such as SAR (successive approximation register) converters are suitable, and may be necessary, to digitize all the EMG signals with a single ADC during the sampling interval because it executes an A/D conversion within a few microseconds. The conversion time and the settling time of the analog multiplexer establish the minimum time between two consecutive A/D conversions.

The total single-channel conversion time is the most important parameter to consider in order to correctly estimate how many EMG signals could be digitized in one conversion cycle (scanning all channels) by a single multiplexed ADC. High-performance ADC devices on the market are equipped with an internal multiplexing stage that allows to digitize a group of input signals (from 4 to 16) without the use of external SH units, saving space and improving the overall performance level at the cost of nonsimultaneous sampling. The outputs of each group of channels are stored in digital registers and then read, by suitable electronics, in the proper sequence.

Sample and hold devices for multiplexing technique are desirable, when the number of multiplexed signals increases, to avoid undesired delays among digitized EMG signals and assure synchronous acquisition. Without using the "hold" feature, the EMG amplitudes of the channels would not be captured by ADC inputs in the same instant. Therefore, the samples would be affected by a variable skew in time (dT_{ch}), a multiple of the ADC conversion time T_{conv}, depending on the channel position pos_{ch} in the multiplexing array ($dT_{ch} = pos_{ch} \times T_{conv}$). This error, if not correctly compensated by offline signal interpolation operations, could affect the accuracy of estimation of some EMG features (e.g., conduction velocity).

To overcome the problems related to multiplexing, an alternative approach is based on the use of single chip, multichannel, simultaneous sampling ADC devices based on high-resolution, parallel $\Delta\Sigma$ modulators. The delta–sigma technique increases the digital resolution with respect to SAR ADC, up to 24 bits or more. This technique requires a noise shaping digital filter to produce the output digital word [37]. The filter has an initial transient that requires to discard a few samples. For this reason the multiplexing technique is not allowed in the case of the $\Delta\Sigma$ system due to the settling time of this digital filter. Simultaneous sampling is guaranteed by multiple $\Delta\Sigma$ converters (as many as the number of EMG channels) which work synchronously within the same monolithic device. The high digital resolution and simultaneous sampling represent two significant quality factors in terms of accuracy in digital recording of many EMG channels. In addition, high-performance $\Delta\Sigma$ ADCs implement an oversampling feature which configures the real sampling of input signals with a clock multiple of the basic sampling ($f_{oversampling}$). This technique drastically reduces the quantization noise within the baseband (BW) and maximizes the real number of noise-free bits of resolution of the digitizing process [effective number of bits (ENoB) parameter] by improving the Signal-to-quantization noise ratio (SNR$_{ADC}$) according to Eqs. (3.11) and (3.12) (see Kester [37] for details):

$$\text{SNR}_{ADC} = \frac{1\text{LSB} \cdot 2^{\text{NBIT}}/2\sqrt{2}}{1\text{LSB} \cdot \sqrt{12}} = 1.76\,\text{dB} + 6.02 \cdot \text{NBIT}\,[\text{dB}] \tag{3.11}$$

$$\text{SNR}_{ADC}^{\Delta\Sigma} = \text{SNR}_{ADC} + 10 \cdot \log_{10}\frac{f_{\text{oversampling}}}{2 \cdot BW}$$

$$= 1.76\,\text{dB} + 6.02 \cdot \text{NBIT} + 10 \cdot \log_{10}\frac{f_{\text{oversampling}}}{2 \cdot \text{BW}}\,[\text{dB}] \tag{3.12}$$

The number of noise-free bits (NFB) is an important parameter which quantifies the quality of the ADC device under selection. It represents the part of the digital word containing the range of amplitude of the input signal greater than the noise floor. The oversampling feature increases this range by improving the signal-to-noise ratio and the noise-free bit as a consequence. For example, in case of $\Delta\Sigma$ ADC with an oversampling factor of 64, and a sampling frequency of 2 ksps, the signal-to-quantization noise ratio for EMG digitizing (BW = 500 Hz) increases by 21 dB (from Eq. (3.11) we obtain $10\log_{10}(128\,\text{ksps}/1000\,\text{Hz}) = 21\,\text{dB}$).

3.4 ASIC SOLUTIONS ON THE MARKET

The use of application-specific integrated circuits (ASIC) for bioelectric signal detection has been rapidly growing since 2001. The biomedical sector offers small-sized, high-cost ASIC devices for biopotential measurements. Three examples of ASIC currently available are described below. The devices presented below provide compact solutions with a excellent performance level in terms of size and power consumption. On the contrary, the analog characteristics (e.g., CMRR, input impedance, gain accuracy, noise floor) delineate a quality level equal to the medium level of EMG amplifier based on discrete component implementations.

3.4.1 IMEC

IMEC produces ASIC devices suitable for biosignal applications (e.g., EEG detection) where reduced size and micro-power features are required (e.g., see references 51, 70, and 71 for details). The proposed amplifiers provide solutions to minimize motion artifact, specifically in the case of dry electrodes. For example, an interesting solution was proposed by Xu et al. [70] by implementing a digitally controlled input impedance boosting circuit that minimizes the voltage dividing effect due to electrode skin impedances mismatch (Fig. 3.3) by means of an array of capacitive loads which are dynamically controlled by a common mode feedback circuit.

3.4.2 Intan Technologies, LLC

Figure 3.13 shows the RHA 2000 family of amplifiers with fixed gain (200 V/V) and externally programmable frequency bandwidth from 100 Hz to 20 kHz by resistor setting. An externally driven fast analog multiplexer (1-MHz maximum switching frequency) canalizes the signals at the output. The typical analog performances are:

- CMRR = 85 dB
- RTI noise floor = $2 \mu V_{RMS}$
- Input impedance $Z_i = 1.3$ GΩ @ 10 Hz and 13 M Ω @ 10 kHz (capacitive)
- Power consumption around 10 mW @ 3.3 V_{cc} (current drain = 3 mA)

New versions with on-board ADC and SPI output are available.

The amplifier stage (see Harrison [34] for details) shows an effective space saving solution (32 channels version occupies only 64 mm^2). Moreover, the single supply voltage, and low power requirement (10 mW at 3.3 V_{DC}) simplifies the design of the power supply. The configurable bandwidth and sampling frequency guarantee versatility and customization of output characteristics. The cost of this device is justified for compact and lightweight multichannel acquisition systems. A full integrated version of this device includes the digitizing section (RHD2000 family). The frame of 24-bit samples is continuously retrieved by means of a standard SPI interface which is used also for the dynamic configuration of the device.

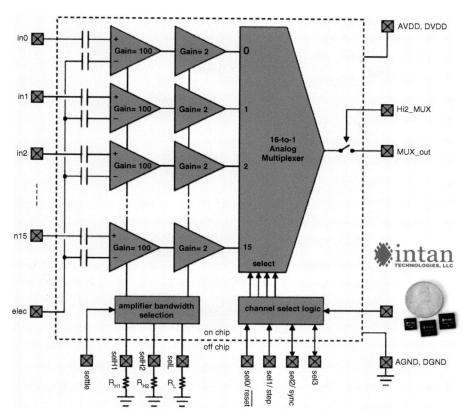

FIGURE 3.13   RHA2000 family, fully integrated, multichannel AFE array with configurable frequency band and fixed voltage gain (www.intantech.com/products_RHA2000.html). Reproduced with permission from Intan Technologies.

3.4.3   Texas Instruments

The ADS1298 device is an eight-channel simultaneous sampling and A/D conversion device with a programmable gain/bandwidth amplifier (PGA: 1 V/V; 232 kHz → 12 V/V; 32 kHz) array with configurable interconnection matrix useful for measurement configuration setting without modifying the external connections with the electrodes (e.g., selection of best patient reference electrode). The high-resolution digitizing section is based on eight parallel $\Delta\Sigma$ converters, with 24 bits of resolution. The digitized samples are simultaneously acquired and retrieved by SPI interface which is also used for device configuration. Additional analog circuits are implemented for ECG/EEG measurements (driven right leg, Wilson node, lead-off arrangement, and on-board oscillator for sampling frequency setting). The typical performances are (see ADS1298 Datasheet 2012 for details): (1) CMRR = 115 dB, (2) referred to input (RTI) noise floor = 1.8 μV_{RMS} with programmable gain amplifier (PGA) configured with gain 12 V/V and bandwidth 2 kHz, (3) input resistance $Z_{iDC} = 1\,G\Omega$, and (4) power consumption $6\,mW$ (low power mode) to $20\,mW$ (high-resolution mode). The device is mainly focused on EEG or ECG applications but could be a possible solution for EMG detection due to its remarkable configurability. Moreover, the system for detection of electrode disconnection from the skin and the driven-right-leg circuits are very useful features. The system provides a compact solution for eight-channel, space-saving ($64\,mm^2$), monopolar or single differential EMG detection system.

3.5   PERSPECTIVES FOR THE FUTURE

While the use of an individual electrode pair per muscle will remain a widely used EMG technique in the clinical environment, the use of electrode arrays and the EMG imaging technique will certainly expand in the future as the technological problems of the electrode–skin interface will be solved (see Chapter 5). Research in this field is strongly interdisciplinary, covering the fields of chemistry, dermatology, textiles, skin-like electronic substrates, organic semiconductors, and material sciences. New gels and/or skin treatments or functional electrode layers must be developed and assessed to improve the critical issue of the electrode–skin interface. Techniques based on new textiles and inks are being investigated [5,72]. Wearable sleeves with dry or capacitive electrodes requiring minimal or no skin treatment will represent an asset for telemedicine and telemonitoring, sport, space, and rehabilitation medicine.

Applications are expected to range from physiopathological investigations (studies of spasticity, co-contraction, work-related disorders, cramps, tremor, etc.), to rehabilitation games, biofeedback applications, and sport training.

REFERENCES

1. Alexe-Ionescu, A. L., G. Barbero, F. C. M. Freire, and R. Merletti, "Effect of composition on the dielectric properties of hydrogels for biomedical applications," *Physiol. Meas.* **31**, S169–S182 (2010).

2. Amer, M., "Novel design of a bioelectric amplifier with minimized magnitude and phase errors," *J. Electronics (China)* **18**, 242–254 (2001).

3. Barbero, M., R. Merletti, and A. Rainoldi, *Atlas of Innervation Zones: Understanding Surface Electromyography and Its Applications*, Springer, Milan, Italy, 2012.

4. Barone, U., and R. Merletti, "Design of a portable, intrinsically safe multichannel acquisition system for high-resolution, real-time processing HD-sEMG," *IEEE Trans. Biomed. Eng.* **60**, 2242–2252 (2013).

5. Beckman, L., C. Neuhaus, G. Medrano, N. Jungbecker, K. M. Walter, T. Gries, and S. Leonhardt, "Characterization of textile electrodes and conductors using standardized measurement setups," *Physiol. Meas.* **31**, 233–247 (2010).

6. Birgersson, U., E. Birgersson, I. Nicander, and S. Ollmar, "A methodology for extracting the electrical properties of human skin," *Physiol. Meas.* **34**, 723–736 (2013).

7. Blok, J. H., J. P. van Dijk, G. Drost, M. J. Zwarts, and D. F. Stegeman, "A high-density multichannel surface electromyography system for the characterization of single motor units," *Rev. Sci. Instrum.* **73**, 1887–1897 (2002).

8. Bronskowski, C., and D. Schroeder, "Systematic design of programmable operational amplifiers with noise-power trade-off," *IET Circuits Devices & Syst.* **1**, 41–48 (2007).

9. Carrillo, J. M., J. L. Ausin, P. Merchan, and J. F. Duque-Carrillo,"Feedback vs. feedforward common-mode control: a comparative study," *Proceedings of the IEEE International Conference on Electronics, Circuits and Systems*, Lisbon, 363–366 (1998).

10. Chang C. W., and J. C. Chiou,"Surface-mounted dry electrode and analog front-end systems for physiological signal measurements," *Proceedings of the IEEE/NIH Life Science Systems and Applications Workshop*, Bethesda, pp. 108–111, 2009.

11. Charles C. T. and R. R. Harrison,"A floating gate common mode feedback circuit for low noise amplifiers," *Proceedings of the Southwest Symposium on Mixed-Signal Design*, Las Vegas, pp. 180–185, 2003.

12. Chi, Y. M., T. P. Jung, and G. Cauwenberghs, "Dry-contact and noncontact biopotential electrodes: methodological review," *IEEE Rev. Biomed. Eng.* **3**, 106–119 (2010).

13. Clippingdale, A., R. Prance, T. Clark, and C. Watkins, "Ultrahigh impedance capacitively coupled heart imaging array," *Rev. Sci. Instrum.* **65**, 269–270 (1994).

14. Dagtekin, M., L. Wentai, and R. Bashirullah, "A multi channel chopper modulated neural recording system," *Proceedings of the 23rd International Conference of the IEEE*, vol. 1, pp. 757–760, 2001.

15. Degen, T., and H. Jackel, "A pseudodifferential amplifier for bioelectric events with DC-offset compensation using two-wired amplifying electrodes," *IEEE Trans. Biomed. Eng.* **53**, 300–310 (2006).

16. Degen, T., and H. Jäckel, "Continuous monitoring of electrode skin impedance mismatch during bioelectric recordings," *IEEE Trans. Biomed. Eng.* **55**, 1711–1715 (2008).

17. Degen, T., S. Torrent, and H. Jackel, "Low-noise two-wired buffer electrodes for bioelectric amplifiers," *IEEE Trans. Biomed. Eng.* **54**, 1328–1332 (2007).

18. Denison, T., K. Consoer, W. Santa, A. T. Avestruz, J. Cooley, and A. Kelly, "A 2 μW 100 nV/$\sqrt{\text{Hz}}$ chopper-stabilized instrumentation amplifier for chronic measurement of neural field potentials," *IEEE J. Solid-State Circ.* **42**, 2934–2945 (2007).

19. Disselhorst-Klug, C., J. Silny, and G. Rau, "Improvement of spatial resolution in surface-EMG: a theoretical and experimental comparison of different spatial filters," *IEEE Trans. Biomed. Eng.* **44**, 567–574 (1997).

20. Dobrev, D. P., T. Neycheva, and N. Mudrov, "Bootstrapped two-electrode biosignal amplifier," *Med. Biol. Eng. Comput.* **46**, 613–619 (2008).

21. Dobrev, D., "Two-electrode low supply voltage electrocardiogram signal amplifier," *Med. Biol. Eng. Comput.* **42**, 272–276 (2004).

22. Drost, G., D. F. Stegeman, B. G. van Engelen, and M. J. Zwarts, "Clinical applications of high-density surface EMG: a systematic review," *J. Electromyogr. Kinesiol.* **16**, 586–602 (2006).

23. Farina, D., and Cescon, C., "Concentric ring electrode system for non-invasive detection of single motor unit activity," *IEEE Trans. Biomed. Eng.* **48**, 1326–1334 (2001).

24. Farina, D., and R. Merletti, "Effect of electrode shape on spectral features of surface detected motor unit action potentials," *Acta Physiol. Pharmacol. Bulg.* **26**, 63–66 (2001).

25. Farina, D., C. Cescon, and R. Merletti, "Influence of anatomical, physical, and detection-system parameters on surface EMG," *Biol. Cybern.* **86**, 445–456 (2002).

26. Farina, D., R. Merletti, and R. M. Enoka, "The extraction of neural strategies from the surface EMG," *J. Appl. Physiol.* **96**, 1486–1495 (2004).

27. Fay, L., V. Misra, and R. Sarpeshkar, "A micropower electrocardiogram amplifier," *IEEE Trans. Biomed. Circuits Syst.* **3**, 312–320 (2009).

28. Fuchs, B., S. Vogel, and D. Schroeder, "Universal application-specific integrated circuit for bioelectric data acquisition," *Med. Eng. Phys.* **24**, 695–701 (2002).

29. Gebrial, W., R. J. Prance, R. J. Harland, and C. J. Clark, "Non-invasive imaging using an array of electric potential sensors," *Rev. Sci. Instrum.* **77**, 063708–063713 (2006).

30. Grimnes, S., "Impedance measurement of individual skin surface electrodes," *Med. Biol. Eng. Comput.* **21**, 750–755 (1983).

31. Harb, A., Y. Hu., M. Sawan, A. Abdelkerim, and M. M. Elhilali, "Low power CMOS interface for recording and processing very low amplitude signals," *Analog. Integr. Circ. S* **39**, 39–54 (2004).

32. Harland, C. J., T. D. Clark, and R. J. Prance, "Electrical potential probes new directions in the remote sensing of the human body," *Meas. Sci. Technol.* **13**, 163–169 (2000).

33. Harland, C. J., T. D. Clark, N. S. Peters, M. J. Everitt, and P. B. Stiffell, "A compact electric potential sensor array for the acquisition and reconstruction of the 7-lead electrocardiogram without electrical charge contact with the skin," *Physiol. Meas.* **26**, 939–950 (2005).

34. Harrison, R. R.,"A versatile integrated circuit for the acquisition of biopotentials," *Proc. IEEE Custom Integrated Circuits Conf*, San Jose, 115–122 (2007).

35. Harrison, R. R., and C. Charles, "A low-power low-noise CMOS amplifier for neural recording applications," *IEEE J Solid-St Circ.* **38**, 958–965 (2003).

36. Hermens, H., B. Freriks, C. Disselhorst-Klug, and G. Rau, "Development of recommendations for SEMG sensors and sensor placement procedures," *J. Electromyogr. Kinesiol.* **10**, 361–374 (2000).

37. Kester, W., *The Data Conversion Handbook*, 3rd ed., Analog Devices Inc., 2004.

38. Lanyi, S., and M. Pisani, "A high input-impedance buffer," *IEEE Trans. Circuits Syst.* **49**, 1209–1211 (2000).

39. Li, J. T., S. H. Pun, M. I. Vai, P. Un Mak, P. In Mak, and F. Wan,"Design of current mode instrumentation amplifier for portable biosignal acquisition system," *Proc. Biomed Circuits and Syst Conf*, Beijing, 9–12 (2009).

40. Lim, Y., K. Kim, and K. Park, "ECG recording on a bed during sleep without direct skin-contact," *IEEE Trans. Biomed. Eng.* **54**, 718–724 (2007).

41. Lindstrom L. and R. Magnusson, "Interpretation of myoelectric power spectra: a model and its applications," *Proc. IEEE* **65**, 653–662 (1977).

42. Marateb, H. R., M. Rojas-Martínez, M. Mansourian, R. Merletti, and M. A. Villanueva, "Outlier detection in high-density surface electromyographic signals," *Med. Biol. Eng. Comput.* **50**, 79–89 (2012).

43. Merletti, R., "The electrode-skin interface and optimal detection of bioelectric signals," *Physiol. Meas.* **31**, 3, editorial (2010).

44. Merletti, R., A. Botter, C. Cescon, M. A. Minetto, and T. M. M. Vieira, "Advances in surface EMG: recent progress in clinical research applications," *Crit. Rev. Biomed. Eng.* **8**, 347–379 (2010).

45. Merletti, R., A. Holobar, and D. Farina, "Analysis of motor units with high-density surface electromyography," *J. Electromyogr. Kinesiol.* **18**, 879–890 (2008).

46. Merletti, R., and P. Parker, *Electromyography: Physiology, Engineering and Non-invasive Applications*, IEEE Press and John Wiley & Sons, Hoboken, NJ, 2004.

47. Merletti, R., M. Aventaggiato, A. Botter, A. Holobar, H. R. Marateb, and T. M. M. Vieira, "Advances in surface EMG: recent progress in detection and processing techniques," *Crit. Rev. Biomed. Eng.* **38**, 305–345 (2010).

48. Merritt, C. R., H. T. Nagle, and E. Grant, "Fabric-based active electrode design and fabrication for health monitoring clothing," *IEEE T Inf. Technol. B* **13**, 274–280 (2009).

49. Mesin, L., R. Merletti, and A. Rainoldi, "Surface EMG: the issue of electrode location," *J. Electromyogr. Kinesiol.* **19**, 719–726 (2009).

50. Metting van Rijn, A., C. Peper, and C. Grimbergen, "The isolation mode rejection ratio in bioelectric amplifiers," *IEEE Trans. Biomed. Eng.* **38**, 1154–1157 (1991).

51. Mitra, S., J. Xu, A. Matsumoto, K. A. A. Makinwa, C. Van Hoof, and R. F. Yazicioglu,"A 700 μW 8-channel EEG/contact-impedance acquisition system for dry electrodes," *Proceedings of the Symposium on VLSI Circuits Digest of Technical Papers*, Honolulu, pp. 68–69, 2012.

52. Nikola, J., P. Ratko, D. Strahinja, and P. B. Dejan,"A novel AC-amplifier for electrophysiology: active DC suppression with differential to differential amplifier in the feedback-loop," *Proceedings of the 23rd Int Conference of the IEEE Eng in Med and Biology Soc*, Istanbul, 3328–3331, 2001.

53. Nishihara, K., Y. Chiba, Y. Suzuki, H. Moriyama, N. Kanemura, T. Ito, K. Takayanagi, and T. Gomi, "Effect of position of electrodes relative to the innervation zone on surface EMG," *J Med Eng Tech* **34**, 141–147 (2010).

54. Parthasarathy, J., A. G. Erdman, A. D. Redish, and B. Ziaie,"An integrated CMOS Biopotential amplifier with a feed-forward DC cancellation topology," *Proceedings of the 28th Int Conference of the IEEE Eng in Med and Biology Soc*, New York, 2974–2977, (2006).

55. Piervirgili, G., F. Petracca, and R. Merletti, "A new method to assess skin treatments for lowering the impedance and noise of individual gelled Ag–AgCl electrodes," *Physiol. Meas.* **35**, 2101–2118 (2014).

56. Prance, H., "Sensor developments for electrophysiological monitoring in healthcare," in G. D. Gargiulo and A. McEwan, eds., *Applied Biomedical Engineering*, InTech, Rijeka, 2011.

57. Prance, R. J., A. Debray, T. D. Clark, H. Prance, M. Nock, C. J. Harland, and A. J. Clippingdale, "An ultra-low-noise electrical-potential probe for human-boy scanning," *Meas. Sci. Technol.* **11**, 291–297 (2000).

58. Prance, R. J., S. T. Beardsmore-Rust, P. Watson, C. J. Harland, and H. Prance, "Remote detection of human electrophysiological signals using electric potential sensors," *Appl. Phys. Lett.* **93**, 033906–033908 (2008).

59. Spinelli, E. M., H. Nolberto, M. Martinez, and A. Mayosky, "A single supply biopotential amplifier," *Med. Eng. Phys.* **23**, 235–238 (2001).

60. Spinelli, E. M., R. Pallas-Areny, and M. A. Mayosky, "AC-coupled front-end for biopotential measurements," *IEEE Trans. Biomed. Eng.* **50**, 391–395 (2003).

61. Spinelli, E., and M. Haberman, "Insulating electrodes: a review on biopotential front ends for dielectric skin–electrode interfaces," *Physiol. Meas.* **31**, S183–S198 (2010).

62. Spinelli, E., M. Haberman, P. García, and F. Guerrero, "A capacitive electrode with fast recovery feature," *Physiol. Meas.* **33**, 1277–1288 (2012).

63. Texas Instruments, "Low-Power, 8-Channel, 24-Bit Analog Front-End for Biopotential Measurements," ADS1298 Datasheet, Jan. 2010 [Revised Jan. 2012].

64. Ueno, A., T. Yamaguchi, T. Iida, Y. Fukuoka, Y. Uchikawa, and M. Noshiro, "Feasibility of capacitive sensing of surface electromyographic potential through cloth," *Sens. Mater.* **24**, 335–346 (2012).

65. Van Helleputte, N., J. M. Tomasik, W. Galjan, A. Mora Sanchez, D. Schroeder, W. H. Krautschneider, and R. Puers, "A flexible system-on-chip (SoC) for biomedical signal acquisition and processing," *Sens. Actuat. A—Phys.* **142**, 361–368 (2008).

66. Wang, C. C., C. C. Huang, J. S. Liou, and K. W. Fang,"A 140-dB CMRR low-noise instrumentation amplifier for neural signal sensing," *Proceedings of the IEEE Asia Pacific Conference Circuits and Systems* Singapore, 696–699 (2006).

67. Winter, B. B., and J. G. Webster, "Driven-right-leg circuit design," *IEEE Trans. Biomed. Eng.* **30**, 62–66 (1983).

68. Winter, B. B., and J. G. Webster, "Reduction of interference due to common mode voltage in biopotential amplifiers," *IEEE Trans. Biomed. Eng.* **30**, 58–62 (1983).

69. Witte, J. F., K. A. A. Makinwa, and J. H. Huijsing, "A CMOS chopper offset stabilized opamp," *IEEE J. Solid-St. Circ.* **42**, 1529–1535 (2007).

70. Xu, J., X. Jiawei, R. F. Yazicioglu, B. Grundlehner, P. Harpe, K. A. Makinwa, and C. Van Hoof, "A 160 μW 8-channel active electrode system for EEG monitoring," *IEEE Trans. Biomed. Circuits Syst.* **5**, 555–567 (2011).

71. Yazicioglu, R. F., P. Merken, R. Puers, and C. Van Hoof, "A 60 μW 60 nV/Hz readout front-end for portable biopotential acquisition systems," *IEEE J. Solid-State Circ.* **42**, 1100–1110 (2006).

72. Yu, Y., J. Zhang, and J. Liu, "Biomedical implementation of liquid metal ink as drawable ECG electrode and skin circuit," *PLoS One* **8**, e58771 (2013).

73. Zwarts, M. J., and D. F. Stegeman, "Multichannel surface EMG: basic aspects and clinical utility," *Muscle Nerve* **28**, 1–17 (2003).

4

SINGLE-CHANNEL TECHNIQUES FOR INFORMATION EXTRACTION FROM THE SURFACE EMG SIGNAL

E. A. Clancy,[1] F. Negro,[2] and D. Farina[2]

[1]*Electrical & Computer Engineering Department; Biomedical Engineering Department, Worcester Polytechnic Institute, Worcester, Massachusetts*
[2]*Department of Neurorehabilitation Engineering, Bernstein Center for Computational Neuroscience, University Medical Center Goettingen, Goettingen, Germany*

4.1 INTRODUCTION

The previous chapter has indicated a number of detection modalities (or electrode montages) for acquiring sEMG signals (monopolar, single or double differential, high density). A raw sEMG signal is usually processed to mainly extract information concerning the "amplitude" of the signal (root mean square value, RMS; average rectified value, ARV; linear envelope, LE) and its power spectral density (through the Fourier or autoregressive approach). The mean and median frequency (MNF and MDF) are obtained from the power spectral density. These parameters provide information, respectively, about the muscle contraction "strength" and about the frequency content of the signal, which is, in turn and in certain conditions, one of the many myoelectric manifestations of muscle fatigue. These parameters are well known in the field and clinically used in movement and rehabilitation sciences. More recently studied features, such as signal entropy and fractal dimension, are less known and not clinically applied. The algorithms to obtain amplitude and spectral parameters have been traditionally applied to single sEMG signals obtained by one pair of electrodes, but can also be used to estimate the spatial distribution of each parameter on the skin

Surface Electromyography: Physiology, Engineering, and Applications, First Edition.
Edited by Roberto Merletti and Dario Farina.
© 2016 by The Institute of Electrical and Electronics Engineers, Inc. Published 2016 by John Wiley & Sons, Inc.

surface, as described in Chapters 5 and 10. Single-channel parameters may vary from place to place in space, on the surface of the skin above a muscle. Finally, most if these parameters are estimated over a signal "epoch" of a given length; others are obtained by low-pass filtering of the rectified or squared signal, which implies the use of past data values. In other words, they have an "integral nature" and the time duration over which they are calculated (or the time constant of the filter) must always be provided together with the feature value, which is not constant in time, even for a stationary signal.

This chapter describes some of the most commonly used techniques for processing single-channel sEMG signals and is largely based on Chapter 6 of the book edited by R. Merletti and P. Parker in 2004 [69]. Processing the signal means to apply algorithms to extract parameters or features to be used for some purpose such as signal classification or quantification of changes. Some basic knowledge of signal theory (the concepts of complex numbers, convolution, Fourier transforms, auto-correlation, and stochastic processes) is assumed [94]. The single-channel techniques described in this chapter address the interference pattern that results from the simultaneous activation of many motor units (MUs). These techniques do not resolve, or decompose, the signal into its individual motor unit action potential trains (MUAPTs) (discussed in Chapter 7); rather they provide a global description of the electric potential observed at the recording site and of its information content. The way to represent such information is usually not directly related to physiological phenomena or events. For example, the description of a signal as a sum of sine waves does not imply that the signal originated as the sum of sine waves by any physiological mechanism. Modeling of a random signal as the output of a filter with random noise at its input neither implies that the signal was generated in such a way nor implies that the filter has a physiological meaning. Nevertheless, these mathematical representations are indeed powerful tools to detect and quantitatively describe the recorded signal and its changes resulting from physiological events. Estimates of the EMG amplitude are used as the control input to myoelectric prostheses (Chapter 20) and as indicators of muscular activity (Chapters 10, 17, and 19). In the frequency domain, the dominant change in the single-channel EMG during sustained, high-effort contractions is a compression (or scaling) of the signal spectrum towards lower frequencies. Measures of this compression are associated with myoelectric manifestations of muscle fatigue in the underlying muscle (Chapter 10).

In the time domain, the dominant changes in the single-channel sEMG are: (a) the modulation of the signal standard deviation (RMS) or of the ARV and (b) the spectral changes due to muscular effort and/or fatigue. As a muscle effort increases, the signal strength (or amplitude) grows. Spectral changes can be evaluated in the time domain with simple techniques based on counting the zero crossings of the signal or on spike analysis. Moreover, in the frequency domain, coherence measures between neural signals (e.g., EEG/MEG-EMG or EMG-EMG) can provide information about shared oscillations in the synaptic inputs to the motor neuron pools [41,97]. This approach has been extensively applied to the study of motor control and movement disorders [38].

The current state of the art of these methods is described in the subsequent sections as well as in Chapter 10. Two sections will introduce these descriptions. The first provides a review of spectral estimation whereas the second describes "traditional" stochastic models applicable to the EMG signal. These models are used to develop, interpret, and test most of the signal processing techniques described in this chapter.

4.2 SPECTRAL ESTIMATION OF DETERMINISTIC SIGNALS AND STOCHASTIC PROCESSES

4.2.1 Fourier-Based Spectral Estimators

The energy spectral density of a finite energy deterministic discrete time signal $m(k)$ is by definition the magnitude squared of the discrete time Fourier transform of the signal $|M(e^{j\omega})|^2$ and represents, as a consequence of Parseval's relation [94], the distribution of signal energy as a function of frequency. The power spectral density (PSD), $S_{mm}(e^{j\omega})$, of a wide-sense stationary (WSS) discrete time stochastic process[1] with zero mean is by definition the discrete time Fourier transform of its auto-correlation sequence:

$$S_{mm}(e^{j\omega}) = \sum_{k=-\infty}^{k=+\infty} r_{mm}(k)e^{-jk\omega} \tag{4.1}$$

where $e^{-jk\omega}$ represents the k th sinusoidal harmonic and $r_{mm}(k)$ is the autocorrelation function defined as $r_{mm}(l) = E[m(k+l)m(k)]$.

In Eq. (4.1), the computation of the autocorrelation sequence implies an expectation whose calculation would require the availability of all the realizations of the process. In practical applications, collection of these data is not possible. However, in the case of ergodic processes, the autocorrelation sequence can be estimated from a single realization by substituting the expectation operation with a temporal average [94]. Given a finite number of samples, the autocorrelation sequence can therefore be estimated as

$$\hat{r}_{mm}(k) = \frac{1}{L} \sum_{l=0}^{L-1-k} m(k+l)m(l), \qquad 0 \le k < L \tag{4.2}$$

where $m(k)$ is a single process realization and L is the number of acquired signal samples. It can be shown that the estimator defined in Eq. (4.2) is a biased estimator of the autocorrelation sequence. Replacing $r_{mm}(k)$ with $\hat{r}_{mm}(k)$ in Eq. (4.1) provides an estimate of the power spectrum of the process. The estimated autocorrelation

---

[1]When both the mean and autocorrelation of a discrete time stochastic process are invariant to time shifts (and the second moment is finite), the process is said to be wide-sense stationary (WSS). In this case, the mean is a single number μ and the autocorrelation can be represented by a sequence of numbers $r_{mm}(l) = E[m(k+l)m(k)]$, where $E[]$ is the expectation operator and k and l are discrete time indices. The EMG signal recorded during isometric constant force contractions can be considered a WWS process, at least for time intervals short enough to exclude fatigue (see below in the text). In the following, we will consider only zero mean WSS stochastic processes.

sequence is generally windowed in order to reduce estimation bias, leading to a class of correlogram-based estimators.

It can be shown that the power spectral estimate based on the discrete Fourier transform of the correlation sequence estimated by Eq. (4.2) is equivalent to the following estimation:

$$\hat{S}_{mm}(e^{j\omega}) = \frac{1}{L}\left|M(e^{j\omega})\right|^2 \tag{4.3}$$

where $\left|M(e^{j\omega})\right|^2$ is the energy spectral density of the finite energy signal obtained by windowing one realization of the stochastic process. The estimator defined in (4.3) is called periodogram. The periodogram is an asymptotically unbiased estimator of the power spectrum (i.e., as $L \to \infty$ the expected value of the periodogram is equal to the true spectrum), but not consistent in the mean square sense since its variance tends to the square of the spectrum value as $L \to \infty$. To reduce the estimation variance, different approaches have been proposed, such as the average of the estimates obtained by different consecutive or partially overlapped signal epochs [109]. For coherence analysis, this averaging is applicable, after the appropriate statistical transformation, even in case of signal epochs recorded in different experiments [2]. This approach is called pooled coherence. Moreover, different window shapes have been introduced.

In a similar way, it is possible to define the coherence function between two discrete time signals as

$$C_{mn}(e^{j\omega}) = \frac{\left|S_{mn}(e^{j\omega})\right|^2}{S_{mm}(e^{j\omega})S_{nn}(e^{j\omega})} = \frac{\left|\sum\limits_{k=-\infty}^{k=+\infty} r_{mn}(k)e^{-jk\omega}\right|^2}{\left(\sum\limits_{k=-\infty}^{k=+\infty} r_{mm}(k)e^{-jk\omega}\right)\left(\sum\limits_{k=-\infty}^{k=+\infty} r_{nn}(k)e^{-jk\omega}\right)} \tag{4.4}$$

where $r_{mn}(k)$ and $S_{mn}(e^{j\omega})$ are, respectively, the cross-correlation and the cross-spectral density between the two signals [44]. The coherence function is a real-valued function that represents the linear relation between the components at different frequencies of the two signals. It is a normalized measure between 0 and 1, where zero indicates completely uncorrelated frequency components and one complete correlation. The square root of Eq. (4.4) is called coherency and provides additional phase information between the two signals. It is worth noticing that, in the case of nonlinear transformations, the information provided by Eq. (4.4) may be inaccurate since the coherence function can only estimate linear relations between stochastic processes.

4.2.2   Parametric-Based Spectral Estimators

Another approach to spectral estimation is based on the methods referred to as parametric- or model-based. The theoretical basis for this class of spectral estimation techniques is the representation of the stochastic process under study as the output of a linear time-invariant (LTI) filter with white noise as its input. If the LTI filter, called the generator model, is identified, the spectrum of the process is also known. The parametric approach is based on estimation of the generator model from the available data. While the Fourier approach implicitly considers the signal to be periodic outside the observation window, the parametric methods propose an estimate that is based on the global process, whose characteristics are estimated from the available data. Thus, in theory, there are no limitations to the frequency resolution; nevertheless some assumptions on the generator model are required. In practical applications the generator model is considered physically realizable, thus the system has to be causal, implying that the LTI filter has a rational transfer function and a finite number of poles:

$$H(e^{j\omega}) = \frac{\sum_{k=0}^{q} b_k e^{jk\omega}}{\sum_{k=0}^{p} a_k e^{jk\omega}} \tag{4.5}$$

It can be shown that the power spectrum of a process generated by filtering white noise with a LTI filter is the multiplication of the power spectrum of the input (a constant in the case of white noise) with the squared magnitude of the filter transfer function. Eq. (4.5) provides the general transfer function which defines a so-called ARMA (autoregressive moving average) model. If $a_i = 0$ $(i = 1, \ldots, p)$ and $a_0 = 1$, a MA (moving average) model results, and, if $b_i = 0$ $(i = 1, \ldots, q)$ and $b_0 = 1$, an AR (autoregressive) model results.

The problem of spectral estimation is thus converted into the problem of estimating a finite number of parameters (from which the term "parametric methods" is derived) which are sufficient to completely describe the entire process from the spectral content point of view. For details about the methods for estimating the model parameters, the interested reader can refer to references 51 and 64.

There are many limitations to the parametric approach. The first issues encountered in dealing with parametric methods are selecting the type (AR, MA, or ARMA) and the order of the model (number of parameters). In the ideal case, the model type should always be the most general one (ARMA) and the order must be chosen larger or equal to the real order: The extra parameters will theoretically be estimated to be zero. In practice, AR parameters are easier to compute than MA parameters; moreover, it is always possible to represent an ARMA or MA model by an infinite AR model [64] that, in practice, will be truncated at a specific order. Thus, AR models are by far the most widely used for spectral estimation. In the following we will therefore refer to AR models.

The choice of an arbitrarily large number of parameters in an AR model is not appropriate since the unnecessary parameters are never estimated as zero. As indicated by Parzen [85], the variance of the AR spectrum estimate for large sample sizes is directly related to the number of parameters and inversely related to the number of samples (available data). Thus, in order to reduce the estimation variance, the model order p should be small relative to the number of data samples. However, a small p may lead to a poor AR approximation of the true spectrum, resulting in increased bias of the estimated spectrum and lower resolution of closely spaced spectral peaks. The trade-off between estimation variance and model order is the counterpart of the trade-off between variance and frequency resolution of the Fourier based methods. Many criteria have been proposed to estimate the appropriate model order from the available data. These criteria include Akaike's final prediction error (FPE) [1], Akaike's information criterion (AIC) [1], Parzen's autoregressive transfer function criterion (CAT) [85], and Rissanen's minimum description length criterion (MDL) [91,92]. However, often, as in the case of the surface EMG signal, the selection of the model order is based on signal simulations and is adapted to the specific application.

4.2.3   Estimation of the Time-Varying PSD of Nonstationary Stochastic Processes

If the statistical properties of a process change over time, the process is said to be nonstationary and spectral analysis with the estimators introduced above may not be appropriate. In particular, the above spectral analysis techniques give frequency information without any time localization. For example, in the case of surface EMG, the signal changes its characteristics as a consequence of changes in muscle effort, muscle fatigue (see Chapter 10) and/or as a consequence of changes in the MU pool. If we consider an EMG signal detected during a prolonged, high-effort muscle contraction of one minute and compute one power spectrum from the entire contraction, we will not obtain any information about the changes which occurred throughout the contraction. Rather, we will get some "averaged" information about the frequency content of the signal during the entire contraction. The simplest approach to obtain both time and frequency information is to divide the signal into many segments (epochs) and estimate a power spectrum for each. If the changes we want to monitor are slow (for example, slower than one second, that is, the signal is quasi-stationary over one second), it would be sufficient to divide a one-minute duration signal into 60 contiguous epochs of one-second duration. In each epoch the signal can be considered as a realization of a WSS stochastic process; thus an estimate of its spectrum is meaningful. This partitioning is the basic idea of the short-time Fourier transform (STFT), from which the spectrogram is defined, which is currently the most widely used method for studying nonstationary signals. In the same manner, a time varying auto-regressive (TVAR) approach results if the spectra of the epochs are estimated with an AR model applied to contiguous signal epochs. The STFT and TVAR approaches can be refined by including epoch overlapping and/or windowing of the data. More advanced time-frequency approaches may be more appropriate in cases of strong nonstationarity of the signals under study [17].

4.3   BASIC SURFACE EMG SIGNAL MODELS

The surface EMG signal detected during voluntary contractions is the summation of the contributions of the recruited MUs that are observed at the recording site

(Chapter 8). A simple analytical model of the generated signal, $m(k)$ (where k is the discrete time index) is the following:

$$m(k) = \sum_{i=1}^{R} \sum_{l=-\infty}^{+\infty} x_{il}(k - \Phi_{i,k}) + \nu(k) \tag{4.6}$$

where R is the number of active MUs, $x_{il}(k)$ the l th MU action potential (MUAP) belonging to MU i, $\Phi_{i,l}$ the occurrence time of $x_{il}(t)$, and $\nu(k)$ an additive noise/interference term [25]. The additive noise/interference represents electrode–electrolyte noise, the noise of the electronic amplifiers, line interference, biological noise, and the interference activity of MUs far from the detection point.

Equation (4.6) is an example of a stochastic process represented by an analytical expression containing parameters which are random variables (the occurrence times of the MU firings). For the case of uncorrelated discharges, it can be shown that the resultant power spectrum is the summation of the power spectra of the MUAP trains. The spectrum of a MUAP train is the product of the spectrum of the MUAP (deterministic finite energy signal) with that of the point process describing the firing pattern (random process). For the case of a Gaussian distributed interpulse interval, the spectrum of the point process is given by [57]

$$S_{\phi}(\omega) = \frac{1}{\mu} \frac{1 - e^{-\sigma^2 \omega^2}}{1 + e^{-\sigma^2 \omega^2} - 2e^{-\sigma^2 \omega^2/2} \cos(\mu\omega)} \tag{4.7}$$

where μ and σ are the mean and standard deviation of the interpulse interval, respectively. As $\omega \to \infty$ the spectrum of the point process tends to a constant value. Substituting into Eq. (4.7) values of the mean and standard deviation of the interpulse interval for normal physiological conditions (for example, a mean firing rate of 8–35 Hz and coefficient of variation of the mean interpulse interval of approximately 15%), one finds that the spectrum of the point process is nonconstant in a rather small frequency region, mainly below 30 Hz. Thus, considering the high-pass analog filters used for surface EMG conditioning (Chapter 3), the influence of the firing patterns of the MUs on the surface EMG power spectrum can be neglected in many applications. The high-frequency range of the EMG power spectrum represents the morphology of the recorded MUAPs, influenced by the relative positions of the MUs with respect to the recording system, the electrode configuration and electrode shape and size, and the conduction velocity (CV) at which each action potential propagates [27,58,103] (see Chapters 5 and 8). Note that in the case of correlated firing patterns, the global EMG spectrum would also contain cross-terms [110]. Dependent firing patterns are due, for example, to short-term synchronization (MUAPs of different MUs firing at approximately the same time more frequently than would be expected by chance alone) or to common drive (the common modulation of firing rates).

For certain applications, the surface EMG signal can be modeled by functional models in a coarser fashion than that provided by Eq. (4.6). Functional models of the EMG seek to capture the observed stochastic behavior of the EMG signal without

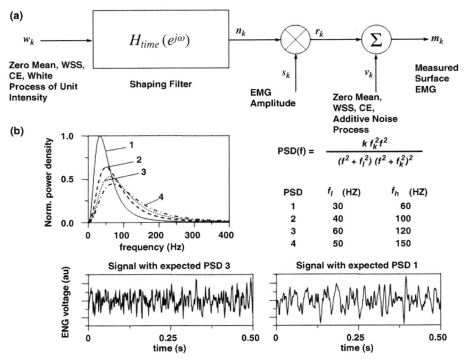

FIGURE 4.1 Simulation of EMG signals by filtering white Gaussian noise. **(a)** The output signal is obtained by filtering white Gaussian noise with a shaping filter [13]. **(b)** Examples of expected spectra obtained from Shwedyk's expression [100] for different values of the two parameters f_h and f_l. Two simulated signals, obtained by filtering white Gaussian noise with the inverse Fourier transform of the square root of PSD(f) are shown to qualitatively simulate non-fatiguing conditions (PSD 3) and myoelectric manifestations of muscle fatigue (PSD 1).

including the complexity that would be involved in modeling the activity of each individual MU (see also Chapter 8). A complete model of this type for a single channel of EMG is shown in Fig. 4.1a. This model produces a surface EMG (m_k) with statistical properties similar to real EMG, during both fatiguing and nonfatiguing contractions. In the model, a zero-mean, WSS, correlation-ergodic (CE), white process of unit variance w_k passes through the stable, inversely stable, linear, time-variant shaping filter $H_{\text{time}}(e^{j\omega})$.

The shaping filter accounts for the spectral shape of the EMG. The signal is then multiplied by the EMG amplitude s_k, which modulates the EMG standard deviation based on the level of muscular activation. The filter $H_{time}(e^{j\omega})$ preserves signal variance so that all modulation in the standard deviation of the noise-free EMG signal (r_k in Fig. 4.1a) is attributed to changes in EMG amplitude. Finally, a zero mean, WSS, CE noise process v_k is added to the signal to form the measured surface EMG m_k. This noise process represents measurement noise (e.g., due to the electrode–skin interface and amplifier circuitry) and cannot be completely eliminated. The processes w_k and v_k are assumed to be uncorrelated with each other.

In Fig. 4.1b, examples of realizations of the stochastic process defined in Fig. 4.1a are reported together with the expected spectra. The shaping filter has been selected as suggested by Shwedyk et al. [100]. The expression of the expected spectrum has two parameters f_h and f_l that allow the shape of the spectrum to be changed. Non-stationarity may be generated by changing these parameters during time, for example to simulate myoelectric manifestations of muscle fatigue. The first moment (mean frequency) of this spectrum can be computed analytically [28]:

$$f_{\text{mean}} = \frac{2f_h}{\pi} \frac{1+\alpha}{1-\alpha} \left[1 - \frac{2\alpha^2}{\alpha^2 - 1} \ln \alpha \right], \qquad \text{with } \alpha = f_l/f_h \qquad (4.8)$$

This model is phenomenological and therefore cannot be used for understanding how physiological events are reflected in the surface EMG signal features. Nevertheless, it assumes importance for the analysis of the statistical properties of estimators of signal features, such as amplitude and frequency content.

4.4 SURFACE EMG AMPLITUDE ESTIMATION

EMG amplitude estimation can be described mathematically as the task of best estimating the time-varying standard deviation of a colored random process in additive noise (refer to the model of Fig. 4.1). This estimation problem has been studied for several decades (see Clancy et al. [16] for a historical review), dating back to Inman et al. in 1952 [50]. Typically, estimators are implemented digitally, with single-channel estimates comprised of the cascade of five sequential processing stages: (1) noise and interference attenuation, (2) whitening, (3) demodulation, (4) smoothing, and (5) relinearization (Fig. 4.2). Noise and interference attenuation seek to limit the adverse effects of motion artifacts, electrode–skin noise, power-line interference, and so on, as described in Chapter 3.

The correlation between neighboring EMG samples is a consequence of the limited signal bandwidth which reflects the actual biological generation of EMG and the low-pass filtering effects of the tissues (see Chapters 2 and 8). Temporal decorrelation— that is, whitening—makes the samples statistically uncorrelated, increasing the "statistical bandwidth" [7], which reduces the variance of amplitude estimation. Demodulation rectifies the whitened EMG and then raises the result to a power (either 1 for mean absolute value (MAV or ARV) processing or 2 for root mean square (RMS) processing). The mean value of the demodulated, relinearized signal is a scaled and noisy version of the standard deviation of the original signal. Smoothing filters this signal whereas relinearization inverts the power law applied during the demodulation stage, returning the signal to units of EMG amplitude. A standard deviation estimate is produced. The quality of EMG amplitude estimates and techniques for implementing the processing stages will be discussed in the following sections.

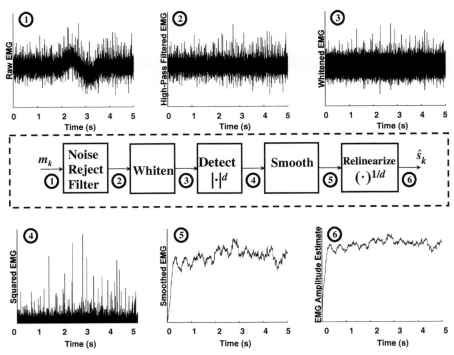

FIGURE 4.2   Cascade of processing stages used to form an EMG amplitude estimate. The acquired EMG signals are assumed to be from bipolar electrodes. The EMG amplitude estimate is $\hat{s}_k$. In the "Detect" and "Relinearize" stages, $d = 1$ for MAV (aka ARV) processing and $d = 2$ for RMS processing.

4.4.1   Measures of Amplitude Estimator Performance

When contraction is isometric, constant-force, and nonfatiguing, it is generally assumed that the EMG is stationary and its amplitude is expected to be a constant. In this case, performance can be measured via a dimensionless signal-to-noise-ratio (SNR) calculated as the mean of a number of amplitude estimates divided by the standard deviation of these estimates (that is, the inverse of the coefficient of variation). This measure does not vary with the gain of the EMG channel and makes no assumption as to any relationship between EMG and muscle force. Better EMG estimators yield higher SNRs. Some authors have used the square of this measure as a performance index.

When force is changing, this SNR is no longer meaningful and alternative performance measures must be used. One approach has been to display a real-time amplitude estimate to the subject as a form of biofeedback. The experimenter generates a target display for the subject to track. The tracking error (e.g., RMS error between the target amplitude and the estimate) serves as a performance measure, with better EMG amplitude estimators providing lower error. This technique also makes no assumption of an EMG–force relationship, but must be evaluated online.

Alternatively, EMG amplitude estimates have been assessed via a common application: the use of surface EMG to estimate joint torques (Chapter 10). Better amplitude estimation is assumed to provide lower EMG–torque errors. Note that the amplitude of EMG is affected by confounding factors other than joint torque, such as the subcutaneous layer thickness, the inclination of the fibers with respect to the detection system, the interelectrode distance selected, joint angle, and so on [27]. As a consequence, EMG-based joint torque estimates must account for these confounding factors—for example, by normalization with respect to a subject specific reference value—and results must be appropriately interpreted.

4.4.2 EMG Amplitude Processing—Overview

Considering the functional model for sampled EMG presented in Fig. 4.1, the goal is to estimate $s(k)$ based on samples of $m(k)$. Formal optimal solutions to the complete model do not exist. However, a simplifying approach can be taken to explain existing solutions (Fig. 4.2).

Stage 1: Noise and Interference Attenuation. The goal of the first stage is to eliminate additive noise, artifacts, and power-line interference that are acquired along with the "true" EMG. Methods to do so are described in Chapter 3. Note that these methods can incorporate proper skin preparation and electrode setup, analog filtering in the amplifier apparatus, digital filtering during post-processing, and so on.

Stage 2: Whitening. Because successive samples of the EMG signal are correlated, direct information extraction from the signal is confounded (in a probabilistic sense). That is, the signal correlation temporally "weights" the information. Whitening resolves this problem by transforming the signal so that successive samples have equal "weight." A frequency domain measure of the degree of correlation in the data is the statistical bandwidth [7]. Hogan and Mann [46,47] showed that as the statistical bandwidth of a signal is increased (and correlation is decreased—for example, via whitening), the SNR of the EMG amplitude estimate (for the constant force case) increases as the square root of the statistical bandwidth. For contractions above 10% MVC, whitening has led to a 63% improvement in the SNR [13] (Fig. 4.2). Laboratory investigations have shown that whitening also improves performance in EMG–force and prosthesis control applications [15,59,60,88].

A whitening filter outputs a theoretically constant, or "whitened" power spectrum in response to an input. This filter is formed by first estimating the PSD of the EMG signal. Then, the inverse of the square root of the PSD is the magnitude of the whitening filter (Fig. 4.3). The phase of the whitening filter is arbitrary, but is selected as causal for a causal EMG processor. For isometric, constant-force, nonfatiguing contractions, it is common to model the EMG as a WSS, amplitude modulated, AR

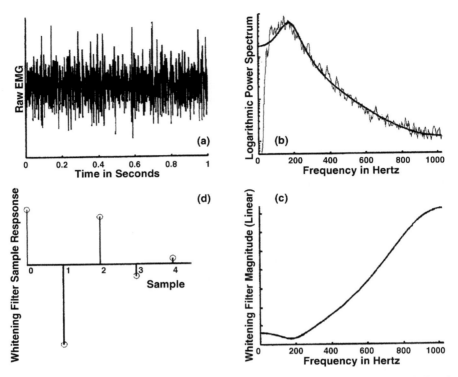

FIGURE 4.3   Fixed whitening filter design. **(a)** A 1-s epoch of an experimental EMG signal. **(b)** The (jagged) thin-line plot is the discrete Fourier transform estimate of the EMG PSD. The thick-line plot is the fourth-order AR estimate of the EMG PSD. **(c)** Magnitude response of the whitening filter obtained from the AR PSD estimate and **(d)** its impulse response. Reproduced from Clancy and Hogan [13] with permission.

process (software for doing so is readily available [89]). With this model, the PSD of EMG, denoted $S_{mm}(e^{j\omega})$, can be written as

$$S_{mm}(e^{j\omega}) = \frac{b_0}{\left|1 - \sum_{i=1}^{p} a_i e^{-ij\omega}\right|^2} \tag{4.9}$$

where the a_i are the AR coefficients, p is the model order, and ω is the angular frequency in rad/s. These coefficients are estimated from a calibration contraction that is typically a few seconds in duration. Once these coefficients have been determined, whitening can be performed on subsequent recordings with a discrete-time MA filter, which operates as indicated by Eq. (4.10) (Fig. 4.3):

$$y(k) = \frac{1}{\sqrt{b_0}}x(k) + \frac{-a_1}{\sqrt{b_0}}x(k-1) + \cdots + \frac{-a_p}{\sqrt{b_0}}x(k-p) \tag{4.10}$$

where $x(k)$ are the generic data input to the whitening filter and $y(k)$ are the whitened output data. During constant-force contractions, parametric model orders of 4–6 have been found sufficient to model the PSD for whitening purposes [13,43,106].

D'Alessio and co-workers [10,19,20] used a generalization of this whitening approach based on the assumption that the PSD of the EMG can vary in a general manner (i.e., not just restricted to an amplitude modulated PSD), and thus the MA whitening filter must do so as well.

Whitening techniques can fail at low contraction levels due to the presence of additive background noise. Clancy and Farry [12] implemented an alternative adaptive whitening technique (again based on an amplitude-modulated PSD model), by incorporating the fact that EMG is invariably acquired in the presence of an additive, broadband background noise. Thus, the whitening filter should be adapted, but the adaptation is not as general as that proposed by D'Alessio et al. [20]. Caution should also be taken when whitening the high-frequency regions of the EMG signal, e.g., frequencies above 500–600 Hz. The relative EMG signal power, compared to noise power, is much lower at these frequencies. Adaptive whitening [12] resolves this issue. Alternatively, signal whitening might only be conducted over a frequency range below approximately 600 Hz [21]—the frequency range over which most of the EMG power is found.

From Fig. 4.3, we note that a whitening filter has low gain at low frequencies and high gain at high frequencies. In fact, the shape of a whitening filter mimics the shape of a *low-order* high-pass filter. Potvin and Brown [88] exploited this fact by implementing a simpler whitening process via low-order FIR high-pass filtering. They produced a filter with this desired shape either by using a first-order high-pass FIR filter with cutoff frequency at 410 Hz or a sixth-order highpass FIR filter with cutoff frequency at 140 Hz. Their filter selection was not "calibrated" to the PSD of each subject; rather it was optimized via an exhaustive search of filter orders and cutoff frequencies. Note that higher-order high-pass filters would likely produce too low of a gain at the lower frequencies, and thus should be avoided. These investigators sampled their EMG signals at 1024 Hz, which inherently limited their whitening bandwidth to 512 Hz. In doing so, they naturally avoided the high-frequency regions of the EMG signal wherein noise can be a detriment to whitening.

Stages 3 and 5: Demodulation and Relinearization.   After the whitening stage, the signal samples are assumed to be noise-free and uncorrelated. In order to estimate the signal standard deviation from the EMG samples, some form of demodulation nonlinearity must be applied to the signal, typically by taking the absolute value or the square of each sample. After demodulation, the signal is smoothed (as discussed in the next section) and then relinearized. Relinearization consists of taking the square root of each sample if squaring had been used in the demodulator; else relinearization is not applied (first-order demodulator). Hence, demodulation and relinearization are considered here together. The demodulated-relinearized signal mean is equal to a scaled standard deviation of the original signal standard deviation.

Theoretically, if the original EMG samples are modeled as conforming to a zero-mean Gaussian probability density function (PDF) [32,46,47], then a squaring demodulator (a.k.a. RMS processor) is the optimal maximum likelihood (ML) estimate. Alternatively, if the original EMG samples conform to the Laplacian PDF (which is more peaked near zero), then the first-power demodulator (computing the MAV or ARV value) gives the optimal estimate. Experimental comparison of MAV and RMS processing indicated that MAV processing led to a higher SNR than RMS processing, but only by 2.0–6.5%, suggesting that, in practical conditions, RMS or MAV processing are nearly indistinguishable [14].

Stage 4: Smoothing. In the smoothing stage, several demodulated samples are time-averaged to form one amplitude estimate. As shown in Fig. 4.2 (signal panel 4), the mean value of the demodulated signal (which now contains the EMG amplitude information) is corrupted by the inherent variability of the EMG signal. A sliding window selects the demodulated samples for each successive amplitude estimate, thereby forming an averaging filter. Because EMG amplitude is, in general, changing during contraction, an appropriate smoothing window length over which the signal is "quasi-stationary" must be selected. Random fluctuations in the EMG amplitude estimate are diminished with a long smoothing window; however, bias (deterministic) errors in tracking the signal of interest are diminished with a short smoothing window. In practice, windows varying in duration from 100–250 ms are common.

Smoothing can alternatively be accomplished with a linear low-pass filter. (The moving average produced by an MAV or RMS filter is, in actuality, a variety of low-pass filter.) Filter cutoff frequencies between 1 and 6 Hz are common. Lower cutoff frequencies are appropriate when the muscle activities under study are constant effort (or slowly force-varying), and higher cutoff frequencies are appropriate for studies including rapid changes in muscle effort. Some studies have dynamically selected the smoothing window length (or linear low-pass filter cutoff) at each data sample [11,16,20,26]. Sanger [98] implemented a nonlinear recursive filter based on Bayesian estimation that responded quickly to fast changes in EMG amplitude while providing more smoothing when the EMG amplitude was more constant.

Finally, note that some EMG–force applications omit the smoothing stage. Instead, a smoothing stage is included as the first step in relating EMG amplitude to force. In this case, the shape (e.g., order and cutoff frequency) of the low-pass filtering can be optimized as part of relating EMG amplitude to force [15,60,105].

4.4.3 Applications of EMG Amplitude Estimation

For many years, myoelectrically controlled upper-limb prosthetics have been a driving force in the development of EMG amplitude estimation algorithms. EMG from remnant muscles has been used to control the operation of a prosthetic elbow, wrist, and/or hand. Most frequently, the electrical activity of two muscle sites is monitored. If biceps and triceps muscles remain, these sites are usually selected. A common scheme is to estimate the EMG amplitude from the two sites, and determine the difference amplitude. If the biceps amplitude is larger, flexion occurs (i.e., elbow flexion, hand closure). If the triceps amplitude is larger, extension occurs (i.e., elbow extension, hand opening). For these applications small variance estimations are required (for a more detailed description of myoelectrically controlled prosthetics see Chapter 20). Clinically, EMG amplitude is used to study muscle coordination and activation intervals. For example, EMG amplitude is used in gait analysis to determine when various muscles are active throughout the gait cycle (see Chapter 16). Finally, surface EMG amplitude is computed as an indicator of muscle fatigue, combined with spectral analysis (see Chapter 10), since it is an indicator of CV decrease, MU pool changes, and other mechanisms occurring with fatigue.

4.5 EXTRACTION OF INFORMATION IN THE FREQUENCY DOMAIN FROM SURFACE EMG SIGNALS

Changes in the power spectrum of the surface EMG signal during muscle contraction were observed for the first time by Piper [87], who detected a decrease in the dominant oscillation of the recorded surface EMG signal during maximal voluntary contractions, as a consequence of muscle fatigue. Subsequent investigators quantified the changes in the frequency content of the EMG signal using various spectral descriptors, such as the centroid frequency [99], the median frequency [102,103], and the high/low-frequency ratio [37]. Moreover, parameters extracted from the signal in the time domain, such as the rate of zero crossings [39,49] or spike or turn properties [33], were proposed as alternative indicators of changes in the surface EMG spectral content.

The theoretical basis for the interpretation of the evolution of the power spectrum of the surface EMG signal during fatigue and for understanding the factors influencing it follow from the works by Lindstrom, De Luca, and Lago [22,57,58]. These works developed the theory needed to interpret frequency changes in the EMG signal with respect to the underlying physiological events. The conclusions by Lindstrom and Magnusson [58], especially those related to the effect of the spatial filter on the PSD of the detected signal, have been extensively validated experimentally. Lindstrom and Magnusson [58] also provided the basis for the interpretation of the changes of the characteristic spectral frequencies as a consequence of changes in MU CV. They indeed proposed the following expression for the PSD of the surface differential signal generated by an intracellular action potential traveling along a muscle fiber parallel to skin:

$$P(f) = \frac{1}{v^2} G\left(\frac{fd}{v}\right) \tag{4.11}$$

where v is the propagation velocity of the action potential, d is the interelectrode distance, and $G(fd/v)$ includes the spectrum of the action potential and the spatial transfer function. This expression implies that the spectrum is scaled by CV, thus its shape does not change but only the frequency axis is scaled when CV changes.

These concepts have been further developed by De Luca [22,103] who definitely clarified the fundamental role of spectral analysis in the study of muscular fatigue. The work by Lago and Jones [57] outlined the effect of the firing pattern of the active MUs on the surface EMG PSD and spectral variables (refer to Eq. (4.7), to be used together with Eq. (4.11) to derive the PSD of a MUAP train).

Since these pioneering works, a large number of studies reported the use of spectral analysis of the surface EMG signal for the investigation of muscle fatigue or MU recruitment strategies. From the experimental evidence provided by Arendt-Nielsen, De Luca, Merletti, et al. [3,4,23,70,71], it was shown that the use of spectral compression (shift in log frequency scale) of the EMG signal as a measure of myoelectric manifestations of muscle fatigue offers a more objective assessment technique compared to the more subjective clinical techniques based on the subsequent manifestations of mechanical fatigue. This observation led, in the early 1980s, to the development of analog and digital instruments for monitoring spectral parameters of the signal [36,86,104].

This section describes the basic body of knowledge concerning EMG spectral analysis (by Fourier and parametric techniques) and how physiological parameters are reflected by surface EMG power spectra. The section is mostly focused on the theoretical basis for the assessment of muscle fatigue during high-intensity isometric constant force contractions since this assessment is by far the most prevalent application of surface EMG spectral analysis. Applications are described in Chapter 10.

4.5.1   Estimation of the PSD of the Surface EMG Signal Detected During Voluntary Contractions

Spectral analysis of EMG signals detected during voluntary constant force isometric contractions is usually performed using the STFT with partially overlapping or nonoverlapping epochs of 0.25 s to 1 s (Fig. 4.4). A few studies in the literature compared the STFT with parametric analysis [28,65,76,84]. The estimation model chosen in these cases was the AR model with an order between 4 and 11.

4.5.2   The Energy Spectral Density of the Surface EMG Signal Detected During Electrically Elicited Contractions

Electrically evoked signals may be considered deterministic and quasi-periodic signals with period determined by the stimulation frequency imposed by the stimulator (see also Chapters 10 and 11). Each M-wave is a finite energy, finite support signal whose frequency content can be described by the energy spectral density. The energy spectral density reflects the properties of the detected MUAPs—in particular their CV, the spread of the MU innervation zones, and the shapes of the intracellular action potentials.

Different strategies can be applied to estimate the spectrum of an electrically elicited EMG signal during a prolonged contraction [74] (Fig. 4.4):

1. The signal can be divided into epochs and the frequency attributes computed over each epoch (which present many M-waves). In this case, the signal is

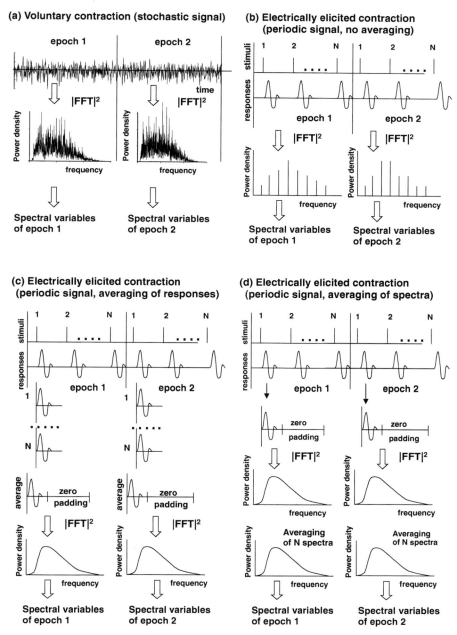

FIGURE 4.4 Estimation of EMG signal spectrum. (**a**) Spectral estimation of EMG recorded during voluntary contractions. The signal is divided into a sequence of epochs during which it is assumed stationary. Spectral variables are obtained from the corresponding spectra. (**b**) Spectral estimation of a signal detected during electrically elicited contractions by dividing the quasi-periodic signal into epochs containing N M-waves (a "line" spectrum estimate is obtained where the "lines" are narrow *sinc* functions whose width is inversely related to the epoch length). (**c**) Spectral estimation of an electrically elicited signal obtained by averaging M-waves within each epoch. (**d**) Spectral estimate obtained by averaging the individual spectral estimates obtained from each M-wave. Reproduced from Merletti et al. [74] with permission.

periodic and the spectral lines (actually, narrow sinc functions whose width is inversely related to the epoch duration) are separated by an interval equal to the stimulation frequency.

2. The spectral features can be computed for each M-wave, which is thus seen as a finite-length, nonperiodic signal. In this case the spacing between spectral lines is determined by the epoch length and can be reduced by zero padding that provides interpolation of an otherwise coarse spectral estimate.

3. The signal can be divided into epochs and the M-waves in each epoch averaged. The spectral features are then computed for each averaged and zero-padded M-wave. Similar considerations as in the previous case can be drawn, but the signal-to-noise ratio is increased by the averaging process and the computational cost is reduced.

The method usually applied is the third one and much more stable estimates of spectral features are obtained with respect to the case of voluntary contractions.

Spectral analysis of electrically evoked EMG signals is mainly used for detecting changes of scale of the M-waves. An alternative approach for estimating scale factors in deterministic signals is to process the signals directly in the time domain. Merletti et al. [75–77] proposed a maximum likelihood approach to solve this problem, while, more recently, Muhammad et al. [80] developed a pseudo joint estimator of the time scale factor and time delay between signals for applications of M-wave analysis during fatigue. These approaches were shown to be, in general, more robust than spectral analysis for estimating the scaling of the M-wave due to fatigue for the case of truncation of the M-wave (when the stimulation frequency is above 30–35 Hz the stimulus interval may be shorter than the total M-wave length and, thus the M-wave is truncated). Alternative approaches have been proposed by Lo Conte et al. [61], who decomposed the M-wave into a particular series of functions, and by Olmo et al. [83], who proposed the Matched Continuous Wavelet Transform to estimate the M-wave scale factor. The cumulative distribution function technique proposed by Rix and Malengé [93] and used in Merletti and Lo Conte [76] can also be directly applied to pairs of M-waves for estimating the scale factor between them.

4.5.3 Descriptors of Spectral Compression

During both voluntary and electrically elicited fatiguing contractions the spectral content of the EMG signal progressively moves towards lower frequencies. This phenomenon can be described as a compression or frequency scaling of the spectrum; that is, the shape of the spectrum does not change but only the scale factor of the frequency axis changes (refer to Eq. (4.11) and to Chapter 10). In the case of a pure scaling, a single parameter would give all the information about the phenomenon of spectral compression, thus any reference spectral frequency can be used as an estimator of spectral compression. One of the possible spectral descriptors is the mean or centroid frequency (MNF) which is defined as

$$f_{\text{mean}} = \frac{\int\limits_0^{f_s/2} f\, S(f)\, df}{\int\limits_0^{f_s/2} S(f)\, df} \tag{4.12}$$

where $S(f)$ is the PSD of the signal and f_s is the sampling frequency.

MNF is the moment of order one, or centroid, of the power spectrum. In general, the central moment of order k is defined as [96]

$$M_{Ck} = \frac{\int\limits_0^{f_s/2} (f - f_{\text{mean}})^k S(f)\, df}{\int\limits_0^{f_s/2} S(f)\, df} \tag{4.13}$$

Merletti et al. [77] proposed a time-domain technique to estimate the central moments of any order, which avoids the Fourier transform and direct estimation of the PSD, and which is particularly suitable for real-time implementation. Other characteristic frequencies f_p are the pth percentile frequencies, indirectly defined as

$$\int\limits_0^{f_p} S(f)\, df = p \int\limits_0^{f_s/2} S(f)\, df, \qquad 0 < p < 1 \tag{4.14}$$

Equation (4.14), with $p = 0.5$, defines the median frequency (MDF), whereas $p = 0.25$ and $p = 0.75$ define the other two interquartile frequencies. Although any percentile frequency can be used to estimate spectral compression, it has been suggested that a better estimate of spectral compression due to CV decrease may be obtained by computing the average change of a number of properly selected percentile frequencies. Each percentile frequency indeed may provide a different indication of fatigue due to the other factors affecting the shape of the PSD of the signal, apart from CV. The analysis of a number of percentile frequencies enables a distinction between the spectral changes due to CV and those due to other factors, which affect spectral shape. It has been shown [62,68] that, indeed, some percentile frequencies are better correlated with CV changes than others. Examples for simulated and real signals are depicted in Fig. 4.5a–d and in Fig. 4.5e, respectively.

In addition to spectral moments and percentile frequencies, all of the model parameters from an AR spectral estimate can be used to characterize spectral changes in the EMG signal [54]. Finally, the frequency at which the first dip introduced by the detection system appears (see Fig. 3.11 and Chapter 5) can be used as a descriptor of spectral changes. In this case the spectral descriptor reflects only the CV of the active MUs [58], but its variance of estimation is much higher with respect to spectral

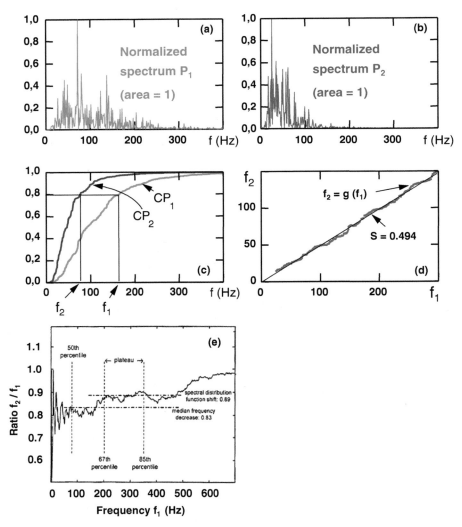

FIGURE 4.5 Cumulative power method. (**a, b**) normalized (unit area) spectra of EMG signals simulated with Shwedyk's model [100] with a scaling factor $s = 0.5$. (**c**) Cumulative power functions (CP1, CP2) of the two spectra depicted in (**a**) and (**b**); the two frequencies f_1 and f_2 correspond to the same percentile of CP1 and CP2. (**d**) If CP1 and CP2 are scaled versions of each other (affine functions), the plot of $f_2 = g(f_1)$ is a straight line through the origin and the ratio f_2/f_1 is an estimate of the scaling factor s, as indicated in (**d**). Reproduced from Merletti and Lo Conte [68] with permission. The displacement of the curve $f_2 = g(f_1)$ from a straight line implies a non constant value of f_2/f_1 and therefore a change of shape of the spectrum in specific frequency bands. (**e**) Plot of the ratio f_2/f_1 for an EMG signal recorded at the beginning and at the end of a 80% MVC contraction of the biceps brachii sustained for 22.5 s. Reproduced from Lowery et al. [62] with permission.

moments [67] and the estimate is blurred by the dispersion of the CV values of different MUs.

MNF and MDF are the most commonly used spectral descriptors. Only a few studies have used higher-order moments to describe the EMG spectrum [70]. In the case of voluntary contractions, spectral descriptors are random variables with particular statistical properties which depend on the nature of the descriptor, epoch signal length, amount of epoch overlapping, type of window and the spectrum estimator adopted. The influence of these parameters can be evaluated by the use of simulation models such as that shown in Fig. 4.1 [6,28,72].

Properties of MNF and MDF. MNF and MDF provide some basic information about the spectrum of the signal and its changes versus time. They coincide if the spectrum is symmetric with respect to its center line, while their difference reflects spectral skewness. A tail in the high-frequency region implies MNF higher than MDF. A constant ratio f_{mean}/f_{med} versus time implies spectral scaling without shape change while a change in this ratio implies a change of spectral skewness or shape. It can be shown that the standard deviation of the estimate of MDF is theoretically higher than that of MNF [6,104], as confirmed in experimental studies [103]. However, it can also be shown that MDF estimates are less affected by additive noise (particularly if the noise is in the high frequency band of the EMG spectrum) [103] and more affected by fatigue (since the spectrum becomes more skewed with fatigue). Because of these pros and cons and because of the additional information that is carried jointly by the two variables, researchers often use both in their reports. However, in cases in which the signal-to-noise ratio may be very low, at least during particular intervals of time (for example at the beginning of a ramp contraction), MDF is often preferred [8,9].

Fourier Versus Parametric Approach. The Fourier and parametric approaches have been compared using simulated EMG signals (model of Fig. 4.1) for different window lengths and degree of non-stationarity [28]. It was found that both in stationary and nonstationary conditions the two approaches lead to similar results, in terms of variance and bias of estimation, for MNF and MDF over a large range of epoch lengths. For very short epochs (below 0.25 s), the parametric approach performs better than the Fourier approach, but the difference can be negligible in practical applications. Figure 4.6 shows the comparison between Fourier and AR estimation of the PSD of experimental surface EMG signals for different epoch lengths. MNF and MDF estimates are also reported.

Window Shape. The type of window shape determines the bias in the power spectrum estimation but it is difficult to predict analytically how the bias in spectral lines is reflected in the bias of spectral descriptors. Again simulation studies provided an evaluation of this effect [72]. In the case of the generation model shown in Fig. 4.1, it was shown that the choice of the window is not critical for MNF or MDF estimation. Despite its greater side lobes, the rectangular window has been used in the majority of experimental studies.

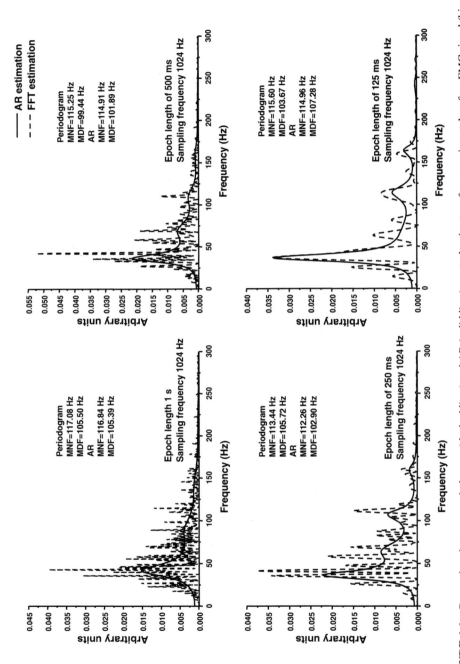

FIGURE 4.6 Comparison between periodogram (*dashed lines*) and AR (*solid lines*) spectral estimates of an experimental surface EMG signal (biceps brachii muscle) for different lengths of the observation window. MNF and MDF are also reported. Reproduced from Farina and Merletti [28] with permission.

Epoch Length and Epoch Overlapping in the Case of Stationary and Nonstationary Conditions. In stationary conditions (model of Fig. 4.1 with fixed parameters), the longer the signal epoch, the lower the variance and bias of estimation of the spectral descriptors. In the case of nonstationary conditions, two sources of bias are present, namely, the bias related to the spectral estimator applied to a finite observation window and that due to the nonstationarity. Bias due to both effects increases with epoch length. On the contrary, variance of MNF and MDF estimates decreases with increased window duration. In the case of isometric, constant force, fatiguing contractions, the signal can be considered stationary for epoch durations of the order of 1–2 s. The spectral descriptors are computed from several sequential (possibly partially overlapping) epochs. Usually the slope of the linear regression which best fits the time changing values of MNF or MDF is used as a fatigue index. A large number of short epochs (that is, of experimental points) in the regression interval will reduce the variance of the slope estimate but will increase the scatter of the experimental points. Finally, an additional factor of interest is the degree of epoch overlapping which allows an increase of the number of experimental points without increasing their scatter, but increasing their statistical dependence.

The standard deviation of the estimated slope of the regression line fitting MNF and/ or MDF versus time in simulated fatiguing contractions (with low to medium nonstationarity) has been found to be minimal for epoch durations between 250 ms and 500 ms, which therefore seem the most suitable for regression line parameter estimation [28]. Epochs shorter than 250 ms lead to high variance and bias of estimation. Epoch overlapping increases the computational load [28] without significant benefits.

4.5.4 Other Approaches for Detecting Changes in Surface EMG Frequency Content During Voluntary Contractions

Another possible indicator of changes in the frequency content of the signal is the rate of zero crossings—that is, the number of sign changes of the signal per unit time. When the signal is noise-free and comprised of only one sinusoidal function, this rate reflects the frequency of the function. For a signal with a Gaussian stationary amplitude distribution [90], the expected number Z of zero crossings per second can be expressed by the following relationship:

$$Z = 2 \left[\frac{\int_0^{f_s/2} f^2 S(f)\, df}{\int_0^{f_s/2} S(f)\, df} \right]^{1/2} \tag{4.15}$$

The distribution of amplitudes of the surface EMG signal is in between Gaussian and Laplacian, thus this hypothesis is almost verified in practical cases. From Eq. (4.15), it is easy to show that Z is scaled by CV (as is MNF and MDF) and can thus be used for evaluating spectral compression. Although the standard deviation of estimation of the zero crossing rate has been shown to be higher than that of MNF and similar to that of the percentile frequencies [45], the technique does not require spectral estimation and it is particularly easy to implement on-line in hardware. Based on this method, Hägg [39] and Inbar et al. [49] developed simple real-time fatigue monitors in the early 1980s. Figure 4.7 shows an example comparison between fatigue analyses performed by MDF and the zero crossing rate.

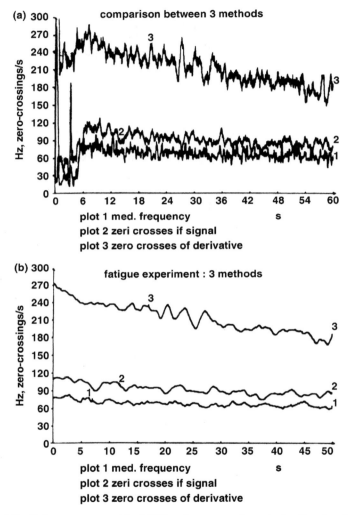

FIGURE 4.7   Fatigue assessment by MDF and zero crossing rate (on the signal and on the signal derivative) during a fatigue experiment on the biceps brachii muscle. **(a)** Unsmoothed raw counts (first few seconds taken during buildup of force). **(b)** same record shifted to the left and smoothed. The correlation between the running zero crossing rate and MDF is 0.76 ($p < 0.001$). The correlation between averaged zero crossing rate and averaged MDF (averages on 6 s) is 0.99 ($p < 0.001$). Reproduced from Inbar et al. [49] with permission.

Another approach to evaluate spectral changes in the surface EMG is based on the automatic detection of spikes in the signal. A spike is defined as a segment of signal shaped by an upward and downward deflection [33]. Both deflections of a spike cross the zero isoelectric baseline and should be at least $100\,\mu V$ in amplitude. The analysis of spike activity has a long history in both clinical neurophysiology [63] and kinesiology [108] but has received less attention for surface EMG analysis in favor of more sophisticated techniques. Mean spike frequency—that is the average number of spikes per unit time—has been shown to be highly correlated to MNF [34]; thus, as the characteristic spectral frequencies, it can be used to monitor spectral changes in the signal. This technique has been indicated as potentially useful in EMG analysis since it does not directly require stationarity of the signal. However, the mean spike frequency, as well as the other spike parameters (see Gabriel [33] for their definitions), implies the use of a signal segment for their computation, thus a trade-off between bias of estimation and variance is implied also with this method.

4.5.5 Applications of Spectral Analysis of the Surface EMG Signal

Spectral analysis of the surface EMG signal has been extensively applied for the study of muscle fatigue both in voluntary and electrically elicited contractions [71] (see Chapters 10 and 11). The preferred application of these techniques has been the analysis of isometric constant force, short-duration contractions of medium–high level. It is now accepted that relative changes in EMG spectral variables reflect fatigue with the possibility of detecting age (or other)-related differences in muscle fiber composition [24,56,73,78], with the limitations discussed in Section 10.7.

Applications of spectral analysis of the EMG signal for fatigue assessment during nonconstant force and dynamic contractions have been proposed in more recent years together with advances in spectral estimation techniques based on time–frequency representations [55]. Nevertheless, many artifacts, mainly related to geometrical and anatomical factors of the EMG generation system, may be associated with these approaches [30]. The relevance of these artifacts on the results shown in the literature is still not clear, thus caution should be taken in extending the considerations drawn for isometric conditions to dynamic exercises, particularly due to the relative movement of the muscle with respect to the surface electrodes and the nonstationarity of the signal. Most studies have investigated relative changes in muscle fatigue, while a few recent studies considered absolute changes [95].

The analysis of the surface EMG PSD has also been applied to the investigation of MU recruitment strategies, in an attempt to extract information about central nervous system (CNS) motor control strategies from a global analysis of the surface EMG signal (i.e., without decomposing the individual MU activities). It was indeed speculated that MNF and MDF should reflect the recruitment of new, progressively larger and faster MUs and increase until the end of the recruitment process. They should then reach a constant value (or decrease) when only rate coding is used to track the desired target force level [8]. The latter hypothesis is based on two theoretical considerations: (1) The CV of a single MUAP scales the power spectrum of that MUAP ([58], Eq. (4.11)), and (2) MU firing rates do not significantly affect the spectral features of the surface EMG signal ([57], Eq. (4.11)). Thus, if recruitment is assumed to progress from the small slow MUs to the large fast MUs as force is increased, then spectral indices should increase when force increases. Observation 1 was experimentally validated within reasonable approximations for the case of isometric constant force fatiguing contractions, as indicated previously. Using intramuscular EMG detection, Solomonow et al. [101] showed experimentally that both observations hold when controlled physiological MU recruitment was obtained by a particular stimulation technique. However, as indicated by the authors, extrapolation of these results to surface EMG was not implied since the detection was very selective. For surface recordings, the volume conductor has a large influence both on the amplitude and on the frequency content of surface-detected MUAPs (see Chapter 2) and could be a powerful confounding factor when inferring conclusions based on surface EMG. Farina et al. [29] indicated that a relationship between force and characteristic spectral frequencies is strongly confounded by anatomical factors. Figure 4.8 shows CV and MNF of simulated EMG signals detected during ramp contractions for different anatomical locations of the active motor units for two recruitment strategies.

Additional methods for quantification of myoelectric manifestations of muscle fatigue, based on recurrence quantification analysis of surface EMG, are described in Chapter 6 of Merletti and Parker [69]. Methods based on two-dimensional electrode arrays are described in Chapter 10.

4.5.6 Correlation and Coherence Analysis of EMG Signals

Time- and frequency-domain correlation methods have been applied for the estimation of neural connectivity between EMG signals or between EEG/MEG and EMG signals. Common synaptic inputs to motor neuron pools increase the probability that action potentials of different motor units are discharged almost at the same time. The phenomenon is called synchronization and can be quantified calculating the cross-correlation in the time domain from pairs of time series of the discharges of concurrently active motor units. Given the difficulty of identifying single motor units during voluntary contractions, the correlation-based approach has been applied directly to the surface EMG signal [35,42,53]. However, Keenan et al. [52] demonstrated that the sensitivity of this approach is relatively low when central and peripheral mechanisms are varied within physiological ranges.

On the other hand, in the frequency domain, corticomuscular and intermuscular coherence measures are used to estimate common features between the outputs of two populations of motor neurons [5,18]. In general, these measures are believed to provide reliable information about the strength and the frequencies of the signals that are transmitted from the motor cortex to the motor neuron pool of a muscle (EEG–EMG coherence) or that are shared between the motor pools of two muscles (EMG–EMG coherence) [79]. However, due to the band-pass nature of the EMG signal (surface or intramuscular), the power spectrum of the EMG signal is highly attenuated in the low-frequency bands; therefore a certain amount of error in the coherence estimation is unavoidable for low frequencies. For these reasons, full wave rectification has been applied as a preprocessing technique to demodulate the neural

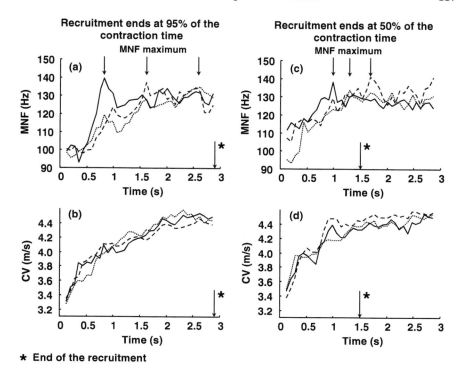

FIGURE 4.8 MNF and CV computed from three signals obtained by simulating ramp contractions with two different recruitment strategies (three line types). CV distribution standard deviation is 0.7 m/s. The three synthetic single differential signals in the two cases have been generated with the same CV distribution, MU sizes and firing pattern; only MU location is different in the three simulations. The end of recruitment is shown in the two cases along with the MNF maximum point of each simulation. In the left column (panels **a** and **b**) the recruitment of new MUs is present from the beginning until almost the end of the contraction, while in the right column (panels **c** and **d**) recruitment ends at 50% of the contraction time. Note that MNF shows a pattern in time very similar in the two recruitment conditions, and the spread of the locations of the maxima of MNF is very large in both cases. Reproduced from Farina et al. [29] with permission.

information from the EMG signal [81]. Unfortunately, since it is nonlinear, this transformation introduces a level of distortion in the coherence estimates. Some studies have demonstrated the usefulness of rectification [40,111], whereas others have strongly criticized its use [66]. Recently, Farina et al. [31] clarified that the level of distortion introduced by the rectification is mathematically related to the amount of amplitude cancellation present in the EMG signal. Amplitude cancellation is defined as the difference between the EMG signal obtained as the sum of the simulated and rectified MUAP trains and the recorded EMG. Therefore, central and pheripheral mechanisms may both bias the coherence estimations calculated from the rectified EMG signals (Fig. 4.9). The innovation provided by multichannel EMG recordings

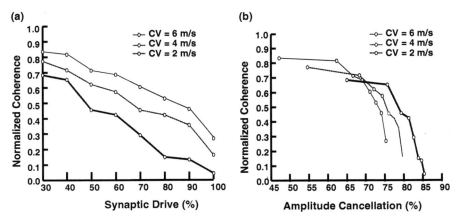

FIGURE 4.9   Effect of amplitude cancellation on EMG–EMG coherence estimation. Simulations were performed using the motor neuron model described in references 31 and 82 and the EMG model described in reference 31. Two motor pools receiving a common synaptic input were simulated. The synaptic input injected into each motor neuron was divided into a common component, shared among all motor neurons, and an independent component. The common component was centered at 20 Hz (1-Hz sidebands) and simulated with a standard deviation equal to 15% of the total standard deviation of the synaptic noise injected into each motor neuron. This level of common input was the same for all the values reported in the plots. The coherence values are normalized with respect to those estimated from the composite spike trains of the two muscles [82]. **(a)** Normalized coherence as a function of the total synaptic current injected into the motor neuron pools. **(b)** Normalized coherence as a function of the EMG amplitude cancellation level. Three scenarios with average conduction velocities of 6, 4, and 2 m/s are shown.

Chapter 5 will probably help to solve the debate on this problem, as demonstrated by recent studies applying principal component analysis (PCA) to coherence analysis [107] and accurate blind source separation techniques for the direct estimation of the neural drive to muscle (Chapter 7) [48].

4.6   CONCLUSIONS

Single-channel processing methods for surface EMG based on the interference pattern of the signal (without aiming at separating the contributions of individual MUs), have been discussed. This material focuses on the establishment of the relationships between the global variables obtained from the signal and the underlying physiological processes. Amplitude and spectral analysis have been used for many years for muscle activity and fatigue assessment. Although they have some limitations (primarily related to the critical issue of electrode location and to the impossibility of extracting information concerning single MU properties), these techniques are simple and useful in a number of basic and clinical research studies. Other global processing techniques, less commonly applied, have also been proven to be potentially useful for muscle assessment and are currently being investigated in experimental and modelling studies.

Recently, extensive research efforts have been devoted to the development of advanced signal detection and processing methods (see Chapters 3, 5, and 7) for more detailed analysis of the surface EMG signal. These methods are mostly focused on the extraction of more localized information on muscle activity (for example, related to a small number of MUs). However, frequency and amplitude analyses are still by far the most widely used methods both in basic and applied studies concerning sports, ergonomics and prevention of neuromuscular disorders. Their advantage with respect to more sophisticated techniques (described in Chapters 5 and 10) is the simpler applicability to a number of experimental conditions and recording systems.

REFERENCES

1. Akaike, H., "A new look at the statistical model identification," *IEEE Trans. Autom. Control* **19**, 716–723 (1974).

2. Amjad, A. M., "An extended difference of coherence test for comparing and combining several independent coherence estimates: theory and application to the study of motor units and physiological tremor," *J. Neurosci. Meth.* **73.1**, 69–79 (1997).

3. Arendt-Nielsen, L., and K. R. Mills, "The relationship between mean power frequency of the EMG spectrum and muscle fibre conduction velocity," *Electroencephalogr. Clin. Neurophysiol.* **60**, 130–134 (1985).

4. Arendt-Nielsen, L., and K. R. Mills, "Muscle fiber conduction velocity, mean power frequency, mean EMG voltage and force during submaximal fatiguing contractions of human quadriceps," *Eur. J. Appl. Physiol.* **58**, 20–25 (1988).

5. Baker, S. N., E. Olivier, and R. N. Lemon, "Coherent oscillations in monkey motor cortex and hand muscle EMG show task-dependent modulation," *J. Physiol.* **501**(Pt1), 225–241 (1997).

6. Balestra, G., M. Knaflitz, and R. Merletti, "Comparison between myoelectric signal mean and median frequency estimates," *Proc. 10th An Conf. IEEE Eng. Med. Biol. Soc.* 1708–1709 (1988).

7. Bendat, J. S., and A. G. Piersol. *Random Data: Analysis and Measurement Procedures*, John Wiley & Sons, New York, pp. 189–193, 277–281, 1971.

8. Bernardi, M., M. Solomonow, G. Nguyen, A. Smith, and R. Baratta, "Motor unit recruitment strategies changes with skill acquisition," *Eur. J. Appl. Physiol.* **74**, 52–59 (1996).

9. Bernardi, M., F. Felici, M. Marchetti, F. Montellanico, M. F. Piacentini, and M. Solomonow, "Force generation performance and motor unit recruitment strategy in muscles of contralateral limbs," *J. Electromyogr. Kinesiol.* **9**, 121–130 (1999).

10. Bonato, P., T. D'Alessio, and M. Knaflitz, "A statistical method for the measurement of muscle activation intervals from surface myoelectric signal during gait," *IEEE Trans. Biomed. Eng.* **45**, 287–299 (1998).

11. Clancy, E. A., "Electromyogram amplitude estimation with adaptive smoothing window length," *IEEE Trans. Biomed. Eng.* **46**, 717–729 (1999).

12. Clancy, E. A., and K. A. Farry, "Adaptive whitening of the electromyogram to improve amplitude estimation," *IEEE Trans. Biomed. Eng.* **47**, 709–719 (2000).

13. Clancy, E. A., and N. Hogan, "Single site electromyograph amplitude estimation," *IEEE Trans. Biomed. Eng.* **41**, 159–167 (1994).

14. Clancy, E. A., and N. Hogan, "Probability density of the surface electromyogram and its relation to amplitude detectors," *IEEE Trans. Biomed. Eng.* **46**, 730–739 (1999).

15. Clancy, E. A., L. Liu, P. Liu, and D. V. Moyer, "Identification of constant-posture EMG–torque relationship about the elbow using nonlinear dynamic models," *IEEE Trans. Biomed. Eng.* **59**, 205–212 (2012).

16. Clancy, E. A., E. L. Morin, and R. Merletti, "Sampling, noise reduction and amplitude estimation issues in surface electromyography," *J. Electromyogr. Kinesiol.* **12**, 1–16 (2002).

17. Cohen, L. "Time–frequency distributions: a review," *Proc. IEEE* **77**, 941–981 (1989).

18. Conway, B. A., D. M. Halliday, S. F. Farmer, U. Shahani, P. Maas, A. I. Weir, and J. R. Rosenberg, "Synchronization between motor cortex and spinal motoneuronal pool during the performance of a maintained motor task in man," *J. Physiol.* **489**(Pt3), 917–924 (1995).

19. D'Alessio, T., N. Accornero, and A. Berardelli, "Toward a real time adaptive processor for surface EMG signals," *Annu. Int. Conf. IEEE Eng. Med. Biol. Soc.* **9**, 323–324 (1987).

20. D'Alessio T., M. Laurenti, and B. Turco, "On some algorithms for the tracking of spectral structure in non-stationary EMG signals," *Proceedings of MIE '87*, Rome, Italy, 1987.

21. Dasog, M., K. Koirala, P. Liu, and E. A. Clancy, "Electromyogram bandwidth requirements when the signal is whitened," *IEEE Trans. Neural Syst. Rehabil. Eng.* **22**, 664–670 (2014).

22. De Luca, C. J., "Physiology and mathematics of myoelectric signals," *IEEE Trans. Biomed. Eng.* **26**, 313–325 (1979).

23. DeLuca, C. J., "Myoelectric manifestations of localized muscular fatigue in humans," *Crit. Rev. Biomed. Eng.* **11**, 251–279 (1984).

24. DeLuca, C. J., "Use of the surface EMG signal for performance evaluation of back muscles," *Muscle Nerve* **16**, 210–216 (1993).

25. Englehart, K. B., and P. A. Parker, "Single motor unit myoelectric signal analysis with nonstationary data," *IEEE Trans. Biomed. Eng.* **41**, 168–180 (1994).

26. Evans, H. B., Z. Pan, P. A. Parker, and R. N. Scott, "Signal processing for proportional myoelectric control," *IEEE Trans. Biomed. Eng.* **31**, 207–211 (1984).

27. Farina D., C. Cescon, and R. Merletti, "Influence of anatomical, physical and detection system parameters on surface EMG," *Biol. Cybern.* **86**, 445–456 (2002).

28. Farina, D., and R. Merletti, "Comparison of algorithms for estimation of EMG variables during voluntary isometric contractions," *J. Electromyogr. Kinesiol.* **10**, 337–350 (2000).

29. Farina, D., M. Fosci, and R. Merletti, "Motor unit recruitment strategies investigated by surface EMG variables. An experimental and model based feasibility study," *J. Appl. Physiol.* **92**, 235–247 (2002).

30. Farina, D., R. Merletti, M. Nazzaro, and I. Caruso, "Effect of joint angle on EMG variables in muscles of the leg and thigh," *IEEE Eng. Med. Biol. Mag.* **20**, 62–71 (2001).

31. Farina, D., F. Negro, and N. Jiang, "Identification of common synaptic inputs to motor neurons from the rectified electromyogram," *J. Physiol.* **591**, 2403–2418 (2013).

32. Filligoi, G. C., and P. Mandarini, "Some theoretical results on a digital EMG signal processor," *IEEE Trans. Biomed. Eng.* **31**, 333–341 (1984).

33. Gabriel, D. A., "Reliability of SEMG spike parameters during concentric contractions," *Electromyogr. Clin. Neurophysiol.* **40**, 423–430 (2000).

34. Gabriel, D. A., J. R. Basford, and K. N. An, "Training-related changes in the maximal rate of torque development and EMG activity," *J. Electromyogr. Kinesiol.* **11**, 123–129 (2001).

35. Gibbs, J., L. M. Harrison, and J. A. Stephens, "Cross-correlation analysis of motor unit activity recorded from two separate thumb muscles during development in man," *J. Physiol.* **499**, 255–266 (1997).

36. Gilmore, L. D., and C. J. De Luca, "Muscle fatigue monitor (MFM): second generation," *IEEE Trans. Biomed. Eng.* **32**, 75–78 (1985).

37. Gross, D., A. Grassino, W. R. D. Ross, and P. T. Macklem, "Electromyogram pattern of diaphragmatic fatigue," *J. Appl. Physiol.* **46**, 1–7 (1979).

38. Grosse P., M. J. Cassidy, and P. Brown, "EEG–EMG, MEG–EMG and EMG–EMG frequency analysis: physiological principles and clinical applications," *Clin. Neurophysiol.* **113.10**, 1523–1531 (2002).

39. Hägg, G., "Electromyographic fatigue analysis based on the number of zero crossings," *Eur. J. Physiol.* **391**, 78–80 (1981).

40. Halliday, D. M., and S. F. Farmer, "On the need for rectification of surface EMG," *J. Neurophysiol.* **103**, 3547 (2010).

41. Halliday, D. M., J. R. Rosenberg, A. M. Amjad, P. Breeze, B. A. Conway, and S. F. Farmer, "A framework for the analysis of mixed time series/point process data: theory and application to the study of physiological tremor, single motor unit discharges and electromyograms," *Prog. Biophys. Mol. Biol.* **64**, 237–278 (1996).

42. Hansen, N. L., S. Hansen, L. O. Christensen, N. T. Petersen, and J. B. Nielsen, "Synchronization of lower limb motor unit activity during walking in human subjects," *J. Neurophysiol.* **86**, 1266–1276 (2001).

43. Harba, M. I. A., and P. A. Lynn, "Optimizing the acquisition and processing of surface electromyographic signals," *J. Biomed. Eng.* **3**, 100–106 (1981).

44. Hinich, M. J., and C. S. Clay. "The application of the discrete Fourier transform in the estimation of power spectra, coherence, and bispectra of geophysical data," *Rev. Geophys.* **6.3**, 347–363 (1968).

45. Hof, A. L. "Errors in frequency parameters of EMG power spectra," *IEEE Trans. Biomed. Eng.* **38**, 1077–1088 (1991).

46. Hogan, N., and R. W. Mann, "Myoelectric signal processing: optimal estimation applied to electromyography—Part I: Derivation of the optimal myoprocessor," *IEEE Trans. Biomed. Eng.* **27**, 382–395 (1980).

47. Hogan, N., and R. W. Mann, "Myoelectric signal processing: optimal estimation applied to electromyography—Part II: Experimental demonstration of optimal myoprocessor performance," *IEEE Trans. Biomed. Eng.* **27**, 396–410 (1980).

48. Holobar, A., M. A. Minetto, A. Botter, F. Negro, and D. Farina. "Experimental analysis of accuracy in the identification of motor unit spike trains from high-density surface EMG," *IEEE Trans. Neural Syst. Rehabil. Eng.* **18**, 221–229 (2010).

49. Inbar, G. F., J. Allin, O. Paiss, and H. Kranz, "Monitoring surface EMG spectral changes by the zero crossing rate," *Med. Biol. Eng. Comput.* **24**, 10–18 (1986).

50. Inman, V. T., H. J. Ralston, J. B. Saunders, B. Feinstein, and E. W. Wright, "Relation of human electromyogram to muscular tension," *Electroencephalogr. Clin. Neurophys.* **4**, 187–194 (1952).

51. Kay, S. M., and L. M. Marple, "Spectrum analysis—A modern perspective," *Proc. IEEE* **69**, 1380–1413 (1981).

52. Keenan, K. G., et al., "Sensitivity of the cross-correlation between simulated surface EMGs for two muscles to detect motor unit synchronization," *J. Appl. Physiol.* **102.3**, 1193–1201 (2007).

53. Kilner, J. M., S. N. Baker, and R. N. Lemon, "A novel algorithm to remove electrical cross-talk between surface EMG recordings and its application to the measurement of short-term synchronisation in humans," *J. Physiol.* **538**, 919–930 (2002).

54. Kiryu, T., C. J. DeLuca, and Y. Saitoh, "AR modeling of myoelectric interference signals during a ramp contraction," *IEEE Trans. Biomed. Eng.* **41**, 1031–1038 (1994).

55. Knaflitz, M., and P. Bonato, "Time–frequency methods applied to muscle fatigue assessment during dynamic contractions," *J. Electromyogr. Kinesiol.* **9**, 337–350 (1999).

56. Kupa, E., S. Roy, S. Kandarian, and C. J. DeLuca, "Effects of muscle fiber type and size on EMG median frequency and conduction velocity," *J. Appl. Physiol.* **78**, 23–32 (1995).

57. Lago, P. J., and N. B. Jones, "Effect of motor unit firing time statistics on e.m.g. spectra," *Med. Biol. Eng. Comput.* **5**, 648–655 (1977).

58. Lindstrom, L., and R. Magnusson, "Interpretation of myoelectric power spectra: a model and its applications," *Proc. IEEE* **65**, 653–662 (1977).

59. Liu, L., P. Liu, E. A. Clancy, E. Scheme, and K. Englehart, "Electromyogram whitening for improved classification accuracy in upper limb prosthesis control," *IEEE Trans. Neural. Syst. Rehabil. Eng.* **21**, 767–774 (2013).

60. Liu, P., L. Liu, F. Martel, D. Rancourt, and E. A. Clancy, "Influence of joint angle on EMG–torque model during constant-posture quasi-constant-torque contractions," *J. Electromyogr. Kinesiol.* **23**, 1020–1028 (2013).

61. Lo Conte, L., R. Merletti, G. V. Sandri, "Hermite expansions of compact support waveforms: applications to myoelectric signals," *IEEE Trans. Biomed. Eng.* **41**, 1147–1159 (1994).

62. Lowery, M. M., C. L. Vaughan, P. J. Nolan, and M. J. O'Malley, "Spectral compression of the electromyographic signal due to decreasing muscle fiber conduction velocity," *IEEE Trans. Rehabil. Eng.* **8**, 353–361 (2000).

63. Magora, A., and B. Gonen, "Computer analysis of the shape of spikes from the electromyograpic interference pattern," *Electromyography* **10**, 261–271 (1970).

64. Makhoul, J., "Linear prediction: a tutorial review," *Proc. IEEE* **63**, 561–581 (1995).

65. Maranzana, M., and M. Fabbro,"Autoregressive description of EMG signals," *ISEK Far East Regional Meeting*, 1981.

66. McClelland, V. M., Z. Cvetkovic, and K. R. Mills, "Rectification of the EMG is an unnecessary and inappropriate step in the calculation of corticomuscular coherence," *J. Neurosci. Methods* **205**, 190–201 (2012).

67. McVicar, G. N., and P. A. Parker, "Spectrum dip estimator of nerve conduction velocity," *IEEE Trans. Biomed. Eng.* **35**, 1069–1076 (1988).

68. Merletti, R., and L. Lo Conte "Surface EMG signal processing during isometric contractions," *J. Electromyogr. Kinesiol.* **7**, 241–250 (1997).

69. Merletti, R., and P. Parker (eds.), *Electromyography: Physiology, Engineering and Noninvasive Applications*, Chapter 6, IEEE Press and John Wiley & Sons, Hoboken, NJ, 2004.

70. Merletti, R., F. Castagno, C. Saracco, G. Prato, and R. Pisani, "Properties and repeatability of spectral parameters of surface EMG in normal subjects," *Rassegna Bioing.* **10**, 83–96 (1985).

71. Merletti, R., M. Knaflitz, C. J. De Luca, "Myoelectric manifestations of fatigue in voluntary and electrically elicited contractions," *J. Appl. Physiol.* **68**, 1657–1667 (1990).

72. Merletti, R., G. Balestra, and M. Knaflitz, "Effect of FFT based algorithms on estimation of myoelectric signal spectral parameters," *11th Annu. Int. Conf. IEEE Eng. Med. Biol. Soc.* **11**, 1022–1023 (1989).

73. Merletti, R., L. R. Lo Conte, C. Cisari, and M. V. Actis, "Age related changes in surface myoelectric signals," *Scand. J. Rehab. Med.* **25**, 25–36 (1992).

74. Merletti, R., M. Knaflitz, and C. J. De Luca, "Electrically evoked myoelectric signals," *Crit. Rev. Biomed. Eng.* **19**, 293–340 (1992).

75. Merletti, R., Y. Fan, and L. Lo Conte, "Estimation of scaling factors in electrically evoked myoelectric signals," *Proc 14th Annu. Conf. IEEE Eng. Med. Biol. Soc.* 1362–1363 (1992).

76. Merletti, R., and L. Lo Conte, "Advances in processing of surface myoelectric signals, Part I and II," *Med. Biol. Eng. Comp.* **33**, 362–384 (1995).

77. Merletti, R., A. Gulisashvili, and L. R. Lo Conte, "Estimation of shape characteristics of surface muscle signal spectra from time domain data," *IEEE Trans. Biomed. Eng.* **42**, 769–776 (1995).

78. Merletti R., D. Farina, M. Gazzoni, and M. P. Schieroni, "Effect of age on muscle functions investigated with surface electromyography," *Muscle Nerve* **25**, 65–76 (2002).

79. Mima, T., and M. Hallett, "Corticomuscular coherence: a review," *J. Clin. Neurophysiol.* **16**, 501–511 (1999).

80. Muhammad, W., O. Meste, H. Rix, and D. Farina, "A pseudo joint estimation of time delay and scale factor for M-wave analysis," *IEEE Trans. Biomed. Eng.* **50**, 459–468 (2003).

81. Myers, L. J., M. Lowery, M. O'Malley, C. L. Vaughan, C. Heneghan, G. A. St Clair, Y. X. Harley, and R. Sreenivasan, "Rectification and non-linear pre-processing of EMG signals for cortico-muscular analysis," *J. Neurosci. Methods* **124**, 57–165 (2003).

82. Negro, F., and D. Farina, "Factors influencing the estimates of correlation between motor unit activities in humans," *PloS One* **7**, e44894 (2012).

83. Olmo, G., F. Laterza, and L. Lo Presti, "Matched wavelet approach in stretching analysis of electrically evoked surface EMG," *Sig. Proc.* **80**, 671–684 (2000).

84. Paiss, O., and G. Inbar, "Autoregressive modelling of surface EMG and its spectrum with application to fatigue," *IEEE Trans. Biomed. Eng.* **34**, 761–770 (1987).

85. Parzen, E., "Some recent advances in time series modeling," *IEEE Trans. Automat. Control* **19**, 723–730 (1974).

86. Petrofsky, S. J., "Filter bank analyzer for automatic analysis of EMG," *Med. Biol. Eng. Comput.* **18**, 585–590 (1980).

87. Piper, H., *Electrophysiologie Menschlicher Muskeln*, Springer-Verlag, Berlin, 1912.

88. Potvin, J. R., and S. H. M. Brown, "Less is more: high pass filtering, to remove up to 99% if the surface EMG signal power, improved EMG-based biceps brachii muscle force estimates," *J. Electromyogr. Kinesiol.* **14**, 389–399 (2004).

89. Press, W. H., W. T. Vetterling, S. A. Teukolsky, and B. P. Flannery, *Numerical Recipes in C: The Art of Scientific Computing*, 2nd ed., Cambridge University Press, Cambridge, UK pp. 572–576, 1994.

90. Rice, R. O., "Mathematical analysis of random noise," In: *Selected Papers on Noise and Stochastic Processes*, N. Wax, ed., Dover, New York, 1945.

91. Rissanen, J., "Modeling by shortest data description," *Automatica* **14**, 465–471 (1978).

92. Rissanen, J., "A universal prior for integers and estimation by minimum description length," *Annu. Stat.* **11**, 416–431 (1983).

93. Rix, H., and J. P. Malengé, "Detecting small variation in shape," *IEEE Trans. Syst. Man. Cybern.* **10**, 90–96 (1980).

94. Roberts, R. A., and C. T. Mullis, *Digital Signal Processing*, Addison Wesley, Reading, MA, 1987.

95. Rogers, D. R., and D. T. MacIsaac, "Training a multivariable myoelectric mapping function to estimate fatigue," *J. Electromyogr. Kinesiol.* **20**, 953–960 (2010).

96. Sachs, L., *Applied Statistics*, Springer-Verlag, Berlin, 1982.

97. Rosenberg, J. R., A. M. Amjad, P. Breeze, D. R. Brillinger, and D. M. Halliday, "The Fourier approach to the identification of functional coupling between neuronal spike trains," *Prog. Biophys. Mol. Biol.* **53**, 1–31 (1989).

98. Sanger, T. D., "Baysian filtering of myoelectric signals," *J. Neurophysiol.* **97**, 1839–1845 (2007).

99. Schweitzer, T. W., J. W. Fitzgerald, J. A. Bowden, and P. Lynne-Davies, "Spectral analysis of human inspiratory diaphragmatic electromyograms," *J. Appl. Physiol.* **46**, 152–165 (1979).

100. Shwedyk, E., R. Balasubramanian, and R. Scott, "A non-stationary model for the electromyogram," *IEEE Trans. Biomed. Eng.* **24**, 417–424 (1977).

101. Solomonow, M., C. Baten, J. Smith, R. Baratta, H. Hermens, R. D'Ambrosia, and H. Shoji, "Electromyogram power spectra frequencies associated with motor unit recruitment strategies," *J. Appl. Physiol.* **68**, 1177–1185 (1990).

102. Stulen, F. B.,"A technique to monitor localized muscular fatigue using frequency domain analysis of the myoelectric signal," Ph.D. thesis, Cambridge, MA, Massachusetts Institute of Technology, 1980.

103. Stulen, F. B., and C. J. De Luca, "Frequency parameters of the myoelectric signal as a measure of muscle conduction velocity," *IEEE Trans. Biomed. Eng.* **28**, 512–522 (1981).

104. Stulen, F. B., and C. J. De Luca, "Muscle fatigue monitor: A noninvasive device for observing localized muscular fatigue," *IEEE Trans. Biomed. Eng.* **29**, 760–769 (1982).

105. Thelen, D. G., A. B. Schultz, S. D. Fassois, and J. A. Ashton-Miller, "Identification of dynamic myoelectric signal-to-force models during isometric lumbar muscle contractions," *J. Biomech.* **27**, 907–919 (1994).

106. Triolo, R. J., D. H. Nash, and G. D. Moskowitz, "The identification of time series models of lower extremity EMG for the control of prostheses using Box–Jenkins criteria," *IEEE Trans. Biomed. Eng.* **36**, 584–594 (1988).

107. van de Steeg, C., A. Daffertshofer, D. F. Stegeman, and T. W. Boonstra, "High-density surface electromyography improves the identification of oscillatory synaptic inputs to motoneurons," *J. Appl. Physiol.* **116**, 1263–1271 (2014).

108. Viitasalo, J. H. T., and P. V. Komi, "Signal characteristics of EMG fatigue," *Eur. J. Appl. Physiol.* **37**, 111–127 (1977).

109. Welch, P. D., "The use of fast Fourier transform for the estimation of power spectra: a method based on time averaging over short, modified periodograms." *IEEE Trans. Electroacoust.* **15**, 70–73 (1967).

110. Weytjens, J. L., and D. van Steenberghe, "The effects of motor unit synchronization on the power spectrum of the electromyogram," *Biol. Cybern.* **51**, 71–77 (1984).

111. Yao, B., S. Salenius, G. H. Yue, R. W. Brown, and J. Z. Liu, "Effects of surface EMG rectification on power and coherence analyses: An EEG and MEG study," *J. Neurosci. Meth.* **159**, 215–223 (2007).

5

TECHNIQUES FOR INFORMATION EXTRACTION FROM THE SURFACE EMG SIGNAL: HIGH-DENSITY SURFACE EMG

R. Merletti,[1] T. M. Vieira,[2] and D. Farina[3]

[1]*Laboratory for Engineering of the Neuromuscular System, Politecnico di Torino, Torino, Italy*
[2]*Laboratory for Engineering of the Neuromuscular System, Politecnico di Torino, Torino, Italy, and Escola de Educação Física e Desportos, Universidade Federal do Rio de Janeiro, Rio de Janeiro, Brasil*
[3]*Department of Neurorehabilitation Engineering, Bernstein Focus Neurotechnology Göttingen, University Medical Center Göttingen, Georg-August University, Göttingen, Germany*

5.1 INTRODUCTION

Bioelectric signals appear on the surface of the body as distributions (images or maps) of electrical potentials. For example, the electrical activity of cortical neural cells is represented by EEG maps and that of heart cells by ECG maps. Similarly, muscle cells generate electrical potentials that result in surface EMG maps which can be detected with electrode grids applied above a muscle. When the grid of electrodes is dense, this EMG representation technique is usually referred to as high-density surface EMG (HDsEMG) or EMG Imaging. The signal intensity under each detection point is represented with false colors.

From an HDsEMG recording, information can be extracted in the form of images. For example, the distribution of electric potential on the skin as detected by an

Surface Electromyography: Physiology, Engineering, and Applications, First Edition.
Edited by Roberto Merletti and Dario Farina.
© 2016 by The Institute of Electrical and Electronics Engineers, Inc. Published 2016 by John Wiley & Sons, Inc.

electrode grid at an instant of time is an instantaneous image; a sequence of these images, taken at multiple time instants, describes how the instantaneous surface potential distribution evolves in time. The color of each pixel (EMG channel) is updated every time sample. Instead of the instantaneous electric potential, one of its features (e.g., average amplitude over an interval of time), may be represented to obtain the image of a specific EMG feature.

This chapter deals with the information that can be extracted from images obtained when electrode grids are applied to the skin above muscles with different architectures. The interpretation of these images, and therefore the information obtainable, depends on the muscle architecture and fiber arrangement (e.g., muscles with fibers parallel to the skin provide images different from those of muscles pinnate in the depth direction). The two cases will be addressed in different sections. It is assumed that the reader has basic knowledge about the concepts of sampling a signal in time and space, has studied the related concepts of Nyquist sampling rate and aliasing in time (see also Chapter 2 and Chapter 3), and is familiar with the relation between time and space for propagating signals. Some concepts described in Chapter 2, Chapter 3, and Chapter 8 are recalled in this chapter for convenience of the reader.

5.2   SPATIAL DISTRIBUTION OF EMG POTENTIAL AND EMG FEATURES IN MUSCLES WITH FIBERS PARALLEL TO THE SKIN

The action potentials propagating along muscle fibers generate electric fields in the surrounding conductive medium (see Chapter 2, Chapter 3, and Chapter 8). These fields produce electric currents which, in turn, generate potential distributions that are travelling on the skin. The following subsections will deal with the surface EMG instantaneous images, with the feature images, and with the spatiotemporal images.

5.2.1   Spatial Distribution of Instantaneous EMG Potential in Muscles with Fibers Parallel to the Skin

1-D Electrode Arrays.   Let us first consider the case of surface electrodes arranged along a line parallel to the fiber direction. These arrays of electrodes correspond to the one-dimensional version of the more general two-dimensional grids. When an action potential moves along a muscle fiber, the surface potential distribution that it generates moves with it along the direction of the electrode array. The two action potentials generated at the neuromuscular junction (NMJ) of a muscle fiber move, respectively, from the NMJ to the extremities of the muscle fiber, in opposite direction.These potential distributions in space evolve in time and their sampled versions are represented in Fig. 5.1 (drawing) and Fig. 5.2 (experimental signals).

The two domains in which the recorded signal can be expressed (time and space) are linked together. At each instant of time, the value of the electric potential at one electrode of the array corresponds to the value at the previous electrode a certain time interval in the past and to the value at the next electrode a certain time interval in the future. If the velocity of propagation (CV) of the source is constant, the two time

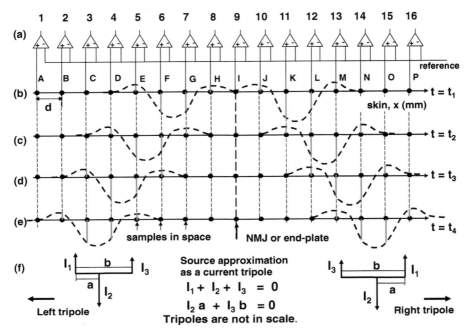

FIGURE 5.1 (a) Sixteen amplifiers in monopolar configuration reading, from 16 electrodes (A–P) with respect to a remote reference. The 16 voltages (1–16 samples in space) present on the skin are shown at time instants t_1, t_2, t_3, and t_4. (b–e) Distribution of voltages along x generated by the propagating tripoles depicted in part f). (f) Tripoles propagating in the two directions from the neuromuscular junction (NMJ) of a single muscle fiber. See Chapter 2 and [47,58]. The tripoles are not in the same scale as the rest of the figure, and their spatial support is 6–10 mm. The black dots are the sampling points in space, and the red lines are the sample values. Modified from Fig. 4.6 of Barbero et al. [1].

intervals are the same and correspond to the time needed by the source to travel from one electrode to the other. In this example, as well as in Fig. 5.1 and Fig. 5.2, we can identify the concept of sampling in time and sampling in space. Indeed, if we record a slowly traveling electric potential at very short intervals of time, we would observe at each instant of time a space-shifted version of the distribution. Collecting several time recordings, we would not only sample the signal in time but we would also be able to reconstruct the distribution of potential in space with fine details (high sampling frequency in space).

As an example, consider a signal in space, propagating from $-\infty$ to $+\infty$ at the velocity $v = 4$ m/s, without changing shape, under a series of electrodes (monopolar detection). Each electrode detects one signal changing in time. Sampling this signal in time at 2 kHz is equivalent to sampling it in space with samples 2 mm apart because this is the distance travelled in the temporal sampling interval ($\Delta t = 0.5$ ms). If the same signal in space would move faster the resulting time signal would have the same shape but shorter duration (v is a time scaling factor). In these ideal conditions, all the

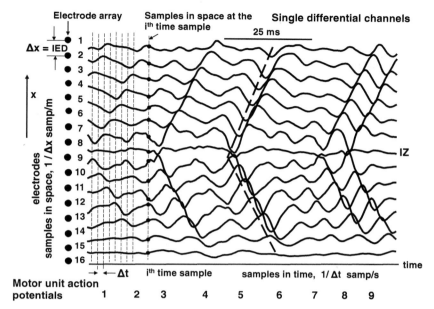

FIGURE 5.2 Differential signals detected during a voluntary contraction of a biceps brachii with an array of 16 electrodes spaced by Δx meters along the fiber direction. All signals are simultaneously sampled in time. Time samples are separated by Δt seconds. Sampling frequencies are $1/\Delta x$ samples/m and $1/\Delta t$ samples/s (e.g., for $\Delta x = 10$ mm the spatial sampling frequency is 100 samples/m and for $\Delta t = 0.5$ ms the temporal sampling frequency is 2000 samples/s or 2 kHz). Nine motor unit action potentials initiating between electrodes 8 and 9 are clearly visible. The potentials are symmetric with respect to the innervation zone (IZ), indicating that the fibers are parallel to the skin and to the electrode array. The slopes of the dashed lines overlapping the motor unit action potential #5 represent the conduction velocity of this motor unit in the two propagation directions (see Section 5.2.5 on conduction velocity estimation).

electrodes of the linear array would detect exactly the same signal, delayed accordingly to the electrode position in the array.

In this example, we can define another sampling in space, which is obtained by the direct reading of the values of the electric potential at each of the electrodes. This spatial sampling has frequency equal to the inverse of the interelectrode distance, and this sampling frequency is unrelated to the indirect spatial sampling associated with the temporal sampling. The concepts of spatial and temporal sampling are valid for any electrode configuration. However, the equivalence of temporal sampling to spatial sampling is valid only in the ideal condition, described above, of propagation along the line defined by the electrode array. For example, this relation does not hold if the signal changes shape as it moves or if it comprises components propagating at different velocities (including $v = \infty$, for "nonpropagating" signals simultaneously present on a group of channels) or in directions not parallel to the skin.

The concepts of spatial and temporal sampling are illustrated in Fig 5.1, where space is along the horizontal axis and time is along the vertical axis (t_1, t_2, t_3, t_4), and in Fig. 5.2, where time is along the horizontal axis and space is along the vertical axis. The question of how close the samples should be, in either space or time, is relevant in order to acquire the full information content of the spatiotemporal signals, especially when the relationship $x - x_o = vt$, linking space and time, does not hold.

2-D Electrode Arrays.   The linear electrode array depicted in Fig. 5.1 and 5.2 can be considered as one column of the 2-D array of Fig 5.3. Each column of a 2-D array would produce a set of signals such as those in Fig 5.3a, where each time sample corresponds to an image. Four such images, corresponding to the time samples t_1, t_2, t_3, t_4, are shown, after spatial interpolation, in Fig. 5.3b, where the color scale

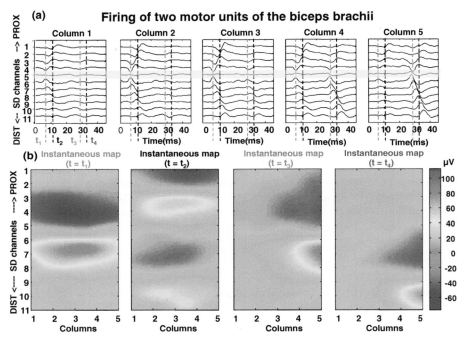

FIGURE 5.3   (a) 2-D electrode array with 12 rows and 5 columns (interelectrode distance: IED = 8 mm) and five sets of single differential (SD) signals, each set obtained from one column. The electrode grid is placed on a biceps brachii, during a mild voluntary isometric contraction, with columns parallel to the fiber direction. Signals are sampled in time at 2000 samples/s; that is, 2000 images per second are obtained. Observation time is 45 ms (90 images). Two motor units discharge within this time interval, one located under columns 1–4, the other under columns 4–5. Four samples (at times t_1, t_2, t_3, t_4) corresponding to the four images in panel b) are indicated by vertical dashed lines. (b) Four instantaneous images, at t_1, t_2, t_3, t_4, are depicted. The images are interpolated with a factor 10 by means of two-dimensional splines and the signal intensity of each pixel is color-coded.

represents the instantaneous signal intensity. Two motor units discharge within the observed 45 ms and can be identified. The first has fibers located under columns 1–4, discharges at t_1, and its motor unit action potential (MUAP) propagates downward from row 5 to 9 and upward from row 5 to 1 at t_2. The second is located under columns 4–5, discharges at t_3, and its MUAP is not as symmetric with respect to the IZ, indicating a slight misalignment between the fibers of the MU and the columns of the array.

The grid operates two spatial samplings in two spatial directions. In the direction of propagation, another spatial sampling related to the temporal sampling under the assumption of pure propagation can be derived as described above. However, in the direction transverse to the muscle fibers, sampling in space is only associated to the electrode spatial locations. Moreover, due to the nonidealities in the propagation of the action potentials along the fiber direction, the sampling by the electrodes located along the muscle fibers is not exactly equivalent to a temporal sampling with scaled sampling frequency, as demonstrated by the slightly different shapes of the MUAPs in different channels in Fig. 5.2 and Fig 5.3a.

As in any digital signal recording, the first question to pose is if the Nyquist limit for sampling the signal is satisfied. If this occurs, then the digital sampling contains all the available information and the digital "movie," representing the evolution of the two-dimensional electric potential over time, is equivalent to the analog (continuous) movie which can be obtained by means of a "reconstruction filter" [52]. While it is simple to determine the Nyquist sampling frequency by analyzing the bandwidth of the signal in the temporal dimension, this is more challenging in the spatial domains The issue of whether or not the Nyquist criterion in space is satisfied is equivalent to asking which interelectrode distance (IED) should be adopted in order not to lose any information on the distribution of electric potential over the skin at each sampling instant and be able to reconstruct the images on the skin and compute their features.

In order to address the issue of spatial sampling in relation to the Nyquist criterion, let us consider that a 2-D signal $f(x,y)$ sampled in space over an area $L_x \times L_y$ has an amplitude spectrum and a power spectrum (defined in spatial frequencies) that are periodic with period $f_{samp} = 1/IED$ where the harmonics are spaced by $1/L_x$ and $1/L_y$ cycles/m. The relation $f_{samp} > 2f_{max}$, where f_{max} is the highest spatial harmonic of the signal, must be satisfied to avoid aliasing (Nyquist limit). Aliasing would not allow the reconstruction of the original analog signal and would affect the estimation of its features [48]. In Fig. 5.4a–c, the same surface potential distribution, produced at a specific time instant by a simulated muscle fiber action potential, is detected over a square support of 0.128×0.128 m, with a grid of point-like electrodes (zero surface) having interelectrode distances of 1 mm, 5 mm, and 10 mm, that correspond to sampling frequencies (f_{samp}) of 1000 samp/m, 200 samp/m, and 100 samp/m, respectively. The magnitudes of the Fourier transforms of these three sampled images are periodic, both along f_x and along f_y, with periods of 1000 cycles/m, 200 cycles/m, and 100 cycles/m respectively, with harmonics (pixels) spaced by 7.8 cycles/m (1.0/ 0.128 m^{-1}). Panels d and e of Fig. 5.4 visualize one cycle (alias) of the amplitude spectrum (for positive frequencies) obtained from the images of panels b and c. Panels

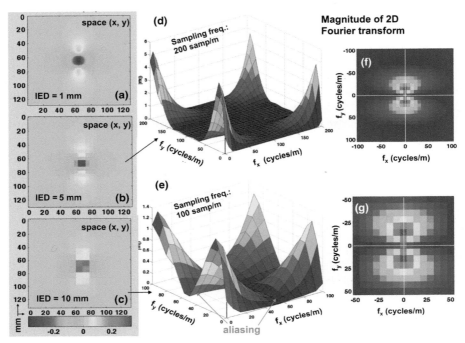

FIGURE 5.4 (a–c) Potential distribution of a simulated monopolar single fiber action potential, beginning to propagate, sampled in space over a surface of 128 mm × 128 mm with three spatial sampling frequencies: 1000 samples/m, 200 samples/m and 100 samples/m— that is, interelectrode distances of 1 mm, 5 mm and 10 mm. The fiber is at depth = 3 mm (1 mm of skin, 1 mm of subcutaneous tissue, 1 mm of muscle). (d, e) Magnitude of the Fourier transform (in space) of the images in parts b and in c. One cycle is represented in panel d for $0 < f_x < 200$ cycles/m and $0 < f_y < 200$ cycles/m and in part e, for $0 < f_x < 100$ cycles/m and $0 < f_y < 100$ cycles/m. Aliasing is evident in panel e as overlapping of spectral repetitions and as nonzero values at $f_{samp}/2$. (f, g) Representations of the same Fourier transforms are shown for positive and negative frequencies (one alias for $-100 < f_x < 100$ cycles/m and $-100 < f_y < 100$ cycles/m in panel f) and one alias $-50 < f_x < 50$ cycles/m and $-50 < f_y < 50$ cycles/m in panel g)). Reproduced from Merletti et al. [48] with permission.

f and g also represent a cycle (alias) for positive and negative frequencies but only the quadrant where $f_x > 0$ and $f_y > 0$ has physical relevance. In this example, it is evident that the "aliases" overlap when IED = 10 mm. This IED is less critical for electrodes of a nonzero surface whose low-pass filtering effect reduces the signal bandwidth in space. This reduction lowers the Nyquist rate and allows lower sampling frequencies—that is, larger IEDs (see Chapter 2 and Section 5.2.4).

The magnitude squared of the 2-D Fourier transform is the energy spectral density of the image. The double integral of this function for $0 < f_x < f_{samp}/2$ and $0 < f_y < f_{samp}/2$ is the energy of the image. Its square root normalized by the size of the image is the root mean square value (RMS) of the image in space.

5.2.2 Spatial Distribution of EMG Features in Muscles with Fibers Parallel to the Skin

Many EMG features (e.g., amplitude or spectral variables) have physiological significance; that is, they reflect physiological mechanisms. They are traditionally estimated from a single (monopolar or differential) EMG channel and associated with the muscle (see Chapter 4). They are also associated with the specific electrode location selected above the muscle and they change when the electrode location changes. When an electrode grid is used, these variables are obtained from each electrode, or electrode pair, and provide images (Figs. 5.5 and 5.6). The large amount of resulting information may be reduced by averaging the selected variables over a region of interest or over the entire image. Other image processing manipulations may be applied, such as interpolation, image rotation, thresholding, segmentation, computation of the centroid, or edge or peak detection.

These variables are estimated over each signal epoch and are functions of time—that is, of the epoch number as well as of the epoch duration (Fig. 5.5b). The finite

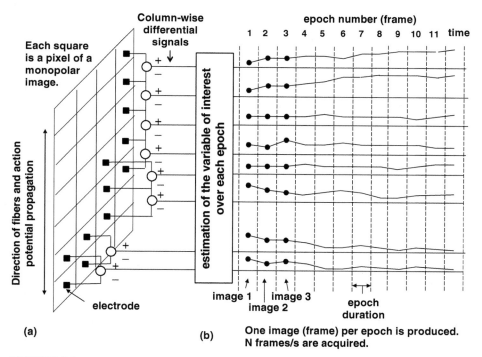

FIGURE 5.5 Example of generation of a feature map from a set of single differential (SD) sEMG signals. Each SD signal is obtained by taking the difference between adjacent monopolar signals along the columns, in the fiber direction. The feature of interest is computed over each epoch of each signal and associated with the corresponding pixel so that a new image is obtained every epoch, representing the evolution in time and space of the feature. The number of epochs per second is usually 1, 2, or 4.

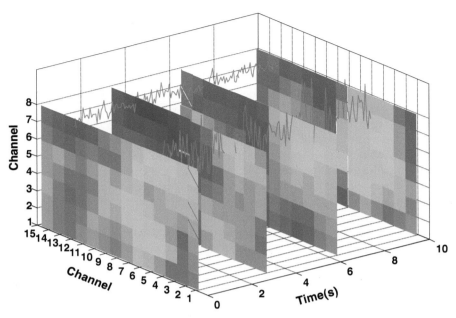

FIGURE 5.6   Four images of ARV (out of 40), each computed over 0.25 s, during a 10-s voluntary contraction of the gastrocnemius muscle. The images are obtained from monopolar signals. Two monopolar signals, corresponding to two pixels, are depicted in red.

duration of the time epoch, due to windowing the signal with a rectangular (or other) window, implies truncation in time and consequent errors in the estimation of features. Because the signal is one realization of a stochastic process, the means and the standard deviations of the estimates of these variables are random variables whose mean and variance depend on the number and duration of the epochs [45].

The same considerations apply to the signals in space (images). Any linear array or grid of electrodes implies truncation in space and error in the estimates of the EMG spatial features. Figure 5.5a shows a grid of electrodes whose pixels are indicated by squares while the instantaneous EMG value is associated to each electrode. Features of interest (e.g., ARV or RMS or MNF or MDF, etc.) are calculated for each epoch duration (see Chapter 4). In this way, a time-evolving image is obtained, which is sampled in time at N frames/s, as indicated in Fig. 5.5b. N is usually 1, 2, or 4 frames/s. In the example of Fig. 5.5, a column-wise differentiation is applied, providing a set of longitudinal single differential (LSD) channels, in the fiber direction, and their features. Figure 5.6 shows images of ARV, each computed over a 0.25-s interval for a total of 40 images in 10 s (only four are shown). Figure 5.7 shows an example of RMS of a simulated motor unit potential distribution estimated with a grid of point-like electrodes whose number is variable along each dimension and equal to the integer part (floor) of the ratio A/IED along x and B/IED along y, for 0.5 mm < IED < 15 mm in steps of 0.5 mm. In this case, $A = B = 127$ mm and the central electrode is fixed. The estimates of the RMS are normalized with respect to the value obtained for

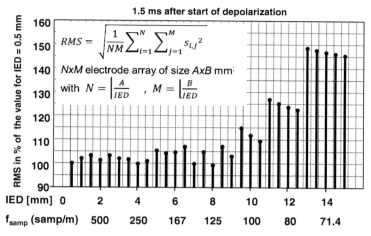

FIGURE 5.7   Estimates of the RMS of the potential distribution in space of a simulated monopolar motor unit action potential [11]. The motor unit has 150 fibers, radius 1.5 mm, depth = 2 mm, width of the innervation zone = 10 mm, L1 = 65 mm, spread1 of fiber-tendon junctions = 8 mm, L2 = 60 mm, spread 2 of fiber–tendon junctions = 10 mm. The RMS of the image is calculated, for a given time instant, over an area of 128×128 mm^2, for interelectrode distance (IED) increasing from 0.5 mm to 15 mm in steps of 0.5 mm. The fluctuations of the estimates are due to truncation effects and aliasing (for IED > 8 mm) as well as to the fact that, as IED increases, the number of rows and columns of the array is progressively decreasing as rows and columns move out of the fixed detection area. IED is increased starting from the electrode in the center of the image which remains fixed. The effect of increasing IED over a fixed area becomes evident, in this example, for IED greater than 8–10 mm. This effect strongly depends on the image morphology, that is, on its frequency content. For a superficial MUAP, the RMS estimation error (in space) usually exceeds 10% for IED > 10 mm but is reduced by the low-pass filtering effect due to the physical size of the electrodes (see Section 5.2.4 and Chapter 2).

IED = 0.5 mm, which is considered as the "true" RMS. It must be emphasized that, as indicated in the figure and in Eqs. (5.1) and (5.2), the RMS is computed by adding the squares of the samples (as is commonly done) and not from the output of a reconstruction filter (as it should be done) [52].

Although the practical implications of surface EMG aliasing in space have not been fully investigated, it is obvious that interpolation in space to increase resolution (such as in Fig. 5.3b) can be applied only if aliasing is absent or minimal. Equations (5.1) and (5.2) provide the continuous and the discrete definition of energy (RMS^2) of an image, or of a region of interest, of area $a \cdot b$ mm^2, sampled in $M \cdot N$ points, showing the mean square value of signals computed over a time interval T, sampled in U equidistant points. Since the integral (or summation) symbols can be exchanged, the result can be seen as the spatial average of RMS^2 in time, or the time average of RMS^2 values in space. The same considerations apply to other features, such ARV and spectral variables.

$$\text{RMS}^2_{a,b,T} = \frac{1}{abT} \int_{-b/2}^{+b/2} \int_{-a/2}^{+a/2} \left[\int_0^T f^2(x,y,t)\, dt \right] dxdy$$

$$= \frac{1}{abT} \int_0^T \left[\int_{-b/2}^{+b/2} \int_{-a/2}^{+a/2} f^2(x,y,t)\, dxdy \right] dt \tag{5.1}$$

$$\text{RMS}^2_{M,N,U} = \frac{1}{NMU} \sum_{i=1}^N \sum_{j=1}^M \sum_{t=1}^U s^2(i,j,t) = \frac{1}{NMU} \sum_{t=1}^U \sum_{i=1}^N \sum_{j=1}^M s^2(i,j,t) \tag{5.2}$$

$\text{RMS}^2_{M,N,U}$ is an approximation of $\text{RMS}^2_{a,b,T}$. If a reconstruction filter is not used (see above), the error depends on the density of samples taken in the time and space dimensions (Fig. 5.7).

The most basic application of maps of EMG, or EMG features, is the study of the location of muscles and of their compartments. Figure 5.8 shows five noninterpolated RMS maps computed over epochs of 1 s from an array of 128 electrodes covering the dorsal side of the forearm. The maps are computed from monopolar signals, and each

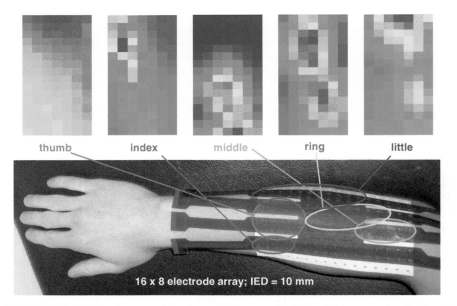

FIGURE 5.8   Electrode array of 8 columns × 16 rows with IED = 10 mm covering the dorsal side of a forearm. The five maps represent the spatial distribution of the RMS of 112 monopolar EMG signal (the most lateral column is not available) computed over a 1-s epoch during extension of each of the fingers [27].

corresponds to the extension of one finger. The most lateral column is not displayed. The figure demonstrates rather abrupt changes in space and the possibility of classification of regions of interest corresponding to the activation of individual muscles. Similar maps have been obtained for the arm muscles by Rojas-Martinez et al. [59] and are reported in Fig. 10.4.

The clinical use of HDsEMG was proposed more than a decade ago [57,70,71] and implemented by many authors for motor unit number estimation [3], investigation of stroke patients [34,69], microgravity studies [50], muscle force estimation [62,63,64], region-specific manifestations of fatigue [67,68], and other applications [8,35,36].

Image processing techniques (such as enhancement, edge and pattern detection) may be applied to spatiotemporal images as well. Figure 5.9a depicts a 100-ms epoch of a 12-channel differential surface EMG detected using a linear electrode array placed on the biceps brachii and parallel to the fiber direction. Figure 5.9b is obtained by representing the signal intensity of the image of Fig. 5.9a in terms of gray levels.

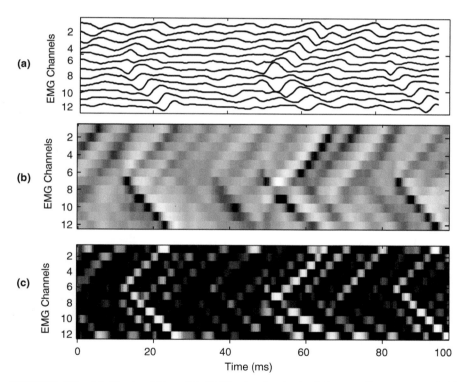

FIGURE 5.9 (a) Single differential EMG signals obtained from a linear array of 12 electrodes placed over the biceps brachii, parallel to the fiber direction. (b) Topographical image of the signals in panel (a) sampled at 2048 samples/s. (c) Same image as in panel (b) after the application of an image enhancing algorithm that facilitates further processing for automatic identification of the innervation zones of the motor units and estimation of their conduction velocities.

Image enhancement techniques can then be applied to obtain Fig. 5.9c from which the conduction velocity and the innervation zone of individual MUAPs can be estimated.

5.2.3   Spatial Filtering

Surface EMG signals can be recorded in monopolar mode or as a linear combination of signals recorded by more than one electrode (Chapters 2 and 3). The most common of these recording modalities is the bipolar (or LSD) derivation which consists in the difference between the signals detected by two electrodes aligned along the fiber direction (Fig. 3.11). The extension of this concept to more than two electrodes leads to the definition of a variety of spatial filters. For example, if a second-order differentiation is applied by making the difference between two adjacent bipolar recordings along an electrode array, a double differential detection is obtained. This procedure approximates the second derivative in the digital spatial domain and enhances the estimation of CV [7,53,54] while reducing the contribution of the nonpropagating components (see Chapter 2).

In a broader sense, one can build a spatial filter of the electric potential distribution by linear combination of monopolar signals from neighboring electrodes, as is usually done for image processing. A classic spatial filter applied to the EMG signal in two dimensions is the Laplacian (or normal double differential, NDD) filter, which is an edge detection filter in image processing. Its spatial transfer function is a high-pass transfer function in both spatial frequency directions. Given an electrode arrangement and the weight associated to each point-like electrode for designing a spatial filter, it is possible to obtain the corresponding transfer function by Fourier transform of a set of delta functions (Chapters 2 and 3). Since the electrodes are located at a certain distance, sampling the skin surface potential, the transfer function of spatial filters obtained in this way is always periodic, with periodicity equal to the spatial sampling frequency—that is, the inverse of the inter-electrode distance. For spatial filters defined as linear combination of electrodes located along the direction of the muscle fiber, there is an obvious association between spatial and temporal filtering. Indeed, for the same reasons discussed above on sampling in the direction of propagation under ideal conditions, filtering in space is equivalent to filtering in time, with the scaling of temporal frequency by the velocity of propagation according to $f_t = f_z \cdot v$, where f_t is a spectral frequency in the time domain and f_z is the corresponding frequency in the spatial domain in the direction of propagation. This equation relates the transfer function of a spatial filter to that of the equivalent temporal filter, in ideal conditions. Figure 5.10a depicts the most common spatial filters, and Fig 5.10b shows one cycle (alias) of the transfer function of two of them, for positive and negative spatial frequencies.

Because the volume conductor is also a spatial filter (see Chapter 2), it is theoretically possible to design a spatial filter for signal detection that would invert the filtering effect of the tissues present between the muscle surface (fascia) and the skin surface. A digital approximation of this analog high-pass filter is described in the work of Farina and Rainoldi [9]. The weights of a truncated (5×5 point-like electrodes) version of the impulse response of this spatial filter, for a specific set

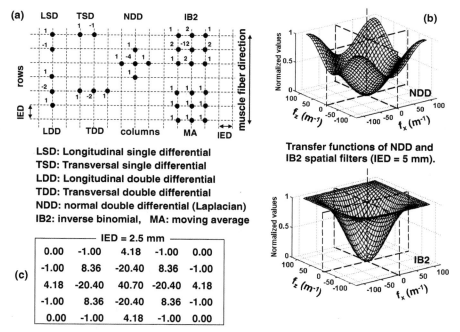

LSD: Longitudinal single differential
TSD: Transversal single differential
LDD: Longitudinal double differential
TDD: Transversal double differential
NDD: normal double differential (Laplacian)
IB2: inverse binomial, **MA:** moving average

Transfer functions of NDD and
IB2 spatial filters (IED = 5 mm).

IED = 2.5 mm				
0.00	-1.00	4.18	-1.00	0.00
-1.00	8.36	-20.40	8.36	-1.00
4.18	-20.40	40.70	-20.40	4.18
-1.00	8.36	-20.40	8.36	-1.00
0.00	-1.00	4.18	-1.00	0.00

FIGURE 5.10   (**a**) Examples of spatial filters. (**b**) Transfer functions, in the spatial frequency domain, of the filters NDD (Laplacian) and IB2 (inverse binomial): The transfer function is periodic in both dimensions, because the surface signal is sampled, and one period is represented for positive and negative frequencies. (**c**) Weights of the monopolar signals detected by a 5×5 grid and producing a spatial filter which compensates for the effect of a given set of skin and subcutaneous tissue properties [9].

of tissue properties, are given in Fig. 5.10c. The properties of the volume conductor (conductivity and thickness of the layers) affect the selectivity of spatial filters [18].

Farina et al. in 2003 [16] used the electrode grid of Fig. 5.11a to detect the monopolar action potentials of a motor unit by combining the signals with a Laplacian (or NDD) spatial filter. The grid was placed on one side of the innervation of the biceps brachii muscle. Twenty-one discharges, detected using six Laplacian Filters centered in positions 1 to 6 of the grid, are superimposed in Fig. 5.11b. The outputs of the 11 transversal single differential (TSD) filters and of the 12 longitudinal single differential (LSD) filters are depicted in Fig. 5.11c, d. The outputs of the six transversal double differential (TDD) filters and of the six longitudinal double differential (LDD) filters are depicted in Fig. 5.11e, f to show the effect of these types of spatial filters on a MUAP.

Visual observation of Fig. 5.11b–f elicits the concept of filter selectivity and the definition of indexes of longitudinal and transversal selectivity which quantify the rate of decay of the filter gain along the longitudinal and the lateral directions. Higher selectivity means greater ability to separate contributions from different motor

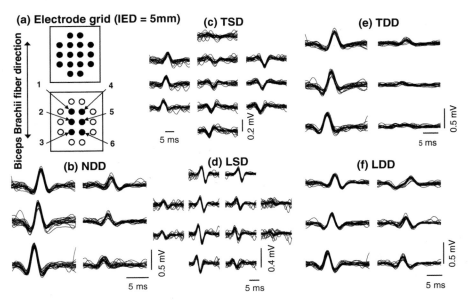

FIGURE 5.11 Examples of responses of five spatial filters applied to the electrode grid indicated in panel a placed on one side of the innervation zone of a healthy biceps muscle. Twenty-one MUAPs of the same motor unit were extracted during a voluntary contraction using the NDD filter. (**a**) Electrode grid and location of the six NDD filters. (**b**) Outputs of the six NDD filters. (**c**) Outputs of the 11 TSD filters. (**d**) Outputs of the 12 LSD filters. (**e**) Outputs of the six TDD filters. (**f**) Outputs of the six LDD filters applied to the central two columns. Filters are defined in Fig. 5.10. Modified from Farina et al. [16].

units [16,60,61]. It is tempting to think that a selective filter would reduce crosstalk from nearby muscles. This is not necessarily so since crosstalk may largely derive from the nonpropagating end-of-fiber effect which decays in space more slowly than the travelling components and may have temporal frequencies higher than the propagating components [6,14,20] (see Chapter 2).

Practical spatial filtering is performed with electrodes having physical dimensions which will act as low-pass spatial filters (they average the electric potential values under the electrode surface) on the detected signal (see Chapter 2). To some degree, it is possible to design the shape of the electrodes to obtain a desired spatial transfer function. For example, the difference between the signals detected by a circular electrode and by a second ring-shaped electrode that surrounds the first corresponds to a two-dimensional high-pass spatial filter with interesting properties (e.g., invariance for rotation) [13].

As a first approximation, the potential distribution in the region surrounding an electrode is assumed to be unchanged. Consider a 2-D signal propagating under the detection area: the electrode implies a moving average filtering—that is, a low-pass filter in space [65] (see Fig. 2.5). The same applies if the signal is not propagating in space but just evolving in time (e.g., the end-of-fiber effect).

The impulse response of a rectangular electrode centered at $x = z = 0$ and with size a along x and b along z is the "box" $h(x, z) = 1/ab$ for $-a/2 < x < a/2$ and $-b/2 < z < b/2$ and zero elsewhere. The transfer function of such electrode is $H(f_x, f_z) = \mathrm{sinc}(\pi a f_x)$ $\mathrm{sinc}(\pi b f_z)$, where $\mathrm{sinc}(y) = (\sin y)/y$. For a square electrode, $H(f_x, f_z) = \mathrm{sinc}(\pi a f_x) \cdot \mathrm{sinc}(\pi a f_z)$ whose first zero is for $f_x = f_z = 1/a$ cycles/m. Therefore, for square electrodes with $a = 5$ mm and $a = 10$ mm, the first zero is at 200 cycles/m and 100 cycles/m, respectively (see Fig. 2.5 for circular electrodes). For a signal propagating with a velocity of 4 m/s, the first zero of the transfer function in the temporal frequency domain is at 800 Hz and 400 Hz, respectively, and the 3-dB cutoff frequency is 360 Hz and 180 Hz, respectively. The low-pass filtering effect, in space and time, due to the averaging of potentials under a physical surface electrode is evident. This effect is not necessarily undesirable if low noise and attenuation of the MUAPs of the most superficial generators are of interest. However, users must be aware of the fact that EMG amplitude and spectral variables are strongly influenced by the electrode configuration (spatial filter), contact area, and interelectrode distance (Chapters 2 and 3).

5.2.4   Estimation of Muscle Fiber Conduction Velocity

The multichannel systems described above detect the surface electric potentials generated by the muscle fiber action potentials. For muscles with fibers parallel to the skin, the resulting surface potentials travel along the fiber direction. The velocity of this propagation is a relevant physiological parameter, as described in Chapter 10. This section addresses the problem of estimating the velocity of propagation (or conduction velocity, CV) from one- or two-dimensional EMG recordings.

The classic way of estimating muscle fiber conduction velocity is based on the estimation of the delay between signals recorded at fixed distance along the direction of propagation and the ratio between such measured distance and the estimated delay. Before addressing this problem, let us once again recall the concepts of spatial and temporal sampling and filtering and their associations.

As previously observed, applying a spatial filter to an EMG signal distribution implies a linear combination of signals from electrodes along the direction of propagation and is equivalent to applying a temporal filter, with a scaling of the frequency axis. The scaling factor is CV, which is unknown. Therefore, theoretically, given a set of coefficients, it is possible to apply the corresponding spatial filter for EMG detection and then apply the equivalent temporal filter to the central monopolar signal. The temporal filter will depend on CV, which would be an unknown parameter of its transfer function. In these conditions, the correct CV value is the value of the unknown parameter in the temporal transfer function such that the spatially and the temporally filtered signals are maximally similar (equal in ideal conditions). On the basis of these considerations, Farina and Merletti [15] proposed an effective method for estimating the CV of action potentials by matching the effects of spatial and temporal filtering. Earlier, a simplified approach based on the same concepts had been proposed and is widely known as the spectral dip approach [39,40]. The spectral dip approach is also based on spatial filtering of an EMG signal and on the scaling between temporal and spatial frequencies by the

CV value. However, contrary to the method of the matched transfer functions, it considers only one specific frequency value—that is, the frequency value at which the spatial transfer function is null (dip frequency). The dip introduced in space is also present in time, and the two frequencies of the dip (in space and time) are associated by the relationship $f_t = f_z \cdot v$. Therefore, once the spectral dip frequency in space is determined (by the design of the spatial filter) and the corresponding spectral dip frequency in time is detected (from the power spectral estimate of the signal in time), the conduction velocity can be estimated. Interestingly, this concept can be extended to multiple dip frequencies [17,22]. This method, as those described in the following, is affected by the spread of the CV values; if the spread is large, the dips are shallow and the estimates of the average delay are affected by greater errors.

Any delay estimator can be used as an approach for the measure of CV. Two electrodes (or electrode systems if a spatial filter is applied) aligned along the direction of propagation will detect the same signal, in ideal conditions, with a delay determined by the interelectrode distance divided by the conduction velocity. Similarly, if more than two signals are available, the delay can be estimated from any signal pair or globally by estimating the interchannel delay that maximally aligns the available signals. Based on these concepts, several methods have been proposed for the delay estimation [10,12]. Rather than describing all of them in detail (see Farina and Merletti [19] for a review), here we will focus on general considerations valid for all these methods and we will then describe representatively only two of these approaches.

The problem of estimating a delay from two signals is trivial when the two signals are identical, continuous, and noise-free. When these conditions are not met, the delay estimation problem is more complex.

If we consider signals recorded at a sampling frequency of 1 kHz, we are limited in time by a resolution equal to the sampling interval, which is 1 ms in this example. Now, let's assume that the CV value to be estimated is 5 m/s and that the electrodes are located at a distance of 5 mm. In this condition, the delay to be estimated and its resolution would be $IED/v = 1$ ms; therefore our relative error (resolution in the estimate divided by the true value) would be 100%, which clearly is not acceptable, despite sampling above the Nyquist rate. In order to improve the estimate, given the associations discussed above, we could either increase the sampling frequency in time (and have much larger raw data files) or increase the spacing between electrodes, or both. For example, by increasing 10 times the sampling frequency in time, we would decrease by a factor 10 the relative error, which would become in the example above equal to 10%. Although the sampling frequency in time can be increased offline without practical limitations (pending that the initial sampling frequency satisfies the Nyquist limit), increasing the interelectrode distance may not be possible because of the constraint that the two systems have to be both within the fiber semi-length where the source propagates in only one direction. Methods that use time interpolation in different ways to solve the sampling issue are described below.

When the two (or more) signals are similar but not identical, the mathematical definition of delay is not unique but depends on the criterion used to compare the signal shapes.

One of the simplest ways to estimate a delay between signals of similar shape is to compute the time lag at which the cross-correlation between the two signals is maximal (or the mean square error is minimum). This approach is not only a practical method for estimating a delay but is also a specific definition of delay between nonidentical signals. This approach is usually applied with interpolation of the cross-correlation function values around the peak.

An alternative approach consists in matching the signals in the frequency rather than in the time domain [37,42,46]. The method is known as spectral matching and consists in finding the delay which minimizes the mean square error between the Fourier transforms of the two signals. The cross-correlation and spectral matching methods, which can be demonstrated to be equivalent, have been extended to the case of more than two signals [21]. Using more than two signals is useful in practice to reduce the effect of additive noise and fluctuations of the MUAP shapes, as indicated in Fig. 5.12.

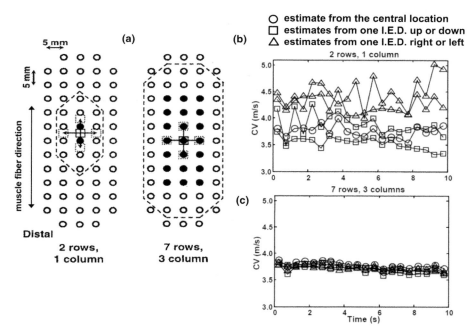

FIGURE 5.12 (a) Electrode array used for a set of measurements of CV. The electrodes used are inside the dashed lines, and the centers of the spatial filters are the black dots. (b) Sequences of CV estimates over 10 s (every 0.5 s). Each estimate is from a pair of double differential filters centered on the central electrode, one electrode below or above it and one electrode to the right or to the left (five locations). (c) Each estimate is from multiple electrode location along seven rows. Multichannel estimates strongly reduce the variability of the estimate caused by small displacements (1 IED up–down or 1 IED right–left). Modified from Farina and Merletti [21] with permission.

Different methods for CV estimation have different sensitivities to the non-propagating signals produced by the generation and extinction of the MUAPs and by crosstalk [25]. In addition, the same method is sensitive to the spatial filter used as described in [16,49,61] and in Fig. 5.12.

5.3   SPATIAL DISTRIBUTION OF EMG POTENTIAL AND FEATURES IN PINNATE MUSCLES

When the muscle fibers are not parallel to the skin, the information contained in the EMG images changes substantially. In this subsection, examples are provided on the information that may be obtained from HD-EMGs detected from muscles with pinnate architecture. Traditionally, surface EMG signals are collected from a specific region of a muscle, typically from its bulk, and their amplitude and spectral descriptors are considered indicative of the degree of activation or fatigue of the entire muscle or most of it, regardless of the architecture of the studied muscle. Recent evidence, obtained with grids of electrodes, suggests however that the amplitude of surface EMGs collected from a specific region of a pinnate muscle is not representative of the entire muscle.

5.3.1   Muscles with Fibers Parallel to the Skin Versus Muscle with Fibers Pinnate in the Depth Direction

An important distinction must be made between (1) pinnate or nonpinnate muscles with fibers parallel to the skin (referred to as *skin parallel-fibered muscles*) and (2) pinnate muscles with fibers inclined in the depth direction, referred to as *muscles pinnate in depth direction*. In pinnate muscles, fibers are arranged into a feather-like structure, with the line of pull of individual fibers directed obliquely with respect to the force-generating axis. They may be classified as unipinnate (e.g., vastus muscle) or multipinnate (gluteus medius muscle [38]).

From an electromyographic perspective, the direction of muscle fibers with respect to the skin is of crucial relevance. Most large leg muscles are *muscles pinnate in the depth direction*. Gastrocnemius muscle fibers, for example, extend from the superficial to the deep aponeurosis, with their insertion angle varying considerably with knee and ankle position, with contraction level, and along the muscle proximo-distal axis [51]. The fibers of thigh muscles, on the other hand, although oriented obliquely with respect to the force-generating axis, are predominantly *skin parallel-fibered muscles*. Notwithstanding some degree of pinnation in depth direction [2], fibers of the vastus medialis muscle are mostly lying in a plane parallel to the skin. These architectural differences between gastrocnemius and vastus medialis dramatically affect the representation of motor unit action potentials in the surface EMGs.

Differences between the representation of MUAPs in *skin parallel-fibered* muscles and in muscles *pinnate in depth direction* are clearly appreciated with a grid of surface electrodes. Figure 5.13 illustrates an example of two MUAPs, one for each of these two muscle architectures. Single-differential EMGs shown in this figure were

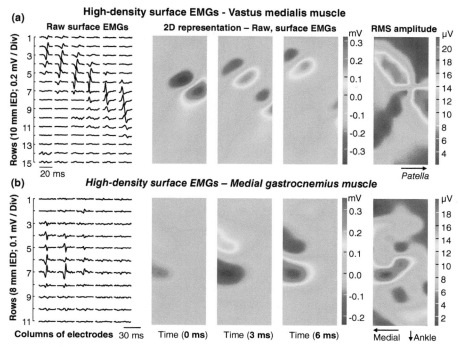

FIGURE 5.13   (a) Single differential action potential of a motor unit of the vastus medialis muscle, detected by a surface array of 16×6 electrodes (10-mm IED). The most lateral column of electrodes was positioned 10 mm medially to a line connecting the center of the patella to the anterior–superior iliac spine, whereas the most distal row was positioned 10 mm proximal to the proximal border of the patella. Raw, surface EMGs are shown in the left panel, whereas the interpolated (by a factor 5), instantaneous amplitude of these raw EMGs is shown in the middle panel, for three different time instants: the action potential generation (0 ms) and then 3 ms and 6 ms later. Amplitude values are represented with the color scale. The distribution of RMS is similarly represented in the right panel. (b) The same graphical representation as in panel (a) is shown here for a motor unit identified in the medial gastrocnemius muscle. Monopolar EMGs were sampled with a grid of 64 electrodes (13 rows $\times$ 5 columns; one missing electrode; 8-mm interelectrode distance). Surface EMG signals detected by electrodes located over the muscle superficial aponeurosis, from row 1 to row 11, are shown. The central column of the grid was positioned 50 mm medial to the junction between gastrocnemius heads. The most proximal row of electrodes was located 20 mm distal to the popliteal fossa.

obtained through decomposition of monopolar, surface EMGs [31], collected from a single subject and interpolated by a factor 10.

Understanding the effect of the relative position of electrodes and muscle fibers is of crucial importance for the interpretation of surface EMGs. For the vastus medialis muscle, electrodes and fibers are located in parallel planes, as suggested by the signals shown in Fig. 5.13a. A first relevant observation prompting from this figure is the local representation of single differential MUAPs, detected by electrodes aligned along an oblique direction with respect to the grid, from row 3 and column 1 to row 9 and column 6. By virtue of such oblique orientation, the location of the innervation zone of this motor unit is not readily clear, however, a phase reversal between potentials detected at the medial–distal half (rows 7–9, columns 5–6) and those detected at the lateral–proximal half (rows 2–6, columns 1–4) is evident. For this particular motor unit, the fiber action potentials are generated at approximately rows 6–7 and columns 4–5 and then propagate in the lateral–proximal and medial–distal directions. Innervation zone location and propagation of these particular action potentials are better observed in terms of the instantaneous distribution of the amplitude of raw EMGs, as shown in the middle panels of Fig. 5.13a. The direction of propagation of these action potentials is also clearly represented in the bidimensional distribution of the RMS amplitude (Fig. 5.13a, right panel). Even though the vastus medialis is a pinnate muscle and its fibers are not parallel to the columns of the grid, features such as the location of innervation zone, the direction of fibers, and the conduction velocity of action potentials may be estimated from HD-EMGs. Mathematical manipulation allows the reorientation of the electrode grid so that columns or rows would be parallel to the fiber direction.

On the other hand, none of these features may be observed or extracted from the HD-EMGs detected from the gastrocnemius muscle, with the exception of the most distal muscle region as explained below. As shown in Fig. 5.13b, the motor unit action potentials in the gastrocnemius muscle are represented locally in the surface EMGs (note that a grid with smaller interelectrode distance was used to sample EMGs from gastrocnemius). The phase reversal observed between rows 5 and 6 cannot be regarded as the location of the innervation zone, given that from this location there is no propagation in any particular direction; potentials disappear at the same location where they appear and are mostly due to the end-of-fiber effect described in Chapter 2 (middle panels in Fig. 5.13b). Differently from the vastus medialis muscle, the distribution of the RMS amplitude on the gastrocnemius surface EMGs does not indicate the innervation zone or the direction of muscle fibers. What information can therefore be obtained from surface EMG detected from muscles *pinnate in depth direction*?

Differently from HD-EMGs collected from *skin parallel-fibered* muscles, HD-EMGs sample from different regions across the physiological cross-sectional area of muscles *pinnate in depth direction*. Because of *in-depth pinnation*, fibers and skin reside in nonparallel planes. As a consequence, for muscles with such architecture, and regardless of whether using grids or electrode pairs, it is not possible to align surface electrodes with the muscle fibers. For any array of consecutive electrodes positioned on skin regions covering *in-depth pinnate* muscles, each electrode is situated in correspondence of the superficial endings of different fibers. The relative position between electrodes and fibers for a muscle *pinnate in the depth direction*, that is, the gastrocnemius muscle, is illustrated in Fig. 5.14. For simplicity, only a few fibers (17 fibers innervated by four different motor neurons) are indicated, extending from the muscle superficial to deep aponeurosis. The superficial extremity of fibers of the most proximal motor neuron (fibers 1, 2, 3, and 5), for example, is located in

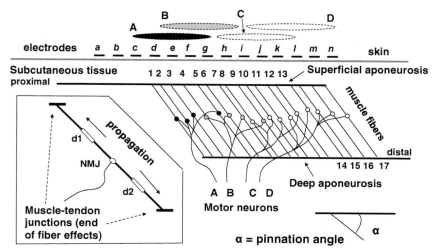

FIGURE 5.14 Schematic illustration of the relative position between fibers and surface electrodes for a muscle pinnate in depth direction. Skin, superficial aponeurosis, and deep aponeurosis are represented with thick, solid lines. Muscle fibers are indicated with thin, solid lines. Seventeen muscle fibers are shown, belonging to four different motor units. Different motor units are distinguished through the gray intensity of the circles denoting the neuro-muscular junctions (NMJs; see inset). Fourteen electrodes, from **a** (most proximal) to **n** (most distal), are indicated. Each of the four ellipses shown in the top (A, B, C, and D) corresponds to the approximate region where the action potentials of each of the four motor units would be predominantly represented in the surface EMGs. Reproduced from Barbero et al. [1] with permission.

proximity of electrodes from **c** to **g**. When this motor unit is active, action potentials are generated at the neuromuscular junctions and propagate toward the deep and superficial aponeurosis, as indicated in the inset of Fig. 5.14. Potentials propagating toward the superficial aponeurosis progressively approach the surface electrodes; that is, the distance between electrodes from **c** to **g** and the moving sources decreases with propagation. For this reason, largest potentials are observed in the surface EMGs when these action potentials reach the superficial aponeurosis. Moreover, as the relative distance between these action potentials and electrodes from **c** to **g** is markedly smaller than that between these action potentials and the other electrodes, the largest surface potentials are detected from electrodes **c** to **g** (cf. black ellipse in Fig. 5.14). For the most distal motor neuron illustrated in the figure, however, the largest surface potentials are detected by electrodes from **i** to **n** (cf. light gray and white ellipses in Fig. 5.14). It is therefore reasonable to expect that, depending on the location and the number of active fibres beneath the electrodes, very different surface potentials may be detected in different skin regions. This is confirmed by the very uneven distribution of surface EMG on different pinnate muscles [26,66–68].

5.3.2   Spatial Distribution of EMG Features in Muscles Pinnate in Depth Direction

Evidence from simulation and experimental studies on the local association between the position of active fibers within the muscle and the amplitude of action potentials in surface EMGs is provided in detail in Chapter 15. Chapter 15 further highlights the relevance of such local association for the study of the anatomy of postural motor units as well as for the study of the functional significance of gastrocnemius regional activation in standing. Here, emphasis is given to whether features typically obtained from surface EMGs detected from *skin-parallel fibered* muscles may be extracted as well for muscles *pinnate in depth direction*. Particular attention is given to the estimation of conduction velocity and to the study of myoelectric manifestation of muscle fatigue (see also Chapter 10). These two features are outlined below exclusively for the pinnate gastrocnemius muscle, given that evidence from HD-EMG for different *in-depth pinnate* muscles is not yet available in the literature.

Notwithstanding the *pinnation in depth direction*, conduction velocity can be estimated from EMGs collected from the distal portion of the gastrocnemius muscle with arrays or grids of electrodes because, in this specific muscle, the distal motor units are almost parallel to the skin.

Figure 5.15 shows the schematic arrangement of the gastrocnemius fibers, two parasagittal echographic images of the muscle, a linear electrode array and two sets of signals. Surface EMGs collected by electrodes over the proximal part of the superficial aponeurosis show neither propagation nor innervation zone; their amplitude distribution on the skin reflects the location and number of active fibers (Fig. 5.14). Conversely, EMGs detected by electrodes positioned over the most distal fibers show propagating action potentials (Fig. 5.15). Physiological estimates of CV and the location of the innervation zone(s) can be obtained exclusively from electrodes positioned over the most distal fibers (Fig. 5.15) [29]. It should be noted that these estimates of CV and innervation zone location cannot be regarded as representative of the whole muscle. Considering the remarkably large gastrocnemius physiological cross section area [51], the representativeness of conduction velocity and innervation zone location from the most distal muscle fibers remains an open issue.

Although surface EMGs detected over the gastrocnemius superficial aponeurosis do not provide reliable estimates of CV, they are sensitive to muscle fatigue mostly because of the widening, in time, of the end-of-fiber effect being reflected by spectral compression (see Chapter 10). Differently from the fatigue induced changes on surface EMGs detected from *skin parallel-fibered* muscles, the myoelectric manifestations of gastrocnemius fatigue are represented markedly locally on the skin. Consider, for example, a large grid of electrodes positioned over the gastrocnemius muscle (Fig. 5.16) while a subject was asked to exert isometric, intermittent plantar flexion contractions until exhaustion (i.e., to alternately keep ankle plantar flexion torque at 50% MVC for 5 s and then at 0% MVC for 5 s [28]. Significant changes in the frequency spectrum of surface EMGs were observed before exhaustion, only for some channels in the grid. Specifically, strong decreases in median frequency were observed in regions A and smaller ones in region B of the gastrocnemius (Fig. 5.16b),

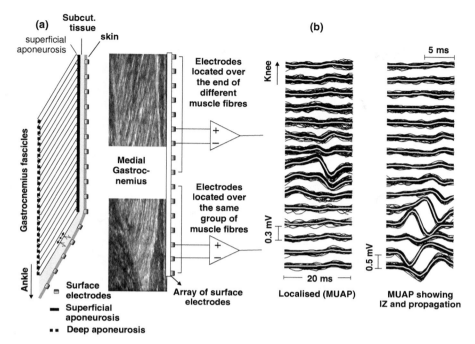

FIGURE 5.15 (a) Schematic representation of the relative position between electrodes and gastrocnemius fibers, according to ultrasound images collected from the muscle proximal and distal regions. (b) Surface representation of multiple action potentials of two motor units identified through decomposition of intramuscular EMGs and spike-triggered averaging of the sEMG. Modified from Vieira et al. [66] with permission.

while increases were observed in region C. Given that single-differential sEMG signals detected by electrodes over the gastrocnemius superficial aponeurosis reflect the activity of fibres in different muscle regions (Fig. 5.14), the spatially localized decreases in median frequency indicate the gastrocnemius regions where myoelectric manifestations of muscle fatigue are most evident. Most interestingly, the region in the grid where greatest decreases in median frequency were observed (region A) was common to all subjects tested [28]. In conclusion, myoelectric manifestations of muscle fatigue based on spectral variables may be well studied from muscles *pinnate in depth direction*.

5.4 CURRENT APPLICATIONS AND FUTURE PERSPECTIVES OF HDsEMG

High-density surface EMG (HDsEMG) techniques have been used in a variety of clinical research applications, ranging from neurophysiology to prostheses and robot control, to biofeedback, to the analysis of back and limb muscles in workers and

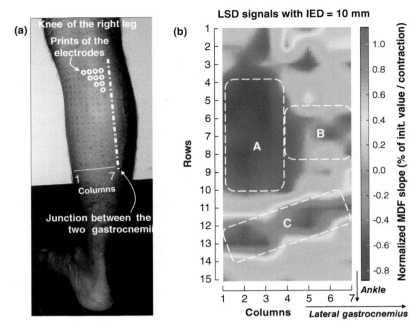

FIGURE 5.16 (a) Position of each of the 128 electrodes (16 rows × 8 columns) used to detect surface EMGs from the medial gastrocnemius muscle. Marks left on the skin upon removal of the grid of electrodes are clearly seen in the picture. (b) Interpolated spatial distribution of median frequency (MDF) slopes. Slopes were calculated through linear regression applied to MDF values obtained for intermittent, 50% MVC plantar flexion efforts (3-s epochs; $N = 18$ plantar flexion contractions), separately for each of the 112 single-differential sEMG (15 × 7 channels; the last column of electrodes was unavailable). Slopes were normalized with respect to the MDF initial values. Modified from Gallina et al. [28] with permission.

musicians, to the targeted application of botulinum toxin. These techniques are based on topographical images and are often globally referred to as "sEMG imaging."

Two main fields of research application of HDsEMG may be identified. The first concerns the extraction, from sEMG, of information which cannot be obtained with a single electrode pair, as described in the preceding sections of this chapter. The second concerns the exploitation of the many surface detection points to provide different "points of view" of the motor units in order to identify them using a decomposition algorithm that extracts their discharge times and action potential templates, as described in Chapter 7.

5.4.1 Surface EMG Imaging

Chapter 10 deals with localized myoelectric manifestations of muscle fatigue mostly observed using sEMG Imaging. Other chapters mention additional applications of sEMG images (e.g., Chapters 13 and 18). A systematic review of HDsEMG applications, presented in 2006 by Drost et al. [8], identified 29 clinical research studies and four reviews, mostly on motor unit studies, muscle fatigue, motor neuron diseases, neuropathies, and myopathies. Since then, the number of publications in the field has considerably increased and 15 review articles are available.

Muscle activity as well as myoelectric manifestations of muscle fatigue in pinnate muscles are localized and mostly reflect end-of-fiber effects of specific motor units whose myoelectric manifestations of fatigue can be monitored locally [28]. In this case, a pair of electrodes placed above an unspecific muscle region may provide very misleading information. An electrode array covering the entire muscle surface (e.g., gastrocnemius) is essential to identify regional activations and myoelectric manifestations of fatigue (see Fig. 5.13b, Fig. 5.16).

Watanabe et al. [67] investigated the sEMG distribution pattern on vastus lateralis in young and elderly men and the region-specific myoelectric manifestations of fatigue in the human vastus lateralis and rectus femoris. They found that the heterogeneity of the spatial EMG patterns (indicated by image entropy and by the correlation coefficient with respect to a reference image) decreases with increasing contraction level, and the change is greater in the case of younger subjects. The same author [68] observed that there are region-specific manifestations of muscle fatigue in the human rectus femoris.

Farina et al. [23] observed that the surface distribution of EMG intensity (RMS) over the trapezius changes during a contraction sustained to the endurance time, and its centroid moves upward, indicating the progressive involvement of motor units in the upper part of the muscle, as indicated in Fig. 13.7. These authors also observed that the subjects showing the largest shifts of the centroid of the EMG activity distribution were those showing longer endurance times (see Chapter 13).

A novel HDsEMG system associated with ultrasound muscle imaging has been proposed by Botter et al. [4]. A layer of silicon rubber, whose acoustic impedance is similar to that of the skin, carries a grid of holes filled with conductive gel where thin stainless steel wires, embedded in the rubber, terminate and make contact with the gel. The system is mildly sticky and can be applied to the skin without air bubbles. The ultrasonic probe is applied above it with ultrasonic gel. The echo given by the rubber–skin interface, the gel, and the wires is negligible. HDsEMG and ultrasonic images can be acquired at the same time to provide simultaneous information about electrical and mechanical muscle activity, as shown in Fig. 5.15.

Readi et al. [55] used grids of 64 electrodes (16 rows × 4 columns) to investigate how activity is distributed along the left and right side of the lumbar spine in sweep rowers. From the distribution of RMS amplitude along the spine, these authors observed that during strokes, right and left erector spinae muscles are activated differently. More interestingly, these asymmetries manifested differently between subjects, suggesting they are not an adaptation of the inherently asymmetric, sweep rowing gesture.

Lumbar muscles are important in relation to back pain and posture investigation. Ringheim et al. [56] investigated these muscle by applying two grids (9 columns by 14 rows each, 4-mm interelectrode distance), one on each side of the lumbar spine during 30-min sitting. They measured the median frequency (MDF) and the root mean square (RMS) for each 1-s epoch as the global average over the channels of each grid, as well as the subjective rate of perceived exertion (RPE) and observed alternating activities between the right and left side.

HDsEMG has been applied by Mulder et al. [50] to assess the time course and the nature of neuromuscular adaptations in the vastus lateralis muscle following eight weeks of bed rest (a simulation of low gravity condition) with and without resistive vibration exercise. They used a 10×13 electrode grid with 5-mm IED and applied a spatial segmentation criterion to estimate RMS, MNF, and muscle fiber CV from specific regions.

The estimation of muscle force based on surface EMG is an open problem (see Chapter 10). Surface EMG reflects the activity of superficial muscles while the torque at a joint is generated by many muscles, some of which are deep and/or outside the detection volume of the electrode array. A contraction generating 40% of the maximal voluntary torque at a joint does not mean that every muscle acting on the joint is contracting at 40% of its maximal force. Unfortunately, many authors made this assumption in their work, and this (together with the variability of electrode types and IED) may partially explain the lack of consensus in the field. This issue has been addressed by Staudenman et al. [62–64] and is discussed in Chapter 10.

Prosthesis and robot control as well as muscle–machine interfaces benefit from the large and redundant information provided by wearable HDsEMG systems. Farina et al. [24] used textile sleeves covering the upper and lower arm with four arrays of 5×5 electrodes each and were able to classify nine tasks with 89.1% accuracy. Further applications are presented in Chapter 20. HDsEMG applications to investigate and monitor neurological disorders are still lagging; however, the works of Kallenberg and Hermens [34] and of Zhang and Zhou [69] are encouraging and show promise.

Finally, applications of HDsEMG in biofeedback and in rehabilitation games are developing rapidly in many fields. Athletes or patients undergoing rehabilitation may benefit from real-time access to images showing when and how much their muscles are activated during a training exercise or a therapeutic protocol. Biofeedback applications involving this technique are addressed in Chapter 18.

5.4.2   Surface EMG Decomposition

A major and growing field of applications of HDsEMG concerns the identification of individual motor unit features (discharge rate, recruitment and derecruitment thresholds, and template shape) [32,43] and their changes in pathologies, during either isometric or dynamic contractions [30]. This application requires the real-time decomposition of the HDsEMG into the constituent MUAP trains. Major contributions to this field are coming from the works of De Luca, Farina, Glaser, Holobar, Zazula, and others, reported and described in Chapter 7. The spatial discrimination of motor units by means of suitable spatial filters boosts the performance of the decomposition algorithms in terms of decomposition accuracy and number of identified motor units.

5.4.3 Future Perspectives

As the number of electrodes and signals increase, the issue of cable and connections between the electrodes and the amplifiers becomes more and more relevant. It is evident that the technology moves toward easy-to-wear elastic sleeves or gloves incorporating the signal conditioning and A/D conversion electronics providing a serial protocol data transmission as output. The options for data transmission are:

1. Thin and flexible plastic optical fibers (very wide bandwidth but physical link).
2. Wi-Fi link to an Internet access point or point-to-point connection to a PC (limited bandwidth).
3. On-board signal compression (either lossless or lossy) and wireless transmission of the compressed information. This option has been investigated by Carotti et al. [5] and Itiki et al. [33].
4. Data logger with on-board memory and wireless transmission of part of the data (e.g., 0.1 s every 1 or 2 s) for online data quality verification.

In addition to these features, the automatic testing for movement artifacts, poor electrode–skin contact quality, and power-line interference will result in indications for skin cleaning and sleeve repositioning in the case of too many poor signals [41].

From the clinical point of view, the expected developments concern more extensive applications in biofeedback and monitoring of the neuromuscular system for prevention purposes, automatic detection of the innervation zone, muscle monitoring during dynamic activities, and sophisticated control of external devices or prosthesis.

REFERENCES

1. Barbero M., R. Merletti, and A. Rainoldi, *Atlas of Muscle Innervation Zones*, Springer-Verlag Italia, Milan, Italy, 2012.
2. Blazevich, A. J., N. D. Gill, and S. Zhou, "Intra- and intermuscular variation in human quadriceps femoris architecture assessed in vivo," *J. Anat.* **209**, 289–310 (2006).
3. Blok, J. H., J. P. Van Dijk, M. J. Zwarts, and D. F. Stegeman, "Motor unit action potential topography and its use in motor unit number estimation," *Muscle Nerve* **32**, 280–291 (2005).
4. Botter, A., T. M. Vieira, I. Loram, R. Merletti, and E. Hodson-Tole, "A novel system of electrodes transparent to ultrasound for simultaneous detection of myoelectric activity and B-mode ultrasound images of skeletal muscles," *J. Appl. Physiol.* **115**, 1203–1214 (2013).
5. Carotti E., J. C. De Martin, R. Merletti, and D. Farina, "Compression of surface EMG signals with algebraic code excited linear prediction," *Med. Eng. Phys.* **29**, 253–258 (2007).
6. Dimitrova, N. A., G. V. Dimitrov, and O. A. Nikitin, "Neither high-pass filtering nor mathematical differentiation of the EMG signals can considerably reduce cross-talk," *J. Electromyogr. Kinesiol.* **12**, 235–246 (2002).

7. Disselhorst-Klug, C., J. Silny, and G. Rau, "Improvement of spatial resolution in surface-EMG: a theoretical and experimental comparison of different spatial filters," *IEEE Trans. Biomed. Eng.* **44**, 567–574 (1997).

8. Drost, G., D. F. Stegeman, G. Baziel, B. G. van Engelen, and M. J. Zwarts, "Clinical applications of high-density surface EMG: a systematic review," *J. Electromyogr. Kinesiol.* **16**, 586–602 (2006).

9. Farina, D., and A. Rainoldi, "Compensation of the effect of sub-cutaneous tissue layers on surface EMG: a simulation study," *Med. Eng. Phys.* **21**, 487–497 (1999).

10. Farina, D., E. Fortunato, and R. Merletti, "Noninvasive estimation of motor unit conduction velocity distribution using linear electrode arrays," *IEEE Trans. Biomed. Eng.* **47**, 380–388 (2000).

11. Farina, D., and R. Merletti, "A novel approach for precise simulation of the EMG signal detected by surface electrodes," *IEEE Trans. Biomed. Eng.* **48**, 637–664 (2001).

12. Farina, D., W. Muhammad, E. Fortunato, O. Meste, R. Merletti, and H. Rix, "Estimation of single motor unit conduction velocity from surface electromyogram signals detected with linear electrode arrays," *Med. Biol. Eng. Comput.* **39**, 225–236 (2001).

13. Farina, D., and C. Cescon, "Concentric ring electrode system for noninvasive detection of single motor unit activity," *IEEE Trans. Biomed. Eng.* **48**, 1326–1334 (2001).

14. Farina, D., R. Merletti, B. Indino, M. Nazzaro, and M. Pozzo, "Surface EMG crosstalk between knee extensor muscles: experimental and model results," *Muscle Nerve* **26**, 681–695 (2002).

15. Farina, D., and R. Merletti, "A novel approach for estimating muscle fiber conduction velocity by spatial and temporal filtering of surface EMG signals," *IEEE Trans. Biomed. Eng.* **50**, 1340–1351 (2003).

16. Farina, D., E. Schulte, R. Merletti, G. Rau, and C. Disselhorst-Klug, Single motor unit analysis from spatially filtered surface electromyogram signals. Part 1: spatial selectivity," *Med. Biol. Eng. Comput.* **41**, 330–337 (2003).

17. Farina, D., L. Arendt-Nielsen, R. Merletti, B. Indino, and T. Graven Nielsen, "Selectivity of spatial filters for surface EMG detection from the tibialis anterior muscle," *IEEE Trans. BME* **50**, 354–364 (2003).

18. Farina, D., L. Mesin, and S. Martina, "Comparison of spatial filter selectivity in surface myoelectric signal detection: influence of the volume conductor model," *Med. Biol. Eng. Comp.* **42**, 114–120 (2004).

19. Farina, D., and R. Merletti, "Methods for estimating muscle fibre conduction velocity from surface electromyographic signals," *Med. Biol. Eng. Comput.* **42**, 432–445 (2004).

20. Farina, D., R. Merletti, B. Indino, and T. Graven-Nielsen, "Surface EMG crosstalk evaluated from experimental recordings and simulated signals. Reflections on crosstalk interpretation, quantification and reduction," *Methods Inf. Med.* **43**, 30–55 (2004).

21. Farina, D., and R. Merletti, "Estimation of average muscle fiber conduction velocity from two-dimensional surface EMG recordings," *J. Neurosci. Methods* **134**, 199–208 (2004).

22. Farina, D., and F. Negro, "Estimation of muscle fiber conduction velocity with a spectral multidip approach," *IEEE Trans. Biomed. Eng.* **54**, 1583–1589 (2007).

23. Farina, D., F. Leclerc, L. Arendt-Nielsen, O. Buttelli, and P. Madeleine, "The change in spatial distribution of upper trapezius muscle activity is correlated to contraction duration," *J. Electromyogr. Kinesiol.* **18**, 16–25 (2008).

24. Farina, D., T. Lorrain, F. Negro, and J. Ning, "High-density EMG E-Textile systems for the control of active prostheses," *Eng. Med. Biol. Soc. (EMBC)*, 3591–3593 (2010).

25. Frahm, K., M. Jensen, D. Farina, and O. Andersen, "Surface EMG crosstalk during phasic involuntary muscle activation in the nociceptive withdrawal reflex," *Muscle Nerve* **46**, 228–236 (2012).

26. Gallina, A., R. Merletti, and M. Gazzoni, "Uneven spatial distribution of surface EMG: what does it mean?" *Eur. J. Appl. Physiol.* **113**, 887–894 (2013).

27. Gallina, A., and A. Botter, "Spatial localization of electromyographic amplitude distributions associated to the activation of dorsal forearm muscles," *Front Physiol.* **13**, 367 (online) (2013).

28. Gallina, A., R. Merletti, and T. M. Vieira, "Are the myoelectric manifestations of fatigue distributed regionally in the human medial gastrocnemius muscle?," *J. Electromyogr. Kinesiol.* **21**, 929–938 (2011).

29. Gallina, A., Ritzel, C. H., Merletti, R., and T. M. Vieira, "Do surface electromyograms provide physiological estimates of conduction velocity from the medial gastrocnemius muscle?" *J. Electromyogr. Kinesiol.* **23**, 319–325 (2013).

30. Glaser, V., A. Holobar, and D. Zazula, "Real-time motor unit identification from high-density surface EMG," *IEEE Trans. Neural Syst. Rehabil. Eng.* **21**, 949–958 (2013).

31. Holobar, A., and D. Zazula, "Multichannel blind source separation using convolution kernel compensation," *IEEE Trans. Signal Process* **55**, 4487–4496 (2007).

32. Holobar, A., M. A. Minetto, A. Botter, F. Negro, and D. Farina, "Experimental analysis of accuracy in the identification of motor unit spike trains from high-density surface EMG," *IEEE Trans. Neural Syst. Rehabil. Eng.* **18**, 221–229 (2010).

33. Itiki, C., S. Furie, and R. Merletti,"Compression of high density EMG signals for trapezius and gastrocnemius muscles" Biomedical Engineering Online 13, http://www.biomedical-engineering-online.com/content/13/1/25 (2014).

34. Kallenberg, L. A., and H. Hermens, "Motor unit properties of biceps brachii in chronic stroke patients assessed with high-density surface EMG," *Muscle Nerve* **39**, 177–185 (2009).

35. Lapatki, B. G., J. P. Van Dijk, I. E. Jonas, M. J. Zwarts, and D. F. Stegeman, "A thin, flexible multielectrode grid for high-density surface EMG," *J. Appl. Physiol.* (1985) **96**, 327–336 (2004).

36. Lapatki, B. G., J. P. van Dijk, B. P. van de Warrenburg, and M. J. Zwarts, "Botulinum toxin has an increased effect when targeted toward the muscle's endplate zone: a high-density surface EMG guided study," *Clin. Neurophysiol.* **122**, 1611–1616 (2011).

37. LoConte, L., and R. Merletti, "Advances in processing of surface myoelectric signals: part 1 and part 2," *Med. Biol. Eng. Comp.* **33**, 362–384 (1995).

38. Lieber, R. L., and J. Fridén, "Functional and clinical significance of skeletal muscle architecture," *Muscle Nerve* **23**, 1647–1666 (2000).

39. Lindstrom, L., R. Magnusson, and R. Petersen, "Muscle fatigue and action potential conduction velocity changes studied with frequency analysis of EMG signals," *Electromyogr. Clin. Neurophysol.* **10**, 341–356 (1970).

40. Lindstrom, L., and R. Magnusson, "Interpretation of myoelectric power spectra: a model and its applications," *Proc. IEEE* **65**, 653–662 (1977).

41. Marateb, H. R., M. Rojas-Martínez, M. Mansourian, R. Merlett, and M. A. Villanueva, "Outlier detection in high-density surface electromyographic signals," *Med. Biol. Eng. Comput.* **50**, 79–89 (2012).

42. McGill, K. C., and L. J. Dorfman, "High resolution alignment of sampled waveforms," *IEEE Trans. Biomed. Eng.* **31**, 462–470 (1984).

43. Merletti, R., A. Holobar, and D. Farina, "Analysis of motor units with high-density surface electromyography," *J. Electromyogr. Kinesiol.* **18**, 879–890 (2008).

44. Merletti, R., M. Knaflitz, and C. J. De Luca, "Myoelectric manifestations of fatigue in voluntary and electrically elicited contractions," *J. App. Physiol.* **69**, 1810–1820 (1990).

45. Merletti, R., and S. Roy, "Myoelectric and mechanical manifestations of muscle fatigue in voluntary contractions," *JOSPT*, **24**, 342–353 (1996).

46. Merletti, R., and L. Lo Conte, Surface EMG signal processing during isometric contractions, *J. Electromyogr. Kinesiol.*, **7**, 241–250 (1997).

47. Merletti, R., L. Lo Conte, E. Avignone, and P. Guglielminotti, "Modeling of surface myoelectric signals, part I: model implementation," *IEEE Trans. Biomed. Eng.* **46**, 819–820 (1999).

48. Merletti, R., B. Afsharipour, and G. Piervirgili, "High density surface EMG technology," in *Converging Clinical and Engineering Research on Neurorehabilitation*, J. Pons, D. Torricelli, and M. Pajaro, eds., Springer-Verlag, Berlin, 2013.

49. Mesin, L., F. Tizzani, and D. Farina, "Estimation of motor unit conduction velocity from surface EMG recordings by signal based selection of the spatial filters," *IEEE Trans. Biomed. Eng.* **53**, 1963–1971 (2006).

50. Mulder, E. R., K. H. Gerrits, B. U. Kleine, J. Rittweger, D. Felsenberg, A. de Haan, and D. F. Stegeman, "High-density surface EMG study on the time course of central nervous and peripheral neuromuscular changes during 8 weeks of bed rest with or without resistive vibration exercise," *J. Electromyogr. Kinesiol.* **19**, 208–218 (2007).

51. Narici, M. V., T. Binzoni, E. Hiltbrand, J. Fasel, F. Terrier, and P. Cerretelli, "In vivo human gastrocnemius architecture with changing joint angle at rest and during graded isometric contraction," *J. Physiol.* **496**, 287–297 (1996).

52. Papoulis, A. *The Fourier Integral and Its Applications*, McGraw–Hill, New York, 1962.

53. Rau, G., and C. Disselhorst-Klug, "Principles of high-spatial-resolution surface EMG (HSR-EMG): single motor unit detection and application in the diagnosis of neuromuscular disorders," *J. Electromyogr. Kinesiol.* **7**, 233–239 (1997).

54. Rau, G., C. Disselhorst-Klug, and J. Silny, "Noninvasive approach to motor unit characterization: muscle structure, membrane dynamics and neuronal control," *J. Biomech.* **30**, 441–446 (1997).

55. Readi, N. G., V. Rosso, A. Rainoldi, and T. Vieira,"Do sweep rowers symmetrically activate their low back muscles during indoor rowing?," *Scand. J. Med. Sci. Sports.* **25**(4), e339–e352 (2015).

56. Ringheim, I., A. Indahl, and K. Roeleveld, "Alternating activation is related to fatigue in lumbar muscles during sustained sitting," *J. Electromyogr. Kinesiol.*, **24**, 380–386 (2014).

57. Roeleveld, K., D. F. Stegeman, H. M. Vingerhoets, and A. Van Oosterom, "The motor unit potential distribution over the skin surface and its use in estimating the motor unit location," *Acta Physiol. Scand.* **161**, 465–472 (1997).

58. Rosenfalck, P., "Intra- and extracellular potential fields of active nerve and muscle fibers. A physico-mathematical analysis of different models," *Acta Physiol. Scand.* (Suppl. 321), 1–168 (1969).

59. Rojas-Martínez, M., M. A. Mañanas, J. F. Alonso, and R. Merletti, "Identification of isometric contractions based on high density EMG maps," *J. Electromyogr, Kinesiol.* **23**, 33–42 (2013).

60. Rodriguez-Falces, J., F. Negro, M. Gonzalez-Izal, and D. Farina, "Spatial distribution of surface action potentials generated by individual motor units in the human biceps brachii muscle," *J. Electromyogr. Kinesiol,* **23**, 766–777 (2013).

61. Schulte, E., D. Farina, G. Rau, G. R. Merletti, and C. Disselhorst-Klug, Single motor unit analysis from spatially filtered surface electromyogram signals. Part 2: conduction velocity estimation. *Med. Biol. Eng. Comput.* **41**, 338–345 (2003).

62. Staudenmann, D., I. Kingma, D. Stegeman, and J. H. van Dieën, "Towards optimal multi-channel EMG electrode configurations in muscle force estimation: a high density EMG study," *J. Electromyogr. Kinesiol.* **15**, 1–11 (2005).

63. Staudenmann, D., I. Kingma, A. Daffertshofer, D. Stegeman, and J. van Dieën. "Improving EMG-based muscle force estimation by using a high-density EMG grid and principal component analysis," *IEEE Trans. Biomed. Eng.* **53**, 712–709 (2006).

64. Staudenmann, D., A. Daffertshofer, I. Kingma, D. Stegeman, and J. van Dieën, "Independent component analysis of high-density electromyography in muscle force estimation," *IEEE Trans. Biomed. Eng.* **54**, 751–754 (2007).

65. van Dijk, J. P., M. Lowery, B. Lapatki, and D. Stegeman, "Evidence of potential averaging over the finite surface of a bioelectric surface electrode," *Ann. Biomed. Eng.* **37**, 1141–1151 (2009).

66. Vieira, T. M., I. D. Loram, S. Muceli, R. Merletti, and D. Farina, "Postural activation of the human medial gastrocnemius muscle: are the muscle units spatially localised?," *J. Physiol.* **589**, 431–443 (2011).

67. Watanabe, K., M. Kouzaki, R. Merletti, M. Fujibayashi, and T. Moritani, "Spatial potential distribution pattern of vastus lateralis muscle during isometric knee extension in young and elderly men," *J. Electromyogr. Kinesiol.* **22**, 74–79 (2012).

68. Watanabe, K., M. Kouzaki, and T. Moritani, "Region-specific myoelectric manifestations of fatigue in human rectus femoris muscle," *Muscle Nerve* **48**, 226–234 (2013).

69. Zhang, X., and P. Zhou, "High-density myoelectric pattern recognition toward improved stroke rehabilitation," *IEEE Trans. Biomed. Eng.* **59**, 1649–1657 (2012).

70. Zwarts, M. J., and G. Drost, Recent progress in the diagnostic use of surface EMG for neurological diseases. *J. Electromyogr. Kinesiol.* **10**, 287–291 (2000).

71. Zwarts, M. J., and D. F. Stegeman, "Multichannel surface EMG: basic aspects and clinical utility," *Muscle Nerve* **28**, 1–17 (2003).

6

MUSCLE COORDINATION, MOTOR SYNERGIES, AND PRIMITIVES FROM SURFACE EMG

Y. P. IVANENKO,[1] A. D'AVELLA,[1,2] AND F. LACQUANITI[1,3,4]

[1]*Laboratory of Neuromotor Physiology, Santa Lucia Foundation, Rome, Italy*
[2]*Department of Biomedical and Dental Sciences and Morphofunctional Imaging, University of Messina, Messina, Italy*
[3]*Center of Space Biomedicine, University of Rome "Tor Vergata", Rome, Italy*
[4]*Department of Systems Medicine, University of Rome "Tor Vergata", Rome, Italy*

6.1 INTRODUCTION

Even the simplest movement requires the coordination of many muscles. For instance, normal locomotion involves activation of tens or even hundreds of muscles. Motor control is distributed across numerous structures of the central nervous system (CNS) integrating motor plans and sensory feedback. However, the *"final common pathway"* [72] is the activation of the α-motoneurons (MNs) and neural control strategies must be understood in terms of the mechanical effect generated by muscle contractions on the skeletal system.

To investigate systematically neural control strategies, muscle activity must be measured during motor behavior. Historically, the first quantitative assessments of muscle activity were obtained by considering the sound produced by contracting individual motor units. A more accurate quantitative measurement came from recording the electrical potential associated with the summation of multiple action potentials propagating along the muscle fibers. Such electromyographic signals

Surface Electromyography: Physiology, Engineering, and Applications, First Edition.
Edited by Roberto Merletti and Dario Farina.
© 2016 by The Institute of Electrical and Electronics Engineers, Inc. Published 2016 by John Wiley & Sons, Inc.

(sEMGs) can be recorded by placing one or more pairs of electrodes on the skin over a muscle and it may be used to estimate the global net firing of spinal MNs innervating that muscle (see Chapter 7). Indeed, the average rectified value (ARV) of the sEMG seems to increase fairly linearly with the net motor unit firing rate [17]. Single or multichannel sEMG thus represents a window that neurophysiologists can use to look at the complex control mechanisms that the CNS has to use for coordinating multiple degrees of freedom.

Recent advances in the investigation of the neural control of movement have led to a reexamination of the mechanisms of sensorimotor integration in the CNS and in the spinal circuitry in particular. The many classes of spinal interneurons may be seen as "functional units" representing different levels of muscle synergies, parts of movements, or even more integrated motor behavior [34]. Thus, the principles that the CNS uses to govern hundreds of muscles to control whole-body movements include a modular organization of the neuronal networks [15,16,24,38,53]. Such modules can capture the structure of the EMG patterns very parsimoniously. In fact, once the modules are given, complex changes in the muscle patterns can be described by the linear combination of a small number N of basic patterns (basis functions), with a small residual error, as is done, for example, in the case of principal or independent component analysis. In the following sections we will consider different approaches used to uncover the modular organization of the motor output in human behaviors such as responding to postural perturbations, reaching with the arm, and locomotion, as well as its plasticity and flexibility in movement disorders.

6.2 MUSCLE SYNERGIES AND SPINAL MAPS

All the approaches presented in this chapter to investigate the strategies that the CNS employs to coordinate the activation of many muscles start from EMG signals recorded simultaneously from many muscles. To unravel muscle coordination strategies, it is essential to monitor a set of muscles large enough to be representative of all the muscles which play a role in the motor behavior being studied. In fact, the control of the many joints involved in every natural motor behavior depends on the net torque generated by multiple muscles on those joints. If muscles with distinctive activations in specific task conditions are missing from the recorded muscle set, the EMG signals may not contain sufficient information to fully uncover the underlying organization of muscle patterns. EMGs are traditionally recorded with bipolar electrodes, band-pass filtered, and amplified before being digitized with an A/D converter. Offline processing involves rectification, low-pass filtering with bidirectional filters (zero group delay), and usually averaging across repetitions of the EMGs recorded in similar movement conditions. Such smoothed sEMG envelopes are closely related to the activation of each muscles [80]. The EMG envelopes recorded over a range of conditions—for example, during different phases of the gait cycle or while reaching to different spatial locations—can be then decomposed into basic building blocks (muscle synergies) or combined to visualize spinal maps of α-MN activation.

6.2.1 EMG Factorization into Muscle Synergies

The fundamental hypothesis motivating sEMG envelope factorization as a tool to investigate the principles of muscle coordination is that the CNS overcomes the complexity inherent to the control of a large number of actuators by operating on a smaller number of muscle synergies. Synergies represent coordinated activations of groups of muscles, with specific activation balances or temporal profiles, that might reflect basic biomechanical properties of the musculoskeletal apparatus or statistical regularities in the muscle activation patterns involved in performing a specific motor behavior. In sensory systems, regularities in the environment, such as edges and bars in the visual scene, represent basic features encoded by the CNS [33] and used as building blocks for perception. In the motor system, muscle synergies might encode regularities in the musculoskeletal system organization and dynamic behavior that can be used as building blocks for action production.

Different muscle synergies models are used to decompose the EMG envelopes using appropriate factorization algorithms. Each model focuses on specific features and thus expresses specific hypothesis on the underlying mechanisms of muscle coordination. Specifically, we consider muscle synergies modeled as either time-invariant, that is, vectors capturing a specific balance of muscle activations [76], or time-varying, that is, a vector sequence capturing a set of specific muscle activation waveforms [15]. In both cases, each muscle pattern $\mathbf{m}(t)$ is generated by a linear combination of the muscle synergies w_i with coefficients c_i. In the time-invariant case

$$\mathbf{m}(t) = \sum_{i=1}^{N} c_i(t)\,\mathbf{w}_i + e(t) \tag{6.1}$$

time-varying muscle patterns $\mathbf{m}(t)$ are generated by time-varying combinations $[c_i(t)]$ of N synergies $\mathbf{w_i}$. In the time-varying case

$$\mathbf{m}(t) = \sum_{i=1}^{N} c_i\mathbf{w}_i(t - t_i) + e(t) \tag{6.2}$$

time-varying muscle patterns are generated by combinations of time-varying synergies, $\mathbf{w}_i(\tau)$, with $\mathbf{w}_i(\tau) = 0$ for $\tau < 0$ and $\tau > T$, where T is the duration of the synergy scaled in amplitude (c_i) and shifted in time (t_i) by scalar coefficients. Since N is finite (the series expansion is truncated) a residual error is implied.

Note that these specific notions and mathematical definitions of muscle synergies, although commonly used in the motor control literature in the last 15 years, differ from those used in the clinical literature and from the definition of "functional synergies" characterized by methods such as the uncontrolled manifold analysis (e.g., [52]). Moreover, the same elements of the time-invariant synergy model are often referred to using a different terminology. In particular, especially in the locomotion literature, as the initial factorization analysis relied on factor analysis [16,37] (see below), muscle synergies are referred to as muscle weights or loadings while the time-varying combination coefficients are referred to as factors or temporal components.

6.2.1.1 Time-Invariant Factorization Algorithms.

Muscle synergies are identified from the sEMG envelopes by optimizing some cost function, typically a reconstruction error or a likelihood function. For time-invariant synergies, the decomposition problem can be expressed as a matrix factorization and dimensionality reduction problem:

$$\mathbf{M} = \mathbf{WC} + \mathbf{E} \qquad (6.3)$$

where $\mathbf{M}$ is a matrix assembled from the EMG envelope of each muscle in each row, $\mathbf{W}$ is a synergy matrix with a number of columns (N) smaller than the number of muscles, $\mathbf{C}$ is a coefficient matrix with N rows and a number of columns (number of samples) equal to the number of columns of $\mathbf{M}$, and $\mathbf{E}$ is the matrix of residual errors to be minimized. Thus, the columns of $\mathbf{W}$ span a N-dimensional subspace of the space of muscle activations (i.e., the set of vectors constructed with simultaneous samples of the EMG envelopes), and the corresponding N rows of $\mathbf{C}$ refer to the basic temporal activation components or patterns.

Standard factorization algorithms, such as non-negative matrix factorization (NMF) [55,56], principal component analysis (PCA) [44], factor analysis (FA) [2], and independent component analysis (ICA) [3,58] have been used. Each algorithm relies on specific assumption on the characteristics of the decomposition and on the statistical properties of the noise affecting the generative models [Eq. (6.1) and (6.2)]. For example, NMF assumes that data ($\mathbf{M}$), synergies ($\mathbf{W}$), and coefficients ($\mathbf{C}$) are non-negative. PCA decomposes mean subtracted data imposing orthogonality between the synergies. FA assumes that the activation coefficients are uncorrelated. ICA assumes that the activation coefficients are statistically independent. However, most algorithms identify very similar synergies [76].

The number of synergies is a free parameter that must be selected according to some criterion. Commonly used criteria are: (1) the minimum number of synergies for which the fraction of data variation explained (R^2) is above a given threshold (often 90%) and (2) the number N of synergies at which the curve of the R^2 value as a function of N has a change in slope.

6.2.1.2 Time-Varying Factorization Algorithms.

According to the time-varying synergy model muscle patterns are generated by combinations of synergies which are scaled in amplitude and shifted in time [see Eq. (6.2)]. Because of the time shift, time-varying synergies are not the linear generators of a subspace in the muscle activation space as time-invariant synergies. Therefore standard matrix factorization algorithms cannot be used. For time-varying synergy factorization, a specific iterative optimization algorithm has been developed [15]. This algorithm uses the reconstruction error as a cost function and a multiplicative rule for updating scaling coefficients and synergies. Because of the additional time-shift operation, at each iteration the time-varying algorithm updates not only synergies and amplitude scaling coefficients but also the onset delay times using a greedy search [59]. However, if the additional assumption is made that synergy time-shifts are fixed across synergies and conditions, it is possible to use standard time-invariant factorization algorithms to identify time-varying synergies by considering the different time samples of each muscles as distinct dimensions [49]. Finally, the time-varying model can be seen as a special case of the anechoic-mixing model in which acoustic signals from sources arrive at an array of microphones with different delays. Thus, an anechoic blind source separation algorithm [65] can be used to identify time-varying synergies by imposing specific relationships between the delays of different sources.

6.2.2 Reconstructing the Spinal Maps of Motoneuron Activation from Surface EMGs

As a complementary approach to statistical methods, it is possible to visualize total segmental α-MN activation by mapping the activity patterns from a large number of simultaneously recorded muscles onto the approximate rostrocaudal location of the MN pools in the human spinal cord [28,38,78]. The implicit assumption is that the rectified EMG provides an indirect measure of the net firing of spinal MNs innervating that muscle [17,79].

In general, each muscle is innervated by several spinal segments, and each segment supplies several muscles [47,71]. To reconstruct the motor-pool output pattern of any given spinal segment Sj of the cervical, thoracic, and lumbosacral segments innervating limb and trunk muscles, all rectified EMG-waveforms corresponding to that segment are averaged using appropriate weighting coefficients:

$$S_j = \frac{\sum_{i=1}^{n_j} k_{ij} \cdot EMG_i}{n_j} \tag{6.4}$$

where n_j is the number of EMG_i corresponding to the jth segment, and k_{ij} is the weighting coefficient for the ith muscle (X and x in Kendall's chart are weighted with $k_{ij} = 1$ and $k_{ij} = 0.5$, respectively [38]). EMGs can be expressed in microvolts or normalized to the maximal voluntary contraction (see references 41 and 57 for details).

In practice, a limited set of leg muscles can be recorded by surface electrodes. Nevertheless, even when using intramuscular recordings or when a slightly different set of muscle recordings was used to generate the maps during walking, the maps remained similar [38,41]. Furthermore, a similar spatiotemporal pattern of the spinal segmental output have been replicated in several different studies [6,9,28,38,39,57,62]. In fact, the maps obtained during human locomotion by excluding any single one of the recorded muscles are generally similar to those obtained from the full set, presumably because the lumbosacral enlargement innervates numerous muscles and each muscle is innervated by several segments. fMRI is now not feasible to study the spinal cord activation during a functional task such as walking; moreover, it is unclear which neurological structures fMRI evaluates precisely. The spinal mapping method may be further refined and compared with other imaging techniques as empirical data on the motor pool mapping in different studies accumulate.

The spinal maps approach provides an interpretation of the motor pool activation at a segmental level rather than at the individual muscle level [28,38,79]. It can be used to characterize network architecture for different gaits by considering relative intensities, spatial extent, and temporal structure of the spinal motor output [39,62]. In particular, in addition to the above-described statistical methods revealing common patterning elements, it provides also spatial information about activity of the whole population of MNs of various muscles at each segmental level.

6.3 MUSCLE SYNERGIES IN POSTURE CONTROL

Balance control is largely modulated by supraspinal (e.g., brain stem) circuits, though it involves numerous brain and spinal structures and multisensory integration. Nevertheless, the motor output during stance perturbations during both standing and walking has a modular spatiotemporal organization. In particular, it has been shown that various stance or support surface perturbations result in direction-dependent activation of specific muscle groups in both healthy subjects and patients with balance disorders [74].

The studies that recorded and analyzed the structure of multi-muscle EMG activity during such tasks are of particular interest. These studies, using different statistical and analytical approaches, converge to a common notion that modulation of a few muscle synergies, or muscle modes, can explain a great part of variation in muscle activity across the tasks studied [35,50,51,61,68,75].

Often, a directionally tuned muscle synergy recruitment is associated with a specific biomechanical function to restore or maintain the center of body mass, even though the control of the specific body segment alignment is also important for the upright posture. The focus on phasic muscle responses to balance perturbations typically dominates in the postural studies (see Ting et al. [74] for a review). Nevertheless, muscle "synergies" or muscle tone underlying the choice and maintenance of the reference body configuration are also essential [30], even though such tone may require smaller muscle activity than phasic responses. Moreover, disturbances in postural tone may result in significant changes and adaptations—for instance, bended postures or even quadrupedal walking in humans (see Ivanenko et al. [42] for a review). These reference-configuration-related muscle "synergies" may not be necessarily the same as those for phasic postural responses.

6.4 MODULAR CONTROL OF ARM REACHING MOVEMENTS

Reaching, moving an effector to a target location, such as reaching for an object with the hand, requires selecting appropriate muscle activation patterns according to proprioceptive and visual information specifying the location of the effector and of the target. Such sensorimotor transformation is thought to be performed by an internal model [46] implemented in the CNS. However, given the complexity of the computations required to select the appropriate activation waveforms of many muscles acting on many articulated body segments, it is not clear what mechanisms allow for an efficient implementation of an internal model. One possibility is that this mapping is simplified by a low-dimensional, modular generation of the muscle patterns. The key idea is that, if all useful muscle patterns can be constructed by the combination of a small number of muscle synergies, selecting the appropriate muscle pattern for a given goal requires only determining how the synergies are combined [4,12].

Recent results from the analysis of muscle patterns recorded during reaching movements support a modular control strategy [11]. Muscle activation waveforms during reaching movements in different directions had been characterized before [21,22,45,77], but the complex dependences of the amplitude and timing of the muscle activation waveforms on movement characteristics had not been directly related to a modular organization. Using a time-varying synergy model, the phasic component of the muscle patterns observed during fast-reaching movements in different directions to targets arranged on two vertical planes have been found to be captured by the combination of four or five time-varying muscle synergies [13]. Phasic components of the muscle activation waveforms [20], responsible for accelerating and decelerating the arm, were identified by subtracting from the sEMG envelopes tonic components, responsible for balancing gravitational forces and for maintaining postural stability, estimated as linear ramps in the muscle activations from movement onset to movement end, preceded and followed by constant activation levels. Each synergy comprised a set of muscle activation waveforms which could be scaled in amplitude and shifted in time as a unit (Fig. 6.1A). The muscle patterns for movements in different directions were reconstructed by modulating the amplitude and timing coefficients of each synergy (Fig. 6.1B). In contrast with the complex directional tuning of individual muscles, in most cases the synergy amplitude coefficients had a simple cosine directional tuning (Fig. 6.1C). These results suggest that synergies provide the CNS with a simple rule for mapping target directions into appropriate muscle patterns.

When the movement speed was varied, the muscle activation patterns for reaching in different directions and with different speeds on a single vertical plane, normalized in time to equal movement duration, were reconstructed by combination of three phasic and three tonic muscle synergies [10]. Phasic synergies were modulated in amplitude and timing by both movement direction and speed (Fig. 6.2), as were the time-varying synergies extracted from the phasic components of fast movements. For slower movements, however, a significant fraction of the sEMG envelopes were not modulated by movement speed and comprised tonic activation waveforms. Instead of

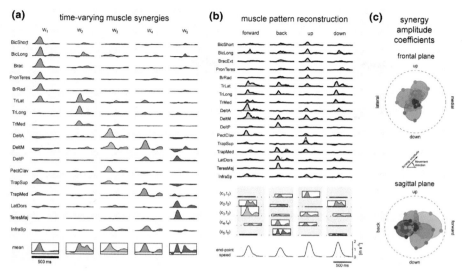

FIGURE 6.1 Muscle synergies for fast arm reaching movements. (**a**) Five time-varying synergies extracted from the phasic muscle patterns recorded during fast (movement duration < 400 ms) point-to-point movements between one central location and eight peripheral locations in the frontal and sagittal planes. Each column illustrates the collection of muscle activation waveform components (rows) comprising a time-varying synergy. The profile inside a rectangular box in the last row represents the mean activation waveforms averaged over all muscle waveforms of each synergy and it is only used to illustrate the gross temporal structure of each synergy. (**b**) The sEMG envelops waveforms of 17 shoulder and arm muscles, averaged across repetitions of the same movement, are reconstructed by multiplying by a scaling coefficient (c_i), shifting in time by an onset delay (t_i), and summing, muscle by muscle, the corresponding activation waveform components of each one of the five synergies. *Top*: The gray area shows the averaged EMG activity and the solid black line its reconstruction by synergy combination. *Middle*: Each scaling coefficient is proportional to the height of a rectangle and the synergy onset is in correspondence of the left edge of the rectangle. *Bottom*: Average end-point tangential velocity. (**c**) The directional dependence of the synergy scaling coefficients is illustrated in polar plots in which the magnitude of the coefficient for each synergy (color-coded as in panel A) for a movement in a specific direction in the frontal (*top*) and sagittal (*bottom*) planes is represented by the radial distance in that direction. Smooth directional tuning curves are obtained by spline interpolations of eight data points in each plot. The directional tuning appears in most cases well captured by a cosine function. *Muscle abbreviations*: biceps brachii short head (BicShort), biceps brachii long head (BicLong), brachialis (Brac), pronator teres (PronTer), brachioradialis (BrRad), triceps brachii lateral head (TrLat), triceps brachii long head (TrLong), triceps brachii medial head (TrMed), deltoid anterior (DeltA), deltoid middle (DeltM), deltoid posterior (DeltP), pectoralis major clavicular (PectClav), trapezius superior (TrapSup), trapezius middle (TrapMid), latissimus dorsi (LatDors), teres major (TeresMaj), infraspinatus (InfraSp). Adapted from d'Avella et al. [13].

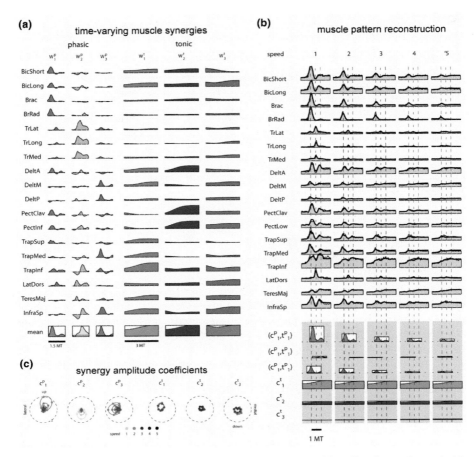

FIGURE 6.2 Modulation of muscle synergies with arm reaching direction and speed. (**a**) Three phasic and three tonic time-varying muscle synergies [see Eq. (6.5)] extracted from the muscle patterns recorded during reaching movements between one central location and eight peripheral locations in the frontal planes at five different speeds. (**b**) Reconstruction of sEMG envelopes, averaged across repetitions with similar movement durations (1, fastest; 5, slowest) for upward movements. (**c**) Directional tuning of synergy amplitude coefficients for different movement speeds illustrated with polar plots as in Fig. 6.1. Different color saturations represents different movement speeds. *Muscle abbreviations*: see legend of Fig. 6.1; pectoralis major sternal (PectInf). Adapted from d'Avella et al. [10].

simply subtracting them, such tonic components were also reconstructed by tonic time-varying muscle synergies that could not be shifted in time and were only modulated in amplitude by direction. Thus, muscle patterns were decomposed into N_p phasic synergies and N_t tonic synergies, according to

$$\mathbf{m}(t) = \sum_{i=1}^{N_p} c_i^p \, \mathbf{w}_i^p(t - t_i^p) + \sum_{i=1}^{N_t} c_i^t \, \mathbf{w}_i^t(t_i) + e(t) \qquad (6.5)$$

where the superscript p refers to phasic and the superscript t to tonic synergies and combination coefficients. Speed modulation by amplitude scaling of phasic synergies can be understood considering one important characteristic of the arm motion equations

$$\mathbf{I}(\mathbf{q})\ddot{\mathbf{q}} + \mathbf{C}(\mathbf{q}, \dot{\mathbf{q}})\dot{\mathbf{q}} + \mathbf{g}(\mathbf{q}) = \boldsymbol{\tau}(t) \qquad (6.6)$$

where $\mathbf{q}$ is a vector on joint coordinates, $\mathbf{I}$ is the inertia matrix, $\mathbf{C}$ is the matrix of Coriolis and centripetal coefficients, $\mathbf{g}$ is the vector of gravity torques, and $\boldsymbol{\tau}(t)$ is the time-varying vector of joint torques generated by muscle contractions. The solutions of the arm motion equations—that is, $\mathbf{q}(t)$ that satisfy Eq. (6.6)—are invariant for time scaling; that is, $\tilde{\mathbf{q}}(t) \triangleq \mathbf{q}(\lambda t)$, where λ is the time scale, is also a solution, if both the gravitational, $\boldsymbol{\tau}_g(t) = \mathbf{g}(\mathbf{q})$, and the dynamic, $\boldsymbol{\tau}_{dyn}(t) = M(\mathbf{q})\ddot{\mathbf{q}} + V(\mathbf{q}, \dot{\mathbf{q}})\dot{\mathbf{q}}$, components of the torque are both scaled in time and only the dynamic component is scaled in amplitude by the square of the time scale, that is, $\tilde{\boldsymbol{\tau}}(t) = c^2 \boldsymbol{\tau}_{dyn}(ct) + \boldsymbol{\tau}_g(ct)$ [1,32]. Thus, the modulation of phasic and tonic synergies might provide a simple rule for controlling the speed of reaching movements.

Finally, superposition and modulation of the synergies identified from point-to-point movement were found to reconstruct the muscle patterns underlying multiphasic movements, such as reaching through a via-point [13] and reaching a target whose location changes after movement initiation [14]. Thus, sequencing of time-varying muscle synergies might constitute an intermittent control strategy that the CNS employs to construct complex movements from simple building blocks.

6.5   MOTOR PRIMITIVES IN HUMAN LOCOMOTION

Locomotion is one of the basic motor behaviors. Its characteristic feature, rhythmicity (alternating limb movements), represents an inherent property of functioning of central pattern generators. It also imposes temporal constraints on muscle activity patterns along with critical gait events such as heel strike and push-off. A line of evidence relating to the nature of the locomotion motor program has come from analyses of EMG activity using statistical approaches (EMG factorization, see above). Below we consider the spatiotemporal organization of the activity patterns of leg and trunk muscles during locomotion.

6.5.1   Spatiotemporal Architecture of Multi-Muscle EMG Activity

A couple of studies in the 1980s and 1990s showed that the activity patterns of leg muscles during locomotion could be accounted for by four to five specific bursts of activation [16,66]. That is, the activity patterns resulting from the modulation of leg muscle activity over a step cycle could be reconstructed for each muscle from a weighted sum of a few basic temporal components (Fig. 6.3). Such functional

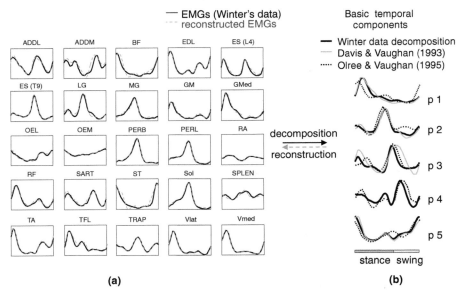

FIGURE 6.3   Linear decomposition of EMG waveforms in human walking. Averaged EMG envelopes recorded from 18 subjects for 25 muscles during a single cycle of overground locomotion at a natural speed (~5 km/h) are plotted on the left. EMG records were filtered with a low-pass cutoff of 3 Hz. Data taken from Winter [78]. Basic temporal components [patterns p1–p5, corresponding to the rows of the coefficient matrix C in Eq (6.3)] derived from these recordings by factor analysis (FA; black traces) are shown on the right. Results are compared with the results, of an FA of EMGs recorded from 16 leg muscles (gray traces, Davis and Vaughan [16]) and from eight muscles in each leg (dotted traces, Olree and Vaughan [64]). Components are designated in a chronological order of their main peak in the cycle beginning with heel strike. Red curves in each of the plots in (a) show the reconstructed EMG waveforms (using five basic temporal components depicted in (b)). *Muscle abbreviations*: adductor longus (ADDL); adductor magnus (ADDM); biceps femoris (BF); extensor digitorum longus (EDL); erector spinae (lumbar) (ES; (L4)); erector spinae (thoracic) (ES (T9)); gastrocnemius lateralis (LG); gastrocnemius medialis (MG); gluteus maximus (GM); gluteus medius (GMed); external oblique lateralis (OEL); external oblique medialis (OEM); peroneus brevis (PERB); peroneus longus (PERL); rectus abdominus (RA); rectus femoris (RF); sartorius (SART); semitendinosus (ST); soleus (Sol); splenius (SPLEN); tibialis anterior (TA); tensor fascia latae (TFL); trapezius (TRAP); vastus lateralis (Vlat); vastus medialis (Vmed).

grouping of muscle activations led to the hypothesis that the total pattern of muscle activity can be controlled by a small set of locomotor activation modules.

A locomotor module is a functional unit—implemented in a neuronal network—that generates a specific motor output by imposing a spatiotemporal structure to muscle activations. Each module involves a basic activation pattern (temporal structure) with variable weights of distribution (spatial structure) to different muscles. It is also worth stressing that tens or even hundreds of muscles are active during locomotion. During the last decade, it has been demonstrated in numerous studies that

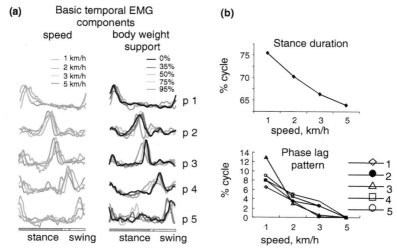

FIGURE 6.4 Effect of walking speed and body weight support on basic temporal components for treadmill locomotion. (**a**) Five common components (p1–p5) derived by factor analysis across conditions are located in a chronological order. (**b**) *Upper plot*: Changes in the relative duration of the stance phase with speed. *Lower plot*: Phase lag required to provide the best fit between each pattern and the pattern determined from the 5 km/h data. Modified from Ivanenko et al. [37].

the EMG activity of trunk and leg muscles can be accounted for by few basic patterns during different locomotion modes, directions, speeds, and body support [5,7,39,60,62,63]. Figure 6.4 illustrates an effect of speed and body weight support on basic EMG patterns for treadmill locomotion. These patterns may be regarded as locomotor primitives in a computational sense, because they are the building blocks from which locomotor activities are constructed.

Correlational analyses [16,38] and biomechanical simulations based on the experimentally derived activation patterns [63] show that the timing of these activity is associated with the major kinematic and kinetic events in the gait cycle, namely (see Fig. 6.5), weight acceptance (p1, involving primarily hip and knee extensors), loading/propulsion (p2, ankle plantar flexors), trunk-stabilization activity during the double support phase and initiation of swing (p3), liftoff (p4, ankle dorsiflexors and hip flexors), and deceleration of the leg in late swing in preparation for heel contact (p5, hamstrings). Thus, it appears that the activation patterns may represent the drive provided by spinal pattern generators and/or sensory feedback.

6.5.2 Segmental Spinal Motor Output

Pattern-generating oscillators are believed to be primarily localized in the cervical and lumbosacral segments of the cord where they control the locomotion movements of the corresponding limbs. To characterize the spatiotemporal organization of the total

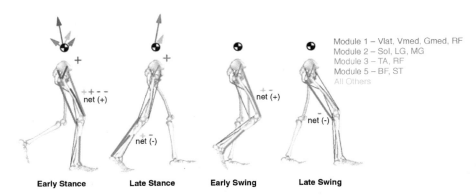

FIGURE 6.5 Contributions of basic patterns (p1–p5 in Fig. 6.4) to walking biomechanics. Contribution of different basic patterns (modules) to early stance (15% of gait cycle), late stance (45%), early swing (70%), and late swing (85%) is shown. Arrows departing from the center of body mass (COM) (*upper plots*) denote the resultant module contributions to the horizontal and vertical ground reaction forces that accelerate the COM providing body support and forward propulsion. Net energy flow by each module to the trunk or leg ("net" total of all modules also presented for leg) is denoted by a + or − for energy increases or decreases, respectively. In early stance, Module 1 contributes largely to body support while acting to decelerate forward motion. In late stance, Module 2 provides both body support and forward propulsion. Muscle abbreviations are as in Fig. 6.4. Reproduced from Neptune et al. [63] with permission.

spinal motor output, the recorded averaged patterns of EMG activity from multiple muscles can be mapped onto the estimated rostrocaudal location of MN pools in the human spinal cord according to the published myotomal charts of segmental innervation in humans [47,71].

Figure 6.6 illustrates spinal segmental output constructed from data recorded during treadmill locomotion at 5 km/h. A simple inspection of the maps suggests that MN activity tends to occur in bursts that are temporally aligned across several spinal segments. These few separate periods of MN activation approximately correspond to the timing of the basic temporal components (Fig. 6.6). Thus, rather than using the statistical techniques, this method independently highlights the major loci of activity of segmental groups of MNs.

In the spinal cord, motor neurons are arranged in columns, with a specific grouping of muscles at each segmental level [67,70]. The functional significance of such motor neuron positioning is largely unknown and has been recently debated [6,43]. Interestingly, the rostrocaudal displacements of the center of bilateral motor neuron activity in the lumbosacral enlargement mirror the changes in the energy of the center-of-body mass motion in human locomotion [6]. These findings suggest that during the evolution of bipedal modes of locomotion the biomechanical mechanisms and global characteristics of whole-body motion, such as the inverted pendulum in walking and the pogo-stick bouncing in running, may be tightly correlated with specific modes of progression of motor pool activity rostrocaudally in the spinal cord.

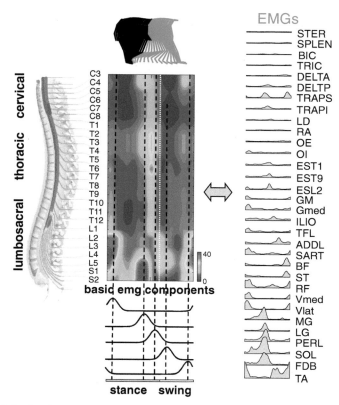

FIGURE 6.6 Spatiotemporal patterns of ipsilateral α-motoneuron activity along the rostrocaudal axis of the spinal cord during walking on a treadmill at 5 km/h. Pattern is plotted versus normalized gait cycle. Output pattern was constructed by mapping the recorded EMG waveform of 32 ipsilateral limb and trunk and shoulder muscles (non-normalized method, adapted from Ivanenko et al.) [38] onto the known charts of segmental localization. Five Gaussian activation components that approximate basic temporal components in Fig. 6.5 (corresponding to five major discrete bursts of activity) and account for ~90% of total variance are shown on the bottom. *Muscle abbreviations*: sternocleido mastoideus (STER); biceps brachii (BIC); triceps brachii (TRIC); deltoideus, anterior and posterior portions (DELTA and DELTP, respectively); trapezius, inferior and superior portions (TRAPS and TRAPI, respectively); latissimus dorsi (LD); internal oblique (OI); flexor digitorum brevis (FDB); other muscle abbreviations are as in Fig. 6.4. The color scale refers to microvolts of corresponding EMG activity.

Application of this technique to different populations of subjects reveals specific characteristics of motor pool activity during development of locomotor patterns from infant to elderly [41,62] and may characterize specific adaptation of the segmental motor output in patients [28].

6.5.3   Bilateral Coordination

A characteristic feature of locomotor behavior, as opposed to arm reaching movements, is a coordinated alternating motion of both limbs. Although each limb controller has a relatively autonomous ability to generate rhythmic pattern (as, for instance, can be observed when walking on a treadmill with split belts; Forssberg et al. [23]), four major activation components are temporally synchronized on both sides of the body. This was first shown by Olree and Vaughan [64], who recorded from leg muscles on both sides of the body and found that two components (1 and 2 in Fig. 6.3) were copies of the other two (3 and 5) but phase shifted by exactly one-half the cycle. Component 4, which explains the least variance and is associated with the ipsilateral foot lift or swing, has no obvious contralateral analog. Components 1 and 2 are predominantly weighted on the ipsilateral leg muscles, whereas components 3 and 5 are more significantly weighted on contralateral leg muscles. It is also evident in the spinal motor pool activation maps (Fig. 6.6) that components 1 and 2 are more prominent in the ipsilateral lumbar and sacral segments.

One might likely expect a kind of symmetry when extracting basic temporal components from the pooled (left and right leg) EMG data because muscle activity on the contralateral side mirrors ipsilateral activity if shifted by one-half the cycle. It is much less obvious, however, that "ipsilateral" components (extracted from only ipsilateral EMG activity, Fig. 6.3) demonstrate the same "symmetry" as bilateral ones [19,64].

This finding may reflect an inherent property of neuronal networks underlying locomotor pattern generation. There is recent evidence from the cat that the spinal circuitry itself encompasses critical periods that may help shape the sequence of muscle activation [69]. The mammalian locomotor CPG is composed of multiple distributed rhythm-generating networks [29] and includes excitatory neurons that are responsible for rhythm generation and glycinergic commissural interneurons that are directly involved in left–right alternation [48,73]. In addition, the bilateral synchronization of activation timings occurs around the heel strike events of both legs and thus the contribution of sensory feedback in left–right alternation may be implicit.

6.5.4   Development of Locomotor Primitives

Uncovering a common underlying neural framework for the modular control of human locomotion and its development represents an interesting avenue for the future work. Motor primitives may reflect in a way how the nervous system develops, by building up or modifying modules as it matures. Some functional units are likely inborn, whereas others may develop later or be dependent on individual body size/ proportions or experience [19,54].

Figure 6.7 illustrates basic patterns of bilateral muscle control at different ages. Muscle co-activation is generally greater in infants leading to only two basic activation patterns in neonates. Through development the two basic patterns of

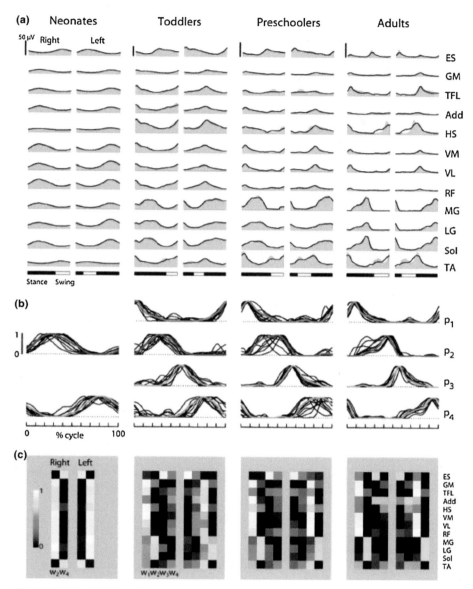

FIGURE 6.7 EMG profiles and derived basic patterns. (**a**) Ensemble-averaged (across all subjects of each group: 39 neonates, 10 toddlers, 10 preschoolers, and 10 adults) EMG profiles during the step cycle, aligned with stance onset in the right leg. Shaded areas are the experimental data, and black traces are the profiles reconstructed as weighted sum of the patterns extracted from the ensemble. (**b**) Basic patterns derived by non-negative matrix factorization from averaged (across steps) EMG profiles in 10 subjects of each group (black). Patterns are from ensemble EMG averages (colored). (**c**) Normalized weights of the ensemble patterns in color scale (the color scale is depicted on the left). Reproduced from Dominici et al. [19].

stepping neonates are retained, augmented by two new patterns first revealed in toddlers (Fig. 6.7). Nevertheless, the underlying neural substrates and how this primitive pattern in neonates evolves in the mature one still require further investigations. Development of motor patterns requires progressively tuning the timing and amplitude of muscle activity to the intrinsic modes of mechanical behavior resulting from the interaction of the limbs/body parts between each other and with the environment, also taking into account the growing body of the child. Some of the results obtained in humans are consistent with comparative studies in vertebrates [54]. Factorization of the EMG of adult rats, cats, macaques, and guinea fowls shows four patterns, closely resembling those found in human toddlers [19].

6.5.5 Reorganization of Motor Patterns in Movement Disorders

Investigating locomotor responses after neurological lesions may have important implications related to the construction of gait rehabilitation technology and exploration of the mechanisms involved in improving locomotor function. Experimental studies performed on individuals with well-identified pathologies have demonstrated distinct adaptations. Due to muscle redundancy, various neuromotor strategies may exist to compensate for impaired neural circuitry, decreased muscle strength, and joint stiffness [27,28,40]. In patients, the mechanisms involved in locomotor improvements may rely on the inherent spatiotemporal organization of neuronal circuitry and its adaptability. The question arises as to whether the rhythmic patterning elements are invariant in conditions where muscle activation patterns can be compromised by spinal cord lesions, brain damage, and other motor disturbances. The findings on the modularity of the motor output in motor-impaired populations are still limited, and some of the conclusions based on statistical methods may possibly be revised as empirical data on the neural substrates underlying normal and pathological gait accumulate. Nevertheless, several recent studies suggest that the building blocks with which the central nervous system constructs the motor patterns can be preserved, even though fewer modules might sometimes be observed to account for muscle activation in motor-impaired population compared with controls [8,18,25,26,31,36].

The application of modern quantitative analyses of the motor patterns discussed here offer a new approach to characterizing the mechanisms underlying control of human locomotion that may potentially benefit the study of pathological gait and the ability of current therapeutic exercises to improve patient outcomes.

6.6 CONCLUSIONS

A modular motor organization may be needed to solve the degrees of freedom problem in biological motor control [24]. Nevertheless, there are still many open questions related to the choice of appropriate modules, their task dependence, and adaptation to the malfunctioning of neuronal networks in the case of different motor pathologies. While many studies succeeded in a decomposition of motor patterns into a few "motor modules," it should be noted that the way in which the central nervous system combines them together, as well as how and where the weighting coefficients are encoded, is not yet fully understood.

In view of task dependence of muscle synergies, it would be interesting also to compare them in different behaviors and examine whether plasticity in muscle patterns originates from adapting these common modules or by creating new muscle synergies. Likely, both strategies represent important compensatory mechanisms in movement disorders or adaptation and learning new motor tasks. It is worth emphasizing though that the spatiotemporal architecture of various motor activities points toward modularity of multi-muscle activation patterns as a hallmark of their neural control. In addition to this general template determined by muscle synergies, a minor but still important part of the motor output consists in *fine tuning* attributable to residual (not explained by muscle synergies) muscle activations.

REFERENCES

1. Atkeson, C. G., and J. M. Hollerbach, "Kinematic features of unrestrained vertical arm movements," *J. Neurosci.* **5**, 2318–2330 (1985).

2. Basilevsky, A. *Statistical Factor Analysis and Related Methods: Theory and Applications*, John Wiley & Sons, New York, 1994.

3. Bell, A. J., and T. J. Sejnowski, "An information-maximization approach to blind separation and blind deconvolution," *Neural Comput.* **7**, 1129–1159 (1995).

4. Bizzi, E., V. C. Cheung, A. d'Avella, P. Saltiel, and M. Tresch, "Combining modules for movement," *Brain Res. Rev.* **57**, 125–133 (2008).

5. Cappellini, G., Y. P. Ivanenko, R. E. Poppele, and F. Lacquaniti, "Motor patterns in human walking and running," *J. Neurophysiol.* **95**, 3426–3437 (2006).

6. Cappellini, G., Y. P. Ivanenko, N. Dominici, R. E. Poppele, and F. Lacquaniti, "Migration of motor pool activity in the spinal cord reflects body mechanics in human locomotion," *J. Neurophysiol.* **104**, 3064–3073 (2010).

7. Chvatal, S. A., and L. H. Ting, "Voluntary and reactive recruitment of locomotor muscle synergies during perturbed walking," *J. Neurosci.* **32**, 12237–12250 (2012).

8. Clark, D. J., L. H. Ting, F. E. Zajac, R. R. Neptune, and S. A. Kautz, "Merging of healthy motor modules predicts reduced locomotor performance and muscle coordination complexity post-stroke," *J. Neurophysiol.* **103**, 844–857 (2010).

9. Coscia, M., V. Monaco, M. Capogrosso, C. Chisari and S. Micera, "Computational aspects of MN activity estimation: a case study with post-stroke subjects," *IEEE International Conference on Rehabilitation Robots*, Zurich, Switzerland 2011.

10. d'Avella, A., L. Fernandez, A. Portone, and F. Lacquaniti, "Modulation of phasic and tonic muscle synergies with reaching direction and speed," *J. Neurophysiol.* **100**, 1433–1454 (2008).

11. d'Avella, A., and F. Lacquaniti, "Control of reaching movements by muscle synergy combinations," *Front. Comput. Neurosci.* **7**, 42 (2013).

12. d'Avella, A., and D. K. Pai, "Modularity for sensorimotor control: evidence and a new prediction," *J. Mot. Behav.*, **42**, 361–369 (2010).

13. d'Avella, A., A. Portone, L. Fernandez, and F. Lacquaniti, "Control of fast-reaching movements by muscle synergy combinations," *J. Neurosci.* **26**, 7791–7810 (2006).

14. d'Avella, A., A. Portone, and F. Lacquaniti, "Superposition and modulation of muscle synergies for reaching in response to a change in target location," *J. Neurophysiol.* **106**, 2796–2812 (2011).

15. d'Avella, A., P. Saltiel, and E. Bizzi, "Combinations of muscle synergies in the construction of a natural motor behavior," *Nat. Neurosci.* **6**, 300–308 (2003).

16. Davis, B. L., and C. L. Vaughan, "Phasic behavior of EMG signals during gait: use of multivariate statistics," *J. Electromyogr. Kinesiol.* **3**, 51–60 (1993).

17. Day, S. J., and M. Hulliger, "Experimental simulation of cat electromyogram: evidence for algebraic summation of motor-unit action-potential trains," *J. Neurophysiol.* **86**, 2144–2158 (2001).

18. Den Otter, A. R., A. C. Geurts, T. Mulder, and J. Duysens, "Gait recovery is not associated with changes in the temporal patterning of muscle activity during treadmill walking in patients with post-stroke hemiparesis," *Clin. Neurophysiol.* **117**, 4–15 (2006).

19. Dominici, N., Y. P. Ivanenko, G. Cappellini, A. d'Avella, V. Mondi, M. Cicchese, A. Fabiano, T. Silei, A. Di Paolo, C. Giannini, R. E. Poppele, and F. Lacquaniti, "Locomotor primitives in newborn babies and their development," *Science* **334**, 997–999 (2011).

20. Flanders, M., and U. Herrmann, "Two components of muscle activation: scaling with the speed of arm movement," *J. Neurophysiol.* **67**, 931–943 (1992).

21. Flanders, M., J. J. Pellegrini, and S. D. Geisler, "Basic features of phasic activation for reaching in vertical planes," *Exp. Brain Res.* **110**, 67–79 (1996).

22. Flanders, M., J. J. Pellegrini, and J. F. Soechting, "Spatial/temporal characteristics of a motor pattern for reaching," *J. Neurophysiol.* **71**, 811–813 (1994).

23. Forssberg, H., S. Grillner, J. Halbertsma, and S. Rossignol, "The locomotion of the low spinal cat. II. Interlimb coordination," *Acta Physiol. Scand.* **108**, 283–295 (1980).

24. Giszter, S. F., C. B. Hart, and S. P. Silfies, "Spinal cord modularity: evolution, development, and optimization and the possible relevance to low back pain in man," *Exp. Brain Res.* **200**, 283–306 (2010).

25. Gizzi, L., J. F. Nielsen, F. Felici, Y. P. Ivanenko, and D. Farina, "Impulses of activation but not motor modules are preserved in the locomotion of subacute stroke patients," *J. Neurophysiol.* **106**, 202–210 (2011).

26. Gizzi, L., J. F. Nielsen, F. Felici, J. C. Moreno, J. L. Pons, and D. Farina, "Motor modules in robot-aided walking," *J. Neuroeng. Rehabil.* **9**, 76–82 (2012).

27. Goldberg, E. J., and R. R. Neptune, "Compensatory strategies during normal walking in response to muscle weakness and increased hip joint stiffness," *Gait Posture* **25**, 360–367 (2007).

28. Grasso, R., Y. P. Ivanenko, M. Zago, M. Molinari, G. Scivoletto, V. Castellano, V. Macellari, and F. Lacquaniti, "Distributed plasticity of locomotor pattern generators in spinal cord injured patients," *Brain* **127**, 1019–1034 (2004).

29. Grillner, S., "Biological pattern generation: the cellular and computational logic of networks in motion," *Neuron* **52**, 751–766 (2006).

30. Gurfinkel, V. S., P. Ivanenko Yu, S. Levik Yu, and I. A. Babakova, "Kinesthetic reference for human orthograde posture," *Neuroscience* **68**, 229–243 (1995).

31. Hayes, H., L. VanHiel, S. Chvatal, L. Ting, K. Tansey, and R. Trumbower "Modularity of muscle activity during robot-assisted locomotion in persons with incomplete spinal cord injury," Presented at *Society for Neuroscience,* Washington, 2011.

32. Hollerbach, M. J., and T. Flash, "Dynamic interactions between limb segments during planar arm movement," *Biol. Cybern.* **44**, 67–77 (1982).

33. Hubel, D. H., and T. N. Wiesel, "Receptive fields of single neurones in the cat's striate cortex," *J. Physiol.* **148**, 574–591 (1959).

34. Hultborn, H., "State-dependent modulation of sensory feedback," *J. Physiol.* **533**, 5–13 (2001).

35. Imagawa, H., S. Hagio, and M. Kouzaki, "Synergistic co-activation in multi-directional postural control in humans," *J. Electromyogr. Kinesiol.* **23**, 430–437 (2013).

36. Ivanenko, Y. P., R. Grasso, M. Zago, M. Molinari, G. Scivoletto, V. Castellano, V. Macellari, and F. Lacquaniti, "Temporal components of the motor patterns expressed by the human spinal cord reflect foot kinematics," *J. Neurophysiol.* **90**, 3555–3565 (2003).

37. Ivanenko, Y. P., R. E. Poppele, and F. Lacquaniti, "Five basic muscle activation patterns account for muscle activity during human locomotion," *J. Physiol.* **556**, 267–282 (2004).

38. Ivanenko, Y. P., R. E. Poppele, and F. Lacquaniti, "Spinal cord maps of spatiotemporal alpha-motoneuron activation in humans walking at different speeds," *J. Neurophysiol.* **95**, 602–618 (2006).

39. Ivanenko, Y. P., G. Cappellini, R. E. Poppele, and F. Lacquaniti, "Spatiotemporal organization of alpha-motoneuron activity in the human spinal cord during different gaits and gait transitions," *Eur J. Neurosci.* **27**, 3351–3368 (2008).

40. Ivanenko, Y. P., R. E. Poppele, and F. Lacquaniti, "Distributed neural networks for controlling human locomotion: lessons from normal and SCI subjects," *Brain Res. Bull.* **78**, 13–21 (2009).

41. Ivanenko, Y. P., N. Dominici, G. Cappellini, A. Di Paolo, C. Giannini, R. E. Poppele, and F. Lacquaniti, "Changes in the spinal segmental motor output for stepping during development from infant to adult," *J. Neurosci.* **33**, 3025–3036a (2013).

42. Ivanenko, Y. P., W. G. Wright, R. J. St George, and V. S. Gurfinkel, "Trunk orientation, stability, and quadrupedalism," *Front Neurol.* **4**, 20 (2013).

43. Jessell, T. M., G. Surmeli, and J. S. Kelly, "Motor neurons and the sense of place," *Neuron* **72**, 419–424 (2011).

44. Jolliffe, I. T., *Principal Component Analysis* 2nd ed., Springer, New York, 2002.

45. Karst, G. M., and Z. Hasan, "Timing and magnitude of electromyographic activity for two-joint arm movements in different directions," *J. Neurophysiol.* **66**, 1594–1604 (1991).

46. Kawato, M., "Internal models for motor control and trajectory planning," *Curr. Opin. Neurobiol.* **9**, 718–727 (1999).

47. Kendall, F. P., E. K. McCreary, P. Provance, M. M. Rodgers, and W. A. Romani, *Muscles: Testing and Function with Posture and Pain*, 5th ed., Williams & Wilkins, Baltimore, 2005.

48. Kiehn, O., "Locomotor circuits in the mammalian spinal cord," *Annu. Rev. Neurosci.* **29**, 279–306 (2006).

49. Klein Breteler, M. D., K. J. Simura, and M. Flanders, "Timing of muscle activation in a hand movement sequence," *Cereb. Cortex* **17**, 803–815 (2007).

50. Klous, M., A. Danna-dos-Santos, and M. L. Latash, "Multi-muscle synergies in a dual postural task: evidence for the principle of superposition," *Exp. Brain Res.* **202**, 457–471 (2010).

51. Klous, M., P. Mikulic, and M. L. Latash, "Two aspects of feedforward postural control: anticipatory postural adjustments and anticipatory synergy adjustments," *J. Neurophysiol.* **105**, 2275–2288 (2011).

52. Krishnamoorthy, V., S. Goodman, V. Zatsiorsky, and M. L. Latash, "Muscle synergies during shifts of the center of pressure by standing persons: identification of muscle modes," *Biol. Cybern.* **89**, 152–161 (2003).

53. Lacquaniti, F., Y. P. Ivanenko, and M. Zago, "Patterned control of human locomotion," *J. Physiol.* **590**, 2189–2199 (2012).

54. Lacquaniti, F., Y. P. Ivanenko, and M. Zago, "Development of human locomotion," *Curr. Opin. Neurobiol.* **22**, 822–828 (2012).

55. Lee, D. D., and H. S. Seung, "Learning the parts of objects by non-negative matrix factorization," *Nature* **401**, 788–791 (1999).

56. Lee, D. D., and H. S. Seung, "Algorithms for non-negative matrix factorisation," in T. K. Leen, T. G. Dietterich, and V. Tresp, eds., *Advances in Neural Information Processing Systems 13* MIT Press, Cambridge, MA, 2001.

57. Maclellan, M. J., Y. P. Ivanenko, G. Cappellini, F. Sylos Labini, and F. Lacquaniti, "Features of hand–foot crawling behavior in human adults," *J. Neurophysiol.* **107**, 114–125 (2012).

58. Makeig, S., T. P. Jung, A. J. Bell, D. Ghahremani, and T. J. Sejnowski, "Blind separation of auditory event-related brain responses into independent components," *Proc. Natl. Acad. Sci. USA* **94**, 10979–10984 (1997).

59. Mallat, S. G., and Z. Zhang, "Matching pursuits with time-frequency dictionaries," *IEEE Trans. Signal Processing* **41**, 3397–3415 (1993).

60. McGowan, C. P., R. R. Neptune, D. J. Clark, and S. A. Kautz, "Modular control of human walking: adaptations to altered mechanical demands," *J. Biomech.* **43**, 412–419 (2010).

61. Milosevic, M., K. M. Valter McConville, E. Sejdic, K. Masani, M. J. Kyan, and M. R. Popovic, "Visualization of trunk muscle synergies during sitting perturbations using self-organizing maps (SOM)," *IEEE Trans. Biomed. Eng.* **59**, 2516–2523 (2012).

62. Monaco, V., A. Ghionzoli, and S. Micera, "Age-related modifications of muscle synergies and spinal cord activity during locomotion," *J. Neurophysiol.* **104**, 2092–2102 (2010).

63. Neptune, R. R., D. J. Clark, and S. A. Kautz, "Modular control of human walking: a simulation study," *J. Biomech.* **42**, 1282–1287 (2009).

64. Olree, K. S., and C. L. Vaughan, "Fundamental patterns of bilateral muscle activity in human locomotion," *Biol. Cybern.* **73**, 409–414 (1995).

65. Omlor, L., and M. A. Giese, "Anechoic blind source separation using Wigner marginals," *J. Machine Learning Res.* **12**, 1111–1148 (2011).

66. Patla, A. E., T. W. Calvert, and R. B. Stein, "Model of a pattern generator for locomotion in mammals," *Am. J. Physiol.* **248**,(4 Pt 2): R484–R494 (1985).

67. Romanes, G. J., "The motor cell columns of the lumbo-sacral spinal cord of the cat," *J. Comp. Neurol.* **94**, 313–363 (1951).

68. Safavynia, S. A., and L. H. Ting, "Task-level feedback can explain temporal recruitment of spatially fixed muscle synergies throughout postural perturbations," *J. Neurophysiol.* **107**, 159–177 (2012).

69. Saltiel, P., and S. Rossignol, "Critical points in the forelimb fictive locomotor cycle and motor coordination: effects of phasic retractions and protractions of the shoulder in the cat," *J. Neurophysiol.* **92**, 1342–1356 (2004).

70. Sharrard, W. J., "The distribution of the permanent paralysis in the lower limb in poliomyelitis: a clinical and pathological study," *J. Bone Joint Surg. Br.* **37-B**, 540–558 (1955).

71. Sharrard, W. J., "The segmental innervation of the lower limb muscles in man," *Ann. R. Coll. Surg. Engl.* **35**, 106–122 (1964).

72. Sherrington, C., *The Integrative Action of the Nervous System*, Charles Scribner's Sons, New York, 1906.

73. Stein, P. S., M. L. McCullough, and S. N. Currie, "Spinal motor patterns in the turtle," *Ann. NY Acad. Sci.* **860**, 142–154 (1998).

74. Ting, L. H., S. A. Chvatal, S. A. Safavynia, and J. L. McKay, "Review and perspective: neuromechanical considerations for predicting muscle activation patterns for movement," *Int. J. Numer Methods Biomed Eng.* **28**, 1003–1014 (2012).

75. Torres-Oviedo, G., and L. H. Ting, "Muscle synergies characterizing human postural responses," *J. Neurophysiol.* **98**, 2144–2156 (2007).

76. Tresch, M. C., V. C. Cheung, and A. d'Avella, "Matrix factorization algorithms for the identification of muscle synergies: evaluation on simulated and experimental data sets," *J. Neurophysiol.* **95**, 2199–2212 (2006).

77. Wadman, W., J. D. v. d. Gon, and R. Derksen, "Muscle activation patterns for fast goal-directed arm movements," *J. Human Movement Studies* **6**, 19–37 (1980).

78. Winter, D. A., The biomechanics and motor control of human gait: normal, elderly and pathological, Waterloo Biomechanics Press, Waterloo (Canada), 1991.

79. Yakovenko, S., V. Mushahwar, G. VanderHorst, G. Holstege, and A. Prochazka, "Spatiotemporal activation of lumbosacral motoneurons in the locomotor step cycle," *J. Neurophysiol.* **87**, 1542–1553 (2002).

80. Zajac, F. E., "Muscle and tendon: properties, models, scaling, and application to biomechanics and motor control," *Crit. Rev. Biomed. Eng.* **17**, 359–411 (1989).

7

SURFACE EMG DECOMPOSITION

A. Holobar,[1] D. Farina,[2] and D. Zazula[1]

[1]Faculty of Electrical Engineering and Computer Science, University of Maribor, Maribor, Slovenia
[2]Department of Neurorehabilitation Engineering, Bernstein Center for Computational Neuroscience, Bernstein Focus Neurotechnology Göttingen, University Medical Center Göttingen, Georg-August University, Göttingen, Germany

7.1 INTRODUCTION

Alpha motor neurons in the spinal cord combine the inputs from supraspinal centers and afferent feedback and transform them into a neural signal which is the drive to muscles. There is a one-to-one correspondence between the discharge of a motor neuron and motor unit action potentials (MUAPs) propagated by the innervated muscle fibers. Thus, the neural drive to a muscle—that is, the cumulative discharge of the active motor neurons innervating the muscle—can be assessed by decomposing electromyogram (EMG) recordings into contributions of individual motor units (MUs). This is beneficial for investigating motor control strategies [24,41,62,63,81,89] and motor unit morphological and functional properties [4,7,16,21,35,60,61,75,102], as well as their adaptation to conditions such as pathology, fatigue, pain, or exercise [3,20,24,32,39,59,87,88,90,106,107,110,116].

In a classic setup, MU discharges are identified from invasively acquired intramuscular EMG. While acceptable and largely established in clinical practice, this procedure has drawbacks in applications such as neurorehabilitation, examination of children, training of athletes, and ergonomics, as well as in many other cases where the acquisition conditions cannot be strictly controlled or where the invasive nature of indwelling EMG prevents its everyday use. As a result, various indirect measures of

Surface Electromyography: Physiology, Engineering, and Applications, First Edition.
Edited by Roberto Merletti and Dario Farina.
© 2016 by The Institute of Electrical and Electronics Engineers, Inc. Published 2016 by John Wiley & Sons, Inc.

neural drive have been proposed in the past, relying on noninvasively acquired surface EMG. Among them, amplitude and median/mean power frequency of the surface EMG have been the most frequently used, especially for estimating muscle force and evaluating the effects of exercise and fatigue [11,66,80]. However, although easily measured, the amplitude and frequency content of the surface EMG are only coarse indicators of the neural drive characteristics and are substantially influenced by many factors, such as MUAP cancelation, muscle anatomy, and low-pass filtering of the subcutaneous tissues [12,19,24,51,55,56]. These factors are difficult to control or assess in experimental conditions. Therefore, they hinder the consistency of observations and are among the most probable reasons for controversy of the results reported by many surface EMG studies in the past.

Over the last two decades, automatic decomposition of multichannel and high-density surface EMG demonstrated relevant progress. Whereas early attempts suffered from the inability to correctly decompose highly interferential MUAP superimpositions [40,42,57,75], a few recent approaches proved to identify complete motor unit discharge patterns, even at maximal contraction forces [49,50,70,94]. The developed decomposition techniques range from template matching to latent component analysis, offering a relatively large palette of performance–cost ratios to choose from. This chapter provides an overview of surface EMG decomposition techniques, along with their basic assumptions, properties, and limitations.

7.2   EMG MIXING PROCESS

Surface electrodes measure the electrical activity of several nearby muscle fibers that are active during a muscle contraction. The electrical activity of each fiber can be described by a single fiber action potential (SFAP) that propagates from the neuromuscular junction towards the tendons (see Chapter 1). With increasing distance from the fiber, the SFAP gets more and more attenuated because of the diffusion process through the interposed subcutaneous tissue that is a passive volume conductor (see Chapter 2). Therefore, due to the distance between the uptake electrode and the muscle fiber, fiber curvature, and anisotropic, and often nonhomogeneous, conductivity of the interposed tissue, the measured SFAP is a function of space on the skin surface.

SFAPs of fibers belonging to the same MU have the same conduction velocity and propagate approximately synchronously along the muscle fibers, depending on the scatter of the neuromuscular junctions (NMJ) within the innervation zone (IZ) of the motor unit. They can therefore be considered phase-locked and mathematically modeled as a common response to a single motor neuron discharge. This is a convenient abstraction that greatly simplifies the EMG analysis as, in every point on the skin surface, the SFAPs of individual MU can be assumed to add up and form a L-sample-long MUAP (Fig. 7.1):

$$\mathbf{h}_{ij} = \begin{bmatrix} h_{ij}(0) & h_{ij}(1) & \cdots & h_{ij}(L-1) \end{bmatrix} \tag{7.1}$$

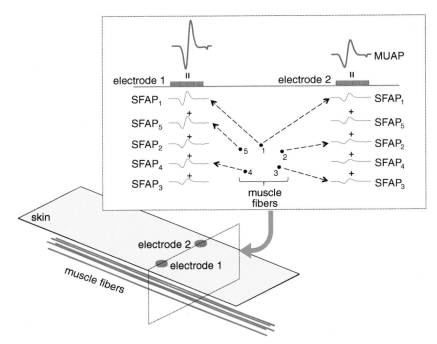

FIGURE 7.1 Schematic representation of five MU fibers belonging to the same motor unit and their corresponding SFAPs as detected by two different surface electrodes. Due to different distances between the fibers and the surface electrodes and the diffusion effect of the interposed tissue, SFAPs appear differently filtered on each surface electrode. In each point in space, the SFAPs can be assumed to add up to form a MUAP.

where $h_{ij}(n)$ is the nth MUAP sample of the jth MU, as detected by the ith uptake electrode.

Being a summation of many spatially filtered SFAPs, the shape of MUAP varies on the skin surface (Fig. 7.2). This is mainly due to inhomogenity and low-pass filtering effect of the volume conductor, and different distances between the individual MU fibers and the uptake electrodes. Although frequently used as a convenient simplification, this spatial MUAP variability cannot be modeled by simple amplitude scaling; that is, MUAP waveforms, as detected by different electrodes, are not simply weighted replicas of each other. This is evident from close inspection of Fig. 7.2.

To quantify the deviation from a simplified model of spatial scaling, it suffices to perform principal component analysis (PCA) of the following matrix of L sample long MUAP waveforms from the jth MU as observed by M uptake electrodes

$$\mathbf{H}_j = \begin{bmatrix} \mathbf{h}_{1j} \\ \mathbf{h}_{2j} \\ \vdots \\ \mathbf{h}_{Mj} \end{bmatrix} = \begin{bmatrix} h_{1j}(0) & \cdots & h_{1j}(L-1) \\ h_{2j}(0) & \cdots & h_{2j}(L-1) \\ \vdots & & \vdots \\ h_{Mj}(0) & \cdots & h_{Mj}(L-1) \end{bmatrix} \qquad (7.2)$$

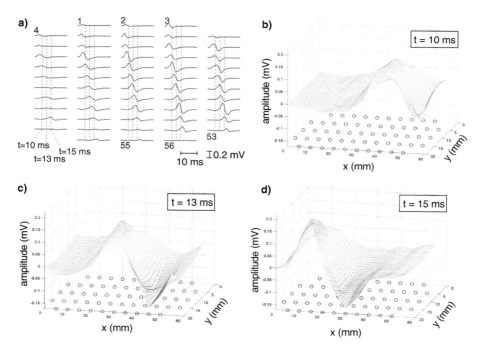

FIGURE 7.2 (**A**) MUAP of an individual MU recorded from the biceps brachii muscle of a healthy man during an isometric contraction sustained at 10% of the maximal voluntary contraction force. The surface EMG signals were detected in single differential (bipolar) mode with an electrode grid of 5 columns and 13 rows (with four missing electrodes at the corners; 56 bipolar recordings; 5-mm interelectrode distance, LISiN, Politecnico di Torino, Italy), with the columns aligned along the fiber direction. (**B–D**) Spatial distribution of MUAP amplitude for all the electrode locations at the time instants indicated by the three vertical lines in part **A**. The single differential signals were detected along the columns. The first and last channel numbers are indicated in part **A**. The innervation zone of the MU is above the matrix in part **A** and to the right of the matrix in parts **B**, **C**, and **D**.

For simplicity reasons, all the observations $\mathbf{h}_{ij}$ in (7.2), associated with MU j and detected by electrode i ($1 < i < M$), are considered to be of equal number of samples L. This is not a limitation as shorter MUAP observations can always be zero padded. Since all the MUAP observations $\mathbf{h}_{ij}$ share the same time support, the nth sample of all $\mathbf{h}_{ij}$, $i = 1, 2, \ldots, M$, is detected simultaneously by the M uptake electrodes.

PCA is a linear mathematical procedure that describes a set of signals as weighted sums of orthogonal basis functions (i.e., principal components), obtained from the signals themselves. If MUAP observations were equal up to a scalar factor, the PCA of $\mathbf{H}_j$ in (7.2) would yield only one principal component with nonzero energy. This principal component would represent the generic MUAP waveform shared by all the uptake electrodes (up to multiplication by a scale factor). In the case of convolutive or even nonlinear relationship among the M different MUAP observations in space, however, the first principal component cannot account for all the MUAP energy. This suggests the following measure to quantify the MUAP spatial variability in terms of energy:

$$\alpha_j = \frac{\lambda_{1j}}{\sum_i \lambda_{ij}} \qquad (7.3)$$

where λ_{ij} stands for ith eigenvalue of the $M \times M$ square matrix $\mathbf{H}_j \mathbf{H}_j^T$ and the M eigenvalues are assumed to be sorted in the descending order. The value of coefficient α_j ranges from 1 to $1/M$. It equals 1 if and only if the spatial MUAP variability can be accounted for by a multiplication by a scale factor. The smaller the α_j, the greater the nonlinearity in, or the greater the convolutive portion of, the MUAP shape variability.

Figure 7.3 depicts the values of α_j, calculated over 500 synthetic MUs that were generated by an advanced simulator described in Farina et al. [26]. The simulated

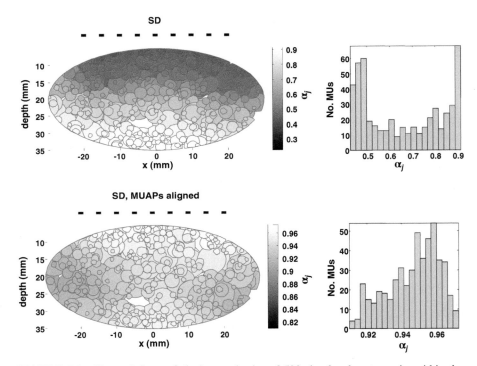

FIGURE 7.3 **Upper left panel** depicts territories of 500 simulated motor units within the muscle cross section. The gray level represents the values of α_j as defined by (7.3) for each motor unit and calculated from 72 MUAPs detected by 9×9 array of electrodes in single-differential configuration (black rectangles, interelectrode distance of 5 mm). MUAPs detected over the innervation zone were discarded and were not considered for α_j calculation. **Lower left panel** depicts again the territories of 500 simulated motor units within the muscle cross section, but the gray level now represents the α_j for MUAPs of each individual MU aligned in time. The right panels depict the distribution of α_j values across all simulated motor units.

acquisition array consisted of 9×9 circular surface electrodes (diameter of 0.5 mm) in single-differential configuration and with an interelectrode distance of 5 mm. The array was centered over the belly of a muscle with an average fiber length of 130 mm and elliptical cross-sectional area of 1413 mm^2. Skin and subcutaneous thickness was set equal to 1 mm and 4 mm, respectively. Other simulation parameters were similar to those in Nielsen et al. [97]. The tendon ending spread and spread of the innervations zone were set equal to 5 mm. The central row of electrodes was located over the innervation zone. Thus, considering the recommendations of the SENIAM project [45], the MUAPs acquired by the central-row electrodes were discarded and excluded from computation of α_j values.

A significant portion of spatial MUAP variability is due to MUAP propagation along the muscle fibers. This can be compensated for by aligning the MUAPs in time [77] and recalculating the α_j values. This considerably increases the α_j values, especially for superficial MUs (Fig. 7.3, lower panels). A similar effect can be achieved by recording the MUAPs with a linear array of surface electrodes oriented perpendicularly to the muscle fibers (Fig. 7.4), but this considerably limits the number of uptake electrodes.

In the case of experimental EMG, the locations of MUs within the muscle are unknown. However, the distribution of α_j values can still be analyzed for all the MUs that are reliably identified from EMG signals. Representative results, calculated over 534 and 384 MUs identified by Convolution Kernel Compensation (CKC) method (see Subsection 7.3.2.2) from surface EMG of tibialis anterior and biceps brachii muscles in young healthy subjects, are depicted in Fig. 7.5. An array of 9×9 circular surface electrodes (diameter of 1 mm and interelectrode distance of 5 mm, LISiN, Politecnico di Torino, Italy) in single-differential configuration was centered over the muscle belly. MUAPs detected over the innervation zone were discarded and were not

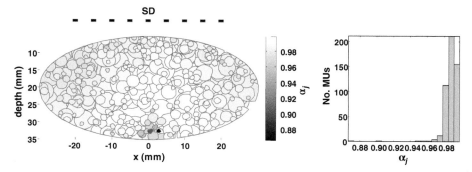

FIGURE 7.4 Territories of 500 simulated motor units within the muscle cross section and corresponding values of α_j as defined by (7.3). The **right panel** depicts the distribution of α_j values calculated across all simulated motor units. Each α_j was calculated from 9 MUAPs as detected by 1×9 array of electrodes (black rectangles, interelectrode distance of 5 mm) located around 30 mm distally from the simulated innervation zone and oriented transversally with respect to muscle fibers.

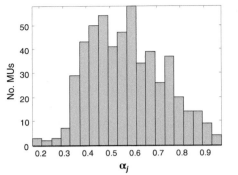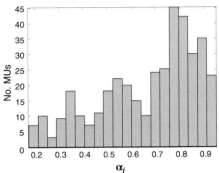

FIGURE 7.5   The distribution of α_j values calculated for 534 and 384 MUs identified from surface EMG of tibialis anterior (**left panel**) and biceps brachii muscle (**right panel**), respectively, detected from 10 healthy young subjects. The surface array was centered over the muscle belly and consisted of 9×9 circular electrodes in single differential configuration with diameter of 1 mm and interelectrode distance of 5 mm. MUAPs detected over the innervation zone were discarded and were not considered for α_j calculations.

considered for calculations of α_j, whereas the detected MUAP shapes were not aligned in time. Recorded levels of isometric muscle contractions ranged from 10% to 70% of maximum voluntary contraction (MVC). In both muscles, the calculated α_j values demonstrated a large variability over the identified MUs, with relatively few MUs with $\alpha_j > 0.9$. This is in agreement with simulation results depicted in Fig. 7.3 as only superficial MUs are usually identified by surface EMG decomposition.

The shape of a MUAP varies also with muscle fatigue and motor unit discharge rate [24] and, in nonisometric contractions, with changes in muscle geometry (Fig. 7.6). As a result, the surface EMG is typically modeled by the following time-varying convolution of MUAP shapes:

$$x_i(t) = \sum_{j=1}^{N}\sum_{k=1}^{K} \mathbf{h}_{ij,\tau_j(k)} * \delta\big(t - \tau_j(k)\big) + \omega_i(t), \quad i = 1,\ldots,M, \quad t = 1,\ldots,T \quad (7.4)$$

where $*$ stands for convolution, $x_i(t)$ is the ith surface EMG channel, $\omega_i(t)$ is an additive noise, $\delta\,(\cdot)$ is the unit-sample pulse, and the kth MUAP of the jth MU appears at time $\tau_j(k)$. The MUAP observation $\mathbf{h}_{ij,\tau_j(k)} = \big[\, h_{ij,\tau_j(k)}(0) \cdots h_{ij,\tau_j(k)}(L-1)\,\big]$ varies in time, modeling all the MUAP changes in time.

In the above model of time-varying MUAP, there are no a priori limitations on the channel responses $\mathbf{h}_{ij,\tau_j(k)}$, except their finite length and energy. Thus, all the anatomic properties of the detected MU (e.g., its location within the muscle, fibers orientation, action potential propagation velocity), volume conductor properties (e.g., thickness of subcutaneous tissue, tissue inhomogeneities and conductivities), as well as the properties of the detection system (e.g., shape and size of electrodes, interelectrode distance and electrode configuration) can be modeled by $\mathbf{h}_{ij,t}$ [51]. Accordingly, the

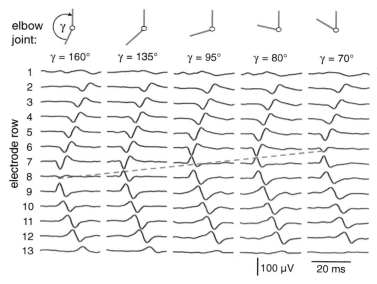

FIGURE 7.6 Changes in MUAPs of an individual MU due to contraction of the biceps brachii muscle as detected by a linear array of 13 surface electrodes with interelectrode distance of 8 mm (LISiN, Politecnico di Torino, Italy). During the contraction, the innervation zone shifted from beneath the 8th to beneath the 6th row of electrodes (dashed line).

convolutive model (7.4) naturally separates the peripheral muscle properties (i.e., MUAP shapes) from the central control strategies (i.e., MU discharge pattern), enabling intuitive interpretation of the decomposition results. In addition, while the MUAP observations vary in space, the MU discharge patterns are modeled as a binary time series occupying the value of 1 when the MU discharge appears and value 0 otherwise. This binary nature of MU discharges can readily be exploited by decomposition approaches, as discussed in Subsection 7.3.2.

Frequently, the surface EMG is assumed to be measured in isometric conditions. This eliminates the variations of detected MUAP shapes due to changes in muscle geometry. By additionally assuming the observed time interval short enough to limit the effect of muscle fatigue, the detected MUAP shapes can be assumed stationary. Under these assumptions, the A/D converted multichannel EMG can be approximated by a linear time-invariant convolutive multiple-input–multiple-output (MIMO) model with the ith model output defined as

$$x_i(t) = \sum_{j=1}^{N}\sum_{l=0}^{L-1} h_{ij}(l)s_j(t-l) + \omega_i(t), \quad i = 1,\ldots,M, \quad t = 1,\ldots,T \tag{7.5}$$

where

$$s_j(t) = \sum_{k=-\infty}^{\infty} \delta\big[t - \tau_j(k)\big], \quad j = 1,\ldots,N \tag{7.6}$$

Although approximate, the assumption of short-term MUAP stationarity proves to be fundamental in practically all decomposition algorithms published so far. Namely, while allowing for gradual changes of MUAP shapes, all known decomposition algorithms search for repeated MUAP occurrences in the signal under examination. Rapid changes of MUAP shapes, such as during explosive muscle contractions, are untraceable.

7.2.1 Redundancy in Multichannel EMG Measurements and the Required Number of EMG Channels

Surface EMG electrodes exhibit relatively low spatial selectivity and detect MUs up to a distance of few centimeters. As a result, nearby electrodes detect the electrical activity of similar populations of MUs. This introduces relatively large amount of redundancy into multichannel surface EMG. However, although redundant, the information provided by larger arrays of surface electrodes proves crucial for discrimination between motor units.

For example, in a simulation study by Farina et al. [31], a single monopolar channel discriminated only 3.4% of the MUs in the simulated population of 200 MUs, whereas an array of 9×9 monopolar channels discriminated 47% of the simulated MUs. Furthermore, the MU identifiability was only moderately related to MU size and location in the muscle. Similar results are demonstrated in Fig. 7.7. In this case, the identifiablility of MUs was studied in the population of 500 MUs, generated by the volume conductor model in [26] (the same simulation setup as in the case of Fig. 7.3). First, the MUs contributing the energy of at least 5% of the maximal detected MUAP energy in at least one EMG channel were classified as detectable. By following the methodology in Farina et al. [31], for each pair of detectable MUs, the mean square difference was computed between their multichannel action potentials and normalized to the mean of the energies of the two action potentials. The MUs with mean square difference below 5% were considered indistinguishable [31]. Finally, only the MUs that were distinguishable from all the other MUs were considered identifiable, whereas all the other detectable MUs were classified as nonidentifiable. In agreement with the results in Farina et al. [31], the results strongly suggest that the larger spatial support of MU sampling increases the number of identifiable MUs, regardless of the decomposition approach. They also suggest that accurate MU identification from only a few surface EMG channels might not be possible.

The results in Fig. 7.7 suggest another advantage of multichannel EMG acquisition. Namely, the acquired multichannel signals can be utilized to build spatial filters that emphasize the contributions of MUs within a selected region of interest and reduce all the contributions of other MUs [15,115] (for details, see Chapter 3). This spatial discrimination of MUs boosts the decomposition performance in terms of accuracy and the number of identified MUs (Fig. 7.7) and has been extensively utilized in the past [31,57,70,94]. For example, in the aforementioned simulation study by Farina et al. [31], a system of 81 Laplacian spatial filters arranged in a 2-D array discriminated 90% of simulated MUs.

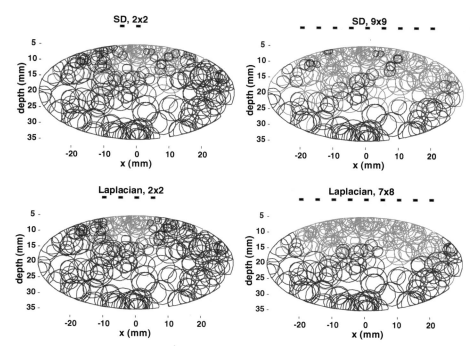

FIGURE 7.7 Representation of identifiable motor units in a simulated population of 500 MUs with fiber length of 120 mm. The territories of all the detectable motor units (i.e., MUs contributing the energy of at least 5% of the maximal detected MUAP energy) are depicted. From this population, the territories of identifiable MUs, that is, the MUs that could be distinguished from all the other MUs based on the criterion of 5% threshold in normalized MUAP energy (see text for details), are depicted in red, whereas the territories of all the other (nonidentifiable) MUs are depicted in black. The results are depicted for longitudinal single differential (upper plots) and Laplacian (lower plots) filtering in the case of 4 (2×2) channels (left panels) and 81 (9×9) and 56 (8×7) channels, respectively (right panels). Black rectangles depict the position of simulated surface electrodes (in transversal direction only), used by aforementioned spatial filters.

In addition to spatial filtering, linear phase temporal high-pass filtering has also been suggested to increase the discriminative power of decomposition algorithms [48,70,78]. The rationale behind this is twofold: (a) high-pass filtering shortens the time support of the detected MUAPs and, consequently, decreases the degree of MUAP superimposition in the surface EMG and (b) it also enhances the residual high-frequency differences between the detected MUAPs that are usually smoothed out by the low-pass filtering effect of the subcutaneous tissue.

7.3 EMG DECOMPOSITION TECHNIQUES

Decomposition of surface EMG aims at reversing the mixing process described in Eq. (7.5) and has been studied for decades. There is large diversity of decomposition techniques that can roughly be categorized either as template matching or latent component analysis (blind source separation) approaches. This section reviews the main methodological principles of both families and provides a brief insight into their performance. A systematic comparison of different approaches is currently still lacking, as well as their standardization. Both issues are crucial for commercial and scientific exploitation of the decomposition algorithms developed so far.

7.3.1   Template Matching Approaches

The template matching approach builds on the methods developed for indwelling EMG decomposition and expands its decomposition algorithms to multichannel surface EMG. The decomposition typically consists of three steps: (a) segmentation of EMG into distinguishable waveforms, (b) identification of MUAP templates, and (c) clustering, that is, matching of the identified MUAP templates to the identified EMG waveforms [42,70].

None of abovementioned steps is trivial in the case of surface EMG, especially at moderate or high muscle contraction levels with several tens of MUAPs detected per second. Typically, the decomposition approaches start with identification of large MUAPs which are then peeled off from the original EMG signal (Fig. 7.8), gradually decreasing the complexity of the measured signal. This procedure has to be applied to each measured channel and is prone to errors in MUAP template estimation. To cope with this limitation, template matching approaches commonly rely on the regularity of the motor unit discharge patterns to predict the time interval in which the next motor unit discharge is most likely to occur [70,78]. In this way, they increase the accuracy and robustness of surface EMG decomposition and, at the same time, decrease the computational complexity. The robustness and efficiency of heuristics used, as well as the level of assumptions made on regularity of MU discharge patterns, vary between published decomposition algorithms [70,78,94]. As a rule of thumb, the stronger the heuristics, the shorter the decomposition time and the stronger the reliance of the decomposition algorithm on the assumed MU discharge properties.

It is noteworthy that not all applications require a full surface EMG decomposition, especially if only peripheral MU properties are of interest, such as MUAP conduction velocity. Many interesting neurophysiologic findings have been based on the results of very fast decomposition algorithms that simply ignore MUAP superimpositions [5,16,42,57,75]. However, more advanced studies on the neural drive [20,22,70,96], MU synchronization [47] and cortico-muscular coupling [95] require a complete and accurate estimation of MU discharge pattern.

7.3.2   Latent Component Analysis

Latent component analysis builds on a mathematical model of the EMG mixing process. Namely, by employing the notation in Eq. (7.5), the EMG mixing process can be written in matrix form:

$$\mathbf{x}(t) = \mathbf{H}\mathbf{\bar{s}}(t) + \mathbf{\omega}(t) \tag{7.7}$$

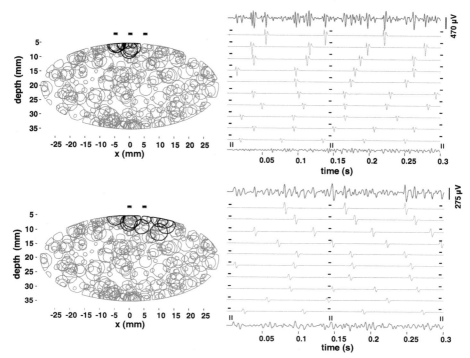

FIGURE 7.8 Schematic representation of the surface EMG decomposition by the template-matching approach in a synthetic EMG channel, filtered by Laplacian (**upper panels**) and single differential spatial filter (**lower panels**). The contraction level of 30% MVC was simulated, resulting in a population of 388 active MUs with the territories depicted in the **left panels**. The synthetic EMG signal (**top trace in the right panels**) comprised the contributions of all 388 active MUs. Large MUAPs were identified (**gray traces in the right panels, contributed by MUs with black territories in the left panels**) and peeled off from the original EMG signal. The **bottom traces in the right panels** depict the residual after subtraction of aforementioned ten MUAP trains from the original EMG signal. In the **left panels**, black rectangles depict the position of simulated surface electrodes (in transversal direction only), used by aforementioned spatial filters.

where $\mathbf{x}(t) = [x_1(t), \ldots, x_M(t)]^T$ is a vector of M surface EMG channels, $\boldsymbol{\omega}(t) = [\omega_1(t), \ldots, \omega_M(t)]^T$ is an additive noise vector, and $\bar{\mathbf{s}}(t) = [s_1(t), s_1(t-1), \ldots, s_1(t-L+1), \ldots, s_N(t), \ldots, s_N(t-L+1)]^T$ stands for vectorized block of L samples from all the MU discharge patterns. The $M \times NL$ mixing matrix $\mathbf{H}$ comprises all the MUAP waveforms that would appear on the electrodes should there be no superimpositions and noise:

$$\mathbf{H} = \begin{bmatrix} \mathbf{H}_1 & \mathbf{H}_2 & \ldots & \mathbf{H}_N \end{bmatrix}$$

The main objective of latent component analysis is to estimate the mixing matrix $\mathbf{H}$ directly from the observations $\mathbf{x}(t)$, without any a priori information on either the mixing process or motor unit discharge patterns. Although contraintuitively, Eq. (7.7) proves to be invertible, mainly due to the sparse nature of MU discharge patterns $s(t)$ and relatively large number of available samples of $x(t)$ [53]. Once the mixing matrix $\mathbf{H}$ is known, the MU discharge patterns can be estimated by applying its (pseudo-)inverse to measured EMG signals:

$$\hat{\mathbf{s}}(t) = \mathbf{H}^{-1}\mathbf{x}(t) \tag{7.8}$$

where $\hat{s}(t)$ stands for estimate of $s(t)$ and $\mathbf{H}^{-1}$ stands for (psudo-)inverse of matrix $\mathbf{H}$.

Estimation (7.8) is completely independent of MU discharge patterns, which makes it an appealing candidate for surface EMG decomposition, especially in pathological conditions like tremor, where the regularity of MU discharge patterns cannot be guaranteed [47]. On the other hand, it builds on overcomplete representation of MU discharge patterns and is, thus, computationally suboptimal, yielding L delayed repetitions of each MU discharge pattern. This property derives from the extension of the source vector $\mathbf{s}(t) = [s_1(t), \ldots, s_N(t)]^T$ in Eq. (7.7) and, thus, from the convolutive nature of the EMG mixing model in (7.4). This computational suboptimality can be compensated by sequential decomposition approaches that extract one MU discharge pattern at a time [10]:

$$\hat{\bar{s}}_j(t) = \mathbf{h}_j^{-1}\mathbf{x}(t) \tag{7.9}$$

where $\mathbf{h}_j^{-1}$ stands for the jth row of $\mathbf{H}^{-1}$. Once $\hat{\bar{s}}_j(t)$ is estimated, the reconstruction of its delayed repetitions may be prevented by employing source deflation approaches presented in Cichocki and Amari [10].

7.3.2.1 *Instantaneous Source Separation.*   As demonstrated by the results in Subsection 7.2, the surface EMG can, to a large extent, be described by an instantaneous multiplicative mixture model, provided that the linear array is used and oriented perpendicularly to the muscle fibers. This cancels out the delays due to MUAP propagation along the muscle fibers, resulting in rather high values of α_j (Fig. 7.4). As a result, the EMG mixing model (7.7) can be rewritten as follows:

$$\mathbf{x}(t) = \mathbf{A}\mathbf{v}(t) + \mathbf{\omega}(t) \tag{7.10}$$

where $\mathbf{v}(t) = [v_1(t), \ldots, v_N(t)]$ and $v_j(t) = \sum_{l=0}^{L-1} h_j(l)s_j(t-l)$ is generic MUAP train of the jth MU [51]. The mixing matrix $\mathbf{A}$ comprises scalar weights projecting the N generic MUAP trains (each train represents one active MU) to M different surface EMG channels:

$$\mathbf{A} = \begin{bmatrix} a_{11} & a_{12} & \cdots & a_{1N} \\ a_{21} & a_{22} & \cdots & a_{2N} \\ \vdots & \vdots & & \vdots \\ a_{M1} & a_{M2} & \cdots & a_{MN} \end{bmatrix}$$

This reduces the dimensionality of model (7.7) by factor L and greatly speeds up the decomposition process. Nevertheless, the results reported by different independent and sparse component analysis of instantaneous mixing model have demonstrated limited success and failed to identify the discharges of individual MUs [40,92,93,111]. The reasons for this failure include relatively large error in MUAP modeling, made by the instantaneous mixing model [51], and, possibly, a relatively low number of surface EMG channels used in these studies.

Recently, an extension of the blind source separation technique based on the second-order blind identification (SOBI) [2] to mixtures of delayed sources has been proposed by Jiang and Farina [54]. This technique reduces the number of unknowns in Eq. (7.5) by transforming the delayed mixture of sources $\bar{s}(t)$ to the mixture of the original sources and their derivatives. As such, it enables the application of instantaneous source separation techniques to EMG signals, regardless of the orientation of the array of surface electrodes. Although not entirely compensating for the convolutive nature of spatial MUAP variability, the proposed technique represents an important step towards simplification of EMG mixing process and EMG decomposition.

Results in Fig. 7.4 suggest another potential application of instantaneous mixing model of the EMG. Namely, when observed with two-dimensional grid of surface electrodes at a given interelectrode distance, MUAP space variability tends to decrease with the distance between the MU territory and the array of uptake electrodes (Fig. 7.3, upper left panel). For far-field potentials the mixing process is relatively close to the instantaneous one, supporting the use of instantaneous ICA approaches for elimination of muscle crosstalk. The latter is caused by poor spatial selectivity of the surface electrodes and represents a vital problem in the field of prosthetics control and muscle activation analysis [25,30,36,109], where the measured electrical activity is often contributed by the muscles not under investigation (see Chapter 2). Results of muscle-crosstalk elimination by SOBI-based separation of linear instantaneous mixtures are presented in [23] and exemplified in Fig. 7.9.

7.3.2.2 Convolutive Source Separation.

Separation of convolutive mixtures is technically much more challenging than separation of instantaneous mixtures, mainly due to increased number of unknowns in (7.7). Moreover, in order to improve the computational conditioning of the EMG mixing process (i.e. condition number of mixing matrix $\mathbf{H}$ in (7.7)), the vector of measurements $\mathbf{x}(t)$ is frequently extended to comprise a block of K samples from each surface EMG channel:

$$\bar{\mathbf{x}}(t) = [x_1(t), x_1(t-1), \ldots, x_1(t-K+1), \ldots, x_M(t), \ldots, x_M(t-K+1)]^T \quad (7.11)$$

with K typically between 10 and 20 [48,50]. This increases the mixing matrix $\mathbf{H}$ up to the size order of 1000×1000, rendering most of instantaneous source separation techniques inefficient. To cope with this problem, Holobar and Zazula [49,50] proposed the so called CKC technique that bypasses the need for direct estimation of the matrix $\mathbf{H}$. The method starts by blindly estimating the cross-correlation vector $\mathbf{c}_{s_j \bar{\mathbf{x}}} = E\left(s_j(t)\bar{\mathbf{x}}(t)\right)$ between the jth MU discharge pattern and the extended vector of

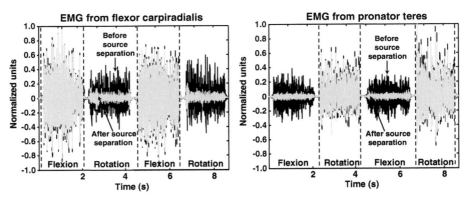

FIGURE 7.9 Example of signals recorded from the sensors located over the flexor carpi radialis and the pronator teres muscles during cyclic wrist flexion/rotation (black lines). A cyclic task, consisting of 2-s flexion and 2-s rotation, both at 50% MVC, is shown. The entire recording lasts 100 s, of which approximately 8 s is shown for clarity. The results shown correspond to the application of the blind-source separation approach to the entire recordings of 100 s. A threshold has been applied to the force signal related to flexion and rotation and the intervals of time in which these forces exceed 15% MVC are shown (limited by dashed lines). The intervals defined by the force signals are those during which the flexor carpi radialis and the pronator teres muscles are active. From the original signals, it is clear that large contributions from the activation of the pronator teres (in the first case) and the flexor carpi radialis (in the second case) during rotation and flexion are present in the two recordings. After source separation, the relative amplitude of the second source in the reconstructed signals (light gray lines) significantly decreases in both cases. Reprinted with permission from Fig. 8 of Farina et al. [23].

measurements $\overline{\mathbf{x}}(t)$, as defined in (7.11), where $E(\cdot)$ stands for mathematical expectation. This can be accomplished either by probabilistic approach presented in Holobar and Zazula [50] or by gradient-based update rule described in Holobar et al. [47]. In the second step, the unknown mixing matrix $\mathbf{H}$ is compensated by calculating the linear minimum mean square error (LMMSE) estimation of the jth MU discharge pattern $s_j(t)$ [50]:

$$\hat{s}_j(t) = \mathbf{c}_{s_j \overline{\mathbf{x}}^T} \mathbf{C}_{\overline{\mathbf{xx}}}^{-1} \overline{\mathbf{x}}(t) \tag{7.12}$$

where $\mathbf{C}_{\overline{\mathbf{xx}}} = E\left(\overline{\mathbf{x}}(t)\overline{\mathbf{x}}^T(t)\right)$ is the correlation matrix of the extended vector of measurements.

In the noiseless case, the estimator (7.12) simplifies to

$$\hat{s}_j(t) = \mathbf{c}_{s_j \overline{\mathbf{s}}^T} \mathbf{H}^T \left(\mathbf{H}\mathbf{C}_{\overline{\mathbf{ss}}}\mathbf{H}^T\right)^{-1} \mathbf{H}\overline{\mathbf{s}}(t) = \mathbf{c}_{s_j \overline{\mathbf{s}}^T} \mathbf{C}_{\overline{\mathbf{ss}}}^{-1} \overline{\mathbf{s}}(t) \tag{7.13}$$

where $\mathbf{C}_{\overline{\mathbf{ss}}} = E\left(\overline{\mathbf{s}}(t)\overline{\mathbf{s}}^T(t)\right)$ and $\mathbf{c}_{s_j \overline{\mathbf{s}}} = E\left(s_j(t)\overline{\mathbf{s}}(t)\right)$ is the vector of cross-correlation coefficients between the jth MU discharge pattern and discharge patterns of all the MUs active in the detection volume.

Equation (7.13) offers a valuable insight into the properties of the CKC method. First, it shows how the CKC method implicitly combines all the second-order information in the measurements $\mathbf{x}(t)$ to cancel out the mixing matrix $\mathbf{H}$. By ignoring the MUAP shapes in $\mathbf{H}$, the CKC method is insensitive either to the anatomy of the muscle or configuration of the uptake electrodes array. Second, the source estimation in (7.12) and (7.13) is also insensitive to MU synchronization, as long as $\mathbf{C_{\overline{ss}}}$ is invertible [47]. This is a significant advantage over other known latent variable analysis approaches, as MU synchronization is rather common, not only in pathologies such as tremor (Fig. 7.10) but also in fatigued healthy muscles.

The robustness of the CKC decomposition approach to MU discharge synchronization is further discussed in [47] and illustrated in Fig. 7.11. The presented results were averaged over 100 decomposition runs of synthetic EMG signals, with $N = 5$ simulated MUs. In each run, random MU discharges (i.e., delta functions in Eq. (7.6)) were generated with the mean interdischarge interval set equal to 0.1 s and the probability of simultaneous MU discharges (with tolerance of ±0.5 sample) ranging from $p = 0.1$ to $p = 0.9$, as in Holobar et al. [47]. With p set equal to 0.9, 90% of all the

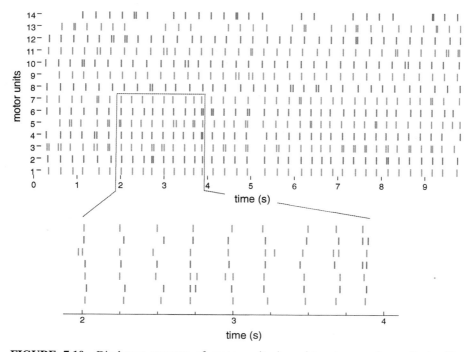

FIGURE 7.10 Discharge patterns of motor units in wrist extensor of a patient with Parkinson's disease during both arms resting on the lap. Each vertical bar represents a discharge of individual motor units as identified by CKC decomposition of 64-channel surface EMG. The zoomed portion of MU discharge patterns (**lower panel**) illustrates high level of MU synchronization.

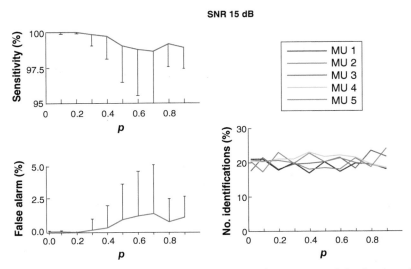

FIGURE 7.11 Average sensitivity, false alarm rate, and percentage of the decomposition runs that converged to specific simulated MU as a function of probability p of MU discharge overlapping (i.e., synchronous MU discharge with the tolerance set to (0.5 samples—see text for details). Vertical bars denote standard deviations. The number of decomposition runs that converged to specific MU was normalized by the number of all decomposition runs. Due to clarity reasons, standard deviations are not reported in this case (**right panel**).

discharges of each simulated MU overlapped with 90% of discharges from other MUs. In each simulation run, multichannel MUAPs of randomly selected identifiable MUs from Fig. 7.7 were used as channel responses $h_{ij}(t)$ and convolved with the simulated MU discharges to produce the observed measurements $\mathbf{x}(t)$. The number of observations M was set equal to 25, simulating the 6×5 array of uptake electrodes (longitudinal single-differential configuration), centered above the simulated muscle (Fig. 7.7). Zero-mean additive noise with 15 dB SNR was added to the measurements. Both sensitivity and false alarm rate in the identification of MU discharges were only slightly correlated with the MU discharge overlapping probability p, whereas the number of identified MUs was independent of the MU synchronization probability p.

In addition, the binary nature of the sources $\mathbf{s}(t)$ in the convolutive model (7.5) guarantees relatively large robustness of CKC to noise and estimation errors. Namely, to suppress the impact of noise, we simply need to threshold source estimations $\hat{s}_i(t)$ in (7.12). The optimal threshold value can be estimated automatically by utilizing threshold segmentation techniques known from the image processing field [117]. Although not necessary, heuristics, similar to the ones used by template matching, can also be applied to speed up the decomposition process and orient the CKC estimation towards specific MU discharge patterns.

Representative results of the CKC-based decomposition of surface EMG of healthy tibialis anterior muscle during its 15-s long isometric contraction at 70%

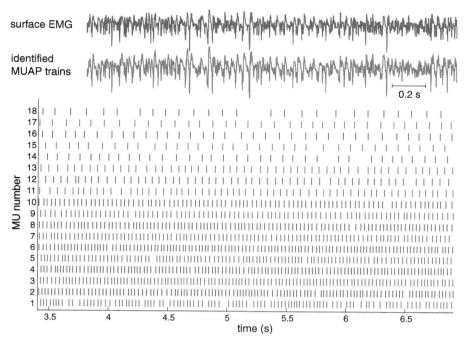

FIGURE 7.12 Surface EMG of tibialis anterior muscle during a 15-s-long isometric contraction at 70% MVC and the sum of MUAP trains identified from surface EMG by CKC decomposition technique (**upper panel**). For clarity reasons, only one out of the 90 (10 rows × 9 columns) acquired EMG channels is depicted. **Lower panel**: Discharge patterns of 18 MUs identified by CKC decomposition technique from surface EMG depicted in the upper panel. Each vertical bar indicates a MU discharge at a given time instant.

MVC are presented in Fig. 7.12. In this case, the surface EMG was recorded by array of 9 × 10 surface electrodes, centered over the muscle belly. For clarity reasons, only one out of 90 acquired surface EMG channels is depicted, along with the corresponding sum of the identified MUAPs.

7.4 VALIDATION OF DECOMPOSITION

Decomposition of surface EMG is a powerful tool enabling noninvasive insight not only into muscle control strategies, but also into peripheral muscle properties. It provides unambiguous information on physiological parameters of individual motor units that can easily be interpreted. This includes MU recruitment and derecruitment thresholds, MU discharge rate, coefficient of variability for MU interdischarge interval, and individual MU conduction velocity [9,28,100,103]. Nevertheless, there are at least two nontrivial questions that have to be answered before the results of EMG decomposition can be put into practical use. The first one is about the accuracy of MU discharge identification. The second one is about representativeness of identified MUs.

7.4.1 Accuracy of Motor Unit Identification

Although under investigation for more than a decade, the methodology of accuracy assessment is still under intense discussion in the field of EMG decomposition. The proposed accuracy measures include indirect measures, such as MUAP shapes, their propagation across the surface EMG channels, and coefficient of variability for MU interdischarge interval [16,42,57,75,99]. These measures support the exclusion of results that fall outside the physiologically established limits, but offer only a coarse insight into the decomposition accuracy, especially in the case of muscle pathologies [47]. In order to surpass this limitation, direct accuracy measures have also been proposed, relaying on advanced surface EMG simulators [26,34] or simultaneous acquisition and decomposition of indwelling and surface EMG [46,48,76,94] and signal-based measures of accuracy [52].

Over the last decade, the progress in the field of surface EMG simulation has been remarkable and the state-of-the-art simulators support realistic simulation of complex muscle properties [6,26,34,68,69,108,109] including variations in tissue conductivity [85], effects of fatigue [66], anatomical variations [82–86], electrode shape and size [14], and crosstalk [67]. Advanced simulators of muscle control strategies have also been proposed for both healthy [37,104] and pathological conditions [13], supporting simulations of both central and afferent components of the neural drive. These simulators provide crucial test-bed environments for systematic testing of decomposition efficiency and accuracy against the muscle anatomy, effect of subcutaneous tissue, number of active MUs and their location within the muscle, MU firing properties, size and shape of surface electrodes, configuration of electrode arrays and signal-to-noise ratio. Nevertheless, while providing a very detailed insight into the behavior of decomposition techniques, the results of simulations cannot trivially be extrapolated to specific experimental conditions.

This limitation can partially be overcome by simultaneously recording of surface and indwelling EMG. Although impractical, invasive, and with double computation costs, this methodology offers a very strong indication of the decomposition accuracy. Namely, it is highly unlikely that two independently developed decomposition techniques applied to different EMG observation modalities (indwelling versus surface EMG) agree on decomposition errors. Therefore, when applied to discharge patterns with at least several tens of MU discharges, the following rate of agreement (RoA) between the decompositions of indwelling and surface EMG can be considered a solid measure of decomposition accuracy for individual MUs detected by both electrode systems (indwelling and surface):

$$\mathrm{RoA}_j = \frac{A_j}{A_j + I_j + S_j} \qquad (7.14)$$

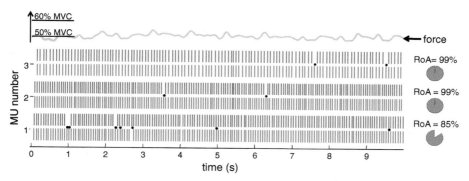

FIGURE 7.13   Discharge patterns for three MUs identified by both CKC from surface EMG (blue lines) and EMGLAB from simultaneously recorded intramuscular EMG (red lines) of the tibialis anterior muscle during an isometric constant force contraction at 50% MVC. Each vertical line indicates a MU discharge at a given time instant. Disagreements on identified motor unit discharges are denoted with black dots. RoA: rate of agreement as defined in Eq. (7.14).

where A_j (agreement), I_j (indwelling), and S_j (surface) are the numbers of discharges of the jth MU that are identified from both surface EMG and indwelling EMG, from indwelling EMG only and from surface EMG only, respectively.

Results of RoA-based analysis of CKC decomposition are exemplified in Fig. 7.13. In this study, the 15-s-long surface EMG and intramuscular EMG signals were acquired simultaneously from the dominant tibialis anterior muscle of a healthy young subject, during a 50% MVC contraction. Surface EMG was acquired in monopolar mode by a bidimensional array of 90 electrodes (LISiN, Politecnico di Torino, Italy, 1 mm in diameter, interelectrode distance of 5 mm, 10 rows $\times$ 9 columns) with the third row centered over the main innervation zone and the columns approximately aligned with the muscle fibers. Indwelling EMG signals were recorded in bipolar mode by three pairs of wire electrodes inserted into the muscle with a 25 G needle. The acquired signals were amplified, band-pass filtered (EMG-USB2 multi-channel amplifier, 3-dB bandwidth 10–500 Hz for surface EMG, 0.1–5 kHz for intramuscular EMG) and decomposed by CKC [50] and EMGLAB decomposition tool [78], respectively.

Although considered as the most reliable way of assessing accuracy of decomposition of the surface EMG, the RoA-based analysis suffers from at least two drawbacks. First, the number of MUs that are detected by both surface and indwelling EMG is usually relatively small [48,76], limiting the strict accuracy assessment to a few MUs per contraction only. Second, the rate of agreement between indwelling and surface EMG decomposition is a conservative underestimate of decomposition accuracy since independent errors may occur in either of the two decomposition techniques [48].

The need for immediate and onsite assessment of decomposition accuracy is substantial. Due to the specific properties of the identified MUs and dynamic interactions among MUAPs, signal artifacts, and noise, this accuracy assessment must be applied not only to each individual MUs, but ideally to the identification of each MU discharge. Such an extensive accuracy assessment has been proposed in the field of indwelling EMG [79]. It utilizes statistical decision theory and a Bayesian framework to integrate the information on MUAP shape and the MU discharge patterns and computes an objective a posteriori measure of confidence in the accuracy of each identified MU discharge. However, this approach is computationally expensive and currently not appropriate for more than eight identified MUs. Thus, it is less suitable for the assessment of surface EMG decomposition.

Another attempt to establish wide-ranging accuracy measure has been proposed by Nawab et al. [94]. The proposed reconstruct-and-test procedure first decomposes the EMG signals into constituent MUAP trains. It then synthesizes the surface EMG signals by summing up the identified MUAP trains and adds additive Gaussian noise to reconstructed EMG signals. Finally, the reconstructed signals are re-decomposed and the results of both decomposition runs mutually compared [94]. This approach, however, only validates the consistency of the decomposition and not the accuracy [27,73].

Another signal-based metric for assessment of accuracy of MU identification from high-density surface EMG, referred to as pulse-to-noise ratio (PNR), was introduced in [52]. This metric is computationally efficient, does not require simultaneous acquisition of indwelling EMG, and can be applied to every MU that is identified by the CKC technique [49,50]. In the experimental and simulated signals, covering the muscle contraction forces from 5% to 70% MVC, the PNR metric correlated significantly with both sensitivity and false alarm rate in identification of MU discharges. All the MUs with PNR > 30 dB exhibited sensitivity > 90% and false alarm rates < 2% [52]. Therefore, a threshold of PNR = 30 dB can be used for selecting MUs that are reliably estimated by CKC decomposition technique. Unfortunately, the PNR metric inherently resides on the results of CKC technique and cannot easily be extended to other decomposition methods.

7.4.2   Representativeness of Identified Motor Units

Surface EMG decomposition identifies from $\sim$5 to $\sim$60 MUs per contraction [16,42,46–48,50,57,70,75,94,114]. This is a relatively small number, especially for high contraction levels with several hundreds of active MUs.

Representativeness of identified MUs has been illuminated for certain applications only. For example, Negro et al. [96] showed that in healthy skeletal muscles, fluctuations in isometric muscle force can reliably be described by a linear projection of smoothed discharge rates of only four MUs per contraction. Similarly, in Gallego et al. [38] the analysis across a wide range of contractions revealed that the use of cumulative spike trains of as little as five randomly selected MUs provides better characterization of central oscillations in pathological tremor than global surface EMG estimates. The simulation study by Negro and Farina [95] also indicated that the corticospinal input can effectively be sampled by a small population of motoneurons.

There are many other applications, where the required number of MUs has not been fully quantified. For example, due to methodological limitations of decomposition and the need for relatively long and accurately identified MU discharge patterns, the studies of MU synchronization have been typically limited to relatively small populations of concurrently active MUs [65,74,98]. Similar conclusions apply to (a) studies on the relationship between the EMG and exerted muscle force [29,71,72], (b) studies on adaptations of central strategies to fatigue, intensive exercise, or pain [1,3,19,20,39,100,112], and (c) diagnosis and pathophysiological studies on neuromuscular disorders such as atrophy, myopathy, and neuropathy [7,8,17,91,101].

It is noteworthy that the problem of representativeness is not limited to decomposition, but extends to surface EMG in general, including global metrics such as EMG amplitude or median spectral frequency. Surface EMG detects superficial MUs, down to the depth of a few centimeters in the muscle tissue. Following the Henneman size principle [44], large high-threshold MUs exhibit larger MUAPs than small low-threshold MUs. In addition, there is physiological evidence of nonhomogeneous spatial distribution of MUs in certain muscles [18,43,58,64,105], suggesting that large MUs are located more superficially. Therefore, at higher contraction levels, surface EMG and its decomposition might be biased towards high-threshold MUs. On the other hand, although based on a few preliminary results and not yet systematically investigated, the surface EMG decomposition has been demonstrated to sample relatively uniformly across MUs recruitment threshold [46,70,94].

The energy accounted for by the identified MUs can be quantified by calculating the energy ratio between the original EMG signal and the sum of identified MUAP trains [48]:

$$\text{SIR}(i) = \left(1 - \frac{E\left[\left(x_i(t) - \sum_j z_{ij}(t)\right)^2\right]}{E\left[x_i^2(t)\right]}\right) \cdot 100\% \tag{7.15}$$

where $x_i(t)$ denotes the ith surface EMG channel and $z_{ij}(t)$ stands for the action potential train of the jth MU, reconstructed from the ith surface EMG signal.

Typically, up to 60% of energy is accounted for by identified MUAP trains [48,52,76]. This is less than with indwelling EMG, where ratios up to 90% or more are not uncommon. However, due to much lower selectivity of surface electrodes, the background physiological noise, consisting of contributions of deep and/or small MUs, is much more evident in surface EMG than in indwelling EMG.

The number of identified MUs can always be increased by subtracting already identified MUs from the original signals and reapplying the decomposition procedure to the subtraction residual. However, such an increase comes at the cost of accuracy in MU identification, as small MUAPs exhibit significantly smaller signal-to-noise ratio than large MUAPs and are thus more receptive to decomposition errors. Another possibility is to combine multichannel indwelling and the surface EMG acquisition technique to sample simultaneously from deep and superficial portions of the muscles [33]. However, such investigations are currently limited to controlled experimental conditions only and are thus suitable mostly for neurophysiologic investigations.

In conclusion, the progress made in the field of surface EMG decomposition has been remarkable. In the last decade, MU identification has been successfully applied to isometric surface EMG signals acquired from skin-parallel-fibered, as well as pennate muscles of highly diverse anatomies [46–50,70,76,94], including facial muscles [103] and anal sphincters (see Chapter 14 and [9]). EMG decomposition algorithms have been validated in healthy controls and in patients with neurodegenerative diseases [47], stroke patients and type II diabetes patients [113] and cleft lip patients [103], to name just a few representative studies. In the near future, the development of surface EMG decomposition algorithms will likely focus on online MU identification, enabling immediate feedback on the quality of acquired surface EMG signals in clinical practice and development of advanced myoelectric control systems. At this time, the decomposition of both surface and indwelling EMG has been limited to isometric muscle contractions only, offering virtually no insight into the dynamics of motor control in humans. Thus, the identification of MU discharge patterns from surface EMG signals, acquired during dynamic muscle contractions, needs to be addressed.

REFERENCES

1. Adam, A., and C. J. De Luca, "Recruitment order of motor units in human vastus lateralis muscle is maintained during fatiguing contractions," *J. Neurophysiol.* **90**, 2919–2927 (2003).

2. Belouchrani, A., K. Abed-Meraim, J. F. Cardoso, and E. Moulines, "A blind source separation technique using second-order statistics," *IEEE Trans. Signal Proces* **45**, 434–444 (1997).

3. Bigland-Ritchie, B., and J. J. Woods, "Changes in muscle contractile properties and neural control during human muscular fatigue," *Muscle & Nerve* **7**, 691–699 (1984).

4. Bischoff, C., E. Stålberg, B. Falck, and K. E. Eeg-Olofsson, "Reference values of motor unit action potentials obtained with multi-MUAP analysis," *Muscle & Nerve* **17**, 842–851 (1994).

5. Blok, J. H., J. P. Van Dijk, M. J. Zwarts, and D. F. Stegeman, "Motor unit action potential topography and its use in motor unit number estimation," *Muscle &Nerve* **32**, 280–291 (2005).

6. Blok, J. H., D. F. Stegeman, and A. van Oosterom, "Three-layer volume conductor model and software package for applications in surface electromyography," *Ann. Biomed. Eng.* **30**, 566–577 (2002).

7. Calder, K. M., D. W. Stashuk, and L. McLean, "Motor unit potential morphology differences in individuals with non-specific arm pain and lateral epicondylitis," *J. Neuroeng. Rehabil.* **5**, 34 (2008).

8. De Carvalho, M., A. Turkman, and M. Swash, "Motor unit firing in amyotrophic lateral sclerosis and other upper and lower motor neurone disorders," *Clin. Neurophysiol.* **123**, 2312–2318 (2012).

9. Cescon, C., L. Mesin, M. Nowakowski, and R. Merletti, "Geometry assessment of anal sphincter muscle based on monopolar multichannel surface EMG signals," *J. Electromyogr. Kinesiol.* **21**, 394–401 (2011).

10. Cichocki, A., and S. Amari, *Adaptive Blind Signal and Image Processing*, John Wiley & Sons, New York, (2002).

11. Clancy, E. A., and N. Hogan, "Probability density of the surface electromyogram and its relation to amplitude detectors," *IEEE Trans. Biomed. Eng.* **46**, 730–739 (1999).

12. Day, S. J., and M. Hulliger, "Experimental simulation of cat electromyogram: evidence for algebraic summation of motor-unit action-potential trains," *J. Neurophysiol.* **86**, 2144–2158 (2001).

13. Dideriksen, J., R. M. Enoka, and D. Farina, "A model of the surface electromyogram in pathological tremor," *IEEE Trans. Biomed. Eng.* **58**, 2178–2185 (2011).

14. van Dijk, J. P., M. M. Lowery, B. G. Lapatki, and D. F. Stegeman, "Evidence of potential averaging over the finite surface of a bioelectric surface electrode," *Ann. Biomed. Eng.* **37**, 1141–1151 (2009).

15. Disselhorst-Klug, C., J. Silny, and G. Rau, "Improvement of spatial resolution in surface-EMG: a theoretical and experimental comparison of different spatial filters," *IEEE Trans. Biomed. Eng.* **44**, 567–574 (1997).

16. Doherty, T. J., and D. W. Stashuk, "Decomposition-based quantitative electromyography: methods and initial normative data in five muscles," *Muscle & Nerve* **28**, 204–211 (2003).

17. Dorfman, L. J., J. E. Howard, and K. C. McGill, "Motor unit firing rates and firing rate variability in the detection of neuromuscular disorders," *Electroencephalogr. Clin. Neurophysiol.* **73**, 215–224 (1989).

18. Elder, G. C., K. Bradbury, and R. Roberts, "Variability of fiber type distributions within human muscles," *J. Appl. Physiol.* **53**, 1473–1480 (1982).

19. Enoka, R. M., "Muscle fatigue—from motor units to clinical symptoms," *J. Biomech.* **45**, 427–433 (2012).

20. Farina, D., L. Arendt-Nielsen, and T. Graven-Nielsen, "Experimental muscle pain reduces initial motor unit discharge rates during sustained submaximal contractions," *J. Appl. Physiol.* **98**, 999–1005 (2005).

21. Farina, D., L. Arendt-Nielsen, R. Merletti, and T. Graven-Nielsen, "Assessment of single motor unit conduction velocity during sustained contractions of the tibialis anterior muscle with advanced spike triggered averaging," *J. Neurosci. Meth.* **115**, 1–12 (2002).

22. Farina, D., L. Arendt-Nielsen, R. Merletti, and T. Graven-Nielsen, "Effect of experimental muscle pain on motor unit firing rate and conduction velocity," *J. Neurophysiol.* **91**, 250–1259 (2004).

23. Farina, D., C. Févotte, C. Doncarli, and R. Merletti, "Blind separation of linear instantaneous mixtures of nonstationary surface myoelectric signals," *IEEE Trans. Biomed. Eng.* **51**, 55–1567 (2004).

24. Farina, D., M. Gazzoni, and F. Camelia, "Low-threshold motor unit membrane properties vary with contraction intensity during sustained activation with surface EMG visual feedback," *J. Appl. Physiol.* **96**, 1505–1515 (2004).

25. Farina, D., R. Merletti, B. Indino, and T. Graven-Nielsen, "Surface EMG crosstalk evaluated from experimental recordings and simulated signals. Reflections on crosstalk interpretation, quantification and reduction," *Method Inform. Med.* **43**, 30–35 (2004).

26. Farina, D., L. Mesin, S. Martina, and R. Merletti, "A surface EMG generation model with multilayer cylindrical description of the volume conductor," *IEEE Trans. Biomed. Eng.* **51**, 415–426 (2004).

27. Farina, D., and R. M. Enoka, "Surface EMG decomposition requires an appropriate validation," *J. Neurophysiol.* **105**, 981–982 (2011).

28. Farina, D., A. Holobar, M. Gazzoni, D. Zazula, R. Merletti, and R. M. Enoka, "Adjustments differ among low-threshold motor units during intermittent, isometric contractions," *J. Neurophysiol.* **101**, 350–359 (2009).

29. Farina, D., A. Holobar, R. Merletti, and R. M. Enoka, "Decoding the neural drive to muscles from the surface electromyogram," *Clin. Neurophysiol.* **121**, 1616–1623 (2010).

30. Farina, D., M. F. Lucas, and C. Doncarli, "Optimized wavelets for blind separation of nonstationary surface myoelectric signals," *IEEE Trans. Biomed. Eng.* **55**, 78–86 (2008).

31. Farina, D., F. Negro, M. Gazzoni, and R. M. Enoka, "Detecting the unique representation of motor-unit action potentials in the surface electromyogram," *J. Neurophysiol.* **100**, 1223–1233 (2008).

32. Farina, D., M. Pozzo, M. Lanzetta, and R. M. Enoka, "Discharge variability of motor units in an intrinsic muscle of transplanted hand," *J. Neurophysiol.* **99**, 2232–2240 (2008).

33. Farina, D., K. Yoshida, T. Stieglitz, and K. P. Koch, "Multichannel thin-film electrode for intramuscular electromyographic recordings," *J. Appl. Physiol.* **104**, 821–827 (2008).

34. Farina, D., and R. Merletti, "A novel approach for precise simulation of the EMG signal detected by surface electrodes," *IEEE Trans. Biomed. Eng.* **48**, 637–646 (2001).

35. Farina, D., D. Zennaro, M. Pozzo, R. Merletti, and T. Läubli, "Single motor unit and spectral surface EMG analysis during low-force, sustained contractions of the upper trapezius muscle," *Eur. J. Appl. Physiol.* **96**, 157–164 (2006).

36. Frahm, K. S., M. B. Jensen, D. Farina, and O. Andersen, "Surface EMG crosstalk during phasic involuntary muscle activation in the nociceptive withdrawal reflex," *Muscle & Nerve* **46**, 228–236 (2012).

37. Fuglevand, A. J., D. A. Winter, and A. E. Patla, "Models of recruitment and rate coding organization in motor-unit pools," *J. Neurophysiol.* **70**, 2470–2488 (1993).

38. Gallego J. A., J. L. Dideriksen, A. Holobar, J. Ibáñez, J. L. Pons, E. D. Louis, E. Rocon, D. Farina, "Influence of common synaptic input to motor neurons on the neural drive to muscle in essential tremor," *J Neurophysiol.* **113**(1), 182–191 (2015).

39. Gandevia, S. C., "Spinal and supraspinal factors in human muscle fatigue," *Physiol. Rev.* **81**, 1725–1789 (2001).

40. García, G. A., R. Okuno, and K. Akazawa, "A decomposition algorithm for surface electrode-array electromyogram. A noninvasive, three-step approach to analyze surface EMG signals," *IEEE Eng. Med. Biol. Mag.* **24**, 63–72 (2005).

41. Garland, S. J., R. M. Enoka, L. P. Serrano, and G. A. Robinson, "Behavior of motor units in human biceps brachii during a submaximal fatiguing contraction," *J. Appl. Physiol.* **76**, 2411–2419 (1994).

42. Gazzoni, M., D. Farina, and R. Merletti, "A new method for the extraction and classification of single motor unit action potentials from surface EMG signals," *J. Neurosci. Methods* **136**, 165–177 (2004).

43. Grotmol, S., G. K. Totland, H. Kryvi, A. Breistøl, B. Essén-Gustavsson, and A. Lindholm, "Spatial distribution of fiber types within skeletal muscle fascicles from Standardbred horses," *Anat. Rec.* **268**, 131–136 (2002).

44. Henneman, E., "Relation between size of neurons and their susceptibility to discharge," *Science* **126**, 1345–1347 (1957).

45. Hermens, H. J., B. Freriks, R. Merletti, D. Stegeman, J. Blok, G. Rau, C. Disselhorst-Klug, and G. Hägg, *"European Recommendations for Surface Electromyography: Results of the Seniam Project,"* Roessingh Research and Development, Enschede, Netherlands (1999).

46. Holobar, A., D. Farina, M. Gazzoni, R. Merletti, and D. Zazula, "Estimating motor unit discharge patterns from high-density surface electromyogram," *Clin. Neurophysiol.* **120**, 551–562 (2009).

47. Holobar, A., V. Glaser, J. A. Gallego, J. L. Dideriksen, and D. Farina, "Non-invasive characterization of motor unit behaviour in pathological tremor," *J. Neural Eng.* **9**, 056011 (2012).

48. Holobar, A., M. A. Minetto, A. Botter, F. Negro, and D. Farina, "Experimental analysis of accuracy in the identification of motor unit spike trains from high-density surface EMG," *IEEE Trans. Neural. Syst. Rehab. Eng.* **18**, 221–229 (2010).

49. Holobar, A., and D. Zazula, "Correlation-based decomposition of surface electromyograms at low contraction forces," *Med. Biol. Eng. Comput.* **42**, 487–495 (2004).

50. Holobar, A., and D. Zazula, "Multichannel blind source separation using convoluion kernel compensation," *IEEE Trans. Signal Proces* **55**, 4487–4496 (2007).

51. Holobar, A., and D. Farina, "Blind source identification from the multichannel surface electromyograms," *Physiol. Meas.* **35**, R143–R165 (2014).

52. Holobar, A., M. A. Minetto, and D. Farina, "Accurate identification of motor unit discharge patterns from high-density surface EMG and validation with a novel signal-based performance metric," *J. Neural Eng.* **11**, 016008 (2014).

53. Hyvärinen, A., J. Karhunen, and E. Oja, *Independent Component Analysis*, John Wiley & Sons, Hoboken, NJ, (2001).

54. Jiang, N., and D. Farina, "Covariance and time-scale methods for blind separation of delayed sources," *IEEE Trans. Biomed. Eng.* **58**, 550–556 (2011).

55. Keenan, K. G., D. Farina, K. S. Maluf, R. Merletti, and R. M. Enoka, "Influence of amplitude cancellation on the simulated surface electromyogram," *J. Appl. Physiol.* **98**, 120–131 (2005).

56. Keenan, K. G., D. Farina, R. Merletti, and R. M. Enoka, "Amplitude cancellation reduces the size of motor unit potentials averaged from the surface EMG," *J. Appl. Physiol.* **100**, 1928–1937 (2006).

57. Kleine, B. U., J. P. van Dijk, B. G. Lapatki, M. J. Zwarts, and D. F. Stegeman, "Using two-dimensional spatial information in decomposition of surface EMG signals," *J. Electromyogr. Kinesiol.* **17**, 535–548 (2007).

58. Knight, C. A., and G. Kamen, "Superficial motor units are larger than deeper motor units in human vastus lateralis muscle," *Muscle & Nerve* **31**, 475–480 (2005).

59. Lanzetta, M., M. Pozzo, A. Bottin, R. Merletti, and D. Farina, "Reinnervation of motor units in intrinsic muscles of a transplanted hand," *Neurosci. Lett.* **373**, 138–143 (2005).

60. Lateva, Z. C., and K. C. McGill, "Estimating motor-unit architectural properties by analyzing motor-unit action potential morphology," *Clin. Neurophysiol.* **112**, 127–135 (2001).

61. Lateva, Z. C., K. C. McGill, and M. E. Johanson, "Electrophysiological evidence of adult human skeletal muscle fibres with multiple endplates and polyneuronal innervation," *J. Physiol. (Lond.)* **544**, 549–565 (2002).

62. LeFever, R. S., and C. J. De Luca, "A procedure for decomposing the myoelectric signal into its constituent action potentials—part I: technique, theory, and implementation," *IEEE Trans. Biomed. Eng.* **29**, 149–157 (1982).

63. LeFever, R. S., A. P. Xenakis, and C. J. De Luca, "A procedure for decomposing the myoelectric signal into its constituent action potentials—part II: execution and test for accuracy," *IEEE Trans. Biomed. Eng.* **29**, 158–164 (1982).

64. Lexell, J., K. Henriksson-Larsén, and M. Sjöström, "Distribution of different fibre types in human skeletal muscles. 2. A study of cross-sections of whole m. vastus lateralis," *Acta Physiol. Scand.* **117**, Issue 1, 115–122 (1983).

65. Logigian, E. L., M. M. Wierzbicka, F. Bruyninckx, A. W. Wiegner, B. T. Shahahi, and R. R. Young, "Motor unit synchronization in physiologic, enhanced physiologic, and voluntary tremor in man," *Ann. Neurol.* **23**, 242–250 (1988).

66. Lowery, M. M., and M. J. O'Malley, "Analysis and simulation of changes in EMG amplitude during high-level fatiguing contractions," *IEEE Trans. Biomed. Eng.* **50**, 1052–1062 (2003).

67. Lowery, M. M., N. S. Stoykov, and T. A. Kuiken, "A simulation study to examine the use of cross-correlation as an estimate of surface EMG cross talk," *J. Appl. Physiol.* **94**, 1324–1334 (2003).

68. Lowery, M. M., N. S. Stoykov, J. P. A. Dewald, and T. A. Kuiken, "Volume conduction in an anatomically based surface EMG model," *IEEE Trans. Biomed. Eng.* **51**, 2138–2147 (2004).

69. Lowery, M. M., N. S. Stoykov, A. Taflove, and T. A. Kuiken, "A multiple-layer finite-element model of the surface EMG signal," *IEEE Trans. Biomed. Eng.* **49**, 446–454 (2002).

70. De Luca, C. J., A. Adam, R. Wotiz, L. D. Gilmore, and S. H. Nawab, "Decomposition of surface EMG signals," *J. Neurophysiol.* **96**, 1646–1657 (2006).

71. De Luca, C. J., and E. C. Hostage, "Relationship between firing rate and recruitment threshold of motoneurons in voluntary isometric contractions," *J. Neurophysiol.* **104**, 1034–1046 (2010).

72. De Luca, C. J., R. S. LeFever, M. P. McCue, and A. P. Xenakis, "Control scheme governing concurrently active human motor units during voluntary contractions," *J. Physiol. (London)* **329**, 129–142 (1982).

73. De Luca, C. J., and S. H. Nawab, "Reply to Farina and Enoka: the reconstruct-and-test approach is the most appropriate validation for surface EMG signal decomposition to date," *J. Neurophysiol.* **105**, 983–984 (2011).

74. De Luca, C. J., A. M. Roy, and Z. Erim, "Synchronization of motor-unit firings in several human muscles," *J. Neurophysiol.* **70**, 2010–2023 (1993).

75. Maathuis, E. M., J. Drenthen, J. P. van Dijk, G. H. Visser, and J. H. Blok, "Motor unit tracking with high-density surface EMG," *J Electromyogr Kinesiol* **18**, 920–930 (2008).

76. Marateb, H. R., K. C. McGill, A. Holobar, Z. C. Lateva, M. Mansourian, and R. Merletti, "Accuracy assessment of CKC high-density surface EMG decomposition in biceps femoris muscle," *J. Neural Eng.* **8**, 066002 (2011).

77. McGill, K. C., and L. J., Dorfman "High-resolution alignment of sampled waveforms," *IEEE Trans. Biomed. Eng.* **31**, 462–468 (1984).

78. McGill, K. C., Z. C. Lateva, and H. R. Marateb, "EMGLAB: an interactive EMG decomposition program," *J. Neurosci. Methods* **149**, 121–133 (2005).

79. McGill, K. C., and H. R. Marateb, "Rigorous a posteriori assessment of accuracy in EMG decomposition," *IEEE Trans. Neural Syst. Rehab. Eng.* **19**, 54–63 (2011).

80. Merletti, R., M. Aventaggiato, A. Botter, A. Holobar, H. Marateb, and T. M. Vieira, "Advances in surface EMG: recent progress in detection and processing techniques," *Crit. Rev. Biomed. Eng.* **38**, 305–345 (2010).

81. Merletti, R., A. Holobar, and D. Farina, "Analysis of motor units with high-density surface electromyography," *J. Electromyogr. Kinesiol.* **18**, 879–890 (2008).

82. Mesin, L., and D. Farina, "Simulation of surface EMG signals generated by muscle tissues with inhomogeneity due to fiber pinnation," *IEEE Trans. Biomed. Eng.* **51**, 1521–1529 (2004).

83. Mesin, L., and D. Farina, "A model for surface EMG generation in volume conductors with spherical inhomogeneities," *IEEE Trans. Biomed. Eng.* **52**, 1984–1993 (2005).

84. Mesin, L., "Simulation of surface EMG signals for a multilayer volume conductor with triangular model of the muscle tissue," *IEEE Trans. Biomed. Eng.* **53**, 2177–2184 (2006).

85. Mesin, L., and D. Farina, "An analytical model for surface EMG generation in volume conductors with smooth conductivity variations," *IEEE Trans. Biomed. Eng.* **53**, 773-779 (2006).

86. Mesin, L., M. Joubert, T. Hanekom, R. Merletti, and D. Farina, "A finite element model for describing the effect of muscle shortening on surface EMG," *IEEE Trans. Biomed. Eng.* **53**, 593–600 (2006).

87. Milner-Brown, H. S., R. B. Stein, and R. G. Lee, "Pattern of recruiting human motor units in neuropathies and motor neurone disease," *J. Neurol. Neurosurg. Psychiatry* **37**, 665–669 (1974).

88. Milner-Brown, H. S., R. B. Stein, and R. G. Lee, "Synchronization of human motor units: possible roles of exercise and supraspinal reflexes," *Electroencephalogr. Clin. Neurophysiol.* **38**, 245–254 (1975).

89. Milner-Brown, H. S., R. B. Stein, and R. Yemm, "The orderly recruitment of human motor units during voluntary isometric contractions," *J. Physiol. (London)* **230**, 359–370 (1973).

90. Minetto, M. A., A. Holobar, A. Botter, and D. Farina, "Discharge properties of motor units of the abductor hallucis muscle during cramp contractions," *J. Neurophysiol.* **102**, 1890–1901 (2009).

91. Mulder, E. R., K. H. L. Gerrits, B. U. Kleine, J. Rittweger, D. Felsenberg, A. de Haan, and D. F. Stegeman, "High-density surface EMG study on the time course of central nervous and peripheral neuromuscular changes during 8 weeks of bed rest with or without resistive vibration exercise," *J. Electromyogr. Kinesiol.* **19**, 208–218 (2009).

92. Nakamura, H., M. Yoshida, M. Kotani, K. Akazawa, and T. Moritani, "The application of independent component analysis to the multi-channel surface electromyographic signals

for separation of motor unit action potential trains: part I-measuring techniques," *J. Electromyogr. Kinesiol.* **14**, 423–432 (2004).

93. Nakamura, H., M. Yoshida, M. Kotani, K. Akazawa, and T. Moritani, "The application of independent component analysis to the multi-channel surface electromyographic signals for separation of motor unit action potential trains: part II-modelling interpretation," *J. Electromyogr. Kinesiol.* **14**, 433–441 (2004).

94. Nawab, S. H., S. S. Chang, and C. J. De Luca, "High-yield decomposition of surface EMG signals," *Clin. Neurophysiol.* **121**, 1602–1615 (2010).

95. Negro, F., and D. Farina, "Decorrelation of cortical inputs and motoneuron output," *J. Neurophysiol.* **106**, 2688–2697 (2011).

96. Negro, F., A. Holobar, and D. Farina, "Fluctuations in isometric muscle force can be described by one linear projection of low-frequency components of motor unit discharge rates," *J. Physiol. (London)* **587**, 5925–5938 (2009).

97. Nielsen, M., T. Graven-Nielsen, and D. Farina, "Effect of innervation-zone distribution on estimates of average muscle-fiber conduction velocity," *Muscle & Nerve* **37**, 68–78 (2008).

98. Nordstrom, M. A., A. J. Fuglevand, and R. M. Enoka, "Estimating the strength of common input to human motoneurons from the cross-correlogram," *J. Physiol. (London)* **453**, 547–574 (1992).

99. Parsaei, H., F. J. Nezhad, D. W. Stashuk, and A. Hamilton-Wright, "A Validating motor unit firing patterns extracted by EMG signal decomposition," *Med. Biol. Eng. Comput.* **49**, 649–658 (2011).

100. Piitulainen, H., A. Holobar, and J. Avela, "Changes in motor unit characteristics after eccentric elbow flexor exercise," *Scand. J. Med. Sci. Sports* **22**, 418–429 (2012).

101. Pino, L. J., D. W. Stashuk, S. G. Boe, and T. J. Doherty, "Probabilistic muscle characterization using QEMG: application to neuropathic muscle," *Muscle & Nerve* **41**, 18–31 (2010).

102. Power, G. A., B. H. Dalton, D. G. Behm, T. J. Doherty, A. A. Vandervoort, and C. L. Rice, "Motor unit survival in lifelong runners is muscle dependent," *Med. Sci. Sport Exer.* **44**, 1235–1242 (2012).

103. Radeke, J., J. P. van Dijk, A. Holobar, and B. G. Lapatki, "Electrophysiological method for examining the muscle fiber architecture of the upper lip in cleft-lip subjects," *J. Orofac. Orthop.* **75**, 51–61 (2013).

104. Revill, A. L., and A. J. Fuglevand, "Effects of persistent inward currents, accommodation, and adaptation on motor unit behavior: a simulation study," *J. Neurophysiol.* **106**, 1467–1479 (2011).

105. Richmond, F. J., K. Singh, and B. D. Corneil, "Marked non-uniformity of fiber-type composition in the primate suboccipital muscle obliquus capitis inferior," *Exp. Brain Res.* **125**, 14–18 (1999).

106. Sauvage, C., M. Manto, A. Adam, R. Roark, P. Jissendi, and C. J. De Luca, "Ordered motor-unit firing behavior in acute cerebellar stroke," *J. Neurophysiol.* **96**, 2769–2774 (2006).

107. Sohn, M. K., T. Graven-Nielsen, L. Arendt-Nielsen, and P. Svensson, "Inhibition of motor unit firing during experimental muscle pain in humans," *Muscle & Nerve* **23**, 1219–1226 (2000).

108. Stegeman, D. F., and W. H. Linssen, "Muscle fiber action potential changes and surface EMG: a simulation study," *J. Electromyogr. Kinesiol.* **2**, 130–140 (1992).

109. Stoykov, N. S., M. M. Lowery, and T. A. Kuiken, "A finite-element analysis of the effect of muscle insulation and shielding on the surface EMG signal," *IEEE Trans. Biomed. Eng.* **52**, 117–121 (2005).

110. Suresh, N., X. Li, P. Zhou, and W. Z. Rymer, "Examination of motor unit control properties in stroke survivors using surface EMG decomposition: a preliminary report," *Conference Proceedings IEEE Engineering in Medicine and Biology Society 2011*, Buenos Aires, pp. 8243–8246 (2011).

111. Theis, F. J., and G. A. García, "On the use of sparse signal decomposition in the analysis of multi-channel surface electromyograms," *Signal Processing* **86**, 603–623 (2006).

112. Tucker, K. J., and P. W. Hodges, "Changes in motor unit recruitment strategy during pain alters force direction," *Eur. J. Pain* **14**, 932–938 (2010).

113. Watanabe K., M. Gazzoni, A. Holobar, T. Miyamoto, K. Fukuda, R. Merletti, and T. Moritani, "Motor unit firing pattern of vastus lateralis muscle in type 2 diabetes mellitus patients," *Muscle & Nerve* **48**, 806–813 (2013).

114. Zaheer, F., S. H. Roy, and C. J. De Luca, "Preferred sensor sites for surface EMG signal decomposition," *Physiol Meas.* **33**, 195–206 (2012).

115. Zhou, P., N. L. Suresh, M. M. Lowery, and W. Z. Rymer, "Nonlinear spatial filtering of multichannel surface electromyogram signals during low force contractions," *IEEE Trans. Biomed. Eng.* **56**, 1871–1879 (2009).

116. Zijdewind, I., and C. K. Thomas, "Firing patterns of spontaneously active motor units in spinal cord-injured subjects," *J. Physiol. (London)* **590**, 1683–1697 (2012).

117. Yoo, T. S., *Insight into Images: Principles and Practice for Segmentation, Registration, and Image Analysis*, A K Peters/CRC Press, Natick, MA (2004).

8

EMG MODELING AND SIMULATION

M. M. Lowery

[1]*School of Electrical & Electronic Engineering, University College, Dublin, Ireland*

8.1 INTRODUCTION

Over the past several decades, mathematical modeling of the surface electromyographic (EMG) signal has become an established tool with which to increase our understanding of EMG signal generation and explore new methods of uncovering muscle structural, physiological, and control properties. To correctly interpret the EMG signal, it is important to understand the underlying processes which influence it and to establish the effect of each individually. This can be difficult to achieve through experimentation alone. The electromyogram is a complex signal, shaped by many factors. Currently it is not possible to measure all of the properties of individual motor units and muscle fibers *in vivo* or *in vitro*. Furthermore, EMG fluctuations during voluntary contractions are not caused by changes in any single parameter alone, but rather a combination of many. An alternative approach is to use computational models in which all of the factors that influence the signal can be directly controlled to simulate synthetic EMG data. In this way, changes that occur as different model parameters are altered, in isolation or in combination, can be explored.

Although a model cannot replicate muscle contraction in all its complexity, it is possible to generate artificial EMG signals that incorporate essential features such as motor unit recruitment and firing patterns, motor unit morphology, transmembrane action potential properties, muscle fiber end-effects, conduction velocities, electrode configuration, volume conduction, and action potential interference. Computational modeling has provided fundamental insights into the mechanisms of generation of the EMG signal, enabling the underlying physiological and physical mechanisms to be

Surface Electromyography: Physiology, Engineering, and Applications, First Edition.
Edited by Roberto Merletti and Dario Farina.
© 2016 by The Institute of Electrical and Electronics Engineers, Inc. Published 2016 by John Wiley & Sons, Inc.

explored. As a result, the influence of factors such as electrode configuration, muscle anisotropy, subcutaneous tissue electrical properties, muscle fiber conduction velocity, fiber orientation, and motor unit firing properties have been quantitatively examined using a range of different EMG models and are now well understood. In addition to applications in scientific and clinical research, EMG models also play an important didactic role.

This chapter outlines the main components that should be considered in the development of physiologically based models of surface EMG. Although the focus will be on surface EMG models, many of the principles considered also apply to the simulation of intramuscular EMG which has been extensively explored, in particular for clinical applications [64,106]. This chapter owes much to the chapter on EMG Modeling and Simulation by Stegeman, Merletti, and Hermens in reference 122 and to earlier reviews by Stegeman et al. [121], and McGill [92]. The discussion of the fundamental components of EMG modeling and simulation are extended here to cover recent developments in areas such as physiological source descriptions, numerical modeling of EMG, and emerging application areas.

8.2 PRINCIPLES OF MODELING AND SIMULATION

Modeling refers to the process by which a system is represented by a series of concepts or laws. The essence of modeling is the formulation of a representation of a physical or physiological system to enhance conceptual understanding of the system and processes involved. Modeling and systems analysis seek to represent observed phenomena by establishing functional relationships between interacting system input and output variables. In studying physiological systems, a model can be used to understand the influence of parameters that can not be accessed in vivo or in vitro. Through the establishment of cause and effect relationships, new methods of estimating physiological and anatomical parameters can be identified. One of the primary purposes of a modeling approach is to identify the mechanisms responsible for experimental observations. A successful EMG model can thereby help the researcher to relate the recorded electrical signal to the underlying processes associated with muscle contraction. An appropriate model can be used to design and test optimal EMG detection systems and processing algorithms for extracting information. The ability of various EMG features to reflect fundamental physical and physiological parameters can also be assessed.

Simulation refers to the application of computational models to the prediction of physical, or physiological, events. Simulations conducted using an appropriate model can help elucidate underlying mechanisms, provided that the inputs to the model can be directly related to established physiological variables. Simulation studies can also provide insight when complex mathematical analysis becomes intractable. In this way, modeling and simulation can help to explain experimental observations and suggest how they should be interpreted. Simulation can also be used to develop novel hypotheses and to propose new experiments with which to test them. An iterative relationship between modeling and experimentation thus exists, where a search for a deeper understanding of experimental observations leads to the development of a conceptual model. The model may lead to the design of new experiments, which in turn results in a refinement of the model, continuing the iterative process and leading to an informed and verified interpretation of a body of experimental observations.

The design and complexity of any model will be determined by the purpose for which it is intended and the extent of its use. A model required to simulate a wide range of muscle structures and physiological phenomena will likely be more complex than one which is very application-specific. The level of detail required of the model is similarly related to the problem to which it is to be applied. A high level of detail may be required of a model used to generate quantitative results—for example, when determining the value of a certain parameter for design purposes—while computational efficiency may be a higher priority in the design of a model which is to be used to explore more qualitative relationships.

Mathematical models may be classified as either phenomenological or structural. In the context of EMG, phenomenological models, also known as 'black-box' models, aim to capture essential features of the statistical properties of the EMG signal without necessarily relating them to the underlying physiology. In contrast, structural or mechanistic models are physiologically-based and aim to simulate the system from a realistic representation of the physiological and physical processes involved. Following the definition outlined by Stegeman et al. [122], the term "structural" model will be used here to define an EMG model that incorporates some or all of the features described.

8.3   PHENOMENOLOGICAL SURFACE EMG MODELS

To avoid the complexities of detailed descriptions of the generation of individual muscle fiber action potentials, at the highest level the EMG signal can be represented as band-limited, Gaussian noise [70,119,124]. The EMG signal may be modeled as zero-mean white noise passed through a linear time-invariant filter to represent the filtering and volume conduction effects that give the power spectrum its characteristic shape. To describe the functional relationship between myoelectric activity and muscle contraction, the properties of the band-limited Gaussian signal may be modulated—for example, as a function of the muscle force or muscle fiber conduction velocity [70,124]. Inherent in this description is an assumption that each motor unit fires independently, and that a sufficiently large number of action potentials are detected at the electrode that the Law of Large Numbers applies, implying that the signal may be assumed to have a band-limited Gaussian probability distribution. Such models are, therefore, not valid at low contraction levels or for EMG signals recorded with very selective electrode configurations where the recorded signal may be sparse. These types of models are phenomenological rather than physiological in nature, providing functional descriptions of the relationship between muscle contraction and EMG activity without accounting for the underlying processes. Phenomenological models have been used mostly in the design of processors to improve signal fidelity and to evaluate EMG amplitude and spectral estimates, as described in

Chapter 4 [13,14,22]. Many have been developed for a particular purpose and tend to be application-specific.

Alternatively, using linear systems theory, each motor unit action potential train can be mathematically represented as a series of unit impulses, or Dirac delta functions, applied to a linear time-invariant, black-box filter, the impulse response of which describes the motor unit action potential (MUAP) waveform [1,17]. The EMG signal is calculated as the linear summation of the signals from all active motor units, as described in Chapter 7 (Fig. 8.1a). By assuming all MUAPs to be identical in shape, the model may be reduced to a single input system which can be described as an amplitude-modulated stochastic process [1,75,108,119] (Fig. 8.1b).

Rather than the factors that determine the shape of individual extracellular action potential waveforms, this type of mathematical model is concerned instead with the manner in which the action potentials combine to produce the EMG signal. This enables the effects of action potential amplitude and, in particular, motor unit firing statistics on the EMG signal and power spectrum to be predicted without the need for complex representations of the muscle architecture or surrounding volume conductor properties. Using this approach, it has been demonstrated that above approximately 40 Hz the Fourier transform of the EMG signal approaches that of the average MUAP, while at lower frequencies it is influenced by the motor unit firing rates which

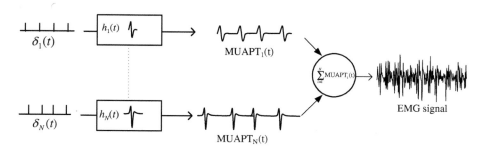

(a) Linear systems surface EMG model

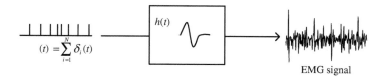

(b) Single-input linear systems surface EMG model

FIGURE 8.1 Models of the generation of surface EMG signals. (**a**) Linear dynamic EMG model similar to that proposed by Agarwal and Gottlieb [1] and De Luca [18]. Each MUAP train is represented as the response of a linear time-invariant filter to a train of unit impulses. (**b**) Simplified single-input linear dynamic model [75,108,119]. All MUAPs are assumed to be equal, and the model reduces to a single input system.

can appear as peaks in the spectrum at the mean firing rate and higher harmonics [1,17,21,75,108].

The classic model of the EMG power spectrum presented by Lindstrom and Magnusson [79] has been widely used in understanding changes related to variation in muscle fiber conduction velocity. The model describes an infinitely long muscle fiber, lying in an isotropic, infinite volume conductor without assuming any particular shape for the transmembrane current source. The filtering effect of bipolar or concentric needle electrodes, differences in fiber action potential arrival times, and muscle fiber conduction velocity were incorporated. Motor unit firing rates were assumed to have little influence on the spectral envelope and, therefore, were not explicitly described by the model. One of the most important results of the Lindstrom model in terms of practical applications has been in the analysis of muscle fatigue [78,80].

Linear systems and spectral models can provide a mathematical understanding of the formation of the EMG signal through the spatiotemporal summation of individual action potentials. Several simplifying assumptions are necessary for analytical tractability. For this reason, the complex physiological, anatomical, physical, and electromagnetic processes that determine the properties of the surface detected action potentials and the instances of motor unit firing are typically not considered. These models have, therefore, been included here, although they are not purely phenomenological but rather are "gray-box" or semiempirical models.

8.4 STRUCTURE-BASED SURFACE EMG MODELS

More detailed representations of the surface EMG signal that capture the physical and physiological process of EMG generation can be obtained by considering the formation and summation of the individual constituent action potentials. When a muscle fiber is stimulated, an action potential propagates in both directions away from the neuromuscular junction until it terminates upon reaching the musculotendon junction. The generation and propagation of the action potential along the muscle fiber gives rise to a change in voltage in the surrounding tissue which can be detected at an electrode located either within the subcutaneous tissue or at the skin surface.

From an appreciation of the underlying biophysics and physiology, the many factors that shape the surface EMG signal can be separated into several components that should be considered when building a structural or physiologically based model of the EMG signal. The first is the *transmembrane muscle fiber action potential* that provides the source of electrical activity in the model. The action potential generated at the muscle fiber is influenced by factors such as muscle fiber geometry, individual ionic currents, the transverse tubular system, and muscle temperature. The next component is the *volume conductor*, which describes the geometrical and electrical properties of the tissues surrounding the muscle fibers. Along with the properties of the transmembrane action potential source, these will determine the shape of each muscle fiber action potential detected at the electrode. In practice, muscle fibers are not activated individually but rather are grouped into motor units. The *morphology of the motor units* of which the muscle is comprised, including innervation ratios, spatial distribution of the motor unit fibers, and fiber properties, should be described. Finally, to represent the EMG "interference" pattern that is formed through the constructive and destructive interference of overlapping motor unit action potentials, the *motor unit recruitment and firing* times must be defined. The recruitment, firing, and synchronization patterns of the motor units relative to one another are governed by the many inputs from around the nervous system received and integrated by the motoneuron pool.

In the following sections, each of these components will be considered to explore how they may be incorporated into EMG models and what information and new insights can thereby be obtained.

8.5 MODELING THE ACTION POTENTIAL SOURCE

8.5.1 The Intracellular Action Potential

The action potential generated by the flow of ionic currents across the fiber membrane is the source of excitation at the center of all EMG models. This electrical source can be represented in terms of the transmembrane potential or the transmembrane current and may be estimated by calculating the individual ionic currents or through the use of a global analytical model.

The simplest way in which the transmembrane current waveform can be represented is as a propagating dipole or tripole source. The tripole source is the combination of two dipole sources. In a dipole or tripole source model, the transmembrane current is concentrated into two or three point sources respectively, representing the main phases of the triphasic transmembrane current waveform [7,49,60,133]. In the case of the tripole model, sources are located at the centroid of each of the three phases, with magnitudes proportional to the area of the phase such that the sum of the point sources and the sum of their moments is equal to zero [61,94] (Fig. 8.2). Dipole and tripole models are computationally efficient, do not require a detailed knowledge of the shape of the transmembrane potential, and, in general, provide a good approximation of the extracellular action potential at relatively large distances from the muscle fiber [3]. The dipole model, however, is unable to replicate extracellular potentials more complex than simple biphasic waveforms [61] and cannot fully capture the effects at action potential generation and extinction. The tripole model, while converging towards the volume conductor model at large distances, does not accurately describe the extracellular potential at short radial distances (<10 mm) from the fiber.

A more accurate representation of the transmembrane action potential, $V_m(z)$, may be obtained by describing the transmembrane current, or potential, as a series of point sources. A convenient method by which to represent the action potential is by means of an analytical model, such as the description of the transmembrane potential as a function of axial distance along the fiber, z, formulated by Rosenfalck [114],

$$V_m(z) = 96z^3 e^{-z} - 90 \qquad (8.1)$$

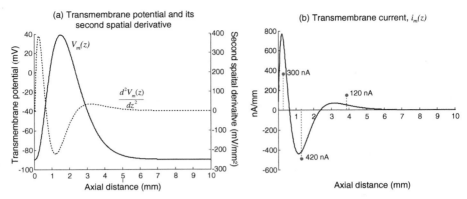

FIGURE 8.2 (a) Simulated transmembrane potential (*solid line*) and its second spatial derivative (*dashed line*). (b) Transmembrane current per unit length, $i_m(z)$, and equivalent tripole source model. Data are based on the modification to Rosenfalck's intracellular potential presented by Nandedkar and Stalberg [104].

and subsequently modified by Nandedkar and Stalberg [104] to provide a closer match to experimental data.

According to the core conductor model, assuming that the transmembrane potential is uniform across a cross section of the fiber, the transmembrane current per unit length, $i_m(z)$, may be assumed to be proportional to the second spatial derivative of $V_m(z)$ [3,15,111] (Fig. 8.2).

$$i_m(z) = \frac{\sigma_{in}\pi d^2}{4} \cdot \frac{d^2 V_m(z)}{dz^2} \tag{8.2}$$

where d is the fiber diameter and σ_{in} is the intracellular conductivity.

In the majority of EMG models proposed in the literature, the volume conductor is assumed to be space invariant in the direction of action potential propagation. Under these conditions, if the conduction velocity and spatial distribution of the action potential along the fiber remain constant, the action potential can be considered as a wave traveling with constant velocity, v. In the case of a fiber of infinite length, the description of the spatially distributed source function, $V_m(z)$, can be interchanged with a temporal description of the transmembrane potential, $V_m(t)$ where $z = v \cdot t$. In reality, slight changes in conduction velocity may occur due to local variations in fiber properties such as fiber diameter and ion channel densities.

Alternative approaches directed at obtaining more realistic representations of the physiological action potential include the use of experimentally measured transmembrane action potentials and currents, extracellularly recorded action potentials [130], and analytical models incorporating the long duration slow repolarization of the transmembrane potential [26,91]. Details of the shape of the extracellularly recorded action potential are sensitive to properties of the intracellular action potential. Using simulations to separate the contribution of the different components of the intracellular action potential, it has been shown that the falling phase of the intracellular action potential determines the shape of the third phase of the monopolar extracellular action potential, while the negative phase is strongly influenced by the duration of the intracellular action potential [112].

The representations of the action potential source in the above models are based on analytical approximations. Changes occurring at the fiber membrane level—for example, due to variations in ion channel properties, ionic concentrations or temperature—are not explicitly included. To incorporate these, an alternative approach using distributed parameter models of the fiber membrane based on Hodgkin–Huxley-type models of the action potential source is required [53]. Ion-channel based models of the muscle fiber action potential that incorporate action potential propagation along the sarcolemma and into the transverse tubular system [132], along with variations in intra- and extracellular ion concentrations and muscle temperature [46], can be coupled to volume conductor models of the surrounding tissue. These types of models can be used to relate parameter variations at the level of the sarcolemma and transverse tubular system, such as the ionic currents, temperature and ion channel properties, directly to the recorded extracellular potential (Fig. 8.3).

8.5.2 Modeling Action Potential Propagation, Generation, and Extinction

Accurate modeling of action potential generation at the neuromuscular junction, or fiber end-plate, and the extinction of the action potential at the musculotendon junction is an important component of the description of source propagation along fibers of finite length. The muscle fiber is effectively insulated at both ends, leading to termination of the action potential upon reaching the musculotendon junction. While an assumption of an infinitely long fiber may suffice if the length of the propagating action potential is short compared to the fiber length and the observation point is located close to the fiber, in shorter muscle fibers, and at greater distances from the fiber, this approximation is unable to replicate recorded extracellular action potentials [113]. To account for the effects at the fiber termination, an approach based on compensation of point current sources has been traditionally adopted [6,30,31,58,81,94,111].

Upon reaching the end of the fiber, each transmembrane current source is assumed to come to a stop and remain stationary, until partly or wholly canceled by the arrival of a source of opposite polarity. This results in the emergence of a stationary (in space), constant latency waveform at the recording electrode, due to the presence of the sources at the fiber termination. The appearance of this nonpropagating waveform depends on the location of the electrode with respect to the end of the fiber, the distance from the fiber, the symmetry of the muscle fiber about its end-plate, and the electrode configuration. The closer the electrodes are to the end of the fiber, the more noticeable this effect will be. If the fiber is exactly symmetrical about the neuro-muscular junction, the effect will be greater, as both action potentials terminate simultaneously. In the case of bipolar detection, the fiber-end effects appear similar at both electrodes and tend to cancel one another out, while in a monopolar configuration the effect is more pronounced. As the stationary component of the

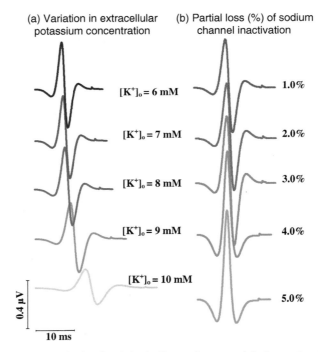

FIGURE 8.3 Variation in simulated single fiber action potentials detected at the skin surface directly above a muscle fiber located 5 mm below the muscle–fat interface in an idealized cylindrical model of radius 43.4 mm, with 1.3-mm-thick skin tissue and 3 mm of fat. Actions were simulated for varying **(a)** extracellular potassium concentration, $[K^+]_e$, and **(b)** partial loss of sodium fast inactivation. To simulate a partial loss of sodium inactivation, similar to that associated with hyperkalemic periodic paralysis [11], in (b) the sodium conductance in the transmembrane action potential was scaled to incorporate a proportion of non-inactivating sodium channels ranging from 1% to 5% of the total sodium inactivation. The action potentials were simulated for bipolar electrodes with an interelectrode distance of 5 mm located 20 mm from the neuromuscular junction with an interelectrode distance of 10 mm, at 37 °C.

potential (end-of-fiber effect), due to the transformation of a propagating double current dipole (tripole) into a simple dipole of decreasing strength, decays with distance more slowly than the propagating component of the action potential, it produces relatively larger action potentials generated from more distant fibers and may constitute an important component of EMG crosstalk [33,42,84].

Action potential generation at the neuromuscular junction can be simulated in a similar manner with all current sources initially positioned at the end plate of the fiber, moving away progressively as the action potential propagation is initiated. Due to the initialization and propagation of two action potentials in opposite directions, the effect at the recording electrode of the action potential generation is much less than the termination effects at the end of the fiber as the effect of the overlapping stationary poles at the neuromuscular junction from one action potential cancels that of the other.

Although the mechanisms underlying the description of action potential generation and extinction in this manner are not fully understood, simulation results are supported by theoretical considerations and experimental observations [74,113].

8.6 MODELS OF VOLUME CONDUCTION AND DETECTION SYSTEMS

The currents generated by excitable tissues give rise to a change in the electrical potential or voltage in the surrounding media. The passive conducting region surrounding an active fiber through which current flows is known as the volume conductor. The geometry and electrical properties of the surrounding tissues play a major role in determining the shape of the single fiber action potential detected at the skin surface. Variations in subcutaneous fat thickness, local tissue in-homogeneities such as blood vessels or bone, fiber location, and recording electrode configuration can all have a substantial influence on the shape of the individual action potential waveforms. To fully capture the variations in individual muscle fiber and motor unit action potential waveforms, therefore, requires a model that can describe the generation, propagation and extinction of the transmembrane action potential and the electrical and geometric properties of the volume conductor in which it is situated (see Fig. 2.3). Understanding how these effects are expressed in the EMG signal is fundamental to a correct interpretation of experimental and clinical data. Often these effects are counterintuitive and can only be correctly interpreted through the use of appropriate models.

A full description of time-varying electric and magnetic fields is obtained through the solution of Maxwell's equations within the appropriate physiological boundary conditions. For simulation of surface EMG signals, it can generally be assumed that the quasi-static approximation holds; that is, at the frequencies of interest, the effects of field propagation through the medium, as well as the inductive and capacitive effects, are negligible and the volume conductor may be assumed to be purely resistive [109]. Assuming piecewise homogeneity of the volume conductor, the relationship between the local electrical potential distribution and the current sources that produce it is described by Poisson's equation,

$$\sigma\nabla^2\phi = -I_v \tag{8.3}$$

reducing further to Laplace's equation if the problem is defined such that the sources lie outside, or at, the boundary of the region of interest.

$$\nabla^2\phi = 0 \tag{8.4}$$

Depending on the level of detail required, volume conductors can range from homogeneous models of infinite extent to numerical models that capture the complex variations of subject-specific geometries by the application of specific boundary conditions. In the following sections, available solution methods for a range of volume conductors of varying complexity will be discussed.

8.6.1 Infinite Volume Conductor Models

The simplest EMG models, and many models to date, assume the muscle fiber to lie within a homogeneous volume conductor of infinite or semi-infinite extent with isotropic or anisotropic conductivity [7,27,61,105,133] (see Section 8.6.3 for additional considerations on conductivity and permittivity). Such models provide a good approximation of the EMG signal detected at the skin surface and capture important features such as spatial filtering of the action potential and electrode configuration. Basic volume conductor theory originally derived for unmyelinated nerve fibers is assumed to be equally valid for muscle fibers as the propagation of action potentials along both is essentially the same (see Chapter 2).

According to the line source model, the total current crossing the fiber membrane may be concentrated into a series of point current sources situated along the center of the muscle fiber. The solution to the Poisson equation in integral form leads to the following expression for the extracellular potential, $\phi(z)$, due to a line of point current sources located at the center of an infinite fiber lying in an isotropic conducting medium of infinite extent, at a point at radial distance r from the fiber [3,114].

$$\phi(z) = \frac{1}{4\pi\sigma_e} \int_{-\infty}^{\infty} \frac{i_m(\zeta)}{\sqrt{r^2 + (z - \zeta)^2}} d\zeta \qquad (8.5)$$

The transmembrane current density, $i_m(z)$, is the current per unit length crossing the fiber membrane and σ_e is the conductivity of the extracellular medium.

For fibers parallel to each other and to the skin, the anisotropic nature of muscle tissue whereby the radial conductivity perpendicular to the direction of the fibers, σ_r, is considerably lower than the axial conductivity along the fiber direction, σ_z, may be accounted for by considering the medium to be cylindrically anisotropic [3,114]:

$$\phi(z) = \frac{1}{4\pi\sigma_r} \int_{-\infty}^{\infty} \frac{i_m(\zeta)}{\sqrt{\sigma_z/\sigma_r r^2 + (z - \zeta)^2}} d\zeta \qquad (8.6)$$

Using the method of images, the potential in a semi-infinite volume conductor with zero conductivity on one side of a plane representing the boundary of the limb can be obtained by multiplying the above expressions by a factor of two [93,117].

The time-dependent extracellular potential, $\phi(t)$, may be calculated at each instant in time as the intracellular action potentials are advanced along the fiber. In many instances, it can be assumed that the shape of the intracellular action potential and the conduction velocity with which it propagates remain unchanged along the fiber. Under these assumptions, a muscle fiber of infinite length may be considered as a linear time-shift invariant system. The formulation for the extracellular potential can then be expressed as the convolution of two time-dependent functions, one representing the intracellular action potential and the other a weighting function representing the impulse response of the system [111]. A simple model providing

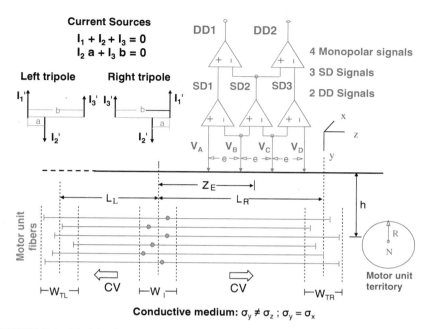

FIGURE 8.4 Model of a motor unit and sEMG detection system for a four-electrode detection array. The motor unit has N fibers uniformly distributed in a cylinder of radius R at depth h. The neuromuscular junctions are uniformly distributed in a region W_I, and the fiber–tendon terminations are uniformly distributed in two regions W_{TR} and W_{TL}. A right-propagating and left-propagating current tripole originate from each neuromuscular junction (NMJ, •) and propagate to the fiber–tendon terminations where they become extinguished. The conduction velocity is the same in both directions and for all fibers of a motor unit but may be different in different motor units. Each of the voltages V_A, V_B, V_C, and V_D is the summation of the contributions from each of the tripoles. Reprinted from Merletti et al. [93] with permission.

four monopolar signals, three single differential signals (SD), and two double differential signals (DD) is depicted in Fig. 8.4 [92,93].

A faster and more concise implementation of action potential generation, propagation and extinction in a fiber of finite length may be obtained by replacing the discrete time and space approach described above with description of the source in terms of the first derivative of the transmembrane potential. This enables the propagating and stationary sources to be combined into a single source function which can be convolved with an impulse response representing the surrounding volume conductor and detection system [27,40].

The convolution function can be formulated in terms of the first derivative of the intracellular action potential, $\frac{dV_m(t)}{dt}$, and an impulse response, $IR(t)$, representing the potential recorded at the electrode due to the propagation of two dipoles in opposite directions away from the end plate [27]. The extracellular potential at an electrode located at coordinates (x_o, y_o, z_o) may be calculated as

$$\phi(t, x_o, y_o, z_o) = \frac{\sigma_z}{\sigma_r} \cdot \frac{dV_m(t)}{dt} * \frac{1}{v} \cdot IR(t) \tag{8.7}$$

where

$$IR(t) = C_1 \frac{d}{dt} \left\{ \frac{1}{\left[\left(x_o - x_{ep} - vt \right)^2 + \frac{\sigma_z}{\sigma_r} (y_o - y_i)^2 + \frac{\sigma_z}{\sigma_r} (z_o + z_i)^2 \right]^{1/2}} \right\}$$

$$+ C_2 \frac{d}{dt} \left\{ \frac{1}{\left[\left(x_o - x_{ep} + vt \right)^2 + \frac{\sigma_z}{\sigma_r} (y_o - y_i)^2 + \frac{\sigma_z}{\sigma_r} (z_o - z_i)^2 \right]^{1/2}} \right\} \tag{8.8}$$

$C_1 = 1$ for $t \leq L_1/v$ and $C_1 = 0$ for $t > L_1/v$; similarly, $C_2 = 1$ for $t \leq L_2/v$ and $C_2 = 0$ for $t > L_2/v$, where L_1 and L_2 are the distances from the end-plate to the proximal and distal ends of the fiber, respectively [27] (see Chapter 2).

8.6.2   Finite Volume Conductor Models

Infinite volume conductor models capture the main morphological features of the MUAP and are computationally efficient, particularly when calculating the potential from to a large numbers of motor units [37,50,81,93]. However, they do not capture the finite dimensions of the surrounding conductive tissues nor allow variations in tissue properties or inhomogeneous structures to be included. In reality, the finite nature of the volume conductor and the different conductivities of the tissues which comprise it will have a substantial influence on the shape of the extracellular single fiber and motor unit action potential.

To incorporate the finite dimensions and geometry of the limb, along with tissues such as fat, skin, and bone, more advanced volume conductor models are required. This can be achieved using either analytical or numerical methods. Analytical models are computationally efficient and provide an exact theoretical solution of the electric field within the tissue. In general, analytical models are limited to problems for which exact theoretical solutions are available such as cylindrical, planar, spherical, or ellipsoidal geometries. To incorporate more complex geometries, including features that cannot be included using traditional analytical methods such as the conducting electrode surface and electrode–tissue interface, irregular inhomogeneities, and asymmetrical tissue geometries, numerical methods such as the finite element method are required. Numerical methods provide an approximation to the exact solution and are generally more computationally demanding than analytical methods. While the majority of finite volume conductor EMG models presented to date have been based on analytical solutions of idealized geometries, in recent years a number of numerical and finite element models have also been presented. Both approaches will be discussed briefly here.

8.6.2.1 Analytical Finite Volume Conductor Models.
The finite dimensions of the limb may be approximated by representing the limb as an idealized cylindrical conductor. Following this approach, a model of an idealized cylindrical limb based on two concentric cylinders representing the muscle tissues and surrounding fat layer was introduced by Gootzen et al. [58]. The approach adopted was based on the solution for the potential in a model of concentric cylinders from classic electrodynamics [72,110] and was later extended to incorporate a third layer representing skin [113] (Fig. 8.5a). When comparing across action potentials simulated using an infinite volume conductor and finite volume conductors consisting of multiple layers, the magnitudes of the different components of the simulated action potential waveforms relative to one another differed across all models, with differences also observed in the rate of action potential decay with increasing distance from the electrode [113] (Fig. 8.5b).

An alternative approach considers the fat and skin layers that overlay the muscle in a planar geometry using a method based on 2-D spatial and temporal filtering of the intracellular action potential [38,40]. Using this approach, the effects of the electrode and volume conductor properties can be interpreted in terms of the application of 2-D filters, with the properties of the transmembrane action potential including extinction and generation of the action potential separately described as a function of space and time

$$
\begin{aligned}
i_m(z,t) = \frac{d}{ds} \Big\{ \psi[\gamma(z - z_1 - vt)] C_1 \big[\gamma(z - z_1 - L_1/t)\big] \\
- \psi[\gamma(-z - z_2 - vt)] C_2 \big[\gamma(z - z_{2-} + L_2/t)\big] \Big\}
\end{aligned}
\tag{8.9}
$$

where $i_m(z,t)$ is the current density, $\psi(z) = \frac{dV_m(z)}{ds}$, z and t are the space and time coordinates, γ is the propagation path, L_1 and L_2 are the semi-fiber lengths, and C_1 and C_2 constants as defined for Eq. (8.8) [44]. This approach has also been generalized to multilayer cylindrical volume conductors (Fig. 8.5c), to simulate action potentials propagating in the longitudinal direction and around the surface of the model to simulate circular muscles, such as the sphincter muscles [43].

8.6.2.2 Numerical Volume Conductor Models.
To address increasing numbers of inhomogeneities and more complex limb anatomies for which analytical solutions are not available, numerical techniques must be employed [68,82,85,86,97,118,123,135]. A number of suitable numerical methods are available, among these the finite element method, a widely used technique for obtaining approximate solutions to boundary-value problems in engineering and mathematics. The method is particularly suited to problems with large numbers of inhomogeneities or complex geometries and involves discretization of the geometry of interest into a large number of very small elements of

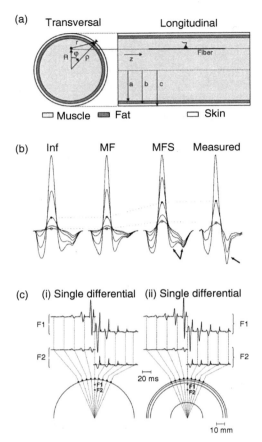

FIGURE 8.5 Examples of analytical, cylindrical, finite volume conductor sEMG models of surface EMG. (**a**) Three-layered cylindrical volume conductor configuration with eccentric source. The cross section on the left is perpendicular to the cylinder axis. The source fiber is shown with its motor end plate halfway along the fiber. Reproduced from Blok et al. [6] with permission. (**b**) Decline of the EMG potential over the circumference of the cylinder in (a), for three of the configurations and for experimental results. Electrodes are spaced 9.8° apart, the first being right above the source. The amplitude of the negative peak of this first signal is used for normalization for absolute amplitude values. The decline with increasing angle is slowest in the three-layer model and is comparable to that observed in experiments. The experimental results are reproduced from Roeleveld et al. [113], for electrodes spaced 12 mm apart. Identical positions 12 mm and 24 mm are marked with "∗"and "o," respectively, in all results, and connected by dotted lines. Reproduced from Blok et al. [6] with permission. (**c**) Examples of simulated monopolar and single differential muscle fiber action potentials in the case of limb muscle with (i) two layers (muscle and air) and (ii) five layers (bone, muscle, fat, skin, and air). The potentials are detected in five (*to the left and right*) locations around the limb circumference. Two fibers (F1 and F2) were simulated. The distances between the sources and the detection points are the same in (i) and (ii). The potentials in (i) and (ii) are normalized with respect to the amplitude of the largest potential in (i) and (ii), respectively. Reprinted from Farina et al. [43] with permission.

variable size, to which electrical properties can be individually assigned. A system of equations describing the behavior of the system are then formulated and solved to obtain the solution to the boundary-value problem. Using this approach, the electric field can be solved throughout a volume conductor of arbitrary shape for a given set of boundary conditions and source functions. This enables models that reflect the complex anatomies of human limbs, including asymmetrical geometries and fibers of arbitrary orientation, to be developed. In addition to models of regular geometries [82], finite-element EMG models have been extended to simulate surface EMG signals in volume conductor models of the upper human arm derived from MRI data [86] (Fig. 8.6) and to simulate the effect of fiber shortening on single fiber action potentials, illustrating variations in action potential amplitude and frequency content in response to changes in muscle geometry [97].

8.6.2.3 *Boundary Conditions.*

In addition to the volume conductor electrical and geometrical properties, the boundary conditions applied to a finite volume conductor should be specified. This can be done by defining the potential on a surface, known as Dirichlet boundary conditions, or by specifying the normal component of the electric field on a surface, known as Neumann boundary conditions [72]. At the interface between any two regions within the volume conductor, the normal component of the current density is continuous, and

$$\sigma_1 E_{1n} = \sigma_2 E_{2n} \tag{8.10}$$

where σ_1 and σ_2 are the conductivities of the first and second region, respectively, and E_{1n} and E_{2n} are the normal components of the electric field in each region. At the skin surface, the conductivity of the surrounding region, air, is assumed to be zero.

In practice, current is not confined within a section of the limb, but is free to flow into the distal and proximal ends. This may be approximated in a number of ways. If the volume conductor is sufficiently long that its finite length does not affect the simulated action potential, the entire volume conductor can be assumed to be surrounded by air [58]. Alternatively, using numerical techniques infinite boundary conditions such as the Bayliss–Turkel approximation [125] can be included to simulate the ability of current to flow outside the model at either end of the limb.

8.6.3 Material Properties

The electrical properties of biological tissues are described by their conductivity and permittivity. Both the conductivity and permittivity of biological tissues are dispersive; that is, they vary as a function of frequency [51,57]. The values of conductivity and relative permittivity reported for biological tissues vary widely, particularly at low frequencies. In the data reviewed by Geddes and Baker [55], conductivity values for muscle in the longitudinal direction vary between 0.093 S/m and 0.8 S/m at 100 Hz. Reported conductivity values for fat similarly range from 0.02 S/m to 0.1 S/m [47,55,57]. Skin is typically simulated as a homogeneous tissue, although in reality it has a laminar structure, with most of the impedance due to the highly resistive stratum

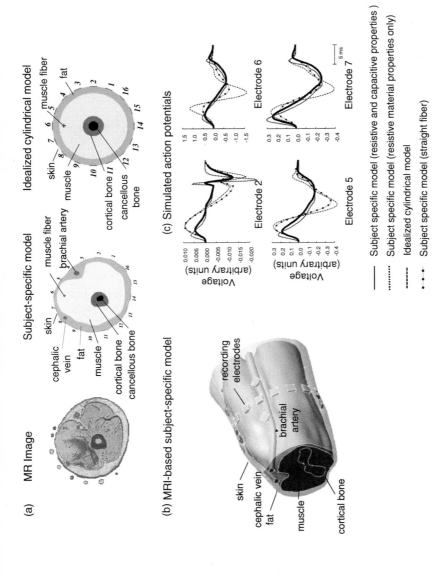

FIGURE 8.6 Subject-specific EMG model. (**a**) MR image of transverse cross section of subject's right arm midway between the bipolar surface electrodes, cross section of subject-specific model at the same location, and cross section of the corresponding idealized cylindrical model. (**b**) Geometry of anatomically based volume conductor model illustrating the tissue geometries and surface electrodes. (**c**) Simulated surface action potentials for a fiber located 14.5 mm below the skin surface beneath electrode 6. Action potentials are presented at different electrode locations for the subject-specific anatomical model with both resistive and capacitive material properties (*solid line*), purely resistive material properties (*dotted line*), a straight muscle fiber (*dot–dashed line*) as well as for the equivalent idealized cylindrical model (*dashed line*). Voltages have been normalized with respect to the peak amplitude of the action potential recorded at electrode 6 in the idealized cylindrical model. Reprinted from Lowery et al. [86] with permission.

corneum which becomes less resistive with increasing depth. Below the highly resistive barrier layer lies deeper granular tissue with material properties closer to those of muscle tissue [137].

8.6.3.1 *Tissue Conductivity and Permittivity.*

While it is generally assumed that capacitive effects, and hence permittivity, may be neglected in EMG modeling, studies using a microscopic network model have indicated that at distances close to the fiber it may be necessary to incorporate membrane capacitance and the frequency-dependent nature of the tissue properties [2,56,115]. At the macroscopic level, inclusion of capacitive effects has been shown to introduce a slight temporal low-pass filtering of simulated surface action potentials, the effect of which is most pronounced on the end-effect components of action potentials detected at locations far from the active fiber [86]. Depending on the combination of electric conductivity and permittivity chosen, capacitive effects in muscle may vary from having a negligible influence on simulated action potentials to having a considerable impact [123]. In general, however, capacitive effects in bioelectric models are deemed negligible and the quasi-static approximation may be assumed to hold. The vast majority of surface EMG models, therefore, assume that, at the frequencies of interest, biological tissues behave as if they are purely conductive materials, with conductivity that is constant with a given tissue and is independent of frequency.

8.6.3.2 *Muscle Anisotropy.*

The conductivity of muscle tissue in the direction parallel to the muscle fibers is higher than in the transverse direction, resulting in preferential current flow in the longitudinal direction and a resulting alteration in the extracellular potential. A wide range of values for muscle anisotropy has been reported, from as low as 1.8 to a maximum of 14.5 [3,10,51,55,57]. Anisotropy, present in both conductivity and permittivity, affects properties of the simulated action potentials and the rate at which the action potential amplitude decays with increasing distance from the fiber [45,82,85]. Increasing the anisotropy ratio, by increasing the conductivity along the longitudinal axis, results in a reduction in the amplitude and frequency content of the simulated action potential [3,15,53,82,93]. It also causes the EMG signal to decay more rapidly around the surface of the volume conductor [82].

If the muscle fibers run parallel to the surface of the skin, the anisotropic nature of muscle conductivity can be incorporated relatively easily by defining the conductivity parallel and perpendicular to the direction of the muscle fibers, as in Eq. (8.6). To incorporate anisotropy in more complex models where fibers are inclined or curved with respect to the detection surface—for example, in bi-pinnate or triangular muscles—requires the conductivity tensor to vary as a function of space. To accommodate this, the conductivity tensor, $\underline{\underline{\sigma}}$, in a region of arbitrary muscle tissue may be defined at each point as

$$\underline{\underline{\sigma}} = \sigma_l \hat{v}_l \hat{v}_l + \sigma_t \hat{v}_{tn} \hat{v}_{tn} + \sigma_t \hat{v}_{tb} \hat{v}_{tb} \qquad (8.11)$$

where σ_l and σ_t are the conductivities parallel and transverse to the fiber direction, and $\hat{v}_l$, $\hat{v}_{tn}$, and $\hat{v}_{tb}$ are vectors in the directions longitudinal, transverse normal, and

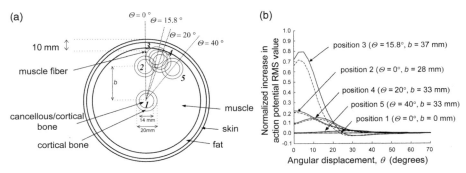

FIGURE 8.7 Influence of inhomogeneous structures **(a)** Cross section of finite element EMG model illustrating simulated bone locations. b denotes the distance from center of the volume conductor to the center of bone, and Θ denotes the angular displacement of the bone from the source. **(b)** Change in surface potential RMS amplitude with the addition of bone at locations 1–5 indicated in part a. Values for the cortical bone are indicated with a solid line, and values for the cortical bone with a core of cancellous tissue are indicated with a dashed line. Values are normalized with respect to the peak value at $\theta = 0°$ without bone. Muscle fiber depth is 10 mm. Reprinted from Lowery et al. [82] with permission.

transverse binomial to the fiber path, respectively. By defining the conductivity tensor based on the direction of source propagation, Mesin and Farina introduced a model of the extracellular potential in a bi-pinnate muscle [95] and in circular muscles [43]. Extending this method, similar analytical models have been used to describe extracellular action potentials generated in non-space-invariant convergent or triangular muscles, such as the deltoid and trapezius [98].

8.6.3.3 Local Inhomogeneities.   The presence of inhomogeneous structures, due to structures such as bone, blood vessels, glands, nerve, fat, or connective tissue, can alter the distribution of the electric field in the volume conductor and hence the properties of the action potential detected at the electrode. A major inhomogeneity such as resistive bone can considerably distort the potential distribution on the skin surface by constraining current flow, particularly when located close to the surface of the volume conductor [82,100], Fig. 8.7. Similarly, inclusion of a highly conductive blood vessel below the skin surface has a shunting effect, causing the potential at the electrode located directly above it to decrease [86,100,116]. Simulations have been used to understand how small inhomogeneities can distort estimated muscle fiber conduction velocity values by altering the delay between action potentials detected along electrode arrays [44,96,118]. In addition to sudden changes in conductivity due to various structures, sEMG models can be used to simulate the effect of conductivity that varies spatially [99] or due to fiber pinnation [95].

8.6.4 Modeling the EMG Electrode

In sEMG modeling, it is generally assumed that the potential detected by a large electrode is equal to the average of the potential over the area beneath the electrode if the electrode were not present [28,32,39,49,60,65]. Two assertions are implicit in this assumption, first that the distribution of the electric potential in the volume conductor, in particular at the skin surface under the electrode, is not altered by the presence of the electrode itself, and secondly that the signal at the electrode is indeed the true average of the potential at the skin surface under the electrode. These assumptions have been tested recently using a combination of experimental and simulation methods [129]. The results indicate that the principle of averaging is valid in practice, if the impedance of the electrical double layer between the electrode and conductive tissue is sufficiently high (see Fig. 8.8 and Chapter 2).

In addition to the properties of the electrode and electrode–skin interface (see Chapters 2 and 3), the configuration of the electrode has a critical role in determining the characteristics of the detected signal. Modeling has been extensively used to examine the effects of interelectrode distance, electrode orientation, electrode size, and configuration, demonstrating how the selectivity of sEMG electrodes can be increased by reducing the interelectrode distance [49,85,88] or through the use of higher-order spatial filters [35].

8.6.5   The Motor Unit Action Potential

When a motoneuron is activated, all muscle fibers belonging to that motor unit are activated almost simultaneously, resulting in the detection of a single motor unit action potential at the electrode. The MUAP is thus the smallest unit of electrical muscle activity that can be recorded by the surface EMG electrode in practice and represents the summation of the single fiber action potentials (SFAPs) generated by all muscle fibers belonging to the motor unit. During voluntary contraction, signals from up to several hundred concurrently active motor units summate linearly at the electrode in accordance with their individual firing patterns to form the sEMG interference signal. This principle of superposition is a function of the linear nature of biological tissues and has been verified experimentally [16]. As the basic building block of EMG, an accurate description of the MUAPs generated by the motor units within a muscle, based on the properties of their composite single fibers, is necessary for complete simulation of the resulting surface-detected signal. Models of MUAPs have also helped in understanding the factors that influence MUAP properties. In the past, simulations of MUAPs have been used, for example, to examine the effects of end-plate dispersion, distance from the electrode, and muscle fiber conduction velocity [7,60,105,133].

The shape of the SFAPs that make up a single MUAP vary due to the different locations of the fibers within the motor unit territory, with fibers located furthest from the electrode generating the smallest action potentials and undergoing the greatest low-pass spatial filtering due to conduction through the volume conductor. The SFAPs are also temporally distributed due to differences in conduction times along the motoneuron terminal branches, random variations in synaptic delays at the end plate (jitter), variations of the axial location of the fiber end plates, and small differences in the conduction velocity of the muscle fibers within the motor unit. To simulate the MUAP, the number and spatial arrangement of fibers within the motor

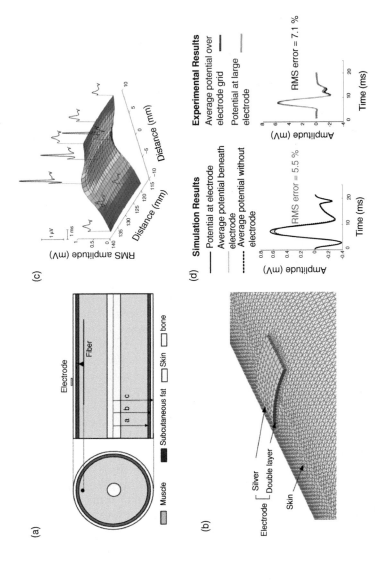

FIGURE 8.8 (a) Schematic diagram of a three-layer volume conductor model with an inner core of bone 5 mm, where $a = 40$ mm, $b = 43$ mm, and $c = 44$ mm. (b) Surface of finite element model showing the thin double layer between the conductive electrode and the skin. (c) Distribution of the electric potential directly beneath the simulated electrode. Examples of single fiber action potentials at different points at the skin surface below the electrode are shown. (d) **Simulation Results:** Simulated muscle-fiber action potential detected at the electrode (*black line*) and average potential beneath the electrode with (*gray line*) and without (*dashed line*) the electrode present. **Experimental Results:** Averaged compound muscle action potential (CMAP) (*gray line*) as recorded by a high-density surface EMG grid. **Black lines** show the CMAP as measured with a large (2.6×3.4 cm^2) electrode over the skin at the covering the same area as the grid. Reproduced from van Dijk et al. [129] with permission.

unit territory must be accounted for, along with the locations of the individual fiber end plates and musculotendon junctions. Based on experimental observations of uniform scattering of fibers throughout the motor unit territory, muscle fibers can be simulated to lie within a uniform random distribution throughout a motor unit territory of typically circular cross-sectional area [7]. The location of the muscle fiber end plates within a motor unit is generally assumed to be randomly located throughout an end-plate region, or innervation zone, of predefined length, often located at the center of the muscle length [7,60,105]. The fiber terminations at the musculotendon junctions can be similarly randomly located. Alternative representations based on experimental observations of anatomical innervation patterns have incorporated a longitudinally convex end-plate zone [133] and biomodal end-plate distributions [105].

In general, sEMG models have assumed fibers belonging to the same motor unit to be parallel and to have the same diameter and conduction velocity. Variations in fiber orientation, conduction velocity, motoneuron terminal branch lengths, and polyneuronal innervation can also be accounted for, and they may be necessary for simulation of muscles with more complex fiber arrangements. Fiber curvature has also been shown to alter the shape of simulated nerve and muscle fiber action potentials [25,86,97,136]. Applications involving more complex geometries include simulation of the fan-shaped genioglossus upper airway muscles [107] and models of the thenar [76] and gastrocnemius muscles [103], all of which illustrated the influence of motor unit anatomical as well as electrophysiological properties. In muscles with nonparallel fiber orientations where anisotropy is defined in the direction of the coordinate system, an error is introduced as the angle between the fibers and coordinate system is increased. To fully incorporate the effect of fiber orientation, the position-dependent nature of the tissue anisotropy should also be included as discussed in Section 8.6.3.2 [95].

8.7   MODELS OF THE SURFACE EMG SIGNAL

8.7.1   Modeling Motor Unit Recruitment, Firing, and Synchronization

During voluntary contraction, motor units fire repetitively in a largely unsynchronized manner. The order of motor unit recruitment during isometric contraction is governed by Henneman's size principle [66], which states that motor units are recruited in increasing order of size. The earliest recruited motor units also maintain a higher mean firing rate than simultaneously active later recruited units ("onion skin" phenomenon) [19,48]. Both properties are at least partly a consequence of the higher susceptibility of smaller, high input impedance motoneurons to discharge in response to synaptic inputs [67]. While evidence of motor unit substitution has been reported [4,134], in addition to limited reversal of recruitment order among motor units of similar thresholds [54,59,62,127] and during dynamic contractions [71,128], the size principle provides a framework for modeling motor unit recruitment that has been broadly verified across a range of contraction types. The influence of additional features such as distributed synaptic inputs which may alter the underlying recruitment pattern may then also be added.

Experimental studies have indicated that motor unit spike trains may be modeled by a stochastic renewal process with Gaussian distribution, where successive interpulse intervals are statistically independent [12]. Based on this, individual motor unit firing times may be modeled by drawing a series of motor unit interpulse intervals from a Gaussian distribution of known mean and standard deviation [50,69]. In addition to the statistical distribution of the firing times of a single motor unit, the distribution of mean firing rates across a population of motor units within the muscle must be simulated. In the model of force and surface EMG presented by Fuglevand et al. [50] (see also Chapter 10), the recruitment thresholds and mean firing rates of a population of motor units are defined as a function of a parameter closely related to synaptic current, termed the "excitatory drive." The mean firing rates of each motor unit at each incremental time interval are determined based on the excitatory drive and assigned motor unit properties, with motor unit size, twitch force, and recruitment threshold distributed exponentially across the population. The model presented by Fuglevand et al. was the first to simulate both the interference EMG signal and the force generated by a muscle based on the underlying physiology. A similar approach has been used as the basis of motor unit firing and recruitment simulation in many subsequent EMG models [24,37,81,84,93].

8.7.1.1 Motor Unit Synchronization.
Simulation of motor unit firing times based on observed statistical firing patterns provides an efficient means of generating physiologically realistic firing times for a large population of motor units. Synchronization of motor unit firings can be incorporated in these types of models by systematically adjusting firing times. Hermens et al. [69] simulated synchronization by aligning firing times according to a Gaussian distribution about a reference motor unit firing time. Using the model of Fuglevand et al. [50], Yao et al. [138] similarly introduced synchronization by adjusting a proportion of the independently generated firing times to align with the firing times of other units within the population. While conceptually simple and computationally efficient, the adjustment of motor unit firing times generated by a renewal point process in this way does not fully capture the physiological mechanisms responsible for the stochastic nature of motor unit firing and synchronization. Rapid changes in excitatory input—for example, due to synaptic noise or common periodic inputs from branched or synchronized presynaptic fibers believed to be responsible for phenomena such as short-term synchronization, common drive, and tremor—can not, therefore, be described. To simulate these effects requires a model that captures the dynamics of individual motoneuron transmembrane potentials [73,87,126].

These types of models, although computationally more intensive, enable factors such as oscillatory inputs to the motoneuron pool, synaptic noise, and motoneuron properties to be simulated and are necessary in order to examine how fluctuations in synaptic inputs can influence the sEMG signal. Models which incorporate the dynamics of the transmembrane potentials of individual motoneurons are particularly suited to simulating synchronization of motor units due to common presynaptic inputs, or effects of synaptic noise. Shared or common inputs arriving at motoneurons that are close to threshold will tend to cause these units to discharge simultaneously or with constant delays, resulting in synchronization.

Depending on the features to be explored, motoneuron models of varying complexity may be incorporated. Kleine et al. [73] examined effects of synchronization on sEMG features as a function of electrode position. Their model was based on the motoneuron model proposed by Matthews [90], which simulates the motoneuron afterhyperpolarization as an exponentially increasing function. Variations in firing rates were simulated by altering the input drive across motoneurons and synchronization was simulated by the addition of a "common" noise signal to subpopulations of the motoneuron pool. Since then, models of varying complexity have been developed to examine the effect of common synaptic inputs of varying amplitude and frequency on correlated motor unit firing and force fluctuations [87,126].

8.7.2 Incorporating Force Generation

In addition to the electrical sEMG activity, simulation of the mechanical output or force generated by the muscle is often required. Simulation of both EMG and force enables to investigate phenomena such as the sEMG–force relationship [24], recruitment, rate coding [50] and tremor [23] to be examined (see Chapter 10). The majority of structural sEMG models that have incorporated force have simulated force as the linear summation of the twitch potentials generated by all motor units within the muscle [37,81,93,139]. Most of these models have used an approach based on that proposed by Fuglevand et al. [50], where the amplitude and duration of the force twitch are assigned as a function of motor unit recruitment index, with twitch force modulated by the interpulse interval. A recent modeling study has adapted the Fuglevand force model for dynamic contractions by adjusting twitch force in accordance with established force–length and force–velocity relationships [23]. Surface EMG models have not yet, however, been extended to incorporate the details of the excitation–contraction coupling process, and the underlying processes such as the calcium dynamics and metabolic processes involved have not been considered. In general, the force component when incorporated in EMG models remains relatively simplified, with a dissociation between the electrical and mechanical activity of muscle.

8.8 MODEL VALIDATION

Validation is a critical step in any modeling process and refers to the procedure by which model outputs are compared with historical data. In the context of sEMG, localization of the origin of the excitation, the active fiber or group of fibers, is particularly challenging as there is currently no in vivo method to reliably determine the number of muscle fibers in a motor unit or the shape and size of the territory through which they are distributed. One of the earliest studies to directly compare experimentally recorded and simulated motor unit action potentials was presented by

Griep et al. [61]. In their study, intramuscularly recorded MUAPs from a single motor unit in the rat extensor digitorum longus were compared with simulated action potentials where the position of the electrodes and activated fibers were determined using histochemical techniques. A high level of similarity was observed between the simulated and recorded MUAPs. Differences were noted in the amplitudes of the MUAPs, due to the high sensitivity to material conductivities which are not well known. In addition, a more rapid rate of action potential increase and decay was observed in the recorded data than in the simulated data at action potential generation and extinction, possibly due to a more dominant effect of fibers located very close to the electrode [61].

More recently, Roeleveld et al. [113] compared surface motor unit action potentials from the human biceps muscle with simulated action potentials generated by muscle fibers located at the electrical center of the experimentally recorded motor units. Motor unit size and depth were estimated using scanning EMG. A three-layer model comprised of skin, fat, and muscle tissue was found to provide the closest agreement between simulated and experimental MUAPs. In all of the models examined, however, MUAP amplitude decreased more rapidly with increasing depth than in the experimental MUAPs. The reason for this discrepancy is not known, but may be due to a combination of factors including single-fiber distribution within the motor unit territory, variations in the true upper arm geometry from an ideal cylinder, incorrect estimates of conductivities, or possible limitations of classic volume conductor theory [113]. A similar observation, of a slower rate of decay (with increasing distance from the source in experimentally detected single-fiber action potential amplitude) than in simulated action potentials, was also reported for intramuscular SFAPs recorded in rat muscle [131]. It may be that to fully capture the volume conductor properties it is necessary to incorporate the microscopic structures of individual fiber membranes and local inhomogeneities which can result in phenomena such as variations in axial and transverse conductivity with frequency and with distance from the current source [2].

In an attempt to isolate the effect of volume conductor properties and avoid the complexities associated with localizing individual fibers, Lowery et al. [86] recorded the voltage distribution around the surface of the upper limb due to an intramuscular current source applied at an intensity below the threshold required for muscle fiber stimulation. The recorded surface potential distribution was compared with data simulated using an anatomically based model for the same subject. In this study, the muscle material properties reported by Gielen et al. [57], which incorporated both conductivity and permittivity, were found to provide the closest agreement between simulated and experimental data in a model based on macroscopic tissue properties.

Model validation remains one of the most challenging areas in EMG modeling. The difficulty of the problem is compounded by the large number of input parameters to even the simplest of sEMG models, in addition to the complex nature of the structures involved and the high level of uncertainty which surrounds many parameters—in particular, the low-frequency electrical properties of biological tissues. This implies that more than one set of (poorly estimated) parameters may result in a good match with experimental observations. As models of increasing accuracy continue to be developed, validation will play a more and more important role in parameter identification and determining critical model properties, particularly when tackling the challenges associated with solving the "inverse" problem and its possible multiple solutions.

8.9 APPLICATIONS OF MODELING

The wide range of areas to which sEMG modeling has been successfully applied is too extensive to summarize here, and many applications have already been mentioned. Rather than attempting a summary of all application areas, this section will instead concentrate on a small number of fields where EMG modeling has provided a fundamental understanding of specific phenomena and has facilitated the establishment of quantitative methods of extracting information from the signal. In addition, a number of selected applications representing the wide range of areas where modeling is now being applied are included.

8.9.1 Modeling Muscle Fatigue

Mathematical modeling has played a major role in understanding the changes in sEMG signal properties that accompany muscle fatigue (see Chapter 10). During sustained fatiguing isometric contractions a progressive decrease in muscle fiber conduction velocity is consistently observed [8,80,89], causing the frequency content of the sEMG signal to decrease and sEMG amplitude to increase as the duration of the extracellular potential increases [79,120]. Mathematical models have been extensively used to examine the relationship between experimentally observed variations in sEMG amplitude and the sEMG power spectrum and concurrent changes in muscle fiber conduction velocity with fatigue [77,94,120]. During voluntary contractions, spectral features including the sEMG median and mean frequencies tend to decrease by a relatively greater amount than predicted by the changes in conduction velocity alone. Model simulation provides a means to explore how additional factors such as changes in action potential shape, firing rates and motor unit synchronization may account for these differences [69,83].

In addition to providing insight into the mechanisms responsible for myoelectric manifestations of fatigue, modeling and simulation studies can provide a means of developing and testing new signal detection and processing algorithms for assessing fatigue. With a model it is possible to directly control parameters related to central and peripheral fatigue, such as conduction velocity, synchronization, motor unit firing rates, and recruitment, either individually or in parallel, and to examine resulting changes in the sEMG signal [34]. Applications have included the development of new nonlinear signal processing tools for estimating changes in conduction velocity and synchronization [41,101], algorithms for estimating individual MUs conduction velocities and amplitudes using sEMG [5], and the use of simulated data to test novel algorithms for quantifying changes in synchronization during fatigue [63].

8.9.2   Understanding Crosstalk

Crosstalk, the detection of unwanted signals from muscles lying in the vicinity of the muscles of interest, is one of the most significant limiting factors associated with surface EMG. This issue is also discussed in Chapter 2. The most direct method of assessing crosstalk is to examine surface signals detected at different sites while selectively activating a single muscle. In practice, this is often difficult to achieve, even with selective electrical stimulation [20]. Using model simulation, however, crosstalk can be directly quantified, its mechanisms explored, and techniques for reducing and removing crosstalk examined. Models enable selective activation of different regions of muscle under conditions in which the inputs to the system, including the locations of the active motor units and the tissue properties, can be controlled.

Using model simulation, Dimitrova and co-workers [29,33] illustrated that action potentials from far-away motor units can be dominated by the fiber-end effects. Their conclusion that temporal high-pass filtering is not a suitable means of reducing crosstalk was confirmed for conditions in which the fiber-end effects dominate the sEMG signal. Simulated signals have also been used to (a) compare the ability of spatial filters to reduce EMG crosstalk [102], (b) understand how the relative contribution of crosstalk to sEMG increases with increasing subcutaneous fat thickness [85], and (c) illustrate how correlation-based methods of estimating cross-talk have limitations [42,84].

8.9.3   Modeling Pathological Conditions

In addition to understanding sEMG signal properties under healthy conditions, modeling has the potential to explore how EMG analysis may be used to derive new biomarkers of neuropathic, myopathic, and central nervous system disorders. Modeling has previously been used to predict the relationship between surface detected motor unit action potentials and characteristic pathological changes in motor unit structure [36]. It has also been used to help understand the factors that contribute to abnormal EMG–force relations of paretic muscles in stroke [140]. There remains considerable scope to develop the clinical potential of sEMG through the simulation of neuromuscular disorders and to use modeling as a tool to provide insight into the mechanisms which underpin them.

8.10   CONCLUSIONS

Substantial progress has been made in the development of physiologically and anatomically accurate models which capture the most important features that determine the characteristics of each unique sEMG signal. Considering the main model components, it is clear that the sEMG signal is an output of a complex system that spans from the cellular to system level. Arising from this are the challenges associated with multiscale and multiphysics modeling including parameter identification, model validation, management of large data sets, incorporation of parameter uncertainty, and integration of simulation methods and measurement systems in real time.

There is now a wide range of models and simulation tools available for researchers to use to answer the questions in which they are interested. When designing the most appropriate model for an application, choices must be made regarding the level of detail that should be accounted for at each stage and the key features which should be included. These will depend on the problem to which a model is to be applied, with a trade-off often necessary between model complexity and computational efficiency. For example, accurate modeling of the limb geometry, asymmetry, tissue properties, and fiber curvature are important when the specific action potential shapes are of interest. However, when the objective is to examine more qualitative features of the surface EMG signal, then an idealized volume conductor model such as a cylindrical model with appropriate tissue properties provides a close enough approximation. The choice of the best model may be viewed as the problem of choosing the appropriate level of detail, often considered the most difficult aspect of modeling [9]. Care should be taken to consider the limitations inherent in any model and the impact that these could have on the results and conclusions. All underlying assumptions should be considered, along with the conditions under which they are valid.

One of the greatest challenges in increasing the accuracy of current sEMG models is the correct identification of the model parameters. Parameter identification is particularly difficult for systems such as this in which it is not possible to accurately measure all of the physiological, anatomical, and physical properties of the system. While anatomical properties such as limb geometries and fiber morphology may be determined through the use of imaging techniques including MRI and ultrasound, methods for determining parameters such as the size and distribution of fibers within each motor unit across the entire population are not currently available. Considerable uncertainty continues to surround the values of the electrical properties of biological tissues, particularly at the low frequencies relevant to sEMG modeling [52].

Although sophisticated models for simulating motor unit firing and recruitment have been developed, these rely heavily on core assumptions and approximations since the exact inputs and outputs of the motoneuron pool are not known. The recent development of sEMG decomposition algorithms will provide a means of determining the firing times and corresponding surface MUAPs of large numbers of motor units from within a single motor unit pool which is not feasible using intramuscular techniques. Data such as these, in combination with advanced imaging techniques, will help to further refine the next generation of sEMG models.

Last but not least, models are powerful teaching tools that enable the user to answer "What if . . . ." type questions, which may not be possible to address through experimental observations. Models can thus play an important role in teaching neuromuscular physiology and muscle electrophysiology, either in the classroom or through online learning. Although material of this type has been used in teaching students from a range of disciplines, including physiology, physical therapy, and movement sciences, there remains a compelling need to expand and disseminate the modeling approach in academic programs.

REFERENCES

1. Agarwal, G. C., and G. L. Gottlieb, "An analysis of the electromyogram by Fourier, simulation and experimental techniques," *IEEE Trans. Biomed. Eng.* **22**, 225–229 (1975).

2. Albers, B. A., W. L. C. Rutten, W. Wallinga-de Jonge, and H. B. K. Boom, "A model study on the influence of structure and membrane capacitance on volume conduction in skeletal-muscle tissue," *IEEE Trans. Biomed. Eng.* **33**, 681–689 (1986).

3. Andreassen, S., and A. Rosenfalck, "Relationship of intracellular and extracellular action-potentials of skeletal-muscle fibers," *CRC Crit. Rev. Bioeng.* **6**, 267–306 (1981).

4. Bawa, P., and C. Murnaghan, "Motor unit rotation in a variety of human muscles," *J. Neurophysiol.* **102**, 2265–2272 (2009).

5. Beck, R. B., C. J. Houtman, M. J. O'Malley, M. M. Lowery, and D. F. Stegeman, "A technique to track individual motor unit action potentials in surface EMG by monitoring their conduction velocities and amplitudes," *IEEE Trans. Biomed. Eng.* **52**, 622–629 (2005).

6. Blok, J. H., D. F. Stegeman, and A. van Oosterom, "Three-layer volume conductor model and software package for applications in surface electromyography," *Ann. Biomed. Eng.* **30**, 566–577 (2002).

7. Boyd, D. C., P. D. Lawrence, and P. J. Bratty, "On modeling the single motor unit action potential," *IEEE Trans. Biomed. Eng.* **25**, 236–243 (1978).

8. Broman, H., G. Bilotto, and C. J. Deluca, "Myoelectric signal conduction-velocity and spectral parameters—influence of force and time," *J. Appl. Physiol.* **58**, 1428–1437 (1985).

9. Brooks, R. J., and A. M. Tobias, "Choosing the best model: Level of detail, complexity, and model performance," *Math Comput Model* **24**, 1–14 (1996).

10. Burger, H. C., and R. van Dongen, "Specific electric resistance of body tissues," *Phys. Med. Biol.* **5**, 431–447 (1961).

11. Cannon, S. C., R. H. Brown, Jr., and D. P. Corey, "Theoretical reconstruction of myotonia and paralysis caused by incomplete inactivation of sodium channels," *Biophys. J.* **65**, 270–288 (1993).

12. Clamann, H. P., "Statistical analysis of motor unit firing patterns in a human skeletal muscle," *Biophys. J.* **9**, 1233–1251 (1969).

13. Clancy, E. A., and N. Hogan, "Single site electromyograph amplitude estimation," *IEEE Trans. Biomed. Eng.* **41**, 159–167 (1994).

14. Clancy, E. A., and N. Hogan, "Probability density of the surface electromyogram and its relation to amplitude detectors," *IEEE Trans. Biomed. Eng.* **46**, 730–739 (1999).

15. Clark, J., and R. Plonsey, "The Extracellular potential field of single active nerve fiber in a volume conductor," *Biophys. J.* **8**, 842–864 (1968).

16. Day, S. J., and M. Hulliger, "Experimental simulation of cat electromyogram: evidence for algebraic summation of motor-unit action-potential trains," *J. Neurophysiol.* **86**, 2144–2158 (2001).

17. De Luca, C. J., and E. J. van Dyk, "Derivation of some parameters of myoelectric signals recorded during sustained constant force isometric contractions," *Biophys. J.* **15**, 1167–1180 (1975).

18. De Luca, C. J., "Physiology and mathematics of myoelectric signals," *IEEE Trans. Biomed. Eng.* **26**, 313–325 (1979).

19. De Luca, C. J., R. S. LeFever, M. P. McCue, and A. P. Xenakis, "Control scheme governing concurrently active human motor units during voluntary contractions," *J. Physiol.* **329**, 129–142 (1982).

20. De Luca, C. J., and R. Merletti, "Surface myoelectric signal cross-talk among muscles of the leg," *Electroencephalogr. Clin. Neurophysiol.* **69**, 568–575 (1988).

21. De Luca, C. J., and J. L. Creigh, "Do the firing statistics of motor units modify the frequency component of the EMG signal during sustained contractions?," in *Biomechanics, International Series on Biomechanics IX-A*, D. A. Winter, eds., Human Kinetics Publishers, Champaign, IL, pp. 358–362, 1985.

22. DeAngelis, G. C., L. D. Gilmore, and C. J. DeLuca, "Standardized evaluation of techniques for measuring the spectral compression of the myoelectric signal," *IEEE Trans. Biomed. Eng.* **37**, 844–849 (1990).

23. Dideriksen, J., R. Enoka, and D. Farina, "A model of the surface electromyogram in pathological tremor," *IEEE Trans. Biomed. Eng.* **58**, 2178–2185 (2011).

24. Dideriksen, J. L., D. Farina, and R. M. Enoka, "Influence of fatigue on the simulated relation between the amplitude of the surface electromyogram and muscle force," *Philos. Trans. R. Soc. A Math. Phys. Eng. Sci.* **368**, 2765–2781 (2010).

25. Dimitrov, G. V., and N. A. Dimitrova, "Modelling of the extracellular potentials generated by curved fibres in a volume conductor," *Electromyogr. Clin. Neurophysiol.* **20**, 27–40 (1980).

26. Dimitrov, G. V., Z. C. Lateva, and N. A. Dimitrova, "Model of the slow components of skeletal muscle potentials," *Med. Biol. Eng. Comput.* **32**, 432–436 (1994).

27. Dimitrov, G. V., and N. A. Dimitrova, "Precise and fast calculation of the motor unit potentials detected by a point and rectangular plate electrode," *Med. Eng. Phys.* **20**, 374–381 (1998).

28. Dimitrov, G. V., C. Disselhorst-Klug, N. A. Dimitrova, A. Trachterna, and G. Rau, "The presence of unknown layer of skin and fat is an obstacle to a correct estimation of the motor unit size from surface detected potentials," *Electromyogr. Clin. Neurophysiol.* **42**, 231–241 (2002).

29. Dimitrov, G. V., C. Disselhorst-Klug, N. A. Dimitrova, E. Schulte, and G. Rau, "Simulation analysis of the ability of different types of multi-electrodes to increase selectivity of detection and to reduce cross-talk," *J. Electromyogr. Kinesiol.* **13**, 125–138 (2003).

30. Dimitrova, N., "Model of the extracellular potential field of a single striated muscle fibre," *Electromyogr. Clin. Neurophysiol.* **14**, 53–66 (1974).

31. Dimitrova, N. A., G. V. Dimitrov, and Z. C. Lateva, "Influence of the fiber length on the power spectra of single muscle fiber extracellular potentials," *Electromyogr. Clin. Neurophysiol.* **31**, 387–398 (1991).

32. Dimitrova, N. A., A. G. Dimitrov, and G. V. Dimitrov, "Calculation of extracellular potentials produced by an inclined muscle fibre at a rectangular plate electrode," *Med. Eng. Phys.* **21**, 583–588 (1999).

33. Dimitrova, N. A., G. V. Dimitrov, and O. A. Nikitin, "Neither high-pass filtering nor mathematical differentiation of the EMG signals can considerably reduce cross-talk," *J. Electromyogr. Kinesiol.* **12**, 235–246 (2002).

34. Dimitrova, N. A., and G. V. Dimitrov, "Interpretation of EMG changes with fatigue: facts, pitfalls, and fallacies," *J. Electromyogr. Kinesiol.* **13**, 13–36 (2003).

35. Disselhorst-Klug, C., J. Silny, and G. Rau, "Improvement of spatial resolution in surface-EMG: a theoretical and experimental comparison of different spatial filters," *IEEE Trans. Biomed. Eng.* **44**, 567–574 (1997).

36. Disselhorst-Klug, C., J. Silny, and G. Rau, "Estimation of the relationship between the noninvasively detected activity of single motor units and their characteristic pathological changes by modelling," *J. Electromyogr. Kinesiol.* **8**, 323–335 (1998).

37. Duchene, J., and J. Y. Hogrel, "A model of EMG generation," *IEEE Trans. Biomed. Eng.* **47**, 192–201 (2000).

38. Farina, D., and A. Rainoldi, "Compensation of the effect of sub-cutaneous tissue layers on surface EMG: a simulation study," *Med. Eng. Phys.* **21**, 487–497 (1999).

39. Farina, D., and R. Merletti, "Effect of electrode shape on spectral features of surface detected motor unit action potentials," *Acta Physiol. Pharmacol. Bulgarica* **26**, 63–66 (2001).

40. Farina, D., and R. Merletti, "A novel approach for precise simulation of the EMG signal detected by surface electrodes," *IEEE Trans. Biomed. Eng.* **48**, 637–646 (2001).

41. Farina, D., L. Fattorini, F. Felici, and G. Filligoi, "Nonlinear surface EMG analysis to detect changes of motor unit conduction velocity and synchronization," *J. Appl. Physiol.* **93**, 1753–1763 (2002).

42. Farina, D., R. Merletti, B. Indino, M. Nazzaro, and M. Pozzo, "Surface EMG crosstalk between knee extensor muscles: experimental and model results," *Muscle Nerve* **26**, 681–695 (2002).

43. Farina, D., L. Mesin, S. Martina, and R. Merletti, "A surface EMG generation model with multilayer cylindrical description of the volume conductor," *IEEE Trans. Biomed. Eng.* **51**, 415–426 (2004).

44. Farina, D., L. Mesin, and S. Martina, "Advances in surface electromyographic signal simulation with analytical and numerical descriptions of the volume conductor," *Med. Biol. Eng. Comput.* **42**, 467–476 (2004).

45. Farina, D., R. Merletti, B. Indino, and T. Graven-Nielsen, "Surface EMG crosstalk evaluated from experimental recordings and simulated signals. Reflections on crosstalk interpretation, quantification and reduction," *Methods Inf. Med.* **43**, 30–35 (2004).

46. Fortune, E., and M. M. Lowery, "Effect of membrane properties on skeletal muscle fiber excitability: a sensitivity analysis," *Med. Biol. Eng. Comput.* **50**, 617–629 (2012).

47. Foster, K. R., and H. P. Schwan, "Dielectric properties of tissues and biological materials: a critical review," *Crit. Rev. Biomed. Eng.* **17**, 25–104 (1989).

48. Freund, H. J., H. J. Budingen, and V. Dietz, "Activity of single motor units from human forearm muscles during voluntary isometric contractions," *J. Neurophysiol.* **38**, 933–946 (1975).

49. Fuglevand, A. J., D. A. Winter, A. E. Patla, and D. Stashuk, "Detection of motor unit action-potentials with surface electrodes—influence of electrode size and spacing," *Biol. Cybern.* **67**, 143–153 (1992).

50. Fuglevand, A. J., D. A. Winter, and A. E. Patla, "Models of recruitment and rate coding organization in motor-unit pools," *J. Neurophysiol.* **70**, 2470–2488 (1993).

51. Gabriel, C., S. Gabriel, and E. Corthout, "The dielectric properties of biological tissues. 1. Literature survey," *Phys. Med. Biol.* **41**, 2231–2249 (1996).

52. Gabriel, C., "Dielectric properties of biological materials," in *Handbook of Biological Effects of Electromagnetic Fields: Bioengineering and Biophysical Aspects of Electromagnetic Fields*, F. S. Barnes, and B. Greenebaum, eds., CRC Press, Boca Raton, FL, (2006).

53. Ganapathy, N., J. W. Clark, and O. B. Wilson, "Extracellular potentials from skeletal-muscle," *Math. Biosci.* **83**, 61–96 (1987).

54. Garnett, R., and J. A. Stephens, "Changes in the recruitment threshold of motor units produced by cutaneous stimulation in man," *J. Physiol.* **311**, 463–473 (1981).

55. Geddes, L. A., and L. E. Baker, "Specific resistance of biological material—a compendum of data for biomedical engineer and physiologist," *Med. Biol. Eng.* **5**, 271–293 (1967).

56. Gielen, F. L., H. E. Cruts, B. A. Albers, K. L. Boon, W. Wallinga-de Jonge, and H. B. Boom, "Model of electrical conductivity of skeletal muscle based on tissue structure," *Med. Biol. Eng. Comput.* **24**, 34–40 (1986).

57. Gielen, F. L. H., W. Wallinga-de Jonge, and K. L. Boon, "Electrical-conductivity of skeletal-muscle tissue—experimental results from different muscles in vivo," *Med. Biol. Eng. Comput.* **22**, 569–577 (1984).

58. Gootzen, T., D. F. Stegeman, and A. van Oosterom, "Finite limb dimensions and finite muscle length in a model for the generation of electromyographic signals," *Electroencephalogr. Clin. Neurophysiol.* **81**, 152–162 (1991).

59. Gorassini, M., J. F. Yang, M. Siu, and D. J. Bennett, "Intrinsic activation of human motoneurons: reduction of motor unit recruitment thresholds by repeated contractions," *J. Neurophysiol.* **87**, 1859–1866 (2002).

60. Griep, P. A., K. L. Boon, and D. F. Stegeman, "A study of the motor unit action potential by means of computer simulation," *Biol. Cybern.* **30**, 221–230 (1978).

61. Griep, P. A. M., F. L. H. Gielen, H. B. K. Boom, K. L. Boon, L. L. W. Hoogstraten, C. W. Pool, and W. Wallinga-de Jonge, "Calculation and registration of the same motor unit action-potential," *Electroencephalogr. Clin. Neurophysiol.* **53**, 388–404 (1982).

62. Grimby, L., and J. Hannerz, "Recruitment order of motor units on voluntary contraction: changes induced by proprioceptive afferent activity," *J. Neurol. Neurosurg. Psychiatry* **31**, 565–573 (1968).

63. Gronlund, C., A. Holtermann, K. Roeleveld, and J. S. Karlsson, "Motor unit synchronization during fatigue: a novel quantification method," *J. Electromyogr. Kinesiol.* **19**, 242–251 (2009).

64. Hamilton-Wright, A., and D. W. Stashuk, "Physiologically based simulation of clinical EMG signals," *IEEE Trans. Biomed. Eng.* **52**, 171–183 (2005).

65. Helal, J. N., and P. Bouissou, "The spatial integration effect of surface electrode detecting myoelectric signal," *IEEE Trans. Biomed. Eng.* **39**, 1161–1167 (1992).

66. Henneman, E., "Relation between size of neurons and their susceptibility to discharge," *Science* **126**, 1345–1347 (1957).

67. Henneman, E., and L. M. Mendell, "Functional organization of motoneuron pool and its inputs," *Comprehensive Physiol.* 423–507 (2011).

68. Heringa, A., D. F. Stegeman, G. J. Uijen, and J. P. de Weerd, "Solution methods of electrical field problems in physiology," *IEEE Trans. Biomed. Eng.* **29**, 34–42 (1982).

69. Hermens, H. J., T. A. M. Vonbruggen, C. T. M. Baten, W. L. C. Rutten, and H. B. K. Boom, "The median frequency of the surface emg power spectrum in relation to motor unit firing and action-potential properties," *J. Electromyogr. Kinesiol.* **2**, 15–25 (1992).

70. Hogan, N., and R. W. Mann, "Myoelectric signal processing: optimal estimation applied to electromyography—part I: derivation of the optimal myoprocessor," *IEEE Trans. Biomed. Eng.* **27**, 382–395 (1980).

71. Howell, J. N., A. J. Fuglevand, M. L. Walsh, and B. Bigland-Ritchie, "Motor unit activity during isometric and concentric-eccentric contractions of the human first dorsal interosseus muscle," *J. Neurophysiol.* **74**, 901–904 (1995).

72. Jackson, J. D., *Classical Electrodynamics*, 3rd ed., Wiley, New York, 1999.

73. Kleine, B. U., D. F. Stegeman, D. Mund, and C. Anders, "Influence of motoneuron firing synchronization on SEMG characteristics in dependence of electrode position," *J. Appl. Physiol.* **91**, 1588–1599 (2001).

74. Kleinpenning, P. H., T. Gootzen, A. Vanoosterom, and D. F. Stegeman, "The equivalent source description representing the extinction of an action-potential at a muscle-fiber ending," *Math. Biosci.* **101**, 41–61 (1990).

75. Lago, P., and N. B. Jones, "Effect of motor-unit firing time statistics on e.m.g. spectra," *Med. Biol. Eng. Comput.* **15**, 648–655 (1977).

76. Lateva, Z. C., K. C. McGill, and C. G. Burgar, "Anatomical and electrophysiological determinants of the human thenar compound muscle action potential," *Muscle & Nerve* **19**, 1457–1468 (1996).

77. Lindstrom, L., R. Magnusson, and I. Petersen, "Muscular fatigue and action potential conduction velocity changes studied with frequency analysis of EMG signals," *Electromyography* **10**, 341–356 (1970).

78. Lindstrom, L., R. Kadefors, and I. Petersen, "Electromyographic index for localized muscle fatigue," *J. Appl. Physiol.* **43**, 750–754 (1977).

79. Lindstrom, L. H., and R. I. Magnusson, "Interpretation of myoelectric power spectra—model and its applications," *Proc. IEEE* **65**, 653–662 (1977).

80. Linssen, W., D. F. Stegeman, E. M. G. Joosten, M. A. Vanthof, R. A. Binkhorst, and S. L. H. Notermans, "Variability and interrelationships of surface emg parameters during local muscle fatigue," *Muscle Nerve* **16**, 849–856 (1993).

81. Lowery, M. M., C. L. Vaughan, P. J. Nolan, and M. J. O'Malley, "Spectral compression of the electromyographic signal due to decreasing muscle fiber conduction velocity," *IEEE Trans. Rehabil. Eng.* **8**, 353–361 (2000).

82. Lowery, M. M., N. S. Stoykov, A. Taflove, and T. A. Kuiken, "A multiple-layer finite-element model of the surface EMG signal," *IEEE Trans. Biomed. Eng.* **49**, 446–454 (2002).

83. Lowery, M. M., and M. J. O'Malley, "Analysis and simulation of changes in EMG amplitude during high-level fatiguing contractions," *IEEE Trans. Biomed. Eng.* **50**, 1052–1062 (2003).

84. Lowery, M. M., N. S. Stoykov, and T. A. Kuiken, "A simulation study to examine the use of cross-correlation as an estimate of surface EMG cross talk," *J. Appl. Physiol.* **94**, 1324–1334 (2003).

85. Lowery, M. M., N. S. Stoykov, and T. A. Kuiken, "Independence of myoelectric control signals examined using a surface EMG model," *IEEE Trans. Biomed. Eng.* **50**, 789–793 (2003).

86. Lowery, M. M., N. S. Stoykov, J. P. Dewald, and T. A. Kuiken, "Volume conduction in an anatomically based surface EMG model," *IEEE Trans. Biomed. Eng.* **51**, 2138–2147 (2004).

87. Lowery, M. M., and Z. Erim, "A simulation study to examine the effect of common motoneuron inputs on correlated patterns of motor unit discharge," *J. Comput. Neurosci.* **19**, 107–124 (2005).

88. Lynn, P. A., N. D. Bettles, A. D. Hughes, and S. W. Johnson, "Influences of electrode geometry on bipolar recordings of the surface electromyogram," *Med. Biol. Eng. Comput.* **16**, 651–660 (1978).

89. Masuda, K., T. Masuda, T. Sadoyama, M. Inaki, and S. Katsuta, "Changes in surface EMG parameters during static and dynamic fatiguing contractions," *J. Electromyogr. Kinesiol.* **9**, 39–46 (1999).

90. Matthews, P. B., "Relationship of firing intervals of human motor units to the trajectory of post-spike after-hyperpolarization and synaptic noise," *J. Physiol.* **492** (Pt2), 597–628 (1996).

91. McGill, K. C., and Z. C. Lateva, "A model of the muscle-fiber intracellular action potential waveform, including the slow repolarization phase," *IEEE Trans. Biomed. Eng.* **48**, 1480–1483 (2001).

92. McGill, K. C., "Surface electromyogram signal modelling," *Med. Biol. Eng. Comput.* **42**, 446–454 (2004).

93. Merletti, R., L. Lo Conte, E. Avignone, and P. Guglielminotti, "Modeling of surface myoelectric signals—part I: model implementation," *IEEE Trans. Biomed. Eng.* **46**, 810–820 (1999).

94. Merletti, R., S. H. Roy, E. Kupa, S. Roatta, and A. Granata, "Modeling of surface myoelectric signals—part II: model-based signal interpretation," *IEEE Trans. Biomed. Eng.* **46**, 821–829 (1999).

95. Mesin, L., and D. Farina, "Simulation of surface EMG signals generated by muscle tissues with inhomogeneity due to fiber pinnation," *IEEE Trans. Biomed. Eng.* **51**, 1521–1529 (2004).

96. Mesin, L., and D. Farina, "A model for surface EMG generation in volume conductors with spherical inhomogeneities," *IEEE Trans. Biomed. Eng.* **52**, 1984–1993 (2005).

97. Mesin, L., M. Joubert, T. Hanekom, R. Merletti, and D. Farina, "A finite element model for describing the effect of muscle shortening on surface EMG," *IEEE Trans. Biomed. Eng.* **53**, 593–600 (2006).

98. Mesin, L., "Simulation of surface EMG signals for a multilayer volume conductor with triangular model of the muscle tissue," *IEEE Trans. Biomed. Eng.* **53**, 2177–2184 (2006).

99. Mesin, L., and D. Farina, "An analytical model for surface EMG generation in volume conductors with smooth conductivity variations," *IEEE Trans. Biomed. Eng.* **53**, 773–779 (2006).

100. Mesin, L., "Simulation of surface EMG signals for a multilayer volume conductor with a superficial bone or blood vessel," *IEEE Trans. Biomed. Eng.* **55**, 1647–1657 (2008).

101. Mesin, L., C. Cescon, M. Gazzoni, R. Merletti, and A. Rainoldi, "A bidimensional index for the selective assessment of myoelectric manifestations of peripheral and central muscle fatigue," *J. Electromyogr. Kinesiol.* **19**, 851–863 (2009).

102. Mesin, L., S. Smith, S. Hugo, S. Viljoen, and T. Hanekom, "Effect of spatial filtering on crosstalk reduction in surface EMG recordings" *Med. Eng. Phys.* **31**, 374–383 (2009).

103. Mesin, L., R. Merletti, and T. M. Vieira, "Insights gained into the interpretation of surface electromyograms from the gastrocnemius muscles: a simulation study," *J. Biomech.* **44**, 1096–1103 (2011).

104. Nandedkar, S. D., and E. Stalberg, "Simulation of single muscle fibre action potentials," *Med. Biol. Eng. Comput.* **21**, 158–165 (1983).

105. Nandedkar, S. D., E. V. Stalberg, and D. B. Sanders, "Simulation techniques in electromyography," *IEEE Trans. Biomed. Eng.* **32**, 775–785 (1985).

106. Nandedkar, S. D., D. B. Sanders, and E. V. Stalberg, "Selectivity of electromyographic recording techniques: a simulation study," *Med. Biol. Eng. Comput.* **23**, 536–540 (1985).

107. O'Connor, C., M., M. M. Lowery, L. S. Doherty, M. McHugh, C. O'Muircheartaigh, J. Cullen, P. Nolan, W. T. McNicholas, and J. O'Malley, "Improved surface EMG electrode for measuring genioglossus muscle activity," *Respir. Physiol. Neurobiol.* **159**, 55–67 (2007).

108. Pan, Z. S., Y. Zhang, and P. A. Parker, "Motor unit power spectrum and firing rate," *Med. Biol. Eng. Comput.* **27**, 14–18 (1989).

109. Plonsey, R., and D. B. Heppner, "Considerations of quasi-stationarity in electrophysiological systems," *Bull. Math. Biophys.* **29**, 657–664 (1967).

110. Plonsey, R., and D. G. Fleming, *Bioelectric Phenomena*, McGraw-Hill, New York, 1969.

111. Plonsey, R., "Active fiber in a volume conductor," *IEEE Trans. Biomed. Eng.* **BM21**, 371–381 (1974).

112. Rodriguez-Falces, J., J. Navallas, L. Gila, A. Malanda, and N. A. Dimitrova, "Influence of the shape of intracellular potentials on the morphology of single-fiber extracellular potentials in human muscle fibers," *Med. Biol. Eng. Comput.* **50**, 447–460 (2012).

113. Roeleveld, K., J. H. Blok, D. F. Stegeman, and A. vanOosterom, "Volume conduction models for surface EMG; confrontation with measurements," *J. Electromyogr. Kinesiol.* **7**, 221–232 (1997).

114. Rosenfalck, P., "Intra- and extracellular potential fields of active nerve and muscle fibres. A physico-mathematical analysis of different models," *Acta Physiol. Scand.* **Suppl 321**, 1–168 (1969).

115. Roth, B. J., F. L. H. Gielen, and J. P. Wikswo, "Spatial and temporal frequency-dependent conductivities in volume–conduction calculations for skeletal-muscle," *Math. Biosci.* **88**, 159–189 (1988).

116. Rutten, W. L. C., B. K. van Veen, S. H. Stroeve, H. B. K. Boom, and W. Wallinga, "Influence of inhomogeneities in muscle tissue on single-fibre action potentials: a model study," *Med. Biol. Eng. Comput.* **35**, 91–95 (1997).

117. Saitou, K., T. Masuda, and M. Okada, "Depth and intensity of equivalent current dipoles estimated through an inverse analysis of surface electromyograms using the image method," *Med. Biol. Eng. Comput.* **37**, 720–726 (1999).

118. Schneider, J., J. Silny, and G. Rau, "Influence of tissue inhomogeneities on noninvasive muscle-fiber conduction-velocity measurements—investigated by physical and numerical modeling," *IEEE Trans. Biomed. Eng.* **38**, 851–860 (1991).

119. Shwedyk, E., R. Balasubramanian, and R. N. Scott, "A nonstationary model for the electromyogram," *IEEE Trans. Biomed. Eng.* **24**, 417–424 (1977).

120. Stegeman, D. F., and W. Linssen, "Muscle-fiber action-potential changes and surface emg—a simulation study," *J. Electromyogr. Kinesiol.* **2**, 130–140 (1992).

121. Stegeman, D. F., J. H. Blok, H. J. Hermens, and K. Roeleveld, "Surface EMG models: properties and applications," *J. Electromyogr. Kinesiol.* **10**, 313–326 (2000).

122. Stegeman, D. F., R. Merletti, and H. J. Hermens, "EMG modeling and simulation," in *Electromyography: Physiology, Engineering, and Noninvasive Applications*, R. Merletti, and P. Parker, eds., Wiley-IEEE Press, Hoboken, NJ, 2004.

123. Stoykov, N. S., M. M. Lowery, A. Taflove, and T. A. Kuiken, "Frequency- and time-domain FEM models of EMG: capacitive effects and aspects of dispersion," *IEEE Trans. Biomed. Eng.* **49**, 763–772 (2002).

124. Stulen, F. B., and C. J. Deluca, "Frequency parameters of the myoelectric signal as a measure of muscle conduction-velocity," *IEEE Trans. Biomed. Eng.* **28**, 515–523 (1981).

125. Taflove, A., and S. C. Hagness, *Computational Electrodynamics: The Finite-Difference Time-Domain Method*, 3rd ed., Artech House, Boston, 2005.

126. Taylor, A. M., and R. M. Enoka, "Quantification of the factors that influence discharge correlation in model motor neurons," *J. Neurophysiol.* **91**, 796–814 (2004).

127. Thomas, C. K., B. H. Ross, and B. Calancie, "Human motor-unit recruitment during isometric contractions and repeated dynamic movements," *J. Neurophysiol.* **57**, 311–324 (1987).

128. van Bolhuis, B. M., W. P. Medendorp, and C. C. Gielen, "Motor unit firing behavior in human arm flexor muscles during sinusoidal isometric contractions and movements," *Exp. Brain Res.* **117**, 120–130 (1997).

129. van Dijk, J. P., M. M. Lowery, B. G. Lapatki, and D. F. Stegeman, "Evidence of potential averaging over the finite surface of a bioelectric surface electrode," *Ann. Biomed. Eng.* **37**, 1141–1151 (2009).

130. van Veen, B. K., H. Wolters, W. Wallinga, W. L. Rutten, and H. B. Boom, "The bioelectrical source in computing single muscle fiber action potentials," *Biophys. J.* **64**, 1492–1498 (1993).

131. van Veen, B. K., E. Mast, R. Busschers, A. J. Verloop, W. Wallinga, W. L. Rutten, P. O. Gerrits, and H. B. Boom, "Single fibre action potentials in skeletal muscle related to recording distances," *J. Electromyogr. Kinesiol.* **4**, 37–46 (1994).

132. Wallinga, W., S. L. Meijer, M. J. Alberink, M. Vliek, E. D. Wienk, and D. L. Ypey, "Modelling action potentials and membrane currents of mammalian skeletal muscle fibres in coherence with potassium concentration changes in the T-tubular system," *Eur. Biophys. J.* **28**, 317–329 (1999).

133. Wani, A. M., and S. K. Guha, "Synthesising of a motor unit potential based on the sequential firing of muscle fibres," *Med. Biol. Eng. Comput.* **18**, 719–726 (1980).

134. Westgaard, R. H., and C. J. de Luca, "Motor unit substitution in long-duration contractions of the human trapezius muscle," *J. Neurophysiol.* **82**, 501–504 (1999).

135. Wood, S. M., J. A. Jarratt, A. T. Barker, and B. H. Brown, "Surface electromyography using electrode arrays: a study of motor neuron disease," *Muscle Nerve* **24**, 223–230 (2001).

136. Xiao, S., K. C. McGill, and V. R. Hentz, "Action potentials of curved nerves in finite limbs," *IEEE Trans. Biomed. Eng.* **42**, 599–607 (1995).

137. Yamamoto, T., and Y. Yamamoto, "Electrical properties of the epidermal stratum corneum," *Med. Biol. Eng.* **14**, 151–158 (1976).

138. Yao, W., R. J. Fuglevand, and R. M. Enoka, "Motor-unit synchronization increases EMG amplitude and decreases force steadiness of simulated contractions," *J. Neurophysiol.* **83**, 441–452 (2000).

139. Zhou, P., and W. Z. Rymer, "Factors governing the form of the relation between muscle force and the EMG: a simulation study," *J. Neurophysiol.* **92**, 2878–2886 (2004).

140. Zhou, P., N. L. Suresh, and W. Z. Rymer, "Model based sensitivity analysis of EMG–force relation with respect to motor unit properties: applications to muscle paresis in stroke," *Ann. Biomed. Eng.* **35**, 1521–1531 (2007).

9

ELECTROMYOGRAPHY-DRIVEN MODELING FOR SIMULATING SUBJECT-SPECIFIC MOVEMENT AT THE NEUROMUSCULOSKELETAL LEVEL

M. Sartori,[1] D. G. Lloyd,[2] T. F. Besier,[3] J. W. Fernandez,[3] and D. Farina[1]

[1]*Department of Neurorehabilitation Engineering, University Medical Center Göttingen, Göttingen, Germany*
[2]*Centre for Musculoskeletal Research, Griffith Health Institute, Griffith University, Gold Coast, QLD, Australia*
[3]*Department of Engineering Science and Auckland Bioengineering Institute, University of Auckland, Auckland, New Zealand*

9.1 INTRODUCTION

Understanding human movement implies decoding the complex dynamics of the neural drive to muscles and reconstructing its link with the associated musculoskeletal function. The neural drive to muscles is comprised of discrete events (i.e., action potentials), generated in central and peripheral systems. These ultimately converge to pools of motor neurons and contribute to recruit muscles and produce subsequent movement [20].

Decoding the dynamics of the neural drive to muscles is crucial to understand how the nervous system and the musculoskeletal system are interfaced with one another. Furthermore, it is crucial for enabling a number of applications with a substantial

Surface Electromyography: Physiology, Engineering, and Applications, First Edition.
Edited by Roberto Merletti and Dario Farina.
© 2016 by The Institute of Electrical and Electronics Engineers, Inc. Published 2016 by John Wiley & Sons, Inc.

impact on the society ranging from the treatment and prevention of musculoskeletal disorders to the control of assistive devices for movement restoration.

To link the neurophysiologic events that underlie human movement with the associated musculoskeletal function, it is necessary to collect electrophysiological data reflecting the neural excitation signals sent to muscles. Then, it is necessary to couple these measurements to physiologically accurate, subject-specific, computational models of the human musculoskeletal system. This enables modeling and simulating the actions of muscles on the skeletal system as controlled by the nervous system [10]. We will refer to this process as "neuromusculoskeletal modeling."

The neural drive to muscles is ultimately determined by action potential trains generated from pools of alpha motor neurons that innervate specific muscles [20]. To record this activity in vivo, it is possible to employ both invasive and noninvasive measurement techniques, which are characterized by different levels of selectivity, stability, and noise contamination. Surface electromyography (EMG) signals indirectly reflect the neural drive to muscles and can be easily recorded in vivo during dynamic motor tasks. Although limited in terms of selectivity and noise contamination, the information content of surface EMG data gives a meaningful insight into the muscle neural excitation level. Therefore, EMG data experimentally recorded from the major superficial muscle groups have been successfully used as a direct input drive to numerical musculoskeletal models of human limbs [10,28,41,46,47,57–60,63,77].

Neuromusculoskeletal modeling driven by surface EMG recordings (i.e., EMG-driven modeling) has demonstrated to be an effective way of predicting muscle dynamics that reflect both experimental EMG data and joint moments in both in healthy [41,46,47,57–60,63] and pathological individuals [6,26,32,38,45,46,52,66,67]. A major benefit is the possibility of determining the subject-specific relationship between patterns of muscle excitation and the resulting force produced in the associated musculotendon units (MTUs) without making any a priori assumption on how muscles are recruited and activated [1,2,71,72]. This opens up the possibility of studying how neuromuscular impairment or adaptation contributes to abnormal movement. It recently enabled investigating motor limitations in subjects with anterior cruciate ligament rupture [67], patellofemoral pain [6], osteoarthritis [26,38], stroke [26,32,66], and upper extremity neuromuscular injuries [45].

Translating EMG-driven modeling to the clinical practice represents a tremendous opportunity for the development of advanced assistive technologies. In this context, EMG-driven modeling enables accurate prediction of an individual's effort and motor capacity. These are key features for implementing myoelectric control of powered orthoses and prostheses in proportion to the predicted user's effort (i.e., also see Chapter 20) [11,25,29,35,48,51,63–65].

The remainder of this chapter will provide a comprehensive description of EMG-driven musculoskeletal modeling methods. It is worth stressing that although the focus will be on the human lower extremity, the same concepts are generalizable to other human extremities. Finally, we will discuss the limitations of current methods and explore relevant applications and future challenges to be addressed in this field.

9.2   MOTION CAPTURING AND BIOMECHANICAL MODELING OF THE HUMAN BODY

EMG-driven modeling requires experimental human motion data to be captured for model calibration and operation (see Section 9.4). Typically, data include recordings of EMG signals, along with three-dimensional joint kinematics and kinetics. The EMG data are usually recorded using multichannel acquisition systems or bipolar electrodes. Joint kinematics and dynamics can be recorded from angle and force sensors placed in correspondence of the joint center of rotation. However, because of the difficulty in identifying the instantaneous center of rotation in human joints, stereophotogrammetry techniques and models of the human skeletal mechanics (i.e., kinematics and dynamics) are often used [13,68,69,78]. In this scenario, retro-reflective markers are placed on the human body and their three-dimensional positions are tracked using optical motion capture cameras. Foot-ground reaction force (GRF) data can be recorded using force plates. In this process, EMG data, marker trajectories, and GRFs are recorded synchronously during the human movement.

A model of the human skeletal mechanics (HSM) is used to extract joint kinematics and kinetics from the experimentally recorded marker trajectories and GRFs. The HSM model is a numerical representation of the anatomical segments (i.e., limbs) interconnected by joints, influenced by forces (i.e., GRFs), and subject to constraints (i.e., joint range of motion) [68]. The anatomical segments are typically treated as rigid bodies that do not undergo deformation when force is exerted on them. The joint interconnecting two segments can be represented by biologically based joints with the center of rotation moving as a function of three-dimensional rotations and translations (Fig. 9.1; also see Section 9.3) [13,69,78].

The marker trajectories recorded during a subject's static standing pose are typically used to scale the HSM model to match the subject's anthropometry. This can be done by including a set of virtual markers in the HSM model that reflects the actual marker set experimentally applied on the real subject. The dimensions of the anatomical segments (i.e., length, width, and depth) are then linearly scaled based on the relative distance between each pair of experimental and corresponding virtual markers. In this process, the mass properties of the subject are also proportionally adjusted in order to match the subject's total measured mass. The adjusted parameters include the segment mass, the location of the segment center of mass, and the mass moment of inertia. The initial values for these parameters are derived from generic templates that represent normative human anthropometry (i.e., also see Section 9.3) [15,16].

The scaled HSM model is then used to reconstruct the three-dimensional joint angles based on the marker trajectories recorded during dynamic motor tasks. In this step, an inverse kinematics (IK) problem is solved. This can be formulated as a least-squares problem that minimizes the squared three-dimensional distances between experimental and virtual markers, subject to the joint constraints and motion ranges of the HSM model [15,43]. As a result, instantaneous joint angle values are computed frame by frame. This is done by positioning the scaled HSM model in a pose that best matches the experimentally recorded marker trajectories sequentially across each time frame [15,43].

(a)   **(b)**

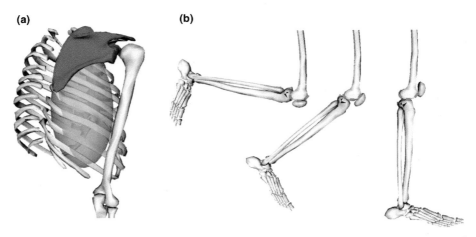

FIGURE 9.1 **(A)** A three degrees of freedom model of the scapula–thoracic joint in the shoulder articulation. The scapula (*highlighted body*) is constrained to move on the surface of an ellipsoid (*shaded object*), which is fixed to the thorax body. **(B)** The knee joint is modeled by the planar sagittal displacement of the tibia body with respect to the femur body. In this, the tibia head is constrained to move on an elliptical curve (*dotted curve*) that defines the femur epicondyles as a function of the knee flexion angle. The position of the knee is here presented at 100 (*left*), 50 (*center*), and 0 (*right*) degrees of knee flexion.

The joint moments needed to track the IK-generated angles are traditionally computed by solving an inverse dynamics (ID) problem [39]. However, traditional ID algorithms do not account for the fact that the IK-generated joint kinematics is not "dynamically" consistent with the measured GRFs. Joint kinematics inconsistencies are the result of limitations in the human movement measuring and modeling process. In this, the recorded marker trajectories are filtered, smoothed, and interpolated (i.e., by using cubic splines) in order to remove recording and movement artifacts and fill marker trajectory gaps occurring when the marker position is not properly recorded. Therefore, by the end of this step, the processed marker trajectories do not entirely reflect the actual recorded human movement. The filtered, smoothed, and interpolated marker trajectories are then used as input to the IK algorithm. This, in turn, will generate a solution that only corresponds to a "best" fit between the scaled HSM model virtual markers and the corresponding experimental ones. As a result, tracking inconsistencies between the virtual HSM model and the actual human movement are introduced and these errors are compounded by the need to double differentiate position data to obtain segmental accelerations. Therefore, the IK-generated joint angle solution is typically not dynamically consistent with the actual subject's performed movement. In this scenario, traditional algorithms solve the ID problem by applying nonphysical forces and moments (i.e., residuals) to a body in the HSM model in order to account for the inherent dynamic inconsistencies between recorded GRFs and derived joint kinematics [39]. However, the preferred solution to account for the joint kinematics and GRFs mismatch is the use of residual reduction analysis

(RRA) [15]. This allows reducing the magnitude of the residuals by slightly adjusting the IK-generated joint angles as well as the HSM model mass properties [15]. This allows obtaining an ID solution with minimized residual forces and moments. In the remainder of this chapter we will refer to the adjusted joint angles and moments produced by this approach as "the experimental" angles and moments.

The steps described in this section can be performed using modeling software packages such as OpenSim (http://opensim.stanford.edu/) [15]. Additional software packages include AnyBody (http://www.anybodytech.com) and Biomechanics of Bodies (http://www.prosim.co.uk/BoB/).

9.3   MUSCULOSKELETAL MODELING

After the kinematics of anatomical joints is estimated, it is possible to determine the resulting kinematics in the multiple MTUs spanning the human joints. This requires the development of a model of the musculoskeletal system that allows representing the complex geometry of MTUs wrapping around bones and spanning multiple joints. We refer to this as "musculoskeletal modeling."

Ideally, a musculoskeletal model is created from medical imaging data of bone and muscle surfaces, such as magnetic resonance imaging (MRI) or computed tomography (Fig. 9.2A) [5]. The scanned images are segmented to obtain a three-dimensional point cloud for each bone (Fig. 9.2B) [5]. The origin and insertion points of major selected MTUs are also segmented from the images. Then, triangulated surfaces are fit to the point cloud and subsequently smoothed in order to create a uniform mesh for each individual bone (Fig. 9.2C) [5]. The shortest line segment connecting the MTU origin point to the MTU insertion point characterizes its path geometry (Fig. 9.2D) [5]. This represents the path geometry of muscle fibers in series with tendons. Depending on the anatomical configuration of the spanned joints, the MTU path (i.e., line segment) can wrap around the modeled bone surfaces. In this case, wrapping points are added to the MTU path when muscle-to-bone contact is detected (Fig. 9.2E). This has been shown to accurately represent the kinematics (i.e., length, velocity, and moment arms) of MTUs, both in the upper [34,56] and lower [4,62] extremities as a function of the three-dimensional joint angles. In this process, regions of bones can also be deformed based on anthropometric data to generate more accurate MTU kinematics estimates [3].

Musculoskeletal models are more commonly created from data averaged across a large number of cadaveric specimens. This results in averaged, generic models that require scaling to match an individual's size and anatomical segment proportions. This is analogous to what described in Section 9.2. It is worth noting that this has the advantage of allowing the application of musculoskeletal modeling to any subject whether or not medical imaging data are available to begin with. The disadvantage of scaling methods is that they might not account for subject-specific bone and muscle geometry, which might be particularly important in pathological populations.

The above-mentioned methodology provides an estimate of the MTU kinematics, but it does not give insight into the kinematics of the muscle fibers. To address this issue, researchers have proposed three-dimensional descriptions of the muscle surface

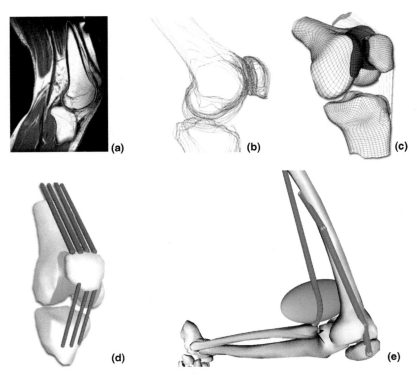

FIGURE 9.2 The steps involved in the creation of a subject-specific musculoskeletal model of the human knee joint [5]. The MRI images are segmented (**A**) and used to represent three-dimensional point clouds (**B**) of bones and muscle origin and insertion points. These are converted into quadrilateral and hexahedral meshes of the bones and cartilage (**C**) and muscle line segments (**D**). The musculotendon line segment (**E**) is defined as the shortest line from the origin to the insertion point. Here, musculotendon line segments can wrap around bony surfaces in the correspondence of muscle-to-bone contact; that is, see vastus lateralis' highlighted contact points in the figure. Furthermore, additional surfaces can be used to better represent the anatomical path of a muscle; that is, see shaded ellipsoidal object constraining the biceps femoris long head in the figure. Part 2A is taken and adapted from Blemker et al. [8]. Parts 2B, 2C, and 2D are taken and adapted from Besier et al. [5].

geometry. In this scenario, finite-element modeling is used to create volumetric representation of muscles from the surface data [8,21,22,74]. This method reveals how muscles change their shape as they interact with neighboring muscles, bones, and other structures as the joints move (Fig. 9.3A) [21]. However, these methods are typically associated with large computational costs that increase with the volumetric surface to be represented. Furthermore, these methods are not easily applicable to a large subject population. This significantly reduces the ability of creating subject-specific large-scale models that comprehend a large number of muscles, bones, and degrees of freedom (DOFs) and validate them on a large subject set.

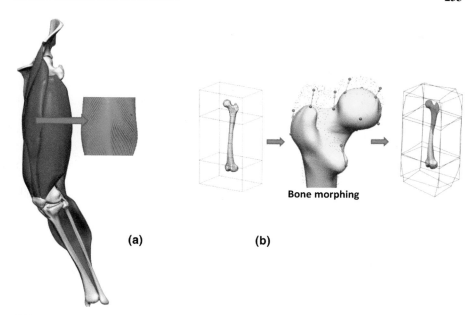

Bone morphing

(a)   (b)

FIGURE 9.3 **(A)** Lower limb model from the Physiome Project repository containing muscle fiber field information from both cadaver dissection and diffusion tensor imaging. The arrow points to an enlarged image of the single rectus femoris muscle with fitted bi-pinnate fiber field. **(B)** Host–mesh morphing of the femoral head using free-form deformation. The femur is embedded inside a host–mesh, which is morphed to minimize the distance between chosen landmark and target control points. The femur, which is embedded inside the host, also experiences the same deformation. Hence a few control points can customize the geometry of many degrees of freedom. Parts A and B are reproduced from Fernandez and co-workers [21,23] with permission.

In this context, the efficiency of extracting muscle anatomy from medical imaging scans can be significantly improved by post processing methods. Fernandez et al. [23] proposed a hybrid approach that allows deforming and scaling an existing high-resolution volumetric finite-element musculoskeletal model to fit a low-resolution model created from MRI data. In this scenario, the complex three-dimensional bone and muscle meshes are individually associated and linked to a simpler three-dimensional geometrical form such as parallelepipeds (i.e., hosting form). A morphing technique is then applied to the simple hosting form rather than to the complex bone or muscle mesh. By deforming the hosting form, the complex musculoskeletal structure within is also morphed accordingly (Fig. 9.3B). This speeds up the process of personalizing high-resolution musculoskeletal model to fit an individual's anatomy. These methods can be performed by using recent, freely available software packages such as the Musculoskeletal Atlas Project (MAP) Client (https://simtk.org/home/map). This has been made available to enable researchers to rapidly generate detailed three-dimensional musculoskeletal models using image fusion, segmentation, and mesh generation methods.

The methods presented here produce models of the musculoskeletal system for the purpose of determining the kinematics of muscles during human movement as well as the resulting deformation due to interaction with neighboring muscular and skeletal structures [22]. However, in order to determine the force produced by muscles during movement, it is still necessary to account for how muscles excite, and develop force and moment around multiple joints and about multiple DOFs simultaneously. To address this problem, experimental measurements reflecting the dynamics of the neural drive (i.e., EMG data) can be used to drive muscles within a computational framework that includes models of muscle activation, kinematics, and contraction dynamics as well as joint mechanics. This framework will be referred to as an EMG-driven musculoskeletal model.

In the following we will refer to an EMG-driven musculoskeletal model in which the MTU kinematics is represented using one-dimensional line segments and wrapping surfaces (Fig. 9.2) [62]. In this context, a muscle can be composed of one or multiple MTUs [58]. For instance, in the methodology described in the following section, the large gluteus maximus muscle is composed of three MTUs defining the main muscular compartments. Pennate muscles such as the rectus femoris are modeled by a single one-dimensional MTU.

9.4 EMG-DRIVEN MUSCULOSKELETAL MODELING AND SIMULATION

EMG-driven musculoskeletal modeling is a forward dynamics approach in which EMG signals experimentally recorded from the major muscle groups are used to predict all the transformations that take place from muscle excitation to MTU force and joint moment production [10,28,41,58,77].

As previously outlined (i.e., Sections 9.2 and 9.3), the musculoskeletal model to be driven via EMG data is previously scaled to match an individual's anthropometry. However, the scaling procedure only adjusts each bone and MTU dimensions and mass properties linearly as a function of the subject's size (Sections 9.2 and 9.3).

In order to model the muscle ability to produce force as a function of EMG data, a number of extra parameters have to be identified (Section 9.4.5). These parameters directly determine the relationship between EMG and force production and as well as the specific muscle properties of an individual. These include the nonlinear EMG-to-force transfer function, the muscle strength, and the tendon and fiber mechanical properties (Section 9.4.5). These parameters vary across individuals and in general they cannot be measured experimentally or derived directly from the literature. As a result, researchers proposed alternative solutions based on optimization procedures [41]. In these procedures, the EMG-driven model is run using an initial set of nominal parameters that do not represent the specific subject being investigated. The generic parameter set is then repeatedly refined until the EMG-driven model output matches the corresponding experimental measure. Ideally, the calibration procedure should reduce the mismatch between predicted and experimental muscle forces. However, direct measurement of *in vivo* muscle force is highly invasive, requiring implantation of buckle transducers or optical fibers inserted in series with tendon [37].

Model validation has been recently made possible at the level of in vivo contact forces recorded by means of joint implants instrumented with multi-axial force transducers. Specifically, instrumented knee implants are providing an opportunity for quantitative evaluation of tibiofemoral contact forces in subjects who have undergone total knee replacement surgery. Since MTU forces are the primary determinants of joint contact forces, the accurate matching of joint contact forces implies physiologically meaningful estimates of MTU forces and the underlying model MTU parameters [27,28,40]. However, in vivo measures of joint contact forces still are limited to patients who have undergone joint replacement surgeries due to degenerative joint diseases, such as osteoarthritis.

Therefore, the viable, noninvasive model calibration is typically based on matching the model predicted joint moments with the externally measured joint moments from inverse dynamics (Section 9.2) [10,41,58,63]. This represents a good rationale of calibration because it relies on the fact that the summed product of predicted MTU force and moment arms (i.e., MTU moment) must equal the measured external joint moment. Therefore, this ensures that the model parameters are calibrated so that the coordinated behavior of the multiple MTUs in the model results in the same experimentally measured joint moment. However, model calibration at the joint moment level may not always ensure a unique muscle force solution for a specific motor task. In this context, model calibration procedures starting from different initial conditions might result in multiple model parameter sets that still match the same experimental joint moment but underlie different muscle force patterns (i.e., also see Section 9.4.5).

Once an optimal, subject-specific parameter set is found, the EMG-driven model is validated by predicting a novel set of trials that were not used for calibration. Validation is based on comparing the EMG-driven model output with respect to the same experimental measure used during calibration (i.e., joint moment). However, it is worth stressing that during validation, the EMG-driven model operates as an open-loop predictive system. That is, it directly determines MTU force and the resulting joint moments as a function of the measured EMG data and joint kinematics, with no need for tracking experimental joint moments.

In this context, calibration can be seen as a supervised learning step. That is, the model parameters and the resulting predictions are continuously refined based on the joint moment fitting error. In this way, the model learns to convert EMG data into MTU force and joint moment during different motor tasks underlying different MTU operations. After calibration the model parameters are fixed and the capacity of the model to generalize to new data is tested in an open-loop, predictive mode.

EMG-driven methods that use calibration have been proposed to create a robust model that only accounts for one selected DOF in the human limbs and for the only associated MTUs spanning that specific DOF. In this, the activity of the selected MTUs is constrained, during calibration, to satisfy the joint moments about the only selected DOF. Such single-DOF models have been designed for the elbow flexion–extension [45,47], knee flexion–extension [6,10,41,63,77] ankle plantar-dorsi flexion [46], and lower back flexion–extension [54] [70], DOFs. When run in open-loop after calibration, single-DOF models perform well at predicting joint moments about the DOF for which the model was calibrated.

However, it was recently shown that single-DOF models do not produce a unique solution to the MTU force estimation problem [58]. Even though they are driven by experimental EMG signals they cannot be successfully applied to predict the MTU dynamics and the resulting joint moment with respect to a different DOF than that used for calibration. Single-DOF calibrations can therefore result in models that predict substantially different force estimates for the same MTU and input data set. To address these limitations, we have developed EMG-driven models that simulate a greater number of MTUs with respect to single-DOF models while providing MTU force solutions that are not specific to an individual DOF only but are generalizable to multiple DOFs and joints.

This section introduces the theory underlying multi-DOF EMG-driven modeling and will refer to a previously presented EMG-driven model of the human lower extremity [58]. The model includes 34 MTUs and accounts for six DOFs including: hip adduction–abduction (HipAA), hip internal–external rotation (HipROT), hip flexion–extension (HipFE), knee flexion–extension (KneeFE), ankle plantar–dorsi flexion (AnkleFE), and ankle subtalar flexion (AnkleSF). The multi-DOF model comprises five main components (Fig. 9.4): musculotendon kinematics, musculotendon activation, musculotendon dynamics, moment computation, and model calibration.

9.4.1 Musculotendon Kinematics

MTU force is related to the level of muscle activation as well as the instantaneous MTU kinematics, which typically include MTU length ℓ^{mt} and moment arms r. In this scenario, ℓ^{mt} is used to determine the individual length of tendons and fibers within a MTU and to account for the fact that the muscle force-generation properties change with fiber length and fiber contraction velocity (Section 9.4.3). The MTU moment arms are subsequently used to project the MTU force onto the multiple DOFs that the MTU spans and to determine the resulting MTU moments (Section 9.4.5).

The MTU kinematics (Fig. 9.4B) can be determined as a function of three-dimensional joint angles using the previously discussed models of the musculo-skeletal kinematics (Section 9.3, Figs. 9.2 and 9.3). As previously stated, these models can represent the musculoskeletal system anatomy and physiology with different levels of accuracy. However, it is worth noting that the associated computational time and memory storage requirements can quickly increase with the complexity of the musculoskeletal model being employed.

In this scenario, it is important to rely on a compact representation of the musculo-skeletal model that allows for fast computation with low storage requirements and no loss of accuracy. This is especially important for applications requiring the real-time computation of ℓ^{mt}, r and the resulting MTU force. Examples of these emerging applications are biofeedback [50], or assistive device control [11,14,25] (also see Section 9.5).

A complex musculoskeletal model can be compactly represented using multi-dimensional spline functions [62]. In this, a set of nominal MTU ℓ^{mt} values for a range

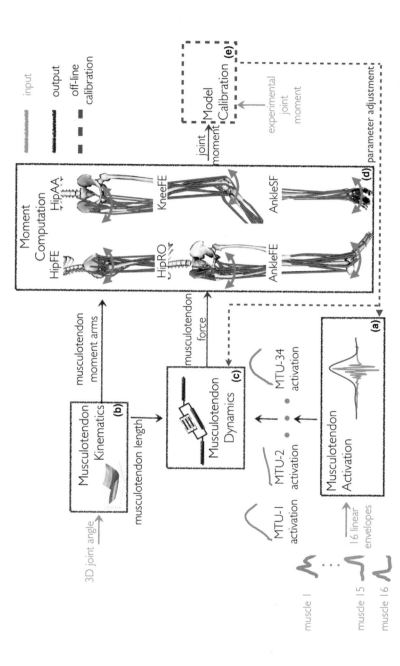

FIGURE 9.4 The schematic structure of the multi-DOF EMG-driven model. It consists of five components: **(A)** musculotendon activation, **(B)** musculotendon kinematics, **(C)** musculotendon dynamics, **(D)** moment computation, and **(E)** model calibration. The multi-DOF EMG-driven model is initially calibrated using the Model Calibration component. After calibration the EMG-driven model is operated in open loop. The force produced by 34 musculotendon units, and the resulting moments are determined as a function of 16 EMG signals and three-dimensional joint angles, without tracking experimental joint moments. Joint moments are predicted with respect to six degrees of freedom (DOFs): hip flexion–extension (HipFE), hip adduction–abduction (HipAA), hip internal–external rotation (HipROT), knee flexion–extension (KneeFE), ankle plantar–dorsi flexion (AnkleFE), and ankle subtalar flexion (AnkleSF).

257

of joint angles are used to create an MTU-specific multidimensional spline function that fits the nominal ℓ^{mt} set. The three-dimensional moment arms r are determined by differentiating the ℓ^{mt} spline function with respect to the joint DOF angle of interest [62].[1]

The MTU-specific multidimensional spline functions can be created from nominal values obtained using musculoskeletal geometry models based on MTU line segments (Fig. 9.2). In this case the nominal ℓ^{mt} reflects the total length of the MTU line segment connecting the insertion to the origin. Alternatively, more realistic ℓ^{mt} values can be obtained using the centroids of the three-dimensional muscle geometries created from experimental imaging data (i.e., MRI) or from volumetric finite-element musculoskeletal models (Fig. 9.3). The remainder of this chapter will assume that the MTU kinematics has been obtained using MTU spline functions created from nominal MTU line segment lengths (Fig. 9.4A).

9.4.2 Musculotendon Activation

This component (Fig. 9.4A) takes the raw EMG signals recorded from 16 major lower-extremity muscle groups and converts them into a measure called "activation" for 32 MTUs in the model. The activation is the ultimate control signal that directly "drives" the MTU contractile component (Section 9.4.3).

The raw EMG signals are collected from muscle groups, including: hip adductors, gracilis, sartorius, rectus femoris, tensor fasciae latae, gluteus medius, gluteus maximus, lateral hamstrings, medial hamstrings, vastus medialis, vastus lateralis, lateral gastrocnemius, medial gastrocnemius, soleus, peroneus group, and tibialis anterior. Raw EMGs are then band-pass filtered (10–450 Hz), then full-wave rectified and low-pass filtered (6 Hz). The resulting linear envelopes are then normalized with respect to the peak processed EMG values obtained from the entire set of recorded trials. In this, the muscle-group-specific EMG linear envelopes are assigned to all MTUs within each group (Table 9.1). In this allocation, the MTUs in deeply located muscle groups such as the gluteus minimus and vastus intermedius are driven by experimental EMG data recorded from a different muscle group that is (1) superficial, (2) that shares the same innervation and, (3) that contributes to the same mechanical action (Table 9.1). EMG signals for the deeply located illiopsoas muscle group cannot be recorded using surface EMG and cannot be in general driven using EMG data recorded from other muscle groups. However, it is still worth modeling their passive elastic force contribution as it is substantially high across motor tasks (i.e., ranging from 200N during walking to 400N during running) [58]. The MTU-allocated linear envelope reflects the level of MTU excitation in response to the neural drive.

The MTU-specific excitation is then processed by a second-order recursive filter [41] that simulates the MTU twitch response to the initial neural excitation: $u(t) = \alpha \cdot x(t - d) - \beta_1 u(t - 1) - \beta_2 u(t - 2)$, where $x(t)$ is the linear envelope at time t, $u(t)$ is the filtered linear envelope, whereas α, β_1, and β_2 are the recursive filtering coefficients. The term d is the electromechanical delay [30,53]. Finally, the MTU twitch response is adjusted to account for the nonlinear relationship between neural

---

[1]Code and documentation are freely available from: http://code.google.com/p/mcbs/

TABLE 9.1   Muscle Groups and the Musculotendon Units (MTUs) within Each Group

Muscle Groups	Experimental EMG[a]	Musculotendon Units
Medial hamstrings	Medial hamstring EMG	Biceps femoris short head (bicfemlh), biceps femoris long head (bicfemsh)
Lateral hamstrings	Lateral hamstring EMG	Semimembranosus (semimem), semitendinosus (semiten)
Rectus femoris	Rectus femoris EMG	recfem
Sartorius	Sartorius EMG	sar
Iliopsoas	No experimental EMG	Iliacus, psoas
Gluteus maximus	Gluteus maximus EMG	gmax1, gmax2, gmax3
Gluteus medius	Gluteus medius EMG	gmed1, gmed2, gmed3,
Gluteus minimus	Gluteus medius EMG	gluteus minimus (gmin1, gmin2, gmin3)
Vastus lateralis	Vastus lateralis EMG	vaslat
Vastus medialis	Vastus medialis EMG	vasmed
Vastus intermedius	(vastus lateralis EMG + vastus medialis EMG)/2	vasint
Gastrocnemius lateralis	gastrocnemius lateralis EMG	gaslat
Gastrocnemius medialis	gastrocnemius medialis EMG	gasmed
Peroneus	peroneus group EMG	Peroneus longus (perlong), peroneus brevis (perbrev), peroneus tertis (perter)
Soleus	soleus EMG	sol
Tibialis anterior	tibialis anterior EMG	tibant
Hip adductors	hip adductor EMG	Adductor magnus (addmag1, addmag2, addmag3), adductor longus (addlong), adductor brevis (addbrev)
Gracilis	gracilis EMG	gra
Tensor fasciae latae	tensof fasciae latae EMG	tfl

[a]This column reports whether or not experimental EMG signals were available for recording from a specific muscle group.

excitation and force [9,44]: $a(t) = \frac{e^{Au(t)}-1}{e^A-1}$, where A is the nonlinear shape factor, which is constrained to $-3 < A < 0$, with 0 being a linear relationship [41]. This results in a damped second-order nonlinear dynamic process controlled by three parameters per MTU [41,58,63]. The final processed signal $a(t)$ is called the MTU activation.

9.4.3   Musculotendon Dynamics

Once the muscle activation level and the three-dimensional kinematics are determined, it is possible to estimate the resulting force produced in the MTU muscle fiber and tendon respectively (Fig. 9.4C).

This requires a model of the muscle fiber contraction and tendon force–strain behavior. Physiological models, such as the Huxley-type model, have been previously proposed to characterize the dynamics of the single actin–myosin cross-bridges underlying fiber contraction [36,75]. In this, the fiber contraction dynamics is governed by multiple ordinary differential equations (ODEs) that need to be numerically integrated in each time frame. This results in computationally time-consuming models that cannot be applied to study the contraction dynamics of multiple MTUs in large-scale musculoskeletal models.

Therefore, most large-scale musculoskeletal models use phenomenological models such as the Hill-type muscle model [33]. That is, the external behavior of the system is characterized rather than its internal physiology. Typically, a Hill-type muscle model is composed of a contractile element, representing the muscle fiber in series with a passive elastic element, representing the tendon. The advantage of this approach is that the MTU dynamics are represented by a single ODE. Therefore, this represents a computationally viable solution to characterize the dynamics of a system of multiple MTUs.

In the Hill-type muscle model, fibers are characterized by an active contractile element in parallel with a passive elastic element. The contractile element is modeled using generic, normalized force–velocity $f(v^m)$, and active force–length $f_A(l^m, a(t))$ curves. The passive element is modeled using a passive force–length $f_P(l^m)$ curve. These curves characterize the ability of the fiber to produce force as a function of the muscle fiber activation $a(t)$, length l^m, and velocity v^m. These functions are normalized with respect to the maximum isometric muscle force $F^{\max}$, optimal fiber length l^m_O, and maximum muscle contraction velocity $v^{\max}$ [80]. The tendon is modeled using a single passive elastic element with a nonlinear force–strain function $f(\varepsilon)$ normalized to $F^{\max}$ [80]. This characterizes the tendon resistive force in response to stretching as a function of the tendon strain ε, that is, the percentage of tendon length l^t variation from its resting slack length l^t_s. Using biomechanical parameters from [16,79], the MTU force F^{mt} is then calculated as a function of $a(t)$, fiber length l^m, and fiber contraction velocity v^m:

$$F^{mt} = F^t = F^m \cos\left(\phi(l^m)\right) = \left[a(t)f_A(l^m, a(t))f(v^m) + f_P(l^m)\right] F^{\max} \cos\left(\phi(l^m)\right) \quad (9.1)$$

where F^t and F^m are the tendon and fiber force, and $\phi(l^m)$ is the pennation angle, that is, the angle of orientation of the fibers with respect to the tendon. This is modeled to change with the instantaneous fiber length assuming the muscle belly has a constant thickness and volume [41].

Using a previously presented method [42], an initial guess of l^m and v^m is determined for the first frame in time. This guess is used to determine the resulting $\phi(l^m)$ and fiber force F^m in (9.1) since all model inputs are available including l^m, v^m, and $a(t)$. Subsequently, the tendon length l^t can be determined since the total MTU length ℓ^{mt} is one of the known inputs of the MTU model, that is, $l^t = \ell^{mt} - l^m \cos\left(\phi(l^m)\right)$. This allows determining the tendon strain ε and the corresponding F^t value by interpolating the $f(\varepsilon)$ curve. This represents the output of the MTU model at the current frame in time. The current values of l^m and $a(t)$ are then used to interpolate $f_A(l^m, a(t))$. The resulting force is subsequently used to interpolate $f(v^m)$ in order to return a new estimate of v^m for the next frame in time. The resulting v^m is numerically integrated using a Runge–Kutta–Fehlberg algorithm to obtain the corresponding l^m value. The procedure is then iterated for each time frame.

It is worth stressing the fact that the integration process is time consuming since the MTU functions are inherently stiff. To reduce the computational time an MTU model based on infinitely-stiff tendons can be used [63]. In this, F^{mt} depends on l^m only because l^t is fixed to its slack length l^t_s. Therefore, F^m (and F^{mt}) can be simply calculated using (9.1), since instantaneous values of ℓ^{mt} and $a(t)$ are known, and if l^t is constant, then l^m is given by

$$l^m = \sqrt{(l^m_O \cdot \sin{(\phi_O)})^2 + (\ell^{mt} - l^t_s)^2} \tag{9.2}$$

where ϕ_O is the pennation angle at optimal fiber length l^m_O, and v^m is found by differentiating (9.2) with respect to time. This simplification is justified by the fact that the tendon is normally very stiff with a strain of only 3.3% when the MTU generates the maximum isometric force [80]. Experimental results demonstrated that this allows speeding up the F^{mt} calculations 250 times with no loss of accuracy in joint moment prediction with respect to using a nonlinear elastic tendon model [63].

This allows running EMG-driven musculoskeletal models on embedded systems with limited computational power that could be used for the EMG-based torque-driven control of powered orthoses or prostheses [11,25,63] (also see Section 9.5).

9.4.4 Joint Moment Computation

After the force produced by a MTU has been calculated, the resulting moment produced about a joint DOF can be predicted. This is done by combining together MTU force with MTU moment arms. That is, the moment computation component (Fig. 9.4D) estimates the joint moments M_x as the sum of the product of F^{mt} and r_X, with $X \in$ (HipAA, HipROT, HipFE, KneeFE, AnkleFE, AnkleSF):

$$M_X = \sum_i F^{mt}_i \cdot r_{X,i} \tag{9.3}$$

where the term i refers to all MTUs having a moment arm with respect to a specific DOF X.

9.4.5 Model Calibration

As previously discussed, the MTU fiber and tendon parameters used for expressing the MTU force (Section 9.4.3) are MTU-specific and person-dependent. These are strongly related to the morphology and anthropometry of the subject. However, once they have been properly identified, their values would not be expected to change over an appreciable time frame. On the other hand, the EMG-to-activation filtering coefficients are used to express the level of muscle activation as a function of the

EMG linear envelope (Section 9.4.1) and have a day-to-day variation due to a number of factors including the electrode placement and position, the skin impedance, and the blood circulation. Therefore, these parameters need to be identified for each recording session.

MTU parameters include strength coefficients δ that are used to scale the MTU-specific F^{max}, thus accounting for the fact that different people have different levels of muscle strength. Initial F^{max} values are chosen from the literature [79]. The scaling coefficients δ are varied between 0.5 and 1.5 and gather MTUs according to their functional action [41,58]. This is done to account for the dimension of the muscle physiological cross-sectional area (PCSA) [41].

Muscle tendon slack lengths l_s^t and optimal fiber lengths l_O^m need to be adjusted as they directly affect the MTU force calculation procedure (Section 9.4.3). However, they do not scale proportionally to the bone length as for ℓ^{mt}, and an alternative approach must be employed. In this, l_s^t and l_O^m can be scaled to maintain the consistency in the normalized fiber length–joint angle relationship between an individual and a generic musculoskeletal model across the joint range of motion [76]. The adjusted values for l_s^t and l_O^m can then be directly used in the EMG-driven model or further adjusted in the calibration algorithm so that $l_s^t = initial\ value \pm 5\%$ and $l_O^m = initial\ value \pm 2.5\%$.

The EMG-to-activation filtering coefficients α, β_1, and β_2 (i.e., Section 9.4.2) are parameterized with two factors that vary between -1 and 1 to realize a stable positive solution and a critically damped impulsive response for the recursive filter. Also, the shape factor A parameter is altered between -3 and 0 to account for the nonlinear EMG-to-force relationship [41].

Calibration (Fig. 9.4E) is an optimization-based procedure that adjusts the model parameters within their ranges until the objective function $f_E = \big(E_{HipAA} + E_{HipROT} + E_{HipFE} + E_{KneeFE} + E_{AnkleFE} + E_{AnkleSF}$ is minimized equally for each DOF. In this, each DOF error term $\big(E_{HipAA}, E_{HipROT}, E_{HipFE}, E_{KneeFE}, E_{AnkleFE}, E_{AnkleSF} \big)$ represents the sum of the root mean square differences between the predicted and experimental joint moments calculated over the calibration trials [58].

The steps presented in this section can be performed by using available EMG-driven modeling software packages such as CEINMS (https://simtk.org/home/ceinms).

9.5 EXPERIMENTAL RESULTS AND APPLICATIONS

This section will demonstrate the use of EMG-driven modeling to predict MTU forces and the resulting joint moments about multiple DOFs during dynamic motor tasks. Furthermore, it will briefly outline the use of EMG-driven modeling for applications in neurorehabilitation technologies. Chapter 20 will provide a more detailed description of EMG-driven modelling applications.

Predicting the force patterns of a large number of MTU spanning multiple DOFs and joints is a challenging problem. This has been previously addressed using optimization methodologies in which the externally measured joint moments are distributed to the relevant MTUs using optimization [1,2,71,72]. However, these methods rely on a priori chosen optimization criteria that cannot be generalized across motor tasks and subjects. Furthermore, dynamic locomotion tasks are characterized by antagonistic muscle co-activation and co-contraction, which cannot be properly predicted using optimization methods [12,31]. In contrast, EMG-driven methodologies can be successfully applied to study dynamic tasks that involve muscle co-contraction [41,58].

The EMG-driven modeling methods so far presented [58] is here applied to one healthy male subject (age: 28 years, height: 183 cm, mass: 67 kg). From the dynamic trials collected, two distinct datasets were created: one for the calibration and the other one for the validation of the EMG-driven musculoskeletal model. The calibration dataset included two repeated trials of four motor tasks including walking (FW), running (RN), sidestepping (SS), and crossover (CO) cutting maneuvers. A different dataset was used to validate the calibrated EMG-driven model. This included 10 repeated novel trials for each of the four considered motor tasks (FW, RN, SS, and CO). None of the trials in the validation dataset were included in the calibration dataset. Therefore there was no intersection of data between the two datasets. Trials were performed at self-selected speeds that ranged on the order of 2.2 ± 0.11 m/s (FW), 4.5 ± 0.3 m/s (RN), 2.5 ± 0.19 m/s (SS), and 2.4 ± 0.19 m/s (CO). The four motor tasks were chosen because (1) they allowed producing substantially high moments (i.e., always greater than 50 Nm) about the six considered DOFs (HipAA, HipROT, HipFE, KneeFE, AnkleFE, and AnkleSF) and (2) they reflected different MTU recruitment strategies and contraction dynamics. This allowed investigating whether the scaled and calibrated model could predict joint moments simultaneously produced about the four considered DOFs while accounting for different MTU operation strategies.

In this, the scaled and calibrated EMG-driven model predicted F^{mt} and M_X, with $X \in$ (HipAA, HipROT, HipFE, KneeFE, AnkleFE, AnkleSF), solely using experimental EMG and joint angle data from the stance phase during the 40 validation trials. Experimental joint moments <u>were not</u> used in the validation stage to refine the joint moment estimation. Therefore, the scaled and calibrated model was operated as an open-loop predictive system.

Data from the same motor task were time-normalized using a cubic spline and averaged across trials, producing motor task-specific ensemble average curves for the predicted F^{mt} and M_X and for the matching experimental joint moments $\bar{M}_X$. The similarity between the predicted and experimental joint moment was quantified using the coefficient of determination R^2 (i.e., square of the Pearson product moment correlation coefficient) and the root mean squared deviation normalized with respect to the range of variation assumed by the experimental joint moment (i.e., NRMSD) [58].

Figure 9.5 shows that the scaled and calibrated EMG-driven model produced joint moment estimates that matched the experimentally measured moments simultaneously produced about the six DOFs during the four considered motor tasks. The NRMSD coefficient ranged from 0.039 to 0.36, while the R^2 coefficient ranged from 0.65 to 0.98. This illustrates the possibility to apply EMG-driven modeling for the study of motor tasks that are substantially different with each other and that underlie different muscle contraction strategies.

Figure 9.6 reports the 34 MTU forces predicted by the EMG-driven model during the 10 validation walking trials that have been chosen as a representative example. As

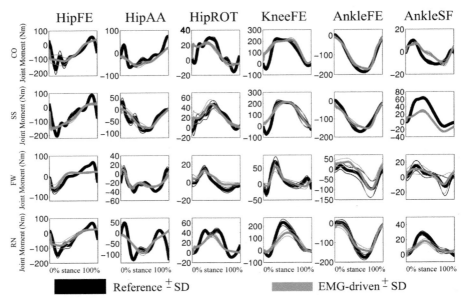

FIGURE 9.5 Ensemble average curves with associated standard deviation (SD) for the experimental joint moments (Reference) about six degrees of freedom (DOF) including: hip flexion–extension (HipFE), hip adduction–abduction (HipAA), hip internal–external rotation (HipROT), knee flexion–extension (KneeFE), ankle plantar–dorsi flexion (AnkleFE), and ankle subtalar flexion (AnkleSF). The reported data are from the stance phase with 0% being heel-strike and 100% toe-off events. The ensemble average curves are also reported for the matching joint moment predicted by the multi-DOF EMG-driven model (EMG-driven). Ensemble average curves are shown for four motor tasks including: walking (FW), running (RN), side-stepping (SS), and cross-over (CO) cutting maneuvers.

previously outlined (Section 9.4), the model mapped the 16 experimentally recorded EMG signals into 34 MTUs. In this, the MTUs within the same group received the same initial excitation pattern (Table 9.1). However, it is worth stressing the fact that the calibrated EMG-driven model was able to produce a force pattern that was specific to each MTU. This was the result of the calibration process. In this, some parameters had a global scope (i.e., the adjusted parameter was applied to all MTUs) while other parameters had a narrower scope (i.e., the adjusted parameter was applied to a group of MTUs or to a single MTU individually). Also note that, although EMG data did not drive the iliacus and psoas MTUs, the EMG-driven model predicted a substantial MTU force contribution during walking for these two MTUs. This is the result of the passive force produced by the MTU component, which was predicted by using (9.1) and setting the MTU activation to zero.

EMG-informed predictions of muscle forces acting on the hip have been also used to improve estimates of bone remodeling stimulus. In a recent study by Fernandez et al. [24], it was shown that EMG-informed estimates of muscle force contributed to redistribute the Von Mises (VM) stress pattern more uniformly and reduce peak stress

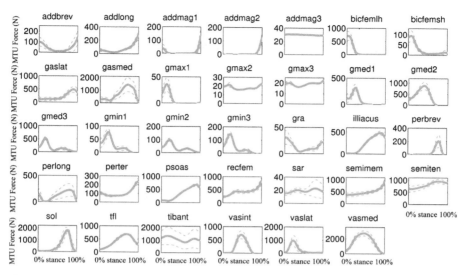

FIGURE 9.6 Ensemble average curves for the force produced by 34 musculotendon units (MTUs) as estimated by the calibrated EMG-driven model. The force data has been averaged over 10 walking trials used for the model validation. The reported data are from the stance phase with 0% being heel-strike and 100% toe-off events. MTU names are defined as in Table 9.1.

throughout the pelvis and within the acetabulum while increasing stress where muscles insert locally, as opposed to traditional Instron compression tests [24]. This suggests that loading ignoring anatomically accurate muscle forces may over- or underestimate bone remodeling stimulus spatially. Here we see an increase in the VM stress on the anterior superior iliac spine (ASIS) due to the insertion of the rectus femoris muscle but reduced peak VM stress in the iliac fossa. At the ASIS, EMG-informed muscle forces increased the mean VM stress from 10.7 ± 2.4 MPa to 11.5 ± 2.7 MPa (i.e., 7.5% increase) and increased the peak stress from 15.6 MPa to 17.0 MPa (i.e., 9% increase). Within the iliac fossa, muscle forces reduced the mean VM stress from 11.9 ± 3.4 MPa to 9.7 ± 2.8 MPa (i.e., 18.5% decrease). This difference was significant ($P < 0.05$) using a one-way ANOVA.

The ability of an EMG-driven model to convert the coordinated activation of MTUs into muscle and joint forces is crucial for enabling a number of applications in neurorehabilitation technologies. The ability of predicting EMG-dependent joint moments during dynamics motor tasks in real-time is important to develop methodologies for the proportional and simultaneous control of multiple DOFs in powered orthoses and prostheses [11,19,25,65]. Furthermore, the ability of predicting the dynamics of individual MTUs is crucial to understand how human muscles continuously modulate joint compliance during locomotion. Predictions of EMG-dependent joint compliance modulation via EMG-driven modeling allow controlling lower limb prostheses to better adapt to the demand of different motor tasks and terrain morphologies [19,55]. Please, refer to Chapter 20 for a more detailed discussion of EMG-driven modeling for neurorehabilitation technologies.

9.6   CONCLUSIONS

This chapter provided a comprehensive description of the methods used to develop physiologically accurate, subject-specific EMG-driven musculoskeletal models for the human lower extremity. The results presented in this chapter suggest that EMG-driven modeling is a powerful tool to gain insights on the form and function of the human musculoskeletal system and for the development of advanced assistive technologies. However, currently available EMG-driven models still have limitations that need to be overcome in the future.

A main limitation of current EMG-driven models is that they, in general, may fail at predicting the high-frequency components of the experimental joint moment waveforms (Fig. 9.5). This is because the procedure of extraction of the EMG linear envelopes requires low-pass filtering, and this is performed using a predefined cutoff frequency. This may not always reflect the actual bandwidth of the muscles neural excitation, which is in fact continuously modulated as a function of multiple variables including the dynamics of the neural drive, the muscle contraction effort, and the dynamics of the task [20]. Another factor limiting the ability of EMG-driven models of matching external moments is the inability of recording the surface EMG activity of deeply located muscles. This may imply that only the MTU passive force can be estimated while the EMG-dependent active force cannot be predicted. In the context of our presented results, the missing illiacus and psoas active force contribution (Fig. 9.6) was clearly reflected in the predicted hip joint moments. During walking (Fig. 9.5), the predicted HipFE moments were smaller than the associated experimental moment during the hip-flexing phase of stance (i.e., 60%–100%).

A potential approach to limit these problems is the use of hybrid EMG-driven/optimization-driven procedures [17,32,59,60]. In this, the EMG-driven model presented in this chapter is combined together with a static optimization-based approach. In this context, experimentally available EMG signals are used to drive the relevant MTUs in the model. Then, the activity of deeply located muscles that cannot be measured is predicted using a static optimization procedure. In this context, the experimentally available EMG envelopes can be further adjusted to account for tracking errors in the high-frequency components of the experimental joint moment waveforms and to account for EMG filtering limitations [59,60].

Alternatively, the theory of muscle synergies [7] can be used as a way to reconstruct the activity of a larger number of muscles from a reduced set of EMG data, thus addressing the problem of accessing deeply located muscles [18,49,61,73]. This may have direct implications in neurorehabilitation technologies where the availability of EMG recording sites may be limited [61].

The use of high-density (HD) surface EMG may provide new viable ways to address limitations in conventional surface EMG selectivity and filtering. With HD-EMG, surface EMG signals are recorded from several closely spaced electrodes arranged in a rectangular grid that is placed in the correspondence of a selected muscle. The ensemble of the multiple differential signals recorded using the high-density grid can be decomposed into the action potentials of the individual motor units active in the specific muscle [20]. As previously outlined, the motor unit action potential reflects the discharge in the axons of the motor neurons that innervate the motor unit and represents a direct measurement of the neural drive to muscles [20]. Therefore, extracting motor unit action potential from HD-EMG allows obtaining richer information on the muscle neural excitation and on how this is continuously modulated. Such information could therefore be used to drive musculoskeletal models of human extremities and predict more physiologically accurate and dynamically consistent muscle force and moment estimates. Although HD-EMG represents an exciting solution to address current limitations in EMG-driven modeling, it is currently only feasible during isometric conditions.

Another factor limiting the use of multi-DOF EMG-driven models is the computational cost associated to the calibration procedure. This can take up to several hours for the calibration of the comprehensive lower extremity model presented in this chapter [58]. Typically, the calibration time increases with the number of MTUs in the model and with the number of MTU parameters to be calibrated. However, experimental studies [25,41] have shown that the MTU-specific morphology parameters (i.e., tendon slack length, optimal fiber length, and maximal isometric force) need to be identified only once for a specific subject. Therefore, across recording sessions it is only needed to readjust the muscle activation parameters to account for different electrode placement and skin preparation (also see Section 9.4). In this scenario, the muscle activation parameters can be assigned a global scope thus speeding up considerably the model recalibration. Furthermore, the use of MTU models that do not require an explicit integration of the MTU dynamics equations could considerably speed up the calibration process further, as was shown in Sartori et al. [63]. Finally, it is worth noting that despite the large computational time associated to calibration, the EMG-driven model open-loop operation (i.e., after calibration) can be performed in a time close to that of the muscles EMD (i.e., between 10 ms and 20 ms), making this suitable for applications with real-time constraints [63].

In conclusion, this chapter presented the methods behind EMG-driven musculoskeletal modeling for the human lower extremity and its applications to assistive technologies. The ability of estimating individual muscle forces as a function of experimental measurements of the neural drive can give insights into neural control and tissue loading and can be helpful for linking events at the neurophysiologic level to events happening at the musculoskeletal level. The ability of translating EMG-driven modeling to the clinical practice is opening new avenues in the design of advanced neurorehabilitation treatments and technologies that can effectively take advantage of a person's movement control strategy without relying on explicit representations of specific models of neuromuscular control.

ACKNOWLEDGMENT

Support for this work was provided by the European Research Council (ERC) via the ERC Advanced Grant DEMOVE (267888).

REFERENCES

1. Anderson, F. C., and M. G. Pandy, "Dynamic optimization of human walking," *J. Biomech. Eng.* **123**, 381–390 (2001).

2. Anderson, F. C., and M. G. Pandy, "Static and dynamic optimization solutions for gait are practically equivalent," *J. Biomech.* **34**, 153–161 (2001).

3. Arnold, A. S., S. S. Blemker, and S. L. Delp, "Evaluation of a deformable musculoskeletal model for estimating muscle-tendon lengths during crouch gait," *Ann. Biomed. Eng.* **29**, 263–274 (2001).

4. Arnold, A. S., S. Salinas, D. J. Asakawa, and S. L. Delp, "Accuracy of muscle moment arms estimated from MRI-based musculoskeletal models of the lower extremity," *Comput. Aided Surg.* **5**, 108–119 (2000).

5. Besier, T. F., G. E. Gold, G. S. Beaupré, and S. L. Delp, "A modeling framework to estimate patellofemoral joint cartilage stress in vivo," *Med. Sci. Sport. Exerc.* **37**, 1924–1930 (2005).

6. Besier, T. F., M. Fredericson, G. E. Gold, G. S. Beaupré, and S. L. Delp, "Knee muscle forces during walking and running in patellofemoral pain patients and pain-free controls," *J. Biomech.* **42**, 898–905 (2009).

7. Bizzi, E., V. C. K. Cheung, A. d'Avella, P. Saltiel, and M. Tresch, "Combining modules for movement," *Brain Res. Rev.* **57**, 125–133 (2008).

8. Blemker, S. S., D. S. Asakawa, G. E. Gold, and S. L. Delp, "Image-based musculoskeletal modeling: applications, advances, and future opportunities," *J. Magn. Reson. Imaging* **25**, 441–451 (2007).

9. Buchanan T. S., and D. G. Lloyd, "Muscle activity is different for humans performing static tasks which require force control and position control," *Neurosci. Lett.* **194**, 61–64 (1995).

10. Buchanan, T. S., D. G. Lloyd, K. Manal, and T. F. Besier, "Neuromusculoskeletal modeling: estimation of muscle forces and joint moments and movements from measurements of neural command," *J. Appl. Biomech.* **20**, 367–395 (2004).

11. Cavallaro, E. E., J. Rosen, J. C. Perry, and S. Burns, "Real-time myoprocessors for a neural controlled powered exoskeleton arm," *IEEE Trans. Biomed. Eng.* **53**, 2387–2396 (2006).

12. Collins, J. J., "The redundant nature of locomotor optimization laws," *J. Biomech.* **28**, 251–267 (1995).

13. Crowninshield, R. D., R. C. Johnston, J. G. Andrews, and R. A. Brand, "A biomechanical investigation of the human hip," *J. Biomech.* **11**, 75–85 (1978).

14. del-Ama, A. J., Á. Gil-Agudo, J. L. Pons, and J. C. Moreno, "Hybrid FES-robot cooperative control of ambulatory gait rehabilitation exoskeleton," *J. Neuroeng. Rehabil.* **11**, 27 (2014).

15. Delp, S. L., F. C. Anderson, A. S. Arnold, P. Loan, A. Habib, C. T. John, E. Guendelman, and D. G. Thelen, "OpenSim: open-source software to create and analyze dynamic simulations of movement," *IEEE Trans. Biomed. Eng.* **54**, 1940–1950 (2007).

16. Delp, S. L., J. P. Loan, M. G. Hoy, F. E. Zajac, E. L. Topp, and J. M. Rosen, "An interactive graphics-based model of the lower extremity to study orthopaedic surgical procedures," *IEEE Trans. Biomed. Eng.* **37**, 757–767 (1990).

17. Demircan, E., O. Khatib, J. Wheeler, and S. L. Delp, "Reconstruction and EMG-informed control, simulation and analysis of human movement for athletics: performance

improvement and injury prevention," *Proceedings of Engineering in Medicine (EMBC), IEEE Conference on*, Minneapolis, MN, pp. 6534–6537, 2009.

18. Duysens, J., F. De Groote, and I. Jonkers, "The flexion synergy, mother of all synergies and father of new models of gait," *Front. Comput. Neurosci.* **7**, 14 (2013).

19. Eilenberg, M. F., H. Geyer, and H. Herr, "Control of a powered ankle-foot prosthesis based on a neuromuscular model," *IEEE Trans. Neural Syst. Rehabil. Eng.* **18**, 164–173 (2010).

20. Farina, D., and F. Negro, "Accessing the neural drive to muscle and translation to neurorehabilitation technologies," *IEEE Rev. Biomed. Eng.* **5**, 3–14 (2012).

21. Fernandez J. W., and P. J. Hunter, "An anatomically based patient-specific finite element model of patella articulation: towards a diagnostic tool," *Biomech. Model. Mechanobiol.* **4**, 20–38 (2005).

22. Fernandez, J. W., and M. G. Pandy, "Integrating modelling and experiments to assess dynamic musculoskeletal function in humans," *Exp. Physiol.* **91**, 371–382 (2006).

23. Fernandez, J. W., P. Mithraratne, S. F. Thrupp, M. H. Tawhai, and P. J. Hunter, "Anatomically based geometric modelling of the musculo-skeletal system and other organs," *Biomech. Model. Mechanobiol.* **2**, 139–155 (2004).

24. Fernandez, J., M. Sartori, D. Lloyd, J. Munro, and V. Shim, "Bone remodelling in the natural acetabulum is influenced by muscle force-induced bone stress," *Int. J. Numer. Method. Biomed. Eng.* **30**, 28–41 (2013).

25. Fleischer C., and G. Hommel, "A human–exoskeleton interface utilizing electromyography," *IEEE Trans. Robot* **24**, 872–882 (2008).

26. Fregly, B. J., "Design of optimal treatments for neuromusculoskeletal disorders using patient-specific multibody dynamic models," *Int. J. Comput. Vis. Biomechnanics* **2**, 145–155 (2009).

27. Fregly, B. J., T. F. Besier, D. G. Lloyd, S. L. Delp, S. A. Banks, M. G. Pandy, and D. D. D'Lima, "Grand challenge competition to predict in vivo knee loads," *J. Orthop. Res.* **30**, 503–513 (2012).

28. Gerus, P., M. Sartori, T. F. Besier, B. J. Fregly, S. L. Delp, S. A. Banks, M. G. Pandy, D. D. D'Lima, and D. G. Lloyd, "Subject-specific knee joint geometry improves predictions of medial tibiofemoral contact forces," *J. Biomech.* **46**, 2778–2786 (2013).

29. Gordon, K. E., and D. P. Ferris, "Learning to walk with a robotic ankle exoskeleton," *J. Biomech.* **40**, 2636–2644 (2007).

30. Heine, R., K. Manal, and T. S. Buchanan, "Using hill-type muscle models and EMG data in a forward dynamic analysis of joint moment: evaluation of critical parameters," *J. Mech. Med. Biol.* **3**, 169–186 (2003).

31. Heintz S., and E. M. Gutierrez-Farewik, "Static optimization of muscle forces during gait in comparison to EMG-to-force processing approach," *Gait and Posture* **26**, 279–288 (2007).

32. Higginson, J. S., J. W. Ramsay, and T. S. Buchanan, "Hybrid models of the neuromusculoskeletal system improve subject-specificity," *Proc. Inst. Mech. Eng. Part H J. Eng. Med.* **226**, 113–119 (2011).

33. Hill, A. V., "The heat of shortening and the dynamic constants of muscle," *Proc. R. Soc. London. Ser. B, Biol. Sci.* **126**, 136–195 (1938).

34. Holzbaur, K. R. S., W. M. Murray, and S. L. Delp, "A model of the upper extremity for simulating musculoskeletal surgery and analyzing neuromuscular control," *Ann. Biomed. Eng.* **33**, 829–840 (2005).

35. Hoover, C., and K. Fite,"A configuration dependent muscle model for the myoelectric control of a transfemoral prosthesis," *Proc Rehabilitation Robotics (ICORR), International Conference on*, Zurich, Switzerland, 2011. pp. 947–952.

36. Huxley H., and J. Hanson, "Changes in the cross-striations of muscle during contraction and stretch and their structural interpretation," *Nature* **173**, 973–976 (1954).

37. Komi, P. V., "Relevance of in vivo force measurements to human biomechanics," *J. Biomech.* **23**, 23–34 (1990).

38. Kumar, D., K. S. Rudolph, and K. T. Manal, "EMG-driven modeling approach to muscle force and joint load estimations: case study in knee osteoarthritis," *J. Orthop. Res.* **30**, 377–383 (2012).

39. Kuo, A. D., "A least-squares estimation approach to improving the precision of inverse dynamics computations," *J. Biomech. Eng.* **120**, 148–159 (1998).

40. Lin, Y., J. P. Walter, S. A. Banks, M. G. Pandy, and B. J. Fregly, "Simultaneous prediction of muscle and contact forces in the knee during gait," *J. Biomech.* **43**, 945–952 (2010).

41. Lloyd, D. G. and T. F. Besier, "An EMG-driven musculoskeletal model to estimate muscle forces and knee joint moments in vivo," *J. Biomech.* **36**, 765–776 (2003).

42. Loan, P. J., *Dynamic Pipeline*, MusculoGraphics, Inc., Evanston, IL, 1992.

43. Lu, T. W., and J. J. O'Connor, "Bone position estimation from skin marker co-ordinates using global optimisation with joint constraints," *J. Biomech.* **32**, 129–134 (1999).

44. Manal K., and T. S. Buchanan, "A one-parameter neural activation to muscle activation model: estimating isometric joint moments from electromyograms," *J. Biomech.* **36**, 1197–1202 (2003).

45. Manal K., and T. S. Buchanan, "Use of an EMG-driven biomechanical model to study virtual injuries," *Med. Sci. Sports Exerc.* **37**, 1917–1923 (2005).

46. Manal, K., K. Gravare-Silbernagel, and T. S. Buchanan, "A real-time EMG-driven musculoskeletal model of the ankle," *Multibody Syst. Dyn.* **28**, 169–180 (2011).

47. Manal, K., R. V. Gonzalez, D. G. Lloyd, and T. S. Buchanan, "A real-time EMG-driven virtual arm," *Comput. Biol. Med.* **32**, 25–36 (2002).

48. McGibbon, C. A., "A biomechanical model for encoding joint dynamics: applications to transfemoral prosthesis control," *J. Appl. Physiol.* **112**, 1600–1611 (2012).

49. McGowan, C., R. Neptune, D. Clark, and S. Kautz, "Modular control of human walking: Adaptations to altered mechanical demands," *J. Biomech.* **43**, 1–15 (2010).

50. Murai, A., K. Kurosaki, K. Yamane, and Y. Nakamura, "Musculoskeletal-see-through mirror: computational modeling and algorithm for whole-body muscle activity visualization in real time," *Prog. Biophys. Mol. Biol.* **103**, 310–317 (2010).

51. Nakamura, Y., K. Yamane, Y. Fujita, and I. Suzuki, "Somatosensory computation for man–machine interface from motion-capture data and musculoskeletal human model," *IEEE Trans. Robot.* **21**, 58–66 (2005).

52. Nikooyan, A. A., H. E. J. Veeger, P. Westerhoff, B. Bolsterlee, F. Graichen, G. Bergmann, and F. C. T. van der Helm, "An EMG-driven musculoskeletal model of the shoulder," *Hum. Mov. Sci.* **31**, 429–447 (2012).

53. Nordez, A., T. Gallot, S. Catheline, A. Guével, C. Cornu, and F. Hug, "Electromechanical delay revisited using very high frame rate ultrasound," *J. Appl. Physiol.* **106**, 1970–1975 (2009).

54. Nussbaum, M. A., and D. B. Chaffin, "Lumbar muscle force estimation using a subject-invariant 5-parameter EMG-based model," *J. Biomech.* **31**, 667–672 (1998).

55. Pfeifer, S., M. Hardegger, and H. Vallery, "Model-based estimation of active knee stiffness," *IEEE Trans. Biomed. Eng.* **59**, 1–6 (2011).

56. Rankin J. W., and R. R. Neptune, "Musculotendon lengths and moment arms for a three-dimensional upper-extremity model," *J. Biomech.* **45**, 1739–1744 (2012).

57. Rao, G., E. Berton, D. Amarantini, L. Vigouroux, and T. S. Buchanan, "An EMG-driven biomechanical model that accounts for the decrease in moment generation capacity during a dynamic fatigued condition," *J. Biomech. Eng.* **132**, 1–9 (2010).

58. Sartori, F., M. Reggiani, D. Farina, and D. G. Lloyd, "EMG-driven forward-dynamic estimation of muscle force and joint moment about multiple degrees of freedom in the human lower extremity," *PLoS One* **7**, 1–11 (2012).

59. Sartori, M., D. Farina, and D. G. Lloyd, "Hybrid neuromusculoskeletal modeling to best track joint moments using a balance between muscle excitations derived from electromyograms and optimization," *J. Biomech.* **47**, 3613–3621 (2014).

60. Sartori, M., D. Farina, and D. G. Lloyd, "Hybrid neuromusculoskeletal modeling," in *Converging Clinical and Engineering Research on Neurorehabilitation*, J. L. Pons, D. Torricelli, and M. Pajaro, eds. Springer, Berlin, pp. 427–430, 2013.

61. Sartori, M., L. Gizzi, D. G. Lloyd, and D. Farina, "A musculoskeletal model of human locomotion driven by a low dimensional set of impulsive excitation primitives," *Front. Comput. Neurosci.* **7**, 1–22 (2013).

62. Sartori, M., M. Reggiani, A. J. van den Bogert, and D. G. Lloyd, "Estimation of musculotendon kinematics in large musculoskeletal models using multidimensional B-splines," *J. Biomech.* **45**, 595–601 (2012).

63. Sartori, M., M. Reggiani, E. Pagello, and D. G. Lloyd, "Modeling the human knee for assistive technologies," *IEEE Trans. Biomed. Eng.* **59**, 2642–2649 (2012).

64. Sawicki G. S., and D. P. Ferris, "Mechanics and energetics of level walking with powered ankle exoskeletons," *J. Exp. Biol.* **211**, 1402–1413 (2008).

65. Sawicki, G. S., and D. P. Ferris, "A pneumatically powered knee–ankle–foot orthosis (KAFO) with myoelectric activation and inhibition," *J. Neuroeng. Rehabil.* **6**, 23 (2009).

66. Shao, Q., D. N. Bassett, K. Manal, and T. S. Buchanan, "An EMG-driven model to estimate muscle forces and joint moments in stroke patients," *Comput. Biol. Med.* **39**, 1083–1088 (2009).

67. Shao, Q., T. D. MacLeod, K. Manal, and T. S. Buchanan, "Estimation of ligament loading and anterior tibial translation in healthy and ACL-deficient knees during gait and the influence of increasing tibial slope using EMG-driven approach," *Ann. Biomed. Eng.* **39**, 110–121 (2011).

68. Sherman, M. A., A. Seth, and S. L. Delp, "Simbody: multibody dynamics for biomedical research," *Procedia IUTAM* **2**, 241–261 (2011).

69. Stokes, V. P., C. Andersson, and H. Forssberg, "Rotational and translational movement features of the pelvis and thorax during adult human locomotion," *J. Biomech.* **22**, 43–50 (1989).

70. Thelen, D. G., A. B. Schultz, S. D. Fassois, and J. A. Ashton-Miller, "Identification of dynamic myoelectric signal-to-force models during isometric lumbar muscle contractions," *J. Biomech.* **27**, 907–919 (1994).

71. Thelen, D. G., and F. C. Anderson, "Using computed muscle control to generate forward dynamic simulations of human walking from experimental data," *J. Biomech.* **39**, 1107–1115 (2006).

72. Thelen, D. G., F. C. Anderson, and S. L. Delp, "Generating dynamic simulations of movement using computed muscle control," *J. Biomech.* **36**, 321–328 (2003).

73. Walter, J., A. L. Kinney, S. A. Banks, D. D'Lima, T. F. Besier, D. G. Lloyd, and B. J. Fregly, "Muscle synergies may improve optimization prediction of knee contact forces during walking," *J. Biomech. Eng.* **136**, 1–9 (2014).

74. Webb, J. D., S. S. Blemker, and S. L. Delp, "3D finite element models of shoulder muscles for computing lines of actions and moment arms," *Comput. Methods Biomech. Biomed. Eng.* **17**, 829–837 (2014).

75. Williams, W. O., "Huxley's model of muscle contraction with compliance," *J. Elast.* **105**, 365–380 (2011).

76. Winby, C. R., D. G. Lloyd, and T. B. Kirk, "Evaluation of different analytical methods for subject-specific scaling of musculotendon parameters," *J. Biomech.* **41**, 1682–1688 (2008).

77. Winby, C. R., D. G. Lloyd, T. F. Besier, and T. B. Kirk, "Muscle and external load contribution to knee joint contact loads during normal gait," *J. Biomech.* **42**, 2294–2300 (2009).

78. Wismans, J., F. Veldpaus, J. Janssen, A. Huson, and P. Struben, "A three-dimensional mathematical model of the knee-joint," *J. Biomech.* **13**, 677–685 (1980).

79. Yamaguchi, G. T., A. G. U. Sawa, D. W. Moran, M. J. Fessler, and J. M. Winters, "A survey of human musculotendon actuator parameters," in *Multiple Muscle Systems: Biomechanics and Movement Organization*, J. M. Winters, and S. Y. Woo, eds. Springer, New York, pp. 717–773, 1990.

80. Zajac, F. E., "Muscle and tendon: properties, models, scaling, and application to biomechanics and motor control," *Crit. Rev. Biomed. Eng.* **17**, 359–411 (1989).

10

MUSCLE FORCE AND MYOELECTRIC MANIFESTATIONS OF MUSCLE FATIGUE IN VOLUNTARY AND ELECTRICALLY ELICITED CONTRACTIONS

R. Merletti,[1] B. Afsharipour,[1] J. Dideriksen,[2] and D. Farina[2]

[1]*Laboratory for Engineering of the Neuromuscular System, Politecnico di Torino, Torino, Italy*
[2]*Department of Neurorehabilitation Engineering, Universitätsmedizin Göttingen, Georg-August-Universität, Göttingen, Germany*

10.1 INTRODUCTION

Muscles produce force under voluntary control or under control of an external drive (e.g., electrical stimulation). Muscles also produce a distribution of electrical potential over the skin (sEMG). During sustained voluntary or electrically stimulated contractions, as well as during intermittent or dynamic contractions, both force and sEMG undergo changes that are referred to as mechanical and myoelectric manifestations of muscle fatigue.

The study of these mechanical and electrical phenomena has many applications in sport (to improve and quantify performance), in rehabilitation medicine (to monitor recovery), in occupational medicine (to prevent and monitor work-related disorders), and in many other fields such as space medicine, prostheses control, and oncology.

Surface Electromyography: Physiology, Engineering, and Applications, First Edition.
Edited by Roberto Merletti and Dario Farina.
© 2016 by The Institute of Electrical and Electronics Engineers, Inc. Published 2016 by John Wiley & Sons, Inc.

While force, torque or EMG voltage are well-defined physical quantities, this is not the case with muscle fatigue.

Many definitions of muscle fatigue exist. Mechanical manifestations of muscle fatigue are defined as an exercise-induced reduction in the maximal force generating capacity of a muscle and may be attributed to (a) changes of muscle drive by the central nervous system (central fatigue) and (b) changes of muscle force generating capability (peripheral fatigue) [56,57]. Myoelectric manifestations of muscle fatigue are defined as changes of features of the sEMG during sustained muscle activity. These features reflect both central and peripheral phenomena leading to mechanical fatigue and are detectable much earlier than mechanical failure.

Since, at this time, the force applied by a muscle to its tendons cannot be measured directly, changes of sEMG have been used for over a century as indirect indicators of muscle force and fatigue [104]. The single-channel sEMG parameters and algorithms used for this purpose have been described in Chapter 4 and will be only briefly recalled in this chapter.

The use of sEMG to "*guess*" the contraction level (force) or the fatigue condition of a muscle can be extremely misleading. Competence in the physiology and biophysics of sEMG generation is required to avoid common traps and misinterpretations. Doctors in occupational, sport, and rehabilitation medicine, as well as other clinicians and therapists, dream of techniques or instruments that could quantify force and fatigue in dynamic conditions during daily life activities, work activities, sport training, or rehabilitation exercises. While most of this is becoming technically feasible, the interpretation of the information derived from the sEMG signals is still difficult (often impossible) because of the large number of phenomena and physical variables reflected by sEMG and of confounding factors [44]. While these factors are being unraveled, relatively simple test conditions can provide indications about the performance of a muscle (or a group of muscles) in specific settings such as isometric constant force contractions, isometric variable force contractions (typically ramps), and some simple dynamic tasks.

This chapter deals with the association between sEMG and force/fatigue and outlines this association, the tools to study it, the many limitations and drawbacks of the available techniques, and the open questions that remain to be solved.

10.2 JOINT TORQUE MEASUREMENT AND MUSCLE FORCE ESTIMATION IN ISOMETRIC CONTRACTIONS

10.2.1 Joint Torque Measurements in Isometric Contractions

Skeletal muscles act on rigid body segments—that is, on bones—that are assumed to be connected by a hinge and rotate around a pivot point (center of rotation). This is an approximation that allows us to define the lever arm of the muscle force and to establish the association between muscle force and torque applied to a joint such as the elbow or knee. Isometric braces lock two body segments, preventing their relative movement, and load cells (with fixed lever arm) or torque meters measure the produced torque. A number of conditions arise concerning these measurements: (a) If torque meters are used, they must be aligned with the center of rotation of the joint, and (b) if a load cell is used, the lever arm of the measured force with respect to the center of rotation must be known to estimate torque. If measurements are performed at a fixed joint angle and only relative torque values (expressed as percentage of the maximum torque) are of interest, these conditions can be somewhat relaxed. The position of the brace and of the body segments may affect the results [37]. For example, the arm (or the thigh) can be either vertical or horizontal and the forearm (or the leg) can be at different angles. The weight of the forearm or leg should be compensated for by adjusting the offset of the measuring instrument in relaxed conditions. A force target is presented to the subject on a computer screen or other display and the subject is asked to match it (constant force isometric conditions) or track it (variable force isometric condition). The target is often a percentage of a previously measured maximal voluntary contraction (MVC). Lack of consensus exists on how MVC should be measured; the average or the peak value of 3–5 short (3–5 s) contractions are often used. This protocol is referred to as a force control task.

Another alternative, for the elbow or knee joints, is to have the subject hold a weight attached to the wrist (arm vertical and forearm horizontal) or to the ankle (trunk horizontal, thigh vertical, and leg horizontal) and hold the required position. This protocol is referred to as a position control task and usually no feedback is given. The neurophysiological control strategy is very different from that of the previous protocol. Among other things, the second protocol leads to a shorter endurance time. For details see the extensive work of Enoka et al. [31,32].

10.2.2   Surface EMG-Based Joint Torque Estimation in Isometric Contractions

Among the many purposes of the investigations described in the previous section one is to establish a relationship between sEMG and muscle force, and another one is to study the sEMG changes reflecting fatigue in isometric conditions with constant or variable torque (ramps or intermittent contractions).

These issues have been investigated for over six decades [65,121,122]; many of them are still open and unsolved, despite new approaches (see Chapter 9). One of these issues concerns the fact that the torque measured with an isometric brace is due to many muscles (agonists, antagonists, mono- or bi-articular). The only exception is the first dorsal interosseous (FDI) which is the only agonist muscle acting on the thumb-index joint: this is why it has been investigated so often [18,19,141,142]. As opposed to force, sEMG, in general, refers mostly to one muscle only. The sEMG amplitude and frequency features depend on the location, size, and interelectrode distance of the electrode pair (see Chapters 2 and 5). In addition, deeply located motor units contribute little to the sEMG features but contribute fully to the produced and measured force/torque. As an extreme

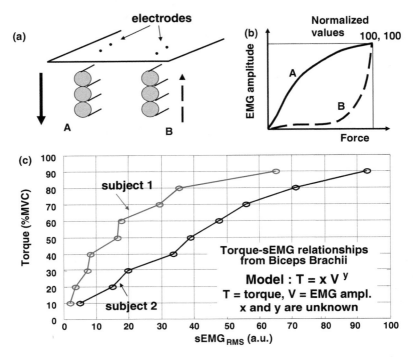

FIGURE 10.1 (a) Schematic extreme examples of two muscles with order of recruitment of the motor units from top-down and from bottom-up. (b) General pattern of the sEMG amplitude versus muscle force for the two cases depicted in part a. (c) Experimental plots of isometric elbow torques versus the average RMS value of the longitudinal single differential sEMG signals obtained from a grid (13 rows × 5 columns, IED = 10 mm) placed, on the biceps brachii, proximally with respect to the innervation zone. The two curves are obtained from two healthy subjects. A possible mathematical model is indicated for the relationship torque = f(sEMG$_{RMS}$). The model implies that the torque is produced only by the biceps brachii.

example, a subject could co-contract agonists and antagonists, producing little or no net torque at the joint but high sEMG from either muscle group.

Figure 10.1a depicts two extreme examples of muscles recruiting motor units from the superficial ones to the deep ones (muscle A) or from the deep ones to the superficial ones (muscle B). Because of the effect of recruitment order (small motor units first) and of distance between the sources and the electrodes, as force increases, the sEMG increases, usually nonlinearly, with the concavity down (A) or up (B) as indicated qualitatively in Fig. 10.1b. These are *not* the only reasons determining the degree of nonlinearity; motor unit size and algebraic summation (cancellation) of motor unit action potentials (MUAP) are additional reasons [49].

Figure 10.1c depicts the relationship between isometric elbow torque and RMS of the sEMG in two healthy subjects. An electrode grid (13 rows and 5 columns with interelectrode distance of 10 mm) was placed on the biceps brachii (proximal with respect to the innervation zone), and the RMSs of all the longitudinal single differential channels were averaged to get a representative sEMG RMS value over a 10-s epoch (see Chapter 5). Voluntary torque was increased in steps of 10% MVC and sustained for 10 s at each level.

Different curves are observed for the two subjects. The two plots of Fig. 10.1b, as well as any monotonously increasing pattern in between these extremes, could be approximated by the mathematical relation described in Eq. (10.1), where T is the torque contributed by the muscle, V is an sEMG amplitude indicator (e.g., RMS or ARV), and x and y are unknown real coefficients defining the curve.

$$T = xV^y \qquad (\text{for } y > 0) \qquad\qquad (10.1)$$

For $0 < y < 1$, concavity is down; for $y = 1$, the relationship is a straight line; for $y > 1$, concavity is up; the sign of x indicates agonist or antagonist muscle [1]. Obviously, T cannot be measured for a single muscle. In the case of the elbow flexion, we have the biceps (short and long head), the brachioradialis, and the brachialis muscles on the agonist side, and we have the three heads of the triceps brachii on the antagonist side. If we consider T as the total measured torque, Eq. (10.1) is not correct since the sEMG is obtained from only one muscle while T is produced by many muscles. To simplify the problem, one may assume that the antagonists are off and the agonists are all pulling at the same percentage of their MVC, but this is not necessarily true. A simple example involving only two muscles is given in Fig. 10.2 and a system of equations based on the mathematical model proposed above is given in Eqs. (10.2–10.5), describing two muscles and four contraction levels. T_i is the total (known) torque at contraction level i ($i = 1, 2, 3, 4$), $V_{i,j}$ is the (known) RMS (or other amplitude parameter) of muscle j at contraction level i, and x_1, y_1, x_2, and y_2 are the unknown

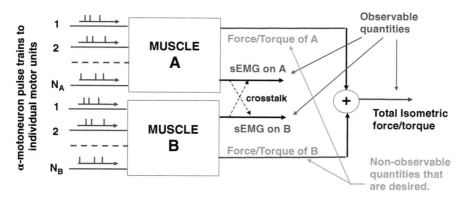

FIGURE 10.2   Input and output variables for two muscles (A and B) acting on the same joint. Inputs are the action potential trains of the N_A and N_B motor neurons. Outputs are the produced force and sEMG amplitude detected from a pair of electrodes or from an electrode grid. Crosstalk may be affecting the signals. Observable and nonobservable quantities are indicated.

model coefficients for muscle 1 and 2.

$$T_1 = x_1 V_{1,1}^{y_1} + x_2 V_{1,2}^{y_2} \tag{10.2}$$

$$T_2 = x_1 V_{2,1}^{y_1} + x_2 V_{2,2}^{y_2} \tag{10.3}$$

$$T_3 = x_1 V_{3,1}^{y_1} + x_2 V_{3,2}^{y_2} \tag{10.4}$$

$$T_4 = x_1 V_{4,1}^{y_1} + x_2 V_{4,2}^{y_2} \tag{10.5}$$

Equations (10.2)–(10.5) provide a system of four nonlinear equations (one for each contraction level), with four unknowns, whose solution(s) provide the unknown model coefficient x_i and y_i which define the torque contributed by each muscle. The x_1 and x_2 unknowns can be obtained from two of the equations and substituted in the other two obtaining a system of two equations and two unknowns (y_1 and y_2 appearing at the exponent) which can be solved either numerically, or by optimization methods by minimizing the mean square error between the measured and the model-predicted torques [right sides of Eqs. (10.2)–(10.5)] [1].

In the given example, the number of contractions and equations could be greater than four when an optimization method is used for finding the solution(s). It is interesting to observe that more than one solution may be obtained, suggesting that more than one load sharing strategy is compatible with the torque outputs. The algorithm can be generalized to more than two muscles (for M muscles at least 2M equations—that is, 2M contraction levels—are needed) and can be applied to subsequent epochs of a fatiguing contraction to observe the changes of the x_i and y_i coefficients and therefore estimate the initial load sharing among the muscles acting on a joint as well as its evolution in time. The algorithm is still under investigation and testing [1].

Figure 10.3 shows an example where the sEMG signals are measured from the two heads of the biceps (as one muscle), the brachioradialis, and two heads of the triceps using electrode grids and arrays and estimating the average envelope of the sEMG RMS over selected channels of each electrode system. The solution is obtained using an optimization algorithm (particle swarm optimization [105]). Because of difficulties in reading the sEMG of the deep muscles (brachialis and medial head of the triceps), the contributions of these muscles cannot be accounted for and are incorporated in those of the others, causing estimation errors. Periodic "parameter estimation"—that is, updating of the x_i and y_i parameters during a test such as that of Fig. 10.3—would provide information about the changes of the torque–sEMG relationship of the individual muscles during a fatiguing exercise.

The model described above is purely mathematical. Its validation is not easy since the contraction force of individual muscles cannot be measured. However, some indirect measurements based on ultrasound technology may become available in the near future [8] and provide means of validation. Other models of force or torque generation offer closer correlation with physiological quantities. Among these is the model proposed by Fuglevand et al. [54] and the more recent and comprehensive model proposed by Dideriksen et al. [23] and applied to the investigation of the

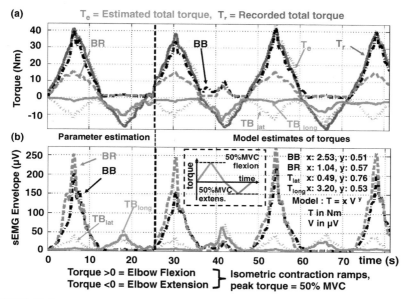

FIGURE 10.3 Example of application of the model described in Eqs. (10.2)–(10.5). The elbow of the subject was locked in an isometric brace and he was tracking a torque target on a computer screen producing a sequence of flexion–extension cycles ranging between 50% MVC flexion and 50% MVC extension (see inset in panel b). Single differential sEMG was detected, with four electrode arrays, from the biceps brachii (BB), the brachioradialis (BR), and the long and lateral heads of the triceps (TBlong, TBlat) (signals from the brachialis and the medial head of the triceps could not be detected). The RMS envelopes were averaged over the channels showing activity. The Particle Swarm Optimization algorithm was used to find the x_i and y_i values that minimized the mean square error between the estimated total torques and the measured ones. The first cycle (25 s, parameter estimation) was used to estimate the x_i and y_i muscle parameters which were then used to estimate the total torque and the contributions of the individual muscles in the subsequent cycles (model estimates of torques). **(a)** Plots of the recorded and estimated torque and of the contributions of the individual muscles considered. Such contributions include those of the not considered muscles (brachialis and medial head of the triceps). **(b)** Plots of RMS envelopes of the considered muscles. The target profile is provided in the inset on the left. The x and y values of the model $T = xV^y$ are provided in the inset on the right for each muscle. It is interesting to notice the co-contraction of agonist and antagonist muscles, which is presumably implemented for better tracking of the force target. The anatomical definition of TBlat and TBlong is according to Saladin, K. S., *Anatomy & Physiology: The Unity of Form and Function*, 3rd ed., McGraw-Hill, New York, 2003.

sEMG–force relation and of fatigue [24]. Dynamic musculoskeletal models are discussed in Chapter 9.

10.2.3 Reading sEMG for Force Estimation

The models described above imply the availability of representative sEMG amplitude values. Although these values can be obtained from pairs of electrodes placed over individual muscles, 2D electrode arrays can provide spatial averages over a region of interest (ROI) defined as a portion of a sEMG map. Electrode grids provide maps of sEMG distribution from which a single indicator of sEMG amplitude (ARV or RMS) can be extracted for each observable muscle and introduced in a force estimation algorithm such as the one defined in Eqs. (10.2)–(10.5). This procedure implies segmentation of the map into ROIs from which the spatial average of ARV or RMS may be obtained, as explained below. Amplitude estimates of individual sEMG channels may be obtained with the algorithms described in Chapter 4.

Figure 10.4 outlines the distribution of single differential sEMG RMS over the arm and forearm of a healthy subject performing various tasks. In this protocol, three

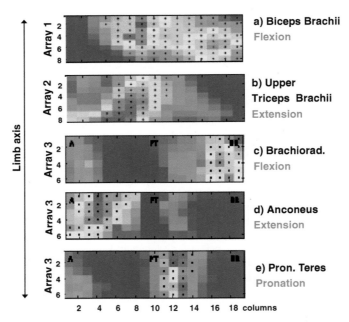

FIGURE 10.4 Examples of single differential sEMG maps obtained from three 2-D arrays, each made with silver-coated eyelets filled with conductive gel, spaced 10 mm apart (8 or 6 rows and 19 columns) and fixed on a cloth. The arrays were applied to the distal part of the biceps brachii (array 1), proximal part of triceps brachii (array 2), and proximal portion of the forearm (array 3), with the columns parallel to the longitudinal direction of the limb. Array 3 covered the anconeus, brachioradialis, and pronator teres muscles (A, PT, BR). Isometric elbow flexion, extension, and forearm pronation were performed at 50% of the MVC of each task. The features of sEMG (amplitude in this case) were estimated over a region of interest, where the signals corresponding to a specific effort are stronger and presumably due to a single muscle. These regions are defined by means of segmentation algorithms and are identified by black dots. Panel a shows the contributions of the two heads of the biceps. The average of the amplitudes, (ARV or RMS) of the pixels marked with a dot provides a sEMG value for that muscle—to be used, for example, in a model such as the one defined by Eqs. (10.2)–(10.5). The same applies to the other panels.

electrode arrays (8 or 6 rows × 20 columns in the fiber direction, IED = 10 mm) were placed respectively across the biceps brachii (array 1), on the triceps brachii (array 2), and on the proximal portion of the forearm (array 3) with the columns aligned with the limb axis. The elbow joint was locked in an isometric brace; efforts of elbow flexion, extension, and forearm pronation were performed at about 50% MVC for 10 s, where MVC was previously measured for each task. The RMS values where computed on a 5-s epoch. Regions of interest (ROI) were identified with an image segmentation method and were indicated with a black dot on each pixel of the ROI [114].

10.3 PHYSIOLOGICAL MECHANISMS OF MUSCLE FATIGUE: A MODELING APPROACH

Different types and definitions of muscle fatigue exist and have been proposed and discussed in the literature. Some are based on the inability to sustain a required force/torque (mechanical manifestations of muscle fatigue), others are based on the force increment elicited by electrical stimulation during a fatiguing contraction (interpolated twitch) [62,84,119], and others are based on the changes that specific features of the sEMG undergo during sustained efforts or tasks [18,19,87,88]. These latter changes may or may not be related to mechanical modifications of performance, take place from the beginning of the contraction, and reflect the muscle drive and the phenomena taking place at the muscle fiber membrane level even in the absence of force changes; they are referred to as myoelectric manifestations of muscle fatigue [19,88,93].

Muscle force is centrally modulated by varying the number of active motor units and their discharge rates [14,20]. Muscle force is also determined by the proper functioning of the neuromuscular junctions, by the sarcolemma excitability, by the excitation–contraction coupling, and by the contractile mechanisms inside the fibers. Some of these mechanisms are affected by blood flow which is blocked at high contraction levels [16,120], causing the muscle to operate in ischemic conditions with progressive accumulation of metabolites [24,25].

All these mechanisms play a role in sustaining force and may fail at different times and in different degrees. These mechanisms have been addressed by De Luca [18,19] and discussed in the book *Fatigue: Neural and Muscular Mechanisms* [56] and in more recent reviews [32]. Although there are still considerable gaps of knowledge concerning the role of these mechanisms and their interactions, it is well known that sEMG features change, during a sustained contraction, without a concurrent change in muscle force and vice versa; therefore the sEMG–force relationship, described in Section 10.2.2, changes as fatigue develops.

Recent modeling efforts contributed to unravel these complex relations. To simulate the motor neuron discharge patterns in fatigue, the model of motor neuron pool behavior and isometric force developed by Fuglevand et al. [54] was adopted and extended by Dideriksen et al. [23–25]. These authors developed a model that accounts for the accumulation of metabolic substances within the muscle, the depletion of energy, and the occlusion of blood flow related to the sustained contraction. The simulated concentration of metabolites determines the degree of inhibitory afferent feedback to the motor neuron pool [58] as well as the changes in the twitch force. The metabolites diffuse, across the cell membranes, into the extracellular space from which they are removed by the bloodstream.

A control algorithm has been implemented [23,24] to (a) simulate sustained contractions at a constant force level while the excitability of the motor neuron pool and its force producing capability change and (b) estimate the descending drive needed to maintain a stable force output. As fatigue progresses, this model adjusts the descending drive to compensate for the neuromuscular changes. At a certain point, the descending drive reaches its maximal level, making it impossible to maintain the target force (task failure).

The intramuscular action potential at different stages of muscle fatigue was described by Dimitrova and Dimitrov [26–28]. According to this description, as fatigue develops, the duration of the intracellular action potential increases and its amplitude decreases. In the Dideriksen model [23], these changes are associated with (i) the metabolite concentration, (ii) the conduction velocity of the muscle fibers of the motor unit, and (iii) the instantaneous discharge rate of the motor unit and its size. Having established the shape of the intracellular action potential, its appearance on the surface is simulated using the model of volume conduction in the muscle, fat, and skin tissue between the single muscle cell and the electrode [45] (see Chapters 2 and 8). In this way, a complete set of motor unit action potentials across different levels of force and fatigue can be generated, the interference sEMG can be simulated and its amplitude (RMS) can be calculated.

Figure 10.5 shows results from a simulated ramp contraction from zero to maximal force. The average motor unit conduction velocity spans the range from no fatigue ($\sim$5 m/s) to severe fatigue ($\sim$3 m/s). The sEMG amplitude corresponding to every combination of muscle activation level and motor unit conduction velocity is determined and represented in a three-dimensional plot.

Figure 10.5 also indicates that an increase in the muscle activity (number of motor unit discharges per second) always involves an increase in the sEMG amplitude. The gain of this relation, however, is slightly higher at low levels of muscle activation, suggesting that the same increase in the muscle activation implies a higher increase in the sEMG amplitudes at low activation levels compared to higher levels. This observation can be explained by a difference in the level of sEMG amplitude cancellation (due to the algebraic summation of MUAPs) of approximately 20% across the range of muscle activations.

As conduction velocity decreases from the nonfatigue values (4–5 m/s), the corresponding sEMG amplitude increases. A peak of sEMG amplitude is reached around conduction velocities of 3.2 m/s, after which the sEMG amplitude decreases because the decrease in amplitude predominates on the increase in the action potential duration [28]. It should be noted that not all the values in the figure are physiologically realistic since the muscle activation level and conduction velocity are not independent variables (for example, conditions yielding the supramaximal amplitudes ($\sim$130%) do not occur in normal conditions). In sustained submaximal contractions at constant force levels, the muscle activation level is expected to start at a submaximal level and

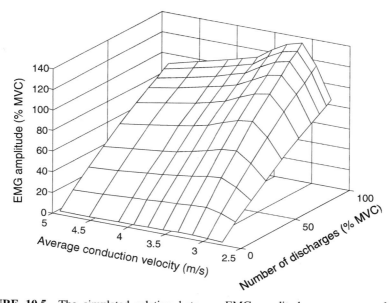

FIGURE 10.5   The simulated relation between EMG amplitude, average muscle fiber conduction velocity, and muscle activation (number of motor unit discharges per second for the entire motor unit population). Adapted from Dideriksen et al. [25].

increase slightly to compensate for the adaptations in motor neuron excitability and contractile properties whereas the conduction velocity decreases. In typical fatigue studies, however, minimal conduction velocities are rarely lower than 3.2 m/s [58] and thus are in the part of the curve where it causes the sEMG amplitude to increase. Therefore, the model predicts that sustained contractions would imply a modest contribution of conduction velocity to the increases in the sEMG amplitude, which is in accordance with experimental observations [81].

Although sEMG and muscle force are direct reflections of the activity of muscles (Figs. 10.1 and 10.2), differences in the adaptations of the action potentials and the force producing capacity may progressively modify their relationship. This was demonstrated in another series of simulations by Dideriksen et al. [24,25] which included a series of sustained contractions at constant force levels, trapezoidal contractions, and repeated ramp contractions. In Fig. 10.6 each thin line represents the time course of the concurrently simulated sEMG amplitude versus force across the different simulation paradigms. The lower dashed, curved line indicates the non-fatigue case, whereas the upper dashed line indicates the relation in severe fatigue. The figure demonstrates that a sEMG amplitude value of 60% of the maximal sEMG value may correspond to forces in the range from 10% to 50% MVC depending on the level of fatigue of the muscle. This very large range indicates that adaptations to fatigue can significantly modulate the sEMG–force relation and that this modulation depends on the characteristics of the task. Therefore, the experimental sEMG–force

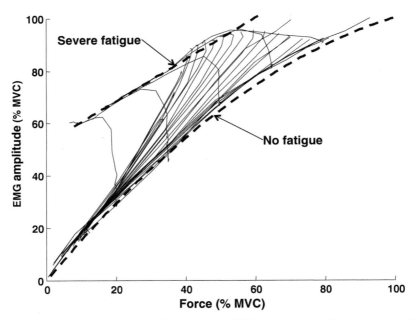

FIGURE 10.6   The simulated relation between EMG amplitude and force across multiple different simulation paradigms. The dashed lines indicate the boundaries for this relation at different levels of fatigue. Reprinted from Dideriksen et al. [24].

relationship indicated in Fig. 10.1c can be estimated only at the beginning of a moderately fatiguing contraction and the muscle parameters x_i and y_i in Eqs. (10.2–10.5) may be changing during a sustained contraction.

Motor unit mechanical twitches produce displacements of the skin above the muscle which can be detected with microphones or accelerometers. The resulting signal, referred to as mechanomyogram, is discussed in Chapter 11 of Merletti and Parker [92] and has been used as indicator of muscle fatigue [66,126].

The following sections of this chapter further deal with myoelectric manifestations of muscle fatigue and expand and update the material presented on this topic in Chapter 9 of Merletti and Parker [92] with specific focus of 2D sEMG.

10.4   MYOELECTRIC MANIFESTATIONS OF MUSCLE FATIGUE IN ISOMETRIC, CONSTANT FORCE, VOLUNTARY CONTRACTIONS

Myoelectric manifestations of muscle fatigue are defined as the changes taking place in the sEMG signal during a sustained muscle activity. As seen in the previous section, the relation between sEMG and muscle force is very complex and affected by many factors which also affect myoelectric manifestations of muscle fatigue. For this reason, myoelectric manifestations of muscle fatigue have been investigated in conditions that eliminate some of these factors and allow the study of the remaining ones, reducing the likelihood of misinterpretation of results. Such conditions are obtained by implementing "bench-test" situations where the muscle(s) of interest operates under controlled conditions (for example, fixed length, fixed force, electrical activation, etc.). Since more than one agonist and more than one antagonist act on a joint (or on two joints), these conditions are difficult to achieve (see Section 10.2). For this reason the first dorsal interosseous (FDI) muscle has been extensively studied because it is the only agonist acting across the thumb–index joint [141,142]. In dynamic conditions, the shifting of the muscle under the skin substantially contributes to sEMG changes under an electrode pair, as explained in Section 10.5, often leading to signal misinterpretations. Further considerations on neurological mechanisms of muscle fatigue are provided in Chapter 12.

10.4.1 Quantification of Muscle Fatigue from sEMG in Isometric Constant Force Voluntary Contractions

The many factors affecting sEMG features, even during the simple experimental paradigm of sustained isometric constant-force contractions, range from anatomical features to the detection system to the estimation algorithm used [38–40,43,44]. The relevance of these factors depends on the produced force and on the joint angle. The two main factors are the decrease of muscle fiber conduction velocity (CV) (see Chapter 5) and the variations of shape and increase of the spatial support and time duration of the transmembrane action potential (intracellular action potential, IAP) [2,3,26,28]. The early work of Lindstrom, Chaffin, and De Luca [11,74,75,124] outlined how CV acts as a scaling factor for the sEMG and for its power spectrum, which is therefore compressed (shifted in logarithmic scale) by a decrease of CV (see Chapter 4).

In the ideal case where a general signal $x_1(\theta)$ is scaled in time (θ) to generate $x_2(\theta) = x_1(k\theta)$, its Fourier transform $X_2(f)$ and its power spectrum $P_2(f) = |X_2(f)|^2$ are scaled in frequency (and in amplitude) so that $X_2(f) = X_1(f/k)/k$ and $P_2(f) = P_1(f/k)/k^2$. This property applies to stochastic signals as well (see Fig. 4.1). Two characteristic frequencies have been used to describe sEMG spectral changes: the mean or centroid (MNF or f_{mean}) and the median (MDF or f_{med}). The latter is the 50th percentile of the power spectrum—that is, the value splitting it into two parts of equal area. Both frequencies are scaled by the same factor k. If the signal in time $x_1(\theta)$ is a sEMG generated by a source (e.g., a motor unit action potential) propagating in space with conduction velocity $CV(t)$ which is slowly decreasing with time (see Chapter 5), then, at time t, $CV(t)$ is a fraction of $CV(0)$, that is, $CV(t) = k\,CV(0)$. The coefficient k ($k < 1$) could be considered an indicator of myoelectric fatigue at time t. As a consequence, $f_{mean}(t) = kf_{mean}(0)$ and $f_{med}(t) = kf_{med}(0)$. If the decrement of CV were the *only* change taking place, the percent decrease of $f_{mean}(t)$ and $f_{med}(t)$ would be identical to the percent decrease of $CV(t)$ and their plots, normalized with respect to the respective initial values $CV(0)$, $f_{mean}(0)$, and $f_{med}(0)$, would overlap. In this case, k could be computed from the f_{mean} or the f_{med} of the power spectrum using a single-channel sEMG signal and could be used as an indicator of fatigue.

It can be shown that the inverse relation holds for the average rectified value of the signal (ARV) and for the root mean square (RMS, which is the square root of the area under the power spectrum). Specifically, $\text{ARV}(t) = \text{ARV}(0)/k$ and $\text{RMS}(t) = \text{RMS}(0)/\sqrt{k}$, indicating an increase of ARV and RMS associated to a decrease of CV. This increase is due to the widening and, therefore, to the increase of the area or the square of the absolute value, of the propagating motor unit action potentials. The plots of normalized CV, f_{mean}, f_{med}, ARV, and RMS (that is, the plots of k, $1/k$, and $1/\sqrt{k}$) are jointly and globally referred to as the "fatigue plot" [92].

A better way to obtain k is to use *all* spectral frequencies, rather than only f_{mean} or f_{med}. This method is referred to as the "cumulative power function method" and was originally proposed by H. Rix and Malengé [112] and applied to sEMG spectra by Merletti and Lo Conte [90]. The method provides the scaling factor between two frequency-scaled power spectra with similar but not identical shapes and outlines the regions where the shapes differ, as discussed in Chapter 4 and presented in Fig. 4.5.

Other factors, such as change of shape of the muscle and thickness of the subcutaneous tissue, also play a role [95]. In addition, the situation is different at different contraction levels. At a low contraction level, only a fraction of the muscle motor units are recruited (mostly small, type I), and dropout and replacement of motor units is possible. At medium levels of contraction, blood flow is almost occluded [119,120] and most motor units are recruited. At high levels of contraction, blood flow is fully occluded, all motor units are recruited, and motor unit substitution is no longer possible.

As indicated above, it should be noted that the rate of decrease of CV is not always matching that of MNF and MDF because:

1. Different motor units have different rates of change of CV. CV values have a statistical distribution that may change in time.

2. The active motor unit pool may change in time because of motor unit dropout and recruitment, and the synchronization of motor unit firings affects spectral variables.

3. The shape and width of the IAP may change.

4. The muscle fibers may not be parallel to the skin (pennate muscles) and CV cannot be measured.

5. The force produced by the muscle of interest may change, even if the total measured torque at the joint remains constant (different sharing of the load among the active muscles, see Section 10.2).

6. The degree of cancellation due to algebraic MUAP summation may change.

An example of sEMG detected with a linear electrode array during a low-level isometric torque (15% MVC sustained for 10 minutes) produced by a biceps brachii is provided in Fig. 10.7 [59]. The array has 8 electrodes, spaced 10 mm apart, and spans most of the muscle, and the signals are longitudinal single differential (LSD, along the fiber direction), thereby clearly showing the innervation zone in the middle of the

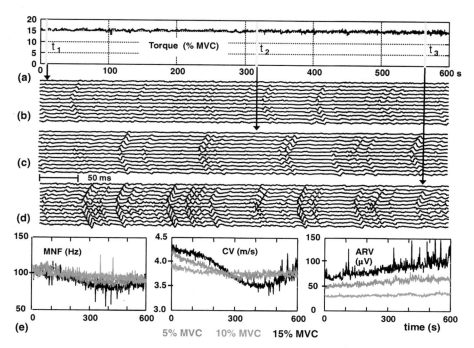

FIGURE 10.7 Isometric contraction of the biceps brachii of a healthy subject, sustained at 15% MVC for 10 minutes. (**a**) Torque plot. (**b–d**) sEMG signals detected with a linear electrode array (IED = 10 mm), centered on the muscle, during three 0.5-s epochs: one at the beginning, one in the middle, and one at the end of the contraction. (**e**) Plots of mean spectral frequency (MNF), muscle fiber conduction velocity (CV), and average rectified value during three 10-min contractions sustained at 5% MVC (green), 10% MVC (red), and 15% MVC (black). Observe the increase of MNF and CV, after 5 min, in the second and third case due to the recruitment of fresh motor units. Reprinted from Gazzoni et al. [59].

muscle. Three signal epochs of 0.5 s are depicted in panels b, c, and d (beginning of contraction, mid-contraction, end of contraction).

Even if there is no change in force (as indicated in panel a), the signals show the progressive recruitment of new motor units, suggesting that those recruited at the beginning are progressively less able to produce the required force. In this situation the definition of myoelectric manifestations of muscle fatigue becomes very question-able and the only meaningful definition is the one concerning the fatigue of individual motor units, which may be investigated after decomposition of the signals into the constituent motor unit action potential trains (Chapter 7). Mean spectral frequency values (MNF or f_{mean}) and the average rectified value (ARV) are computed for channels on either side of the innervation zone and averaged, and conduction velocity is computed using a group of channels on one side of the innervation zone (see Chapter 5). Figure 10.7e shows the time course of MNF, CV, and ARV corresponding to three contraction levels (5% MVC, 10% MVC, 15% MVC). While the patterns of MNF and CV show a small decrement for the contraction level of 5% MVC, they show a decrement followed by an increment for 10% MVC and 15% MVC, reflecting fatigue of the initial motor unit pool followed by recruitment of fresh motor units.

The case described in Fig. 10.7 shows how carefully the changes of sEMG signals should be interpreted. For example, selecting a group of channels that include the innervation zone would provide erroneous results since CV cannot be defined, MNF and MDF have abnormally high values, and RMS and ARV have abnormally low values [44].

As discussed in Chapter 5, sEMG features can be defined in two dimensions, as maps above the skin. Figure 10.8 provides three maps of RMS and three maps of

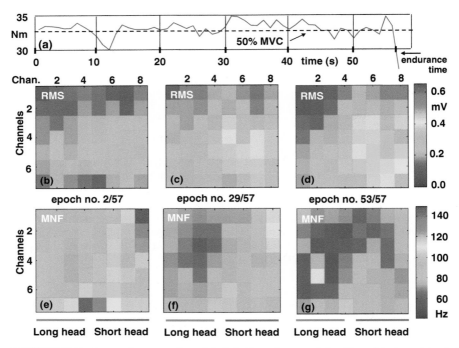

FIGURE 10.8 Isometric contraction of the left elbow flexors of a healthy subject, sustained at 50% MVC to endurance time (57 s). An electrode grid of 8×8 electrode (IED $= 10$ mm) is placed just proximal with respect to the innervation zone of the biceps brachii and sEMG detection is single differential along the columns, in the fiber direction. RMS and MNF are estimated on 57 one-second epochs. The left half of the grid is on the long head and the right part of the grid is on the short head of the muscle. No interpolation is applied. (a) Torque plot. (b–d) three maps of the RMS spatial distribution at epochs 2, 29 and 53; (e-g) three maps of the MNF spatial distribution at epochs 2, 29, and 53. The short head shows a marked increase of RMS values and a marked decrease of MNF values. The long head shows smaller initial values of the two variables and slightly smaller variations with time. No information is available about the torque contributed by these two muscles, and by other muscles acting on the elbow, to the total torque.

MNF obtained from a left biceps brachii using an electrode grid of 8×8 electrodes with IED $= 10$ mm (no interpolation). An isometric, constant flexion torque is produced at the elbow at 50% MVC and sustained to the endurance time (57 s) as indicated in Fig. 10.8a. For each longitudinal single differential signal the RMS and MNF values are computed over 1-s-long epochs and displayed as colors of the pixels. The columns of the grid are parallel to the fiber direction and the innervation zone of the muscle is just below the bottom row. The RMS and MNF maps are relative to epoch 2, 29, and 53. Panels b, c, and d show a progressive increase of RMS, particularly in the region of the short head. Panels e, f, and g show higher initial values of MNF and greater MNF decrease in the region of the short head. This figure demonstrates the possibility of studying localized myoelectric manifestations of muscle fatigue in individual muscle compartments.

Figure 10.9 shows maps of ARV and MNF estimated over 0.5-s epochs of single differential sEMG signals detected on the upper trapezius muscle during an isometric

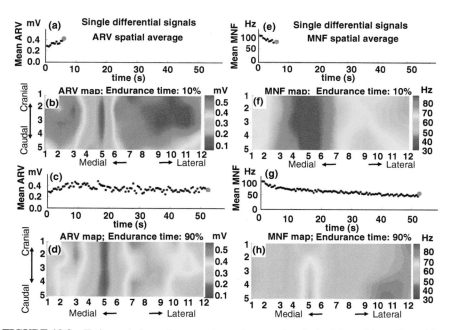

FIGURE 10.9 Fatigue study on the upper trapezius muscle of a healthy subject. The subject is pulling up a handle fixed to a load cell, with the straight vertical right arm, at 50% MVC. The effort is sustained to endurance (55 s). Single differential signals in the fiber direction are obtained using a 13×5 grid with IED $= 8$ mm. ARV and MNF are computed every 0.5 s and their maps are interpolated with a factor 10. **(a)** spatial average of ARV over the entire map every 0.5 s. **(b)** ARV map at the sixth second. **(c)** Spatial average of the ARV map at the 52nd second, few seconds before failure. **(d)** ARV map at the 52nd second. **(e–h)** Same plots and maps for MNF. Observe an important muscle innervation zone between rows 4 and 6 showing a valley in the ARV distribution and a peak in the MNF distribution.

contraction at 50% MVC sustained until endurance (55 s). The maps are obtained from a 13×5 grid with IED = 8 mm and are interpolated with a factor 10. The spatial averages of ARV and MNF, computed over a 1-s epoch and over the entire grid, are depicted in panels a and c for the 6th second of contraction and in panels e and g for the 52nd second of contraction, just before force failure. The valley of ARV and the peak of MNF on the innervation zone are evident. The initial increase of the spatially averaged ARV, followed by a decrease, and the progressive decrease of the spatially averaged MNF are also evident. No compartments or particular changes of image patterns can be clearly identified in this case. Similar results have been reported for monopolar and bipolar sEMG detection by other investigators [70,121,122] using smaller surface electrode arrays. The maps in Figs. 10.8 and 10.9 also show the importance of a topographic recording and point out how different bipolar sEMG signal could be when detected in different locations.

Other experiments demonstrated that topological changes of EMG distribution of the trapezius muscle take place during an isometric sustained contraction and that the entropy of the RMS image (degree of uniformity) decreases while the centroid of the image moves cranially [48]. These authors demonstrated that the subjects with more heterogeneous activity and larger shift of the image centroid towards the cranial direction could sustain the required isometric contraction force longer than subjects with more uniform activity and smaller centroid shift.

Other indices of fatigue have been investigated searching for those most sensitive, most repeatable or most focusing on specific central or peripheral mechanisms. Some examples are provided by Dimitrov et al. [26] who used spectral moments, other than the first, and their ratio as indicated in Eq. (10.6):

$$FI_k = \frac{\int_{f_1}^{f_2} f^{-1} P(f) \, df}{\int_{f_1}^{f_2} f^k P(f) \, df} \qquad \text{for} \quad k = 2, 3, 4, 5 \qquad (10.6)$$

where $P(f)$ is the normalized power spectrum of the signal (spectrum with unit area), and f_1 and f_2 define the bandwidth over which the fatigue index FI is calculated. The integral at the denominator in Eq. (10.6) defines the moment of order k and the numerator is the moment of order -1. The moment of order zero is the power of the signal—that is, its RMS^2.

Other authors proposed indices based on the fractal dimension (FD) [61,139] of the sEMG signal and the percentage of determinism (%DET, obtained from recurrence quantification analysis) [42,51,53]. Mesin et al. [94] compared the variability and the sensitivity of a number of fatigue indicators, applied to simulated stationary and nonstationary interferential signals, with respect to CV and motor unit synchronization changes. The coefficient of variation (COV) for stationary signals, computed over 60 epochs of 0.5 s each, turned out to be less than 0.5% for the estimated conduction velocity (ECV), between 1.5% and 4% for RMS, ARV, MNF, and MDF, between 4.5% and 9% for the FI idexes defined in Eq. (10.6), about 1% for entropy (S),

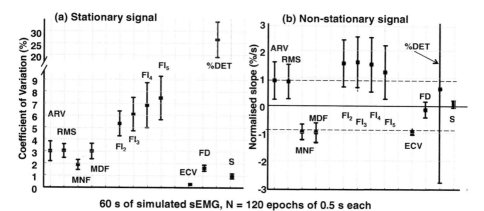

FIGURE 10.10 (a) Mean ± st. dev. of the coefficient of variation (σ/mean in %) of a number of fatigue indicators calculated over 60 epochs of 0.5 s each of a simulated stationary EMG signal with no synchronization among motor units. An array of eight electrodes was simulated and the spatial mean of each feature was computed for each epoch. (b) Mean ± st. dev. of the normalized slope (% of the initial value per second) of the same fatigue indicators calculated from nonstationary signals simulating a decrease of CV from 4 m/s to 3 m/s (0.83% of the initial value per second, indicated by the dashed lines) and an increase of synchronization from 9% to 20% in 30 s. The muscle fibers are parallel to the skin, and the array is on one side of the innervation zone. ARV and RMS: average rectified value and root mean square, MNF and MDF: mean and median frequency of the power spectrum, F_2, F_3, F_4, and F_5, are defined in Eq. (10.6). ECV, estimated muscle fiber conduction velocity; FD, fractal dimension; S, signal entropy; %DET, percentage of determinism obtained from the recurrence quantification analysis [42,53]. See text and Mesin et al. [94] for further details. Modified from [94] with permission.

between 1.5% and 2% for FD, and greater than 20% for %DET. Results from Mesin et al. [94] are indicated in Fig. 10.10a. The nonstationary signals were simulated (for fibers parallel to the skin) by imposing a distribution of motor unit CV values whose mean decreased linearly from 4 m/s to 3 m/s in 30 s (0.83% of the initial value per second) while synchronization (see definition in Mesin et al. [94]) increased linearly from 0% to 20% in 30 s. Results reported in Fig. 10.10b show (a) a nearly correct value of the mean slope of spectral variables (negative slope) and of amplitude variables (positive slope), (b) a higher sensitivity, associated with a greater variability, of the FI indexes defined in Eq. (10.6), (c) an estimate of CV (ECV) with small variability and slight bias, (d) small sensitivity of FD, and (e) a very large variability of %DET (see Mesin et al. [94] for details). The sensitivity of the estimated CV (ECV) is high with respect to variations of CV and low with respect to variations of motor unit synchronization, while the sensitivity of FD is low with respect to variations of CV and high with respect to variations of synchronization. These authors, therefore, proposed to define a fatigue index as a vector with two components, ECV and FD. The separate quantification of myoelectric manifestations of central and peripheral fatigue in clinical settings deserves further research.

The results reported above are relative to skin-parallel-fibered muscles and cannot be extrapolated to in-depth-pinnate muscle where most of the signal is due the end-of-fiber effect. In the in-depth-pinnate muscles spectral variables decrease with fatigue because of the progressive widening in time of the end-of-fiber effect consequent to the decreasing values of muscle fiber CV (which cannot be measured directly).

If an in-depth-pinnate muscle sustains an isometric constant force contraction, changes of spectral parameters may appear locally and correspond to the fiber-aponeurosis terminations of the active motor units. Gallina et al. [55] observed these localized myoelectric manifestations of muscle fatigue in the medial gastrocnemius muscle during a sequence of 5 s ON and 5 s OFF voluntary isometric contractions. Results are reported in Fig. 10.11.

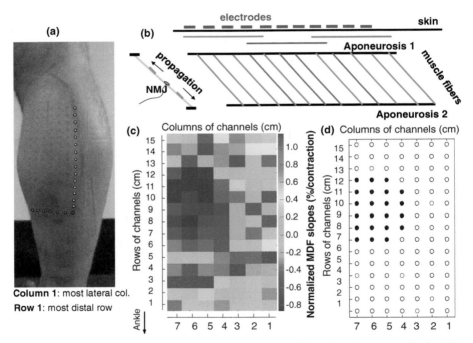

FIGURE 10.11   Localized myoelectric manifestations of fatigue in a pinnate muscle (medial gastrocnemius) observed by means of an electrode grid (16×7 electrodes, IED = 10 mm) during isometric intermittent contractions (5 s ON and 5 s Off). (a) Medial gastrocnemius region after removal of the grid, showing the locations of the electrodes. (b) Schematic arrangement of the fibers in the medial gastrocnemius. Three motor units are depicted, as an example, with fibers represented in red, blue, and green. Their signals would appear mostly in the red, blue, and green regions indicated under the electrodes. (c) Color map of the fatigue index "rate of decrement of MNF" defined as the slope (% of initial value per contraction) of the linear regression of MNF versus time for each electrode pair (longitudinal differential detection). (d) Segmented portion of the image showing high negative slope of MNF (see also Chapter 5).

The analysis of myoelectric manifestations of muscle fatigue in pinnate muscles requires considerable competence and understanding of the muscle structure and is an open field of research.

10.5 MYOELECTRIC MANIFESTATIONS OF MUSCLE FATIGUE IN DYNAMIC CONTRACTIONS

In dynamic conditions, the shifting of the muscle under the skin substantially contributes to sEMG changes under an electrode pair, as indicated in Fig. 10.12. These changes should *not* be interpreted as myoelectric manifestations of muscle fatigue. They are due to the change of muscle position under the skin, as demonstrated by the shift of the innervation zone (IZ). For this and other reasons, most studies have been carried out in isometric conditions or by reading the sEMG obtained at a fixed joint angle during cyclic

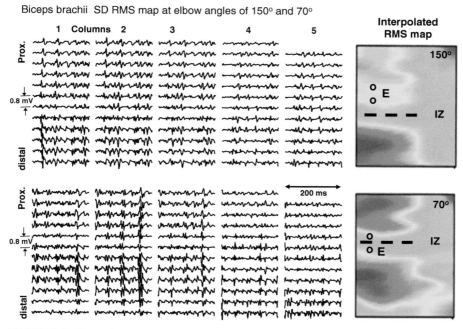

FIGURE 10.12   Effect of elbow angle on biceps EMG signals while the subject is holding a 3-kg weight in his hand and flexing the forearm with respect to the arm. Surface EMG signals were acquired in longitudinal single differential (LSD) mode from a biceps brachii muscle using a grid of 13 rows and 5 columns (columns along the limb direction, grid IED = 8 mm) at two elbow angles (150° and 70°). The image of RMS values is interpolated by a factor 10. As the elbow flexes from 150° to 70° the biceps shortens under the electrode grid and the innervation zone (thick dashed line) moves proximally by about 16 mm. The signals detected by the same pair of electrodes E in the two conditions are different because of the anatomical and geometrical changes rather than because of changes of muscle activation or force.

movements [5,6,99]. The muscle movement with respect to the skin and the electrodes is an important cause of signal change, in particular when a single electrode pair is used and the innervation zone moves under it (or away from it) during the movement. In the case of muscles with fibers parallel to the skin, an electrode array or a grid provides means to track the shift of the innervation zone and to read signals at a fixed distance from it [95].

The main problems arising in the study of fatigue in dynamic contractions are related to (a) the nonstationarity of the signal and (b) the relative movement of the muscle with respect to the electrodes. In the special case of cyclic movements (e.g., pedaling) short EMG epochs can be selected for feature extraction at the same fraction of each cycle.

In general, the study of nonstationary signals is addressed using the "time frequency representations" (TFR) which can be imagined as a short time spectrum evolving in time (time along the x axis, frequency along the y axis, and power represented by false colors) and providing a description of the time evolution of the "instantaneous" spectrum of the signal. Short bursts and explosive contractions can be analysed either with this method [5,6,68,99,119] or by the estimation of CV of individual multichannel MUAPs recorded during short sEMG bursts [106].

The Fourier expansion of a signal in series of basis functions (sinusoids) of infinite time support is appropriate for stationary signals with wide time support (high-frequency resolution and low-time resolution). A series expansion using basis functions with short time support is better suited to describe waves of short duration (motor unit action potential, M-waves) or signals that change rapidly in time (bursts). A family of expansions is based on a single shape (mother wavelet) from which the basis functions are obtained as scaled and shifted versions of the mother wavelet. Another family of expansions is based on non identical waveforms derived from the product of Gaussian functions and sinusoids (Gabor "wavelets") or from the product of Gaussian functions and Hermite polynomials (Hermite "wavelets") [79]. Applications to sEMG signals have been and are still being investigated [12,15,17,67,68].

10.6 MYOELECTRIC MANIFESTATIONS OF FATIGUE IN ELECTRICALLY ELICITED CONTRACTIONS

Mechanical and myoelectric manifestations of muscle fatigue are very relevant in functional electrical stimulation when electrically elicited contractions should be functionally efficient for a relatively long time. This section, however, deals mostly with myoelectric manifestations of muscle fatigue as investigation tools. Further discussions are presented in Chapter 11.

When a train of electric pulses is applied to a muscle motor point or to a motor nerve, the activated motor units discharge synchronously, producing a deterministic quasi-stationary signal, referred to as M-wave or compound motor action potential (CMAP), described in Fig. 4.4, as well as a twitch force. The number of recruited motor units depends on the amplitude and duration of the pulse according to the intensity–duration curve of the nerve fibers or of their terminal branches, as well as on

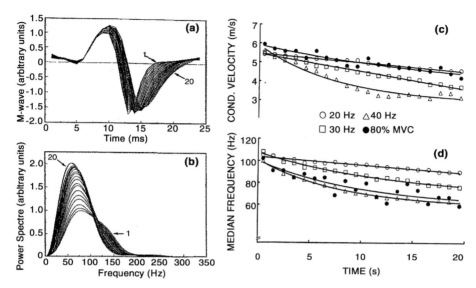

FIGURE 10.13 (a, b) Time evolution of a single differential M-wave elicited in the tibialis anterior (TA) muscle by motor point stimulation at 30 Hz* for 20 s (0.2-ms pulse duration). Each of the 20 waves is the average of 30 responses (1 s) and the power spectra are obtained by zero-padding to 1 s of each average M-wave. (c, d) Time course of conduction velocity (CV) and median frequency (MDF) during 20 s of stimulation of the TA at 20 Hz, 30 Hz, and 40 Hz. Black dots show the patterns detected in the same conditions during a voluntary contraction of the TA at 80% MVC. Note the widening of the M-wave and the compression of the spectrum, with increase of its area. Note the values of CV during the voluntary contraction being similar to those corresponding to 20 Hz stimulation, while the values of MDF decrease more, being sensitive to other factors such as the change of shape of the sources. Reproduced from [85] with permission. *: The currently used unit for stimulation frequency is "Hz" but "pulses/s" would be more appropriate.

geometrical factors (distance between such fibers and the electrodes) determining the current density at the nerve fiber level. Pulses are usually produced by current generators, rather than by voltage generators, to reduce the effect of changing electrode-skin impedance. Afferent fibers are activated as well; and the effect of their stimulation, which is relevant at low amplitudes and long pulse durations, is under investigation (see Chapters 11 and 12).

Figures 10.13a and 10.13b show the widening of the M-wave, along with the associated compression of the power spectral density, during an electrically elicited contraction of the tibialis anterior muscle (TA) sustained for 20 s at 30 Hz. Figures 10.13c and 10.13d show the decreasing pattern of conduction velocity (CV) and spectral median frequency (MDF) during 20 s of stimulation of the TA at 20 Hz, 30 Hz, and 40 Hz. For comparison, the same investigation included a voluntary contraction at 80% MVC, performed in the same conditions, whose results are shown as black dots in Figs. 10.13c and 10.13d [85,89].

It seems reasonable to think that each pulse of a train would produce a "quantum" of fatigue and therefore the fatigue-related changes of sEMG variables should be plotted

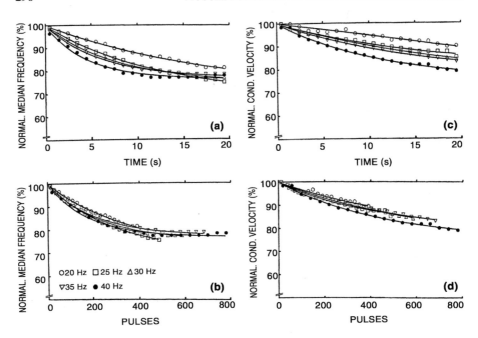

FIGURE 10.14 (a) Median frequency (normalized with respect to its initial value) of the power spectrum of the single differential sEMG detected on the tibialis anterior (TA) muscle during 20 s of electrical stimulation at 20 Hz, 25 Hz, 30 Hz, 35 Hz, and 40 Hz. (b) Same plot where the time scale has been replaced by the number of delivered pulses. (c) Normalized plot of conduction velocity (CV) in the same conditions as a). (d) Same plot as in part c, where the time scale has been replaced by the number of delivered pulses. The contractions are performed in isometric conditions and ankle torque shows a slight increase during each 20-s stimulation train due to widening of the mechanical twitch response (not shown in the figure). Reproduced from [89] with permission.

versus the number of delivered pulses rather than versus time, resulting in the same plot regardless of stimulation frequency. This approach is depicted in Figs. 10.14 and 10.15 for two subjects showing different patterns. Figures 10.14a and 10.14c show the percentage decrements of MDF and CV versus time for the TA of Subject 1 being stimulated at 20 Hz, 25 Hz, 30 Hz, 35 Hz, and 40 Hz.

In Figs. 10.14b and 10.14d the time scale has been replaced by the number of delivered pulses. The same presentation is provided in Fig. 10.15 for Subject 2 whose myoelectric manifestations of muscle fatigue are much greater than those of Subject 1. Although the curves of normalized MDF and CV versus the number of delivered pulses are much closer than those versus time, some difference exists, possibly due to the frequency effect on muscle vascularization, ischemia level, or other factors. The plots of Figs. 10.14 and 10.15 suggest that the differences between the two subjects might reflect differences in fiber constituency or vascularization of the respective muscles [89].

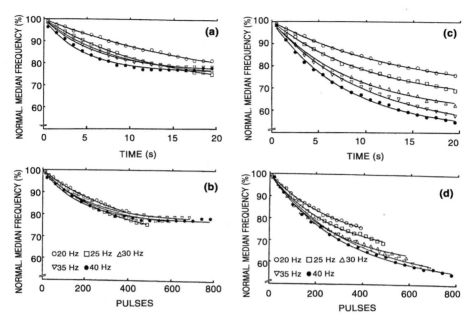

FIGURE 10.15   Same data reported in Fig. 10.14 but for a different subject showing greater myoelectric manifestations of muscle fatigue. Reproduced from [89] with permission.

A further issue under investigation is the use of pulses with different amplitude and duration attempting to activate different nerve fibers (and therefore different motor unit groups) having different intensity–time curves as proposed by Neyroud et al. [100].

10.7   EMG POWER SPECTRUM AND FIBER-TYPING; A CONTROVERSIAL ISSUE

Muscle are made of fibers which can be roughly classified, according to their histologic and metabolic properties, into three main types: slow oxidative fibers (SO or type I, slow twitch), fast oxidative glycolitic fibers (FOG or type IIa), and fast glycolitic fibers (FG or type IIb, fast twitch). These properties (together with others, such as fiber diameter) are reflected into sEMG features describing fatigue [60,76,77,111]. A classic result, obtained from in vitro neuromuscular preparations of three rat muscles with different fiber type distribution, is depicted in Fig. 10.16 which shows different decrease rates of median frequency (MDF) obtained from electrically elicited EMG signals detected with an electrode pair placed on the muscle in an in vitro preparation [72]. After the data collection the three muscles were fiber typed and showed different percentages of the three types of fibers (Fig. 10.16). The percentage of cross-sectional area occupied by different fiber types was clearly related to the rate of change of MDF. However, this association is not

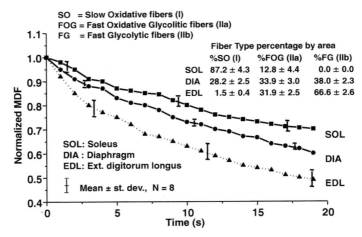

FIGURE 10.16   In vitro study of myoelectric manifestations of muscle fatigue. Normalized values of median frequency (MDF) were obtained from the soleus (SOL), diaphragm (DIA) and extensor digitorum longus (EDL) muscles of eight Wistar rats (mean ± st. dev.). The neuromuscular preparations were stimulated at 40 Hz for 20 s. Data for each muscle groups are normalized with respect to their initial values. The rate of change of MDF in the three muscles is correlated with the percentage of muscle cross section occupied by fast glycolytic fiber. Redrawn from [72] with permission.

straight-forward in vivo and in humans because of many other confounding factors, as discussed in Section 10.4 and in the following.

The conduction velocity with which an action potential propagates along a muscle fiber determines the duration of the generated waveform. Faster conduction velocity corresponds to shorter action potentials, and vice versa. Because the duration of an action potential is inversely associated with the bandwidth of its power spectrum, variables (or their changes) that are directly influenced by the spectral bandwidth of the sEMG signals have been traditionally associated to muscle fiber conduction velocity [74,75,80,123]. These considerations are at the basis of the observed myoelectric manifestations of muscle fatigue, as discussed in Chapter 4 and in this chapter. In addition to the association between *relative* decrease of conduction velocity and *relative* decrease of spectral variables (expressed as percentage change per second with respect to the initial value) commonly studied in the sEMG-based studies of fatigue, many researchers have tried to find an association between absolute values of spectral variables and the percentages of fibers of different type in a muscle. This association is based on the relations between conduction velocity and power spectrum and between fiber type and fiber size. The latter depends on the assumption that generally type I fibers have smaller diameter than type II fibers and therefore lower CV, resulting in lower MNF and MDF. Both associations, however, are only partially valid, as indicated in the following.

The association between fiber typing and fiber diameter is not direct. The distribution of conduction velocities of the fibers in a muscle is not bimodal but presents a single peak [129] where it is not possible to distinguish different populations of fibers. Basically all other physiological properties of the muscle fibers follow a unimodal distribution [30,33]. Moreover, the association between type I fibers and small-diameter fibers is absent in back muscles [82,83].

The power spectrum of the sEMG is not influenced *only* by the muscle fiber conduction velocity. Other factors of influence are anatomical (e.g., subcutaneous layer thickness) and physiological (e.g., duration and shape of the intracellular action potentials) and are related to the recording method (e.g., type of derivation, electrode size, and interelectrode distance). These factors may vary between subjects and muscles and therefore make it questionable to compare absolute values of spectral variables for estimating the relative percentage of fiber typing [46,47,50]. Conversely, the relative changes of spectral variables over time with respect to their initial values are less influenced by the above mentioned factors, so that the rate of change of these variables have been used as a relatively robust index of fatigue [88].

For the above physiological and technical reasons, extracting indexes that characterize the percentage of fiber types in human muscles [83] or the proportion of fiber types used in a task [47,111,132,133] is a questionable approach.

10.8 REPEATABILITY OF MEASUREMENTS AND APPLICATIONS OF RESULTS

10.8.1 Repeatability of Measurements

Clinical applications of sEMG posit the reliability and repeatability testing of the techniques to be used for diagnosis and monitoring of patients. The repeatability of sEMG measurements and fatigue indicators has been tested by many researchers. Analysis of variance (ANOVA), coefficient of variation (CoV), standard error of the mean (SEM), and intraclass correlation coefficient (ICC) are widely used tools and indices. A critical issue concerning features of single-channel sEMG, obtained in different tests or days, is the repeatability of electrode position and interelectrode distance [44,73,96,108].

Rainoldi et al. [107] investigated the repeatability of amplitude (ARV), spectral parameters (MNF), and muscle fiber conduction velocity (CV) at four contraction levels in the biceps of 10 healthy subjects. Good inter- and intrasubject repeatability was found, in particular for CV values, although the rate of decrement (normalized with respect to the initial value) of MNF and CV showed differences among subjects, probably reflecting individual differences.

Fauth et al. [52] tested the reliability of RMS estimates of sEMG from the quadriceps and hamstring muscles, during isometric and ballistic activities, in 24 healthy subjects performing three repetitions. They found ICC mostly greater than 90%, intrasubjects CoV in the range of 11.5% to 49.3% and intersubject CoV in the range of 5.4% to 148%. They concluded that RMS is a reproducible quantity.

The long-term repeatability of RMS of shoulder and neck muscles was investigated by Røe et al. [113], who performed measurements over three years in 26 healthy subjects finding considerable intraindividual and interyear variations. Ollivier et al. [102] compared the repeatability of estimates of RMS, MNF, and CV obtained from the biceps brachii of 10 healthy subjects using a bipolar pair of electrodes and a Laplacian (normal double differential) during isometric contractions at different levels. Not surprisingly the repeatability of the features obtained from one electrode pair was higher than that obtained with the Laplacian arrangement because of the higher spatial selectivity and smaller detection volume of the second method. Castroflorio et al. [10] tested the reproducibility of sEMG amplitude and spectral variables of jaw elevator muscles. Three voluntary contractions were performed at 80% MVC and the protocol was repeated in three days. ICC >70% was obtained only with an interelectrode distance of 30 mm—that is, with a very large detection volume. Zaman et al. [140] studied the decline of MNF of the biceps brachii during five weekly fatigue tests (sustained isometric contractions) performed on 11 participants and found the 95% confidence interval of the CoV to be 13.18% to 21.85%. Similar values were found for cyclic contractions.

Other reliability/repeatability studies were carried out on the quadriceps muscle [71,87], the back muscles [22,101,130], the biceps [107], the sternocleido-mastoid, and anterior scalene muscles [34].

In general, the reproducibility of estimates of the sEMG features in isometric or dynamic conditions is not excellent; for this reason, its value in assessing the effectiveness of treatments or training procedures has been questioned. Only major changes can be detected. This observation is, in part, the result of the persisting lack of standards in the field. Despite the efforts of the European Project on "Surface Electro-myography for Noninvasive Assessment of Muscles (SENIAM)" [63], most operators are not aware of the effects of changing electrode position and interelectrode distance on the estimates of sEMG features. Minor differences of this type in the repetition of an experimental protocol or task may have considerable impact on the results [21].

A substantial contribution to the issue of sEMG reliability and reproducibility may come, in the near future, from the use of electrode grids and from the automatic identification of regions of interest from which more reliable/repeatable signals can be extracted and fatigue indexes can be computed (Figs. 10.4, 10.8, and 10.9). It will also be possible to assess if the pool of the recruited motor units is the same or not in repeated contractions (see Chapter 7), thereby resolving one of the possible reasons of variability.

10.8.2   Applications of Results

Current findings on repeatability of results are still unsatisfactory, mostly due to the lack of standards and the inconsistent methodology. Nevertheless, sEMG has been used in a large number of clinical applications ranging from work-related disorders (see Chapter 13) to sport medicine, to studies of aging individuals, to "cancer fatigue," and to a variety of pathologies (stroke, Parkinson disease, multiple sclerosis, diabetes). An excellent review concerning fatigue in central and peripheral nervous system disorders is provided in the work of Zwartz et al. [143].

A few examples of clinical applications are summarized below. In 2006, Schulte et al. [117] observed, in eight individuals with painful trapezius muscle, an increased fatigue-related recruitment of motor units and a less pronounced decrement of CV, with respect to healthy subjects. Falla et al. [34–36] investigated myoelectric manifestations of muscle fatigue in neck flexor muscles comparing the painful and nonpainful sides in patients with chronic unilateral neck pain. They revealed greater estimates of the initial value and slope of the MNF for both the sternoclei-domastoid and anterior scalene muscles on the side of the patient's neck pain at 25% and 50% of MVC. Minetto et al. [97] conducted a single-blind, placebo-controlled study on the elbow flexors, knee extensors, and tibialis anterior of 20 men who received dexamethasone (8 mg/d) or placebo for one week. They observed an increase of voluntary muscle force and a smaller rate of decrease of MNF during voluntary contractions after treatment.

More recently, Minetto et al. [98] demonstrated that the levels of circulating muscle proteins, as well as CV and MNF of the sEMG, were significantly decreased in subjects affected by Cushing's disease. Differences between the CV values estimated in 10 patients and 30 controls were of the order of 23% to 26% for the knee extensors and 11.6% for the tibialis anterior. This association between biochemical and electrophysiological markers is of great interest for the follow up of myopatic patients and should be further investigated.

Watanabe et al. [134–136] investigated the knee extensors of nine type 2 diabetes mellitus patients al low-level voluntary contractions (10% MVC and 20% MVC) and found a smaller motor unit firing modulation and less spatial variability (lower entropy) of sEMG amplitude with respect to a healthy control group.

Rainoldi et al. [110] recorded sEMG signals from the vastus medialis longus (VML), the vastus medialis obliquous (VMO), and the vastus lateralis (VL) muscles during isometric knee extension contractions at 60% and 80% of the maximum voluntary contraction (MVC). They observed different myoelectric manifestations of fatigue that might be useful to noninvasively describe functional differences between the vasti muscles. The issue of load sharing, discussed in Section 10.2.2, may be relevant in this case, since the observed differences may be due to a different sharing of the load among these muscles in different subjects. This fact may be relevant in sport medicine.

It is reasonable to expect that a relation should exist between tissue oxygenation and myoelectric manifestations of muscle fatigue. Taelman et al. [125] found a strong correlation between MNF of the sEMG detected on the biceps brachii and the near-infrared spectroscopy results. Since both techniques provide information from the most superficial portion of a muscle, the combined information could be relevant with respect to such portion. Botter et al. [7] investigated the response of the vastus lateralis (VL) and vastus medialis longus (VML) to electrical stimulation finding that VL is more fatigable than VML and that motor units tend to be recruited, by increasing electrical stimulation intensity, in order of increasing CV in both muscles (see Chapter 11). Other authors investigated the effect of age and gender [86,129], cancer [69], and physical training [9] on myoelectric manifestations of muscle fatigue.

The agreement between sEMG-based estimation of fatigue and perceived exertion (assessed with the Borg scale), at least in isometric fatiguing contractions, is of considerable interest in occupational medicine [64,127].

In conclusion, while the modest reproducibility of sEMG features in repeated tests must be investigated and improved, the currently available techniques are applicable in many situations where relatively large differences (greater than about 20%) are observed in sEMG features or in their rate of change with fatigue. The recent development of the sEMG imaging technology is very promising, in this regard, for the near future [55,70,136].

REFERENCES

1. Afsharipour, B.,"Estimation of load sharing among muscles acting on the same joint and applications of surface electromyography," Ph.D. dissertation, Department of Electronics and Telecommunications, Politecnico di Torino, Torino, Italy, 2014.

2. Andreassen, S., and L. Arendt-Nielsen, "Muscle fibre conduction velocity in motor units of the human anterior tibial muscle: a new size principle parameter," *J. Physiol.* **391**, 561–571 (1987).

3. Arendt-Nielsen, L., and K. R. Mills, "The relationship between mean power frequency of the EMG spectrum and muscle fibre conduction velocity," *Electroencephalogr. Clin. Neurophysiol.* **60**, 130–134 (1985).

4. Bilodeau, M., A. B. Arsenault, D. Gravel, and D. Bourbonnais, "EMG power spectrum of elbow extensors: a reliability study," *Electromyogr. Clin. Neurophysiol.* **34**, 149–158 (1994).

5. Bonato, P., "From the guest editor - recent advancements in the analysis of dynamic EMG data," *Eng. Med. Biol. Mag. IEEE* **20**, 29–32 (2001).

6. Bonato, P., S. H. Roy, M. Knaflitz, and C. J. De Luca, "Time-frequency parameters of the surface myoelectric signal for assessing muscle fatigue during cyclic dynamic contractions," *IEEE Trans. Biomed. Eng.* **48**, 745–753 (2001).

7. Botter, A., F. Lanfranco, R. Merletti, and M. A. Minetto, "Myoelectric fatigue profiles of three knee extensor muscles," *Int. J. Sports Med.* **30**, 408–417 (2009).

8. Bouillard, K., A. Nordez, P. W. Hodges, C. Cornu, and F. Hug, "Evidence of changes in load sharing during isometric elbow flexion with ramped torque," *J. Biomechanics* **45**, 1424–1429 (2012).

9. Casale, R., A. Rainoldi, J. Nilsson, and P. Bellotti, "Can continuous physical training counteract aging effect on myoelectric fatigue? A surface electromyography study application," *Arch. Phys. Med. Rehabil.* **84**, 513–517 (2003).

10. Castroflorio, T., K. Icardi, B. Becchino, E. Merlo, C. Debernardi, P. Bracco, et al., "Reproducibility of surface EMG variables in isometric sub-maximal contractions of jaw elevator muscles," *J. Electromyogr. Kinesiol.* **16**, 498–505 (2006).

11. Chaffin, D. B., "Localized muscle fatigue—definiton and measurement," *J. Occupational Environ. Med.* **15**, 346–354 (1973).

12. Chowdhury, S. K., A. D. Nimbarte, M. Jaridi, and R. C. Creese, "Discrete wavelet transform analysis of surface electromyography for the fatigue assessment of neck and shoulder muscles," *J. Electromyogr. Kinesiol.* **23**, 995–1003 (2013).

13. Cifrek, M., V. Medved, S. Tonković, and S. Ostojić, "Surface EMG based muscle fatigue evaluation in biomechanics," *Clin. Biomech.* **24**, 327–340 (2009).

14. Contessa, P., and C. J. De Luca, "Neural control of muscle force: indications from a simulation model," *J. Neurophysiol.* **109**, 1548–1570 (2013).

15. Coorevits, P., L. Danneels, D. Cambier, H. Ramon, H. Druyts, J. S. Karlsson, et al., "Test–retest reliability of wavelet- and Fourier-based EMG (instantaneous) median frequencies in the evaluation of back and hip muscle fatigue during isometric back extensions," *J. Electromyogr. Kinesiol.* **18**, 798–806 (2008).

16. Crenshaw, A., S. Karlsson, B. Gerdle, and J. Friden, "Differential responses in intra-muscular pressure and EMG fatigue indicators during low versus high level isometric contractions to fatigue," *Acta Physiol. Scand.* **160**, 353–361 (1997).

17. da Silva, R. A., C. Lariviere, A. B. Arsenault, S. Nadeau, and A. Plamondon, "The comparison of wavelet- and Fourier-based electromyographic indices of back muscle fatigue during dynamic contractions: validity and reliability results," *Electromyogr. Clin. Neurophysiol.* **48**, 147–162 (2008).

18. De Luca, C. J., "Myoelectrical manifestations of localized muscular fatigue in humans," *Crit. Rev. Biomed. Eng.* **11**, 251–279 (1984).

19. De Luca, C. J., "The use of surface electromyography in biomechanics," *J. Appl. Biomech.* **13**, 135–163 (1997).

20. De Luca, C. J., and E. C. Hostage, "Relationship between firing rate and recruitment threshold of motoneurons in voluntary isometric contractions," *J. Neurophysiol.* **104**, 1034–1046 (2010). Erratum in *J. Neurophysiol.* **107**(5):1544(2012).

21. De Nooij, R., L. A. C. Kallenberg, and H. J. Hermens, "Evaluating the effect of electrode location on surface EMG amplitude of the m. erector spinae p. longissimus dorsi," *J. Electromyogr. Kinesiol.* **19**, e257–e266 (2009).

22. Dedering, Å., M. Roos af Hjelmsäter, B. Elfving, K. Harms-Ringdahl, and G. Németh, "Between-days reliability of subjective and objective assessments of back extensor muscle fatigue in subjects without lower-back pain," *J. Electromyogr. Kinesiol.* **10**, 151–158 (2000).

23. Dideriksen, J. L., D. Farina, M. Baekgaard, and R. M. Enoka, "An integrative model of motor unit activity during sustained submaximal contractions," *J. Appl. Physiol.* **108**, 1550–1562 (2010).

24. Dideriksen, J. L., D. Farina, and R. M. Enoka, "Influence of fatigue on the simulated relation between the amplitude of the surface electromyogram and muscle force," *Philos. Trans. A Math. Phys. Eng. Sci.* **368**, 2765–2781 (2010).

25. Dideriksen, J. L., R. M. Enoka, and D. Farina, "Neuromuscular adjustments that constrain submaximal EMG amplitude at task failure of sustained isometric contractions," *J. Appl. Physiol.* **111**, 485–494 (2011).

26. Dimitrov, G. V., T. I. Arabadzhiev, J. Y. Hogrel, and N. A. Dimitrova, "Simulation analysis of interference EMG during fatiguing voluntary contractions. Part II—changes in amplitude and spectral characteristics," *J. Electromyogr. Kinesiol.* **18**, 35–43 (2008).

27. Dimitrov, G. V., Z. C. Lateva, and N. A. Dimitrova, "Effects of changes in asymmetry, duration and propagation velocity of the intracellular potential on the power spectrum of extracellular potentials produced by an excitable fiber," *Electroencephalogr. Clin. Neurophysiol.* **28**, 93–100 (1988).

28. Dimitrova, N. A., and G. V. Dimitrov, "Interpretation of EMG changes with fatigue: facts, pitfalls, and fallacies," *J. Electromyogr. Kinesiol.* **13**, 13–36 (2003).

29. Enoka, R. M., S. C. Gandevia, A. J. McComas, D. G. Stuart, C. K. E. Thomas, and P. A. Pierce, *Fatigue: Neural Control and Muscular Mechanisms*, vol. **384**, Plenum Press, New York, 1995.

30. Enoka, R. M., "Neural adaptations with chronic physical activity," *J. Biomech.* **30**, 447–455 (1997).

31. Enoka, R. M., and J. Duchateau, "Muscle fatigue: what, why and how it influences muscle function," *J. Physiol.* **586**, 11–23 (2008).

32. Enoka, R. M., S. Baudry, T. Rudroff, D. Farina, M. Klass, and J. Duchateau, "Unraveling the neurophysiology of muscle fatigue," *J. Electromyogr. Kinesiol.* **21**, 208–219 (2011).

33. Enoka, R. M., "Muscle fatigue—from motor units to clinical symptoms," *J. Biomech.* **45**, 427–433 (2012).

34. Falla, D., P. Dall'Alba, A. Rainoldi, R. Merletti, and G. Jull, "Repeatability of surface EMG variables in the sternocleidomastoid and anterior scalene muscles," *Eur. J. Appl. Physiol.* **87**, 542–549 (2002).

35. Falla, D., A. Rainoldi, R. Merletti, and G. Jull, "Myoelectric manifestations of sterno-cleidomastoid and anterior scalene muscle fatigue in chronic neck pain patients," *Clin. Neurophysiol.* **114**, 488–495 (2003).

36. Falla, D., G. Jull, A. Rainoldi, and R. Merletti, "Neck flexor muscle fatigue is side specific in patients with unilateral neck pain," *Eur. J. Pain* **8**, 71–77 (2004).

37. Farina, D., R. Merletti, A. Rainoldi, M. Buonocore, and R. Casale, "Two methods for the measurement of voluntary contraction torque in the biceps brachii muscle," *Med. Eng. Phys.* **21**, 533–540 (1999).

38. Farina, D., and R. Merletti, "Comparison of algorithms for estimation of EMG variables during voluntary isometric contractions," *J. Electromyogr. Kinesiol.* **10**, 337–349 (2000).

39. Farina, D., and R. Merletti, "Effect of electrode shape on spectral features of surface detected motor unit action potentials," *Acta Physiol. Pharmacol. Bulg.* **26**, 63–66 (2001).

40. Farina, D., R. Merletti, M. Nazzaro, and I. Caruso, "Effect of joint angle on EMG variables in leg and thigh muscles," *Eng. Med. Biol. Mag. IEEE* **20**, 62–71 (2001).

41. Farina, D., R. Merletti, B. Indino, M. Nazzaro, and M. Pozzo, "Surface EMG crosstalk between knee extensor muscles: experimental and model results," *Muscle & Nerve* **26**, 681–695 (2002).

42. Farina, D., L. Fattorini, F. Felici, and G. Filligoi, "Nonlinear surface EMG analysis to detect changes of motor unit conduction velocity and synchronization," *J. Appl. Physiol.* **93**, 1753–1763 (2002).

43. Farina, D., P. Madeleine, T. Graven-Nielsen, R. Merletti, and L. Arendt-Nielsen, "Standardising surface electromyogram recordings for assessment of activity and fatigue in the human upper trapezius muscle," *Eur. J. Appl. Physiol.* **86**, 469–478 (2002).

44. Farina, D., C. Cescon, and R. Merletti, "Influence of anatomical, physical, and detection-system parameters on surface EMG," *Biol. Cybern.* **86**, 445–456 (2002).

45. Farina, D., L. Mesin, S. Martina, and R. Merletti, "A surface EMG generation model with multilayer cylindrical description of the volume conductor," *IEEE Trans. Biomed. Eng.* **51**, 415–426 (2004).

46. Farina D., R Merletti, and R. Enoka, "The extraction of neural strategies from the surface EMG," *J. Appl. Physiol. (1985)* **96**, 1486–1495 (2004).

47. Farina, D., "Counterpoint: spectral properties of the surface EMG do not provide information about motor unit recruitment and muscle fiber type," *J. Appl. Physiol. (1985)* **105**, 1673–1674 (2008).

48. Farina, D., F. Leclerc, L. Arendt-Nielsen, O. Buttelli, and P. Madeleine, "The change in spatial distribution of upper trapezius muscle activity is correlated to contraction duration," *J. Electromyogr. Kinesiol.* **18**, 16–25 (2008).

49. Farina, D., C. Cescon, F. Negro, and R. M. Enoka, "Amplitude cancellation of motor-unit action potentials in the surface electromyogram can be estimated with spike-triggered averaging," *J. Neurophysiol.* **100**, 431–440 (2008).

50. Farina D., R. Merletti, and R. Enoka, "The extraction of neural strategies from the surface EMG: an update," *J. Appl. Physiol. (1985)* **117**, 1215–1230 (2014).

51. Fattorini, L., F. Felici, G. C. Filligoi, M. Traballesi, and D. Farina, "Influence of high motor unit synchronization levels on non-linear and spectral variables of the surface EMG," *J. Neurosci. Methods* **143**, 133–139 (2005).

52. Fauth, M. L., E. J. Petushek, C. R. Feldmann, B. E. Hsu, L. R. Garceau, B. N. Lutsch, et al., "Reliability of surface electromyography during maximal voluntary isometric contractions, jump landings, and cutting," *J. Strength Cond. Res.* **24**, 1131–1137 (2010).

53. Filligoi, G., and F. Felici, "Detection of hidden rhythms in surface EMG signals with a non-linear time-series tool," *Med. Eng. Phys.* **21**, 439–448 (1999).

54. Fuglevand, A. J., D. A. Winter, and A. E. Patla, "Models of recruitment and rate coding organization in motor-unit pools," *J. Neurophysiol.* **70**, 2470–2488 (1993).

55. Gallina, A., R. Merletti, and T. M. M. Vieira, "Are the myoelectric manifestations of fatigue distributed regionally in the human medial gastrocnemius muscle?," *J. Electromyogr. Kinesiol.* **21**, 929–938 (2011).

56. Gandevia, S., R. Enoka, A. McComas A., D. Stuart, and C. Thomas, *Fatigue: Neural and Muscular Mechanisms*, Plenum Press, New York, 1995.

57. Gandevia, S. C., "Spinal and supraspinal factors in human muscle fatigue," *Physiol. Rev.* **81**, 1725–1789 (2001).

58. Gazzoni, M., F. Camelia, and D. Farina, "Conduction velocity of quiescent muscle fibers decreases during sustained contraction," *J. Neurophysiol.* **94**, 387–394 (2005).

59. Gazzoni, M., D. Farina, and R. Merletti, "Motor unit recruitment during constant low force and long duration muscle contractions investigated with surface electromyography," *Acta Physiol. Pharmacol. Bulg.* **26**, 67–71 (2001).

60. Gerdle, B., K. Henriksson-Larsén, R. Lorentzon, and M. L. Wretling, "Dependence of the mean power frequency of the electromyogram on muscle force and fibre type," *Acta Physiol. Scand.* **142**, 457–465 (1991).

61. Gitter, J. A., and M. J. Czerniecki, "Fractal analysis of the electromyographic interference pattern," *J. Neurosci. Methods* **58**, 103–108 (1995).

62. Herbert, R. D., and S. C. Gandevia, "Twitch interpolation in human muscles: Mechanisms and implications for measurement of voluntary activation," *J. Neurophysiol.* **82**, 2271–2283 (1999).

63. Hermens, H., B. Freriks, R. Merletti, D. Stegeman, J. Blok, G. Rau, C. Disselhorst-Klug, G. Hägg,"European recommendations for surface electromyography: results of the SENIAM Project," Roessing Research and Development, Enschede, The Netherlands, www.seniam.org, 1999.

64. Hummel, A., T. Läubli, M. Pozzo, P. Schenk, S. Spillmann, and A. Klipstein, "Relationship between perceived exertion and mean power frequency of the EMG signal from the upper trapezius muscle during isometric shoulder elevation," *Eur. J. Appl. Physiol.* **95**, 321–326 (2005).

65. Inman, V. T., H. J. Ralston, J. B. De C. M. Saunders, M. B. Bertram Feinstein, and E. W. Wright Jr, "Relation of human electromyogram to muscular tension," *Electroencephalogr. Clin. Neurophysiol.* **4**, 187–194 (1952).

66. Islam, M. A., K. Sundaraj, R. B. Ahmad, and N. U. Ahamed, "Mechanomyogram for muscle function assessment: a review," *PLoS One* **8**, e58902 (2013).

67. Karlsson, S., Y. Jun, and M. Akay, "Enhancement of spectral analysis of myoelectric signals during static contractions using wavelet methods," *IEEE Trans. Biomed. Eng.* **46**, 670–684 (1999).

68. Karlsson, S., Y. Jun, and M. Akay, "Time-frequency analysis of myoelectric signals during dynamic contractions: a comparative study," *Trans. Biomed. Eng. IEEE* **47**, 228–238 (2000).

69. Kisiel-Sajewicz, K., V. Siemionow, D. Seyidova-Khoshknabi, M. P. Davis, A. Wyant, V. K. Ranganathan, et al., "Myoelectrical manifestation of fatigue less prominent in patients with cancer related fatigue," *PLoS One* **8**, e83636 (2013).

70. Kleine, B., N. Schumann, D. Stegeman, and H. Scholle, "Surface EMG mapping of the human trapezius muscle: the topography of monopolar and bipolar surface EMG amplitude and spectrum parameters at varied forces and in fatigue," *Clin. Neurophysiol.* **111**, 686–693 (2000).

71. Kollmitzer, J., G. R. Ebenbichler, and A. Kopf, "Reliability of surface electromyographic measurements," *Clin. Neurophysiol.* **110**, 725–734 (1999).

72. Kupa, E. J., S. H. Roy, S. C. Kandarian, and C. J. De Luca, "Effects of muscle fiber type and size on EMG median frequency and conduction velocity," *J. Appl. Physiol.* **79**, 23–32 (1995).

73. Li, W., and K. Sakamoto, "The influence of location of electrode on muscle fiber conduction velocity and EMG power spectrum during voluntary isometric contraction measured with surface array electrodes," *Appl. Hum. Sci.* **15**, 25–32 (1996).

74. Lindstrom, L., R. Magnusson, and I. Petersen, "Muscular fatigue and action potential conduction velocity changes studied with frequency analysis of EMG signals," *Electromyography* **10**, 341–356 (1970).

75. Lindstrom, L. H., and R. I. Magnusson, "Interpretation of myoelectric power spectra: a model and its applications," *Proc. IEEE* **65**, 653–662 (1977).

76. Linssen, W. H. J. P., M. Jacobs, D. F. Stegeman, E. M. G. Joosten, and J. Moleman, "Muscle fatigue in McArdle's disease: muscle fiber conduction velocity and surface EMG frequency spectrum during ischaemic exercise," *Brain* **113**, 1779–1793 (1990).

77. Linssen, W. H., D. F. Stegeman, E. M. Joosten, R. A. Binkhorst, M. J. Merks, H. J. ter Laak, et al. , "Fatigue in type I fiber predominance: a muscle force and surface EMG study on the relative role of type I and type II muscle fibers," *Muscle Nerve* **14**, 829–837 (1991).

78. Lo Conte, L. R., and R. Merletti, "Advances in processing of surface myoelectric signals: Part 2," *Medical and Biological Engineering and Computing* **33**, 373–384 (1995).

79. Lo Conte, L. R., R. Merletti, and G. V. Sandri, "Hermite expansions of compact support waveforms: applications to myoelectric signals," *IEEE Trans. Biomed. Eng.* **41**, 1147–1159 (1994).

80. Lowery, M. M., C. L. Vaughan, P. J. Nolan, and M. J. O'Malley, "Spectral compression of the electromyographic signal due to decreasing muscle fiber conduction velocity," *IEEE Transactions on Rehabilitation Engineering* **8**, 353–361 (2000).

81. Maluf, K., M. Shinohara, J. Stephenson, and R. Enoka, "Muscle activation and time to task failure differ with load type and contraction intensity for a human hand muscle," *Exp. Brain Res.* **167**, 165–177 (2005).

82. Mannion, A. F., G. A. Dumas, R. G. Cooper, F. J. Espinosa, M. W. Faris, and J. M. Stevenson, "Muscle fibre size and type distribution in thoracic and lumbar regions of erector spinae in healthy subjects without low back pain: normal values and sex differences," *J. Anat.* **190**(Pt4), 505–513 (1997).

83. Mannion, A. F., G. A. Dumas, J. M. Stevenson, and R. G. Cooper, "The influence of muscle fiber size and type distribution on electromyographic measures of back muscle fatigability," *Spine (Phila Pa 1976)* **23**, 576–584 (1998).

84. McKenzie, D. K., B. Bigland-Ritchie, R. B. Gorman, and S. C. Gandevia, "Central and peripheral fatigue of human diaphragm and limb muscles assessed by twitch interpolation," *J. Physiol.* **454**, 643–656 (1992).

85. Merletti, R., "Surface electromyography: possibilities and limitations," *J. Rehabil. Sciences* **7**, 25–34 (1994).

86. Merletti, R., D. Farina, M. Gazzoni, and M. P. Schieroni, "Effect of age on muscle functions investigated with surface electromyography," *Muscle & Nerve* **25**, 65–76 (2002).

87. Merletti, R., A. Fiorito, L. R. Lo Conte, and C. Cisari, "Repeatability of electrically evoked EMG signals in the human vastus medialis muscle," *Muscle & Nerve* **21**, 184–193 (1998).

88. Merletti, R., M. Knaflitz, and C. J. De Luca, "Myoelectric manifestations of fatigue in voluntary and electrically elicited contractions," *J. Appl. Physiol.* **69**, 1810–1820 (1990).

89. Merletti, R., M. Knaflitz, and C. J. DeLuca, "Electrically evoked myoelectric signals," *Crit. Rev. Biomed. Eng.* **19**, 293–340 (1992).

90. Merletti, R., and L. R. Lo Conte, "Surface EMG signal processing during isometric contractions," *J. Electromyogr. Kinesiol.* **7**, 241–250 (1997).

91. Merletti, R., L. R. Lo Conte, and D. Sathyan, "Repeatability of electrically-evoked myoelectric signals in the human tibialis anterior muscle," *J. Electromyogr. Kinesiol.* **5**, 67–80 (1995).

92. Merletti, R., and P. A. Parker, eds., *Electromyography: Physiology, Engineering, and Noninvasive Applications*, IEEE Press and John Wiley & Sons, Hoboken, NJ, 2004.

93. Merletti, R., and S. Roy, "Myoelectric and mechanical manifestations of muscle fatigue in voluntary contractions," *J. Orthop. Sports Phys. Ther.* **24**, 342–353 (1996).

94. Mesin, L., C. Cescon, M. Gazzoni, R. Merletti, and A. Rainoldi, "A bi-dimensional index for the selective assessment of myoelectric manifestations of peripheral and central muscle fatigue," *J. Electromyogr. Kinesiol.* **19**, 851–863 (2009).

95. Mesin, L., M. Joubert, T. Hanekom, R. Merletti, and D. Farina, "A finite element model for describing the effect of muscle shortening on surface EMG," *IEEE Trans. Biomed. Eng.* **53**, 593–600 (2006).

96. Mesin, L., R. Merletti, and A. Rainoldi, "Surface EMG: the issue of electrode location," *J. Electromyogr. Kinesiol.* **19**, 719–726 (2009).

97. Minetto, M. A., A. Botter, F. Lanfranco, M. Baldi, E. Ghigo, and E. Arvat, "Muscle fiber conduction slowing and decreased levels of circulating muscle proteins after short-term dexamethasone administration in healthy subjects," *J. Clin. Endocrinol. Metab.* **95**, 1663–1671 (2010).

98. Minetto, M. A., F. Lanfranco, A. Botter, G. Motta, G. Mengozzi, R. Giordano, et al., "Do muscle fiber conduction slowing and decreased levels of circulating muscle proteins represent sensitive markers of steroid myopathy? A pilot study in Cushing's disease," *Eur. J. Endocrinol.* **164**, 985–993 (2011).

99. Molinari, F., M. Knaflitz, P. Bonato, and M. V. Actis, "Electrical manifestations of muscle fatigue during concentric and eccentric isokinetic knee flexion–extension movements," *IEEE Trans. Biomed. Eng.* **53**, 1309–1316 (2006).

100. Neyroud, D., D. Dodd, J. Gondin, N. A. Maffiuletti, B. Kayser, and N. Place, "Wide-pulse-high-frequency neuromuscular stimulation of triceps surae induces greater muscle fatigue compared with conventional stimulation," *J. Appl. Physiol.* **116**, 1281–1289 (2014).

101. Ng, J. K. F., and C. A. Richardson, "Reliability of electromyographic power spectral analysis of back muscle endurance in healthy subjects," *Arch. Phys. Med. Rehabil.* **77**, 259–264 (1996).

102. Ollivier, K., P. Portero, O. Maïsetti, and J.-Y. Hogrel, "Repeatability of surface EMG parameters at various isometric contraction levels and during fatigue using bipolar and Laplacian electrode configurations," *J. Electromyogr. Kinesiol.* **15**, 466–473 (2005).

103. Pedrinelli, R., L. Marino, G. Dell'Omo, G. Siciliano, and B. Rossi, "Altered surface myoelectric signals in peripheral vascular disease: correlations with muscle fiber composition," *Muscle & Nerve* **21**, 201–210 (1998).

104. Piper, H. E., *Electrophysiologie, Menschlicher Muskeln*, J. Springer, Berlin, 1912.

105. Poli, R., J. Kennedy, and T. Blackwell, "Particle swarm optimization," *Swarm Intell.* **1**, 33–57 (2007).

106. Pozzo, M., E. Merlo, D Farina, G. Antonutto, R. Merletti, and P. Di Prampero, "Muscle fiber conduction velocity estimated from surface EMG signals during explosive dynamic contractions," *Muscle & Nerve* **29**, 823–833 (2004).

107. Rainoldi, A., G. Galardi, L. Maderna, G. Comi, L. Lo Conte, and R. Merletti, "Repeatability of surface EMG variables during voluntary isometric contractions of the biceps brachii muscle," *J. Electromyogr. Kinesiol.* **9**, 105–119 (1999).

108. Rainoldi, A., M. Nazzaro, R. Merletti, D. Farina, I. Caruso, and S. Gaudenti, "Geometrical factors in surface EMG of the vastus medialis and lateralis muscles," *J. Electromyogr. Kinesiol.* **10**, 327–336 (2000).

109. Rainoldi, A., J. E. Bullock-Saxton, F. Cavarretta, and N. Hogan, "Repeatability of maximal voluntary force and of surface EMG variables during voluntary isometric contraction of quadriceps muscles in healthy subjects," *J. Electromyogr. Kinesiol.* **11**, 425–438 (2001).

110. Rainoldi, A., D. Falla, R. Mellor, K. Bennell, and P. Hodges, "Myoelectric manifestations of fatigue in vastus lateralis, medialis obliquus and medialis longus muscles," *J. Electromyogr. Kinesiol.* **18**, 1032–1037 (2008).

111. Rainoldi, A., M. Gazzoni, and G. Melchiorri, "Differences in myoelectric manifestations of fatigue in sprinters and long distance runners," *Physiol. Meas.* **29**, 331–340 (2008).

112. Rix, H., and J. P. Malengé, "Detecting small variations in shape," *IEEE Trans. Syst. Man Cybern.* **10**, 90–96 (1980).

113. Røe, C., Ó. A. Steingrímsdóttir, S. Knardahl, E. S. Bakke, and N. K. Vøllestad, "Long-term repeatability of force, endurance time and muscle activity during isometric contractions," *J. Electromyogr. Kinesiol.* **16**, 103–113 (2006).

114. Rojas-Martínez, M., M. A. Mañanas, J. F. Alonso, and R. Merletti, "Identification of isometric contractions based on high density EMG maps," *J. Electromyogr. Kinesiol.* **23**, 33–42 (2013).

115. Roy, S. H., C. J. De Luca, and J. Schneider, "Effects of electrode location on myoelectric conduction velocity and median frequency estimates," *J. Appl. Physiol.* **61**, 1510–1517 (1986).

116. Sadoyama, T., T. Masuda, H. Miyata, and S. Katsuta, "Fibre conduction velocity and fibre composition in human vastus lateralis," *Eur. J. Appl. Physiol. Occup. Physiol.* **57**, 767–771 (1988).

117. Schulte, E., O. Miltner, E. Junker, G. Rau, and C. Disselhorst-Klug, "Upper trapezius muscle conduction velocity during fatigue in subjects with and without work-related muscular disorders: a non-invasive high spatial resolution approach," *Eur. J. Appl. Physiol.* **96**, 194–202 (2006).

118. Shield, A., and S. Zhou, "Assessing voluntary muscle activation with the twitch interpolation technique," *Sports Med.* **34**, 253–267 (2004).

119. Singh, V. P., D. K. Kumar, B. Polus, and S. Fraser, "Strategies to identify changes in SEMG due to muscle fatigue during cycling," *J. Med. Eng. Technol.* **31**, 144–151 (2007).

120. Sjøgaard, G., G. Savard, and C. Juel, "Muscle blood flow during isometric activity and its relation to muscle fatigue," *Eur. J. Appl. Physiol.* **57**, 327–335 (1988).

121. Staudenmann, D., A. Daffertshofer, I. Kingma, D. F. Stegeman, and J. H. van Dieen, "Independent component analysis of high-density electromyography in muscle force estimation," *IEEE Trans. Biomed. Eng.* **54**, 751–754 (2007).

122. Staudenmann, D., J. H. van Dieën, D. F. Stegeman, and R. M. Enoka, "Increase in heterogeneity of biceps brachii activation during isometric submaximal fatiguing contractions: a multichannel surface EMG study," *J. Neurophysiol.* **111**, 984–890 (2014).

123. Stulen, F. B., and C. J. De Luca, "The relation between the myoelectric signal and physiological properties of constant-force isometric contractions," *Electroencephalogr. Clin. Neurophysiol.* **45**, 681–698 (1978).

124. Stulen, F. B., and C. J. De Luca, "Frequency parameters of the myoelectric signal as a measure of muscle conduction velocity," *IEEE Trans. Biomed. Eng.* **28**, 515–523 (1981).

125. Taelman, J., J. Vanderhaegen, M. Robijns, G. Naulaers, A. Spaepen, and S. Van Huffel, "Estimation of muscle fatigue using surface electromyography and near-infrared spectroscopy," in *Oxygen Transport to Tissue XXXII*, vol. **701**, J. C. LaManna, M. A. Puchowicz, K. Xu, D. K. Harrison, and D. F. Bruley, eds., Springer, New York, pp. 353–359, 2011.

126. Tarata, M. T., "Mechanomyography versus electromyography, in monitoring the muscular fatigue," *Biomed. Eng. Online* **2**, 3 (2003).

127. Troiano, A., F. Naddeo, E. Sosso, G. Camarota, R. Merletti, and L. Mesin, "Assessment of force and fatigue in isometric contractions of the upper trapezius muscle by surface EMG signal and perceived exertion scale," *Gait & Posture* **28**, 179–186 (2008).

128. Troni, W., R. Cantello, and I. Rainero, "Conduction velocity along human muscle fibers in situ," *Neurology* **33**, 1453–1459 (1983).

129. Tsuboi, H., Y. Nishimura, T. Sakata, H. Ohko, H. Tanina, K. Kouda, et al., "Age-related sex differences in erector spinae muscle endurance using surface electromyographic power spectral analysis in healthy humans," *Spine J.* **13**, 1928–1933 (2013).

130. van Dieën, J. H., and P. Heijblom, "Reproducibility of isometric trunk extension torque, trunk extensor endurance, and related electromyographic parameters in the context of their clinical applicability," *J. Orthop. Res.* **14**, 139–143 (1996).

131. Viitasalo, J. H. T., and P. V. Komi, "Signal characteristics of EMG with special reference to reproducibility of measurements," *Acta Physiol. Scand.* **93**, 531–539 (1975).

132. von Tscharner, V., and B. M. Nigg, "Point: spectral properties of the surface EMG can characterize/do not provide information about motor unit recruitment strategies and muscle fiber type," *J. Appl. Physiol. (1985)* **105**, 1671–1674 (2008).

133. Wakeling, J. M., S. A. Pascual, and B. M. Nigg, "Altering muscle activity in the lower extremities by running with different shoes," *Med. Sci. Sports Exerc.* **34**, 1529–1532 (2002).

134. Watanabe K., T. Miyamoto, Y. Tanaka, K. Fukuda, and T. Moritani, "Type 2 diabetes mellitus patients manifest characteristic spatial EMG potential distribution pattern during sustained isometric contraction," *Diabetes Res. Clin. Pract.* **97**, 468–473 (2012).

135. Watanabe, K., M. Gazzoni, A. Holobar, T. Miyamoto, K. Fukuda, R. Merletti, and T. Moritani, "Motor unit firing pattern of vastus lateralis muscle in type 2 diabetes mellitus patients," *Muscle & Nerve* **48**, 806–813 (2013).

136. Watanabe, K., M. Kouzaki, and T. Moritani, "Region-specific myoelectric manifestations of fatigue in human rectus femoris muscle," *Muscle & Nerve* **48**, 226–234 (2013).

137. Westgaard, R. H., and C. J. De Luca, "Motor unit substitution in long-duration contractions of the human trapezius muscle," *J. Neurophysiol.* **82**, 501–504 (1999).

138. Yao, W., R. J. Fuglevand, and R. M. Enoka, "Motor-unit synchronization increases EMG amplitude and decreases force steadiness of simulated contractions," *J. Neurophysiol.* **83**, 441–452 (2000).

139. Yassierli, and M. A. Nussbaum, "Utility of traditional and alternative EMG-based measures of fatigue during low-moderate level isometric efforts," *J. Electromyogr. Kinesiol.* **18**, 44–53 (2008).

140. Zaman, S. A., D. T. MacIsaac, and P. A. Parker, "Repeatability of surface EMG-based single parameter muscle fatigue assessment strategies in static and cyclic contractions," in *Engineering in Medicine and Biology Society, EMBC, 2011 Annual International Conference of the IEEE*, pp. 3857–3860, 2011.

141. Zijdewind, I., and D. Kernell, "Fatigue associated EMG behavior of the first dorsal interosseous and adductor pollicis muscles in different groups of subjects," *Muscle & Nerve* **17**, 1044–1054 (1994).

142. Zijdewind, I., M. J. Zwarts, and D. Kernell, "Fatigue-associated changes in the electromyogram of the human first dorsal interosseous muscle," *Muscle & Nerve* **22**, 1432–1436 (1999).

143. Zwartz M., G. Bleijenberg, and B. van Engelen, "Clinical neurophysiology of fatigue," *Clin. Neurophysiol.* **119**, 2–10 (2008).

11

EMG OF ELECTRICALLY STIMULATED MUSCLES

A. Botter and R. Merletti

Laboratory for Engineering of the Neuromuscular System (LISiN) Politecnico di Torino, Torino, Italy

11.1 ELECTRICAL STIMULATION OF THE PERIPHERAL NERVOUS SYSTEM

Neuromuscular electrical stimulation (NMES) involves the application of a train of stimuli to a nerve branch or a superficial skeletal muscle, with the objective of triggering muscle contractions through the activation of intramuscular nerve branches [41]. The stimulus strength (either current or voltage) determines the number of recruited motor units and the stimulus frequency determines their discharge rate. All the recruited motor units are activated synchronously, and the sEMG signal is deterministic rather than stochastic. The factors of variability are reduced with respect to voluntary contractions.

NMES is attracting increasing attention because of the possibilities it offers for a number of research and clinical applications. NMES is used as a valid research tool for in vivo assessment of neuromuscular function of healthy and impaired muscles in both fresh and fatigued conditions [41,52,84]. As compared to voluntary contractions, it allows the evaluation of the contractile function of intact muscles in a more standardized way (e.g., fatigue during constant frequency stimulation, investigation of the force–frequency relationship). Moreover, it is possible to evaluate the level of voluntary activation with the "interpolated twitch" technique, a joint analysis of stimulated and voluntary muscle contractions [31,40].

Surface Electromyography: Physiology, Engineering, and Applications, First Edition.
Edited by Roberto Merletti and Dario Farina.
© 2016 by The Institute of Electrical and Electronics Engineers, Inc. Published 2016 by John Wiley & Sons, Inc.

In clinical settings, NMES is used for muscle training in normal subjects and athletes [2,22,28,77], prevention of disuse atrophy and activation of denervated muscles (with specific stimulation parameters) [2,44,71], functional control of paralyzed extremities [16,47,72,74], rehabilitation of patients following stroke [32,66] or spinal cord injury [4,15], and many other fields ranging from geriatric to cardiovascular medicine [49].

Despite the wide use in basic research, clinical applications, and recently also in daily life, many controversies about methodological aspects concerning NMES are still open. As an example, motor unit (MU) recruitment order during electrically elicited contraction as well as the effect of different stimulation parameters on the muscle response are still a topic of debate.

A great variety of stimulation waveforms, protocols, testing procedures, and methods are reported in the literature, as well as many approaches to describe and quantify changes induced by electrical stimulation. The combination of variability in stimulation and assessment modalities makes it very difficult to compare results reported by different laboratories and therefore enhances the above-mentioned controversies [59].

The main topics addressed in this chapter are: stimulation techniques, the issue of the stimulation electrode positioning, MU activation order, and spinal involvement in electrically elicited contractions. This section provides a brief review on methodological and physiological aspects related to NMES. Methodological issues concerning the acquisition of surface EMG signals during neuromuscular electrical stimulation (e.g., M-wave analysis and stimulation artifact reduction) are discussed in Section 11.2.

11.1.1 Stimulation Techniques

Two stimulation techniques are commonly used: they are referred to as monopolar and bipolar electrode configuration (Fig. 11.1). The differences between these two methods concern the geometry and relative position of the stimulation electrodes.

In either case, constant voltage or constant current stimulators may be used where the term "constant" refers to the voltage or current being insensitive to the load impedance between the electrodes.

The term monopolar refers to the fact the electrical stimulation takes place in the proximity of only one of the two stimulation electrodes and not at the other, as indicated in Fig. 11.1a [61]. The stimulation electrode has small dimensions (usually a few square centimeters) and is located either near a nerve (nerve stimulation) or above a muscle motor point (motor point stimulation). The second electrode is much larger than the stimulation electrode (around tens of square centimeters) and is generally placed over the antagonist muscle. With this electrode configuration, for a certain current level, the current density in the proximity of the stimulation electrode may exceed the excitation level of the nerve branches, whereas the large dimension of the second electrode assures that the current density under such electrode remains below the excitation threshold. Therefore, this technique allows the stimulation of localized populations of superficial motor units.

In the bipolar arrangement (Fig. 11.1b), two electrodes of similar dimensions are applied over the muscle. With respect to monopolar stimulation, the current

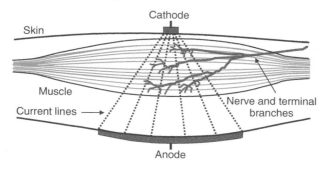

a) Monopolar stimulation

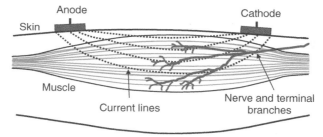

b) Bipolar stimulation

FIGURE 11.1 Schematic representation of monopolar and bipolar stimulation electrode configuration.

distribution is more confined in space and the current density is more uniform along the current path between the electrodes. This implies a less localized stimulation that may take place under either electrode or along the current path between the electrodes.

In both monopolar and bipolar stimulation, the position of the stimulation electrodes is a critical issue: a proper positioning is necessary in order to optimize the stimulation paradigm to maximize force while avoiding uncomfortable levels of stimulation current.

Bipolar stimulation is commonly used in rehabilitation and training protocols. In these cases, the objective is to maximize the force output while minimizing injected current and discomfort. This objective is achieved by placing the stimulation electrodes close to the proximal and distal tendons (Fig. 11.1b). In this way, current lines run trough the longitudinal direction of the muscle, thus maximizing the number of axons activated and therefore the force output.

Monopolar stimulation is mainly used in research fields. It is generally associated with EMG or MMG (mechanomyogram) detection in order to characterize the peripheral properties of MUs. In this case, the optimization of the stimulation paradigm means a selective and stable activation of few MUs with the lowest injected current (that minimizes artifacts in EMG recordings and subject discomfort). This objective is achieved by placing the stimulation electrode over the main muscle motor point. Muscle motor point represents the location where the motor branches of a nerve enter the muscle belly. It can be noninvasively identified by electrical stimulation as the skin area above the muscle where the excitability threshold is the lowest.

A precise localization of the muscle motor points is paramount not only for a proper positioning of the stimulation electrodes and maximization of the output force [33], but also for improving therapeutic effectiveness of anesthetic or neurolytic motor nerve blocks [43]. The motor point and the innervation zone of a muscle do not necessarily coincide but are often close to each other [1,3,8].

11.1.2   MU Recruitment During NMES: Direct Activation of Motor Axons

The involvement of motor units during NMES contractions is considerably different from that underlying voluntary activation. Besides the obvious differences in MUAP discharges, which are asynchronous in voluntary contractions and synchronous during NMSE, the activation order of MUs during NMES is still a controversial issue. In contrast to the order in which MUs are activated during voluntary contractions, in stimulated contractions the classic view is that large-diameter axons, having stimulation threshold lower than small-diameter axons, are more easily excited by imposed electrical fields [76]. However, in the case of transcutaneous stimulation, the size of motor neuron axon is not the only factor that plays a role in determining the motor unit recruitment order. Other factors, such as size of the axonal branches, distance of the axonal branches from the stimulation electrode, orientation of the axonal branches with respect to the current field, distribution of the MUs within the muscle, and changes in axonal excitability with trains of stimuli, may play a role in determining the motor unit activation order. Many investigators [4,25,26,30] studied electrical stimulation of the muscle motor point and reported that an orderly activation is more likely than a reverse one. Jubeau et al. [42] investigated the recruitment order of motor units during voluntary and electrically induced contractions of the quadriceps muscle. They concluded that transcutaneous stimulation would result neither in motor unit recruitment according to Henneman's size principle nor in a reversal of the orderly recruitment. This conclusion provided support to the hypothesis, originally formulated by Gregory and Bickel [34], that MU activation during electrical stimulation is nonselective—that is, without any order related to fiber features. On the contrary, Hennings et al. [37] found an inverse (from large to small) activation order of ulnar nerve fibers with transcutaneous electrical stimulation of the nerve trunk: They detected muscle action potentials from the abductor digiti minimi and adopted a modified Hopf's collision technique to determine the distributions of conduction velocity (NCV) of ulnar nerve fibers and of muscle fibers activated by electrical stimuli eliciting 20%, 50%, and 80% of the maximal muscle response.

The comparison between the results of the studies of Jubeau et al. [42] and Hennings et al. [37] and those of other studies [25,26,46] highlights that the mode of transcutaneous stimulation represents an additional factor that may play a role in determining the MU activation order: direct stimulation of the nerve trunk with surface electrodes produced random [42] and orderly inverse [37] motor unit recruitment, whereas the electrical stimulation of the muscle motor point produced an orderly motor unit activation, from low to high conduction velocity (CV), with increasing stimulation current [25,26,46]. Since, during muscle motor point stimulation, progressively higher current intensities depolarize fibers located at a progressively greater distance from the electrode (i.e., deeper MUs), it may be assumed that activation progressed from superficial small-diameter muscle fibers to deep large-diameter ones. Consistently, Henriksson-Larsen et al. [38] found, in tibialis anterior muscle, that both type I and II fibers have a larger diameter in the deep than in the superficial muscle layers. Similar conclusions were drawn by Lexell and Taylor [48], who showed that fibers in the deep parts of the vastus lateralis muscle were larger than the more superficial ones.

Overall, NMES activates motor axons without obvious sequencing related to motor unit type or size, but related to excitability threshold in the case of nerve stimulation and to geometrical distribution of axonal branches in the case of muscle motor point stimulation.

11.1.3   MU Recruitment During NMES: Reflexive Activation of Spinal Motor Neurons

The considerable involvement of different neural structures, such as reflex circuitry, during NMES, is well known [24,78] and partially described in Chapter 12. It has been suggested that NMES provides a multimodal "bombardment" of the central nervous system [49], which results in spinal motor neuron facilitation [12,14] and increased cortical activity and corticospinal excitability [51,82] (see also Chapter 12).

It has been proposed that short and long-term neural adaptations can be triggered by NMES protocols (NMES-induced plasticity), based on the following findings:

a. Persistence of the NMES benefits after the stimulation is turned off, as demonstrated by Vodovnik in 1973 (republished in 2003 in his collected works [82]) and referred to as "carryover effect" (e.g., temporary functional improvements during and after ankle dorsiflexion stimulation in stroke patients with drop foot);

b. Increases in muscle strength after only a few sessions of NMES training, with no concurrent variations in muscle fiber size and functional/enzymatic properties.

c. Increases in muscle strength for the homologous contralateral muscle after unilateral NMES training (a phenomenon referred to as "contralateral limb effect" or "cross-education").

Interestingly, not only "chronic" adaptations may be triggered, but also "acute" neural involvement may occur during a single NMES session. The degree of "acute" involvement of the spinal motor neurons occurring during NMES seems to be related either to the properties of the stimulation train (e.g., frequency, duration) [12,17] or to the features of the stimulation pulse (width, amplitude, charge) [7,12,58].

Conventional NMES involves pulse widths close to the nerve cronaxy (rectangular pulse duration corresponding to the minimal stimulation energy), between 100 and 400 µs and frequencies between 20 and 50 pps. This stimulation paradigm activates motor axons without obvious sequencing related to motor unit type [34]. During a sustained bout of conventional NMES, the torque first increases, due to the lengthening of the mechanical twitches, and then progressively decreases because of mechanical fatigue, while a "slowing" of the surface EMG signals occurs (myoelectric manifestations of muscle fatigue) [58]. This "slowing" is described by a progressive widening of the M-wave, due to progressive decrease of muscle fiber conduction velocity and change of the intracellular action potential shape (see Chapter 10). This results in a progressive compression of the EMG power spectrum, changes of its moments, and decrease of characteristic frequencies [19].

On the contrary, during wide-pulse (1 ms) high-frequency (50–100 pps) NMES, torque increases significantly beyond that elicited by conventional NMES and is sustained for a longer time during the course of the stimulation. This phenomenon has been referred to as "central torque" and is described in Fig. 11.2 [13,14]. Collins et al. demonstrated that the underlying mechanism is the reflexive recruitment of spinal motor neurons (which activates motor units in the normal physiological recruitment order); in fact, the "extra" force could not be elicited following peripheral nerve block [13,14]. Interestingly, it has been observed that a weak involuntary contraction (associated with EMG activity) often persists even after the wide-pulse high-frequency NMES is turned off, thus indicating that the electrically elicited sensory volley may trigger either bistable discharge of motor neurons (i.e., already recruited motor neurons keep firing following the burst) or self-sustained discharge of motor neurons (i.e., recruitment and sustained discharge in previously silent motor units). Bistable and self-sustained discharge are consistent with the development of persistent inward currents (PIC) in spinal motor neurons and interneurons. Further, a possible role for the cortex in the development of the central torque cannot be excluded, as well as a presynaptic mechanism related to the potentiation of neurotransmitter release consequent to afferent discharges.

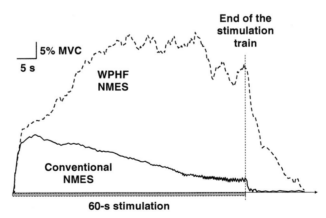

FIGURE 11.2 Representative example of force traces detected from the biceps brachii muscle of a healthy subject during conventional NMES (*continuous line*) and wide-pulse high-frequency NMES (*dashed line*). WPHF (wide-pulse high-frequency) NMES: 1-ms pulse width, 100 pps, 6 mA. Conventional NMES: 100-μs pulse width, 25 pps, 50 mA. MVC: maximal voluntary contraction. Reproduced from [56] with permission.

Independently of the physiological mechanism(s) underlying the "extra" force that can be elicited during and after stimulation, the use of a wide-pulse high-frequency stimulation paradigm is promising for rehabilitation medicine as it seems to recruit primarily slow motor units which are not preferentially activated when conventional NMES is used. The preferential recruitment of slow motor units (if proven) could allow us to prevent or reduce the slow-to-fast transition of motor units that is associated with immobilization and disuse. The reflexive motor neuron recruitment underlying the "central torque" phenomenon may also be involved in NMES-induced muscle cramps, a possible side effect of NMES that takes place if the electrically excited muscles shorten during the stimulation [5,65] (see Chapter 12).

11.2 SURFACE EMG DETECTION DURING ELECTRICAL STIMULATION

11.2.1 M-Waves and Stimulation Artifact

Surface EMG (sEMG) signals can be detected during selective electrical stimulation of a nerve branch or of a motor point of a muscle. The resultant sEMG signal is a compound motor action potential (CMAP), also termed as M-wave. Since the M-wave represents the sum of the potentials of the concurrently activated MUs, its change is generally assumed to reflect changes either in the number or in the sarcolemmal properties of activated motor units (MUs) (this topic is addressed later in this chapter).

A major issue in M-wave studies is the stimulation artifact. In fact, during an electrical stimulus, the current field extends to a large tissue volume which usually includes the detection volume of the sEMG electrodes [57,60,61]. The stimulation artifact is typically spike-shaped and is followed by an exponential decay whose amplitude and time constant depend on several factors such as stimulus intensity and shape, type of the stimulator output stage, relative geometry of the stimulation and detection electrodes, and properties of the signal amplifier input stage [50,53]. The stimulation artifact affects the sEMG temporal and spectral features since it constitutes a high-frequency signal component. Since the artifact is narrow and synchronous in all the sEMG channels (Fig. 11.3), it biases conduction velocity estimations and spectral parameters toward high values and it can overlap with the M-wave and modify its shape.

The issue of the stimulation artifact reduction has been addressed with three main approaches: (a) hardware techniques (dealing with the stimulator output stage and the front end electronics of the EMG amplifier), (b) software methods (mainly based on neural networks designed to reconstruct the shape of the stimulation artifact and to subtract it from the recorded M-wave), and (c) methods oriented to the optimization of the experimental setup (stimulation pulse features, electrode–skin interface and electrode configuration) [50]. A brief description of the hardware and software techniques for the reduction of the stimulation artifact is reported below [42,50,57,61].

Front-End Electronics. Since stimulation current causes a common mode excitation to the first stage of the front-end amplifier, a high CMRR (>100 dB) and a high

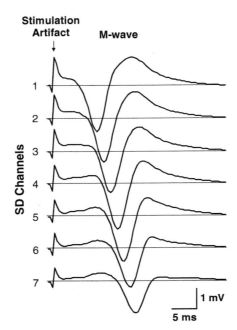

FIGURE 11.3 Surface EMG signals recorded with a linear array of eight electrodes (5 mm apart) aligned with the muscle fibers during monopolar electrical stimulation of the biceps brachii muscle. Each curve is the average of the 25 responses obtained during one second of stimulation (pulse frequency = 25 pps). Stimulation artifacts (synchronous across the channels of the detection system) and M-waves are indicated. It can be observed that for the first three channels (closest to the active stimulation electrode) the stimulation artifact and the M-wave are partially superimposed.

input impedance ($>100\,\text{M}\Omega$) are desirable to reduce artifact amplitude (see Chapter 3). A typical sEMG amplifier input stage is configured with a high-pass filter with cutoff frequency between 10 and 20 Hz in order to reduce the effect of unequal half-cell DC potential, galvanic skin potentials, and slow movement artifacts. The stimulation pulse generates a slow exponential transient in the output of such a filter, thus extending the duration of the artifact.

A slew rate limiter (SRL) may be used to reduce the fast phase of the stimulation artifact. This circuit takes advantage of the higher slew rate (i.e., the the maximal rate of increase or decrease of the signal) that characterizes the stimulation artifact with respect to the M-wave. The circuit limits the maximal slope of the output, leaving the M-wave unchanged while limiting the amplitude of the stimulation artifact and preventing saturation. A second circuit, which can be associated with the SRL, is the blanking circuit which forces the detected signal to zero in a defined time window in which the stimulation artifact is present. The circuit is usually controlled by the stimulator and produces a binary output that is active for the width of the blanking window and forces the output of the blanking circuit to zero. In the case of superimposition between stimulation artifact and M-wave, the blanking circuit not only deletes the artifact but also a portion of the M-wave. Both techniques are described in [45,50]. Other artifact-reducing circuits have been proposed, including a circuit for the automatic identification of the artifact [23].

Stimulator Output Stage. The features of the stimulator output stage play an important role in the duration of the stimulation artifact. The exponential decay of the artifact depends on the discharge of the electrode and tissue capacitances charged during the stimulation pulse.

If the stimulation output stage is current-controlled, its output impedance is at least one order of magnitude higher than the interelectrode impedance. Therefore, between stimuli, the electrode and tissue capacitance finds a high impedance discharge path yielding a slow decay across the stimulation electrodes (τ_2 in Fig. 11.4).

In a voltage-controlled output stage, the output impedance is much lower than the inter-electrode impedance and, between stimuli, the output stage behaves like an active short circuit. This leads to a fast decay of the stimulation artifact due to the fact that electrode and tissue capacitance can discharge through a low impedance path (τ_1 in Figure 11.4). On the other hand, in a current-controlled output stage, the stimulation current is not affected by the interelectrode impedance. In a voltage-controlled output stage, instead, the constant voltage maintained between the stimulation electrodes may generate unpredictable variation of amplitude and shape of the stimulation current due to variation of interelectrode impedance (Fig. 11.4). The compromise between these two stimulation methods was originally proposed in 1978 by Del Pozo and Delgado [18] and optimized 10 years later by Knaflitz and Merletti [45]. These authors designed a hybrid output stage, combining the advantages of the current-controlled stimulation with those of the voltage-controlled mode between stimuli. The main feature of this output stage is its ability to function as a current generator during stimuli and as an active short circuit between stimuli. Figure 11.4 (hybrid configuration) shows that this approach provides a constant (not affected by interelectrode impedance) current level during a stimulation pulse and a fast discharge of the interelectrode capacitance through a low impedance path (time constant: τ_1) that minimizes the duration of the stimulation artifact.

Software Techniques. Software techniques for the offline artifact removal have also been developed. Most of them are based on the estimation of the artifact shape and its subtraction form the recorded signal(s) [9,35,39,50,67]. Grieve et al. [35] and Boudreau et al. [9] proposed a neural network adaptive filter designed to reconstruct the shape of the stimulation artifact and to subtract it from the recorded M-wave contaminated by artifact. In these methods, the shape of the stimulation artifact was estimated from subthreshold responses and used as the reference input of the neural network. Mandrile et al. [50] proposed a neural network based method to be used concomitantly with arrays of sEMG electrodes. The algorithm was designed to estimate the artefact and M-wave from the channels where they are separated in time and use them to reconstruct the M-wave on the channels where stimulation artifact and M-wave overlap.

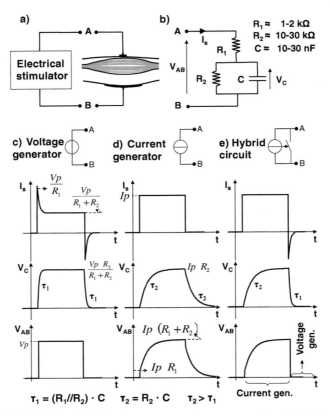

FIGURE 11.4 **(a)** Electrical stimulator connected to a monopolar electrode system. **(b)** Approximate equivalent electrical circuit of the load. **(c)** Current and voltage waveforms present in the equivalent circuit when a rectangular pulse is applied by a voltage generator. **(d)** Current and voltage waveforms present in the equivalent circuit when a rectangular pulse is applied by a current generator. **(e)** Current and voltage waveforms present in the equivalent circuit when a rectangular pulse is applied by a hybrid generator. A hybrid generator is a current generator for the duration of the stimulus and turns into a short circuit (zero voltage generator) in between stimuli.

11.2.2 Myoelectric Manifestations of Muscle Fatigue During NMES

The analysis in the time and frequency domains of the sEMG signal detected during fatiguing muscle contractions is a tool to assess myoelectric manifestations of localized muscle fatigue, which are useful for noninvasive muscle characterization (see Merletti et al. [58,60] and Chapter 10). Myoelectric manifestations of fatigue during voluntary contractions appear to be a multifactorial phenomenon, involving different physiological processes (from the excitatory drive to the contraction mechanisms of the fiber) which evolve simultaneously. As indicated in Chapter 10, they can be grouped under the headings of (1) central fatigue, (2) fatigue of the neuromuscular junction, and (3) muscle fatigue [58].

NMES combined with surface EMG provides interesting experimental paradigms to "isolate" the contributions of peripheral fatigue because it gives the experimenter control of motor unit firing frequency and recruitment. For these reasons, estimates of myoelectric signal variables show less variability in electrically elicited fatigue than in voluntary fatigue [46,61] as pointed out in Chapter 10.

The reduction of muscle fiber conduction velocity due to fatigue causes a compression of the power spectrum and a decrease of spectral features [such as mean and median frequency (MNF and MDF, respectively)] as well as an increase in amplitude features [such as the average rectified value (ARV) or root mean square value (RMS)] due to widening of the M-wave (Fig. 11.5).

In some cases, ARV and RMS show a curvilinear behavior with downward concavity constituted by a linear increase in the first part of the curve (due to an initial

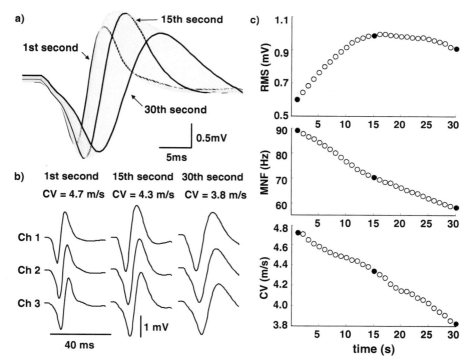

FIGURE 11.5 (a) M-wave time evolution over 30 s of sustained stimulation at 30 pps of the biceps brachii muscle. Each curve is the average of the 30 responses obtained during one second epoch. The 1st, the 15th, and 30th (averaged) M-waves are highlighted. (b) M-waves detected by three consecutive channels (corresponding to the 1st, the 15th, and 30th second of stimulation) and relative conduction velocity (CV) estimates. (c) Time course of the EMG variable estimations during the 30 s of electrically elicited contraction. Black dots correspond to the estimations of variables of the M-waves highlighted in panel a). See also Chapter 10. Reproduced from Piitulainen et al. [68] with permission.

M-wave widening that is related to a decrease in conduction velocity of the activated muscle fibers) followed by a slow decrease (see [6,19] and Fig. 11.5).

This decrease may have many explanations among which are the variation in the amplitude of muscle fiber action potentials and the alterations in neuromuscular junction transmission [30,36], During sustained stimulated contractions, the presence of different MU pools with different fatigue profiles may lead to the progressive separation of the M-wave in sub-M-waves, as represented in Fig. 11.6. Classic EMG

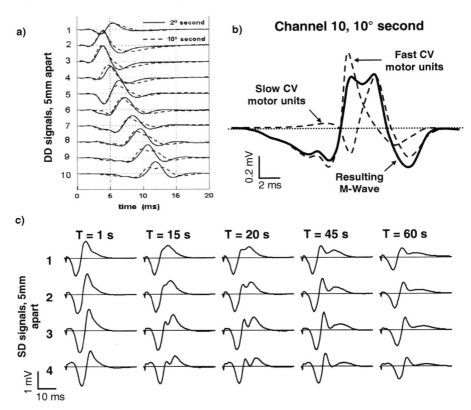

FIGURE 11.6 **(a)** Ten channels of double differential (DD) M-waves detected from the biceps brachii in one subject stimulated for 10 s at 30 pps. The continuous line shows the array of (averaged) M-waves detected during the 2nd second while the dotted line shows the array of (averaged) M-waves detected during the 10th second. The second array shows that some MUs maintained (or slightly increased) their conduction velocity while others decreased it. **(b)** Possible interpretation of the contributions of the two pools of motor units. CV: conduction velocity. **(c)** Four channels of single differential (SD) M-waves (5 mm IED) detected from a biceps brachii stimulated for 60 s at 20 pps. The five columns show the time course of the M-waves (averaged over 1-s epochs) throughout the fatiguing contraction. Propagation of the M-wave along the electrode array can be detected from channel 1 to channel 4 in each panel. A clean-cut separation of two pools of MUs with different conduction velocities is evident after 15–20 s from the stimulation onset.

variables (conduction velocity, amplitude, and frequency indicators) are not sufficient to describe and interpret these complex multichannel signals. The analysis of these signals requires mathematical models and new methods for multichannel M-wave deconvolution [64].

At stimulation frequencies above 30–35 pps the duration of an M-wave is longer than the interstimulus interval and M-wave truncation takes place. This nonlinear event alters the spectrum of the signal and its characteristic frequencies are no longer usable. Alternative techniques, based on estimation of scaling factors and on wavelet expansion, have been developed to quantify M-wave scaling using the portion before truncation [62,63].

11.2.3 Incremental M-Waves

The M-wave represents the sum of the action potentials of the concurrently activated MUs. In a nonfatigued muscle a change of M-wave amplitude is generally assumed to reflect a change in the number of activated MUs. In fact, the M-wave amplitude increases with the stimulation current intensity up to a plateau level, which is reached when all the MUs excitable from a given stimulation point and detectable by the EMG electrode system are activated. The difference between M-waves detected using two different stimulation levels provides the response of the MUs whose activation thresholds are in between the two stimulation levels. Indeed, the difference (ΔM-wave) between a multichannel M-wave elicited by a current intensity I_2 and that elicited by a lower current intensity I_1 provides the multichannel M-wave of the MUs incrementally recruited when current is increased from I_1 to I_2. This process allows the characterization of small groups of MUs and therefore the investigation of MU recruitment order during electrical stimulation (Fig. 11.7). The number of new MUs recruited by an increment of stimulation current ΔI depends on the amplitude of the current increment and on the excitability thresholds of the nerve fibers or their axonal branches, depending on the site of stimulation. Therefore it is theoretically possible for very small ΔI (lower than tenths of mA) to study the incremental recruitment of very few MUs.

In addition, when M-waves are detected with bi-dimensional EMG recording systems, (see Ch. 5) the spatial distribution of incremental M-waves can be investigated. In muscles where the surface representation of motor unit action potential is localized (such as in pinnate muscles), this approach provides information about the localization and the size of the territories of the newly recruited MUs as indicated in Fig. 11.8 and [81].

11.2.4 Incremental Stimulation for Estimating Motor Unit Number

The EMG response to incremental nerve stimulation has been used to estimate the number of motor units in a muscle. The technique is termed motor unit number estimation (MUNE). The estimation of the number of active motor neurons finds important clinical applications in the assessment of the degree of denervation as well as in the monitoring of neurodegenerative processes of motor neuron diseases and

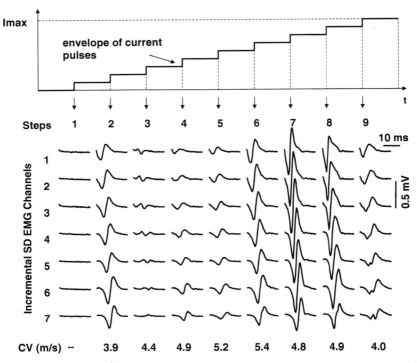

FIGURE 11.7   Nine steps of incremental M-waves (and relative CV estimates) generated by progressively recruited MUs during an electrically elicited contraction (staircase amplitude increase, stimulation frequency: 20 pps). Each set of seven signals represents the seven bipolar recordings obtained from the linear array of eight electrodes (5 mm apart) placed on the biceps brachii muscle. Reproduced from Botter et al. [7] with permission.

peripheral neuropathies [10,27,69,80]. The original technique, proposed by McComas in 1971 [54,55], and further developed by other investigators [12,69,70] consists in increasing the stimulation amplitude from a subthreshold value until a number (a dozen) of increments are observed in the muscle electrical response. The average amplitude of the increments (assumed to represent the average MUAP amplitude) is obtained dividing the amplitude of the last response by the number of observed increments. Motor unit number is then estimated through the ratio between the supramaximal compound action potential (CMAP, or maximal M-wave) and the average amplitude of the increments. This method is based on three main assumptions [20,21,54,55]: (i) each incremental response is due to the additional excitation of a single MU, (ii) the MU sample obtained with the incremental stimulation at low amplitude is representative of the global MU population, and (iii) no cancellation between incrementally stimulated MUAPs occurs.

The issue of possible subtractions between motor unit contributions (cancellation) can be minimized, in the case of a single and localized muscle innervation zone, by careful electrode placement [54]. To avoid possible confounding factors due to

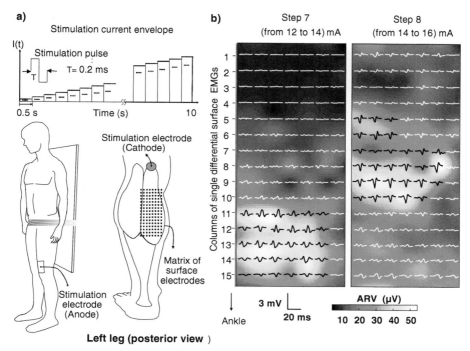

FIGURE 11.8 (a) Schematic representation of the stimulation current envelope and of the experimental setup. The positions of the 128 electrode grid, of the stimulation electrode and of the reference electrode used to close the stimulation current loop (monopolar stimulation of the nerve) are shown. (b) Incremental M-waves and interpolated maps of their averaged rectified value (ARV; 15 ms epochs from each stimulus) for two successive stimulation steps. Localized M-waves changes can be observed by comparing the two maps.

cancellation, some author proposed to use muscle mechanical response (always positive) rather than action potentials [75]. Whether the sample of MU detected during the incremental protocol is representative of the global population of muscle MUs (assumption ii) is a question that involves, among others, the issue of recruitment order during electrically induced contractions. Indeed, if a random recruitment occurs, a group of a few (tens) MUs might be representative of the global population. However, if large-diameter axons (and motor units) are recruited first, the average MUAP amplitude will be overestimated, thus leading to an underestimation of the total motor unit number. Although the MUNE error decreases with the number of stimulated MUs, an incremental protocol from subthreshold to high stimulation levels is not always viable in clinical settings as it is time-consuming and not well-tolerated by patients. The most important limitation seems to be related to the first assumption, that is the excitation of single MUs during the incremental stimulation. Indeed, the activation threshold of a single axon cannot be defined as a fixed value above which the excitation always takes place. It is, instead, a range of stimulus intensities over which the excitation probability increases from 0 to 1 [11,54]. Therefore, for the same stimulation level a MU can or cannot be activated according to a certain probability function. When the thresholds of two (or more) axons overlap, the same stimulus intensity could generate a set of different M-wave increments due to different combinations of MUs that can respond at that given stimulus intensity. This

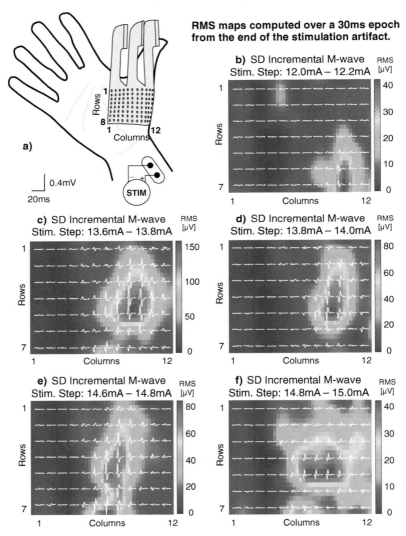

FIGURE 11.9 (a) Two-dimensional M-wave detection over the surface of the abductor and flexor pollicis during electrical stimulation of the median nerve. The five color maps (**b–f**) are interpolated with a factor 4 and represent the distribution of single differential incremental M-waves due to 0.2-mA increments of the stimulation current. M-waves of individual channels are superimposed in white.

phenomenon, also known as alternation, can lead to errors in the estimation of the number of motor units.

To deal with this problem, several modifications of the original method have been proposed [20,73,83]. In 2008 van Dijk et al. [79] introduced a modified version of this method based on high-density EMG detection.

By sampling the two-dimensional potential distribution from multiple locations over the skin surface with an electrode array, it is possible to obtain spatiotemporal profiles of potentials (Fig. 11.9). Because of the added spatial dimension, superimposition of elicited potentials can be solved and alternation can be easily recognized. This approach results in larger sample sizes and in an overall improvement of MUNE accuracy.

REFERENCES

1. Barbero, M., R. Merletti, and A. Rainoldi, *Atlas of Innervation Zones: Understanding Surface Electromyography and Its Applications*, Springer, Milano, Italy, 2012.

2. Bax, L., F. Staes, and A. Verhagen, "Does neuromuscular electrical stimulation strengthen the quadriceps femoris? A systematic review of randomised controlled trials," *Sports Med.* **35**, 191–212 (2005).

3. Behringer, M., A. Franz, M. McCourt, and J. Mester, "Motor point map of upper body muscles," *Eur. J. Appl. Physiol.* **114**, 1605–1617 (2014).

4. Belanger, M., R. B. Stein, G. D. Wheeler, T. Gordon, and B. Leduc, "Electrical stimulation: can it increase muscle strength and reverse osteopenia in spinal cord injured individuals?," *Arch. Phys. Med. Rehabil.* **81**, 1090–1098 (2000).

5. Bertolasi, L., D. De Grandis, L. G. Bongiovanni, G. P. Zanette, and M. Gasperini, "The influence of muscular lengthening on cramps," *Ann. Neurol.* **33**, 176–180 (1993).

6. Botter, A., F. Lanfranco, R. Merletti, and M. A. Minetto, "Myoelectric fatigue profiles of three knee extensor muscles," *Int. J. Sports Med.* **30**, 4008–4017 (2009).

7. Botter, A., R. Merletti, and M. A. Minetto, "Pulse charge and not waveform affects M-wave properties during progressive motor unit activation," *J. Electromyogr. Kinesiol.* **19**, 564–573 (2009).

8. Botter, A., G. Oprandi, F. Lanfranco, S. Allasia, N. A. Maffiuletti, and M. A. Minetto, "Atlas of the muscle motor points for the lower limb: implications for electrical stimulation procedures and electrode positioning," *Eur. J. Appl. Physiol.* **111**, 2461–2471 (2011).

9. Boudreau, B. H., K. B. Englehart, A. D. Chan, and P. A. Parker, "Reduction of stimulus artifact in somatosensory evoked potentials: segmented versus subthreshold training," *IEEE Trans. Biomed. Eng.* **51**, 1187–1195 (2004).

10. Bromberg, M. B., "Updating motor unit number estimation (MUNE)," *Clin. Neurophysiol.* **118**, 1–8 (2007).

11. Brown, W. F., and H. S. Milner-Brown, "Some electrical properties of motor units and their effects on the methods of estimating motor unit numbers," *J. Neurol. Neurosurg. Psychiatry* **39**, 249–257 (1976).

12. Brown, W.F., M. J. Strong, and R. Snow. "Methods for estimating numbers of motor units in biceps-brachialis muscles and losses of motor units with aging," *Muscle Nerve* **11**, 423–432 (1988).

13. Collins, D. F., "Central contributions to contractions evoked by tetanic neuromuscular electrical stimulation," *Exerc. Sport Sci. Rev.* **35**, 102–109 (2007).

14. Collins, D. F., D. Burke, and S. C. Gandevia, "Large involuntary forces consistent with plateau-like behavior of human motoneurons," *J. Neurosci.* **21**, 4059–4065 (2010).

15. Crameri, R. M., A. R. Weston, S. Rutkowski, J. W. Middleton, G. M. Davis, and J. R. Sutton, "Effects of electrical stimulation leg training during the acute phase of spinal cord injury: a pilot study," *Eur. J. Appl. Physiol.* **83**, 409–415 (2000).

16. Creasey, G. H., C. H. Ho, R. J. Triolo, D. R. Gater, A. F. DiMarco, K. M. Bogie, and M. W. Keith, "Clinical applications of electrical stimulation after spinal cord injury," *J. Spinal Cord Med.* **27**, 365–375 (2004).

17. Dean, J. C., L. M. Yates, and D. F. Collins, "Turning on the central contribution to contractions evoked by neuromuscular electrical stimulation," *J. Appl. Physiol.* **103**, 170–176 (2007).

18. Del Pozo, F., and J. M. Delgado, "Hybrid stimulator for chronic experiments," *IEEE Trans. Biomed. Eng.* **25**, 92–94 (1978).

19. Dimitrova, N. A., J. Y. Hogrel, T. I. Arabadzhiev, and G. V. Dimitrov. "Estimate of M-wave changes in human biceps brachii during repetitive stimulation," *J. Electromyogr. Kinesiol.* **15**, 341–348 (2005).

20. Doherty, T. J., and W. F. Brown, "The estimated numbers and relative sizes of thenar motor units as selected by multiple point stimulation in young and older adults," *Muscle Nerve* **16**, 355–366 (1993).

21. Doherty, T., Z. Simmons, B. O'Connell, K. J. Felice, R. Conwit, K. M. Chan, T. Komori, T. Brown, D. W. Stashuk, and W. F. Brown, "Methods for estimating the numbers of motor units in human muscles," *J. Clin. Neurophysiol.* **12**, 565–584 (1995).

22. Duchateau, J., and K. Hainaut, "Training effects of sub-maximal electrostimulation in a human muscle," *Med Sci Sports Exerc* **20**, 99–104 (1988).

23. Eck, U., and G. Vossius, "A method for analysis of muscle activity during electric stimulation," *Biomed. Tech.* **47**, 517–520 (2002).

24. Enoka, R. M., "Activation order of motor axons in electrically evoked contractions," *Muscle Nerve* **25**, 763–764 (2002).

25. Farina, D., A. Blanchietti, M. Pozzo, and R. Merletti, "M-wave properties during progressive motor unit activation by transcutaneous stimulation," *J. Appl. Physiol.* **97**, 545–555 (2004).

26. Feiereisen, P., J. Duchateau, and K. Hainaut, "Motor unit recruitment order during voluntary and electrically induced contractions in the tibialis anterior," *Exp. Brain Res.* **114**, 117–123 (1997).

27. Felice, K. J., "A longitudinal study comparing thenar motor unit number estimates to other quantitative tests in patients with amyotrophic lateral sclerosis," *Muscle Nerve* **20**, 179–185 (1997).

28. Fitzgerald, G. K., and A. Delitto, "Neuromuscular electrical stimulation for muscle strength training," in A. Rainoldi, M. A. Minetto, and R. Merletti, eds., *Biomedical Engineering in Exercise and Sports*, Edizioni Minerva Medica, Torino, Italy, 2006.

29. Fitzgerald, G. K., S. R. Piva, and J. J. Irrgang, "A modified neuromuscular electrical stimulation protocol for quadriceps strength training following anterior cruciate ligament reconstruction," *J. Orthop. Sports Phys. Ther.* **33**, 492–501 (2003).

30. Fuglevand, A. J., K. M. Zackowski, K. A. Huey, and R. M. Enoka, "Impairment of neuromuscular propagation during human fatiguing contractions at submaximal forces," *J. Physiol.* **460**, 549–572 (1993).

31. Gandevia, S. C., "Spinal and supraspinal factors in human muscle fatigue," *Physiol. Rev.* **81**, 1725–1789 (2001).

32. Glinsky, J., L. Harvey, and P. Van Es, "Efficacy of electrical stimulation to increase muscle strength in people with neurological conditions: a systematic review," *Physiother. Res. Int.* **12**, 175–194 (2007).

33. Gobbo, M., P. Gaffurini, L. Bissolotti, F. Esposito, and C. Orizio, "Transcutaneous neuromuscular electrical stimulation: influence of electrode positioning and stimulus amplitude settings on muscle response," *Eur. J. Appl. Physiol.* **111**, 2451–2459 (2011).

34. Gregory, C. M., and C. S. Bickel, "Recruitment patterns in human skeletal muscle during electrical stimulation," *Phys. Ther.* **85**, 358–364 (2005).

35. Grieve, R., P. A. Parker, B. Hudgins, and K. Englehart, "Nonlinear adaptive filtering of stimulus artifact," *IEEE Trans. Biomed. Eng.* **47** (3), 389–395 (2000).

36. Griffin, L., B. G. Jun, C. Covington, and B. M. Doucet, "Force output during fatigue with progressively increasing stimulation frequency," *J. Electromyogr. Kinesiol.* **18**, 426–433 (2008).

37. Hennings, K., E. N. Kamavuako, and D. Farina, "The recruitment order of electrically activated motor neurons investigated with a novel collision technique," *Clin. Neurophysiol.* **118**, 283–291 (2007).

38. Henriksson-Larsen, K., J. Friden, and M. L. Wretling, "Distribution of the fibre sizes in human skeletal muscle. An enzyme histochemical study in m tibialis anterior," *Acta Physiol. Scand.* **123**, 171–177 (1985).

39. Hines, A. E., P. E. Crago, G. J. Chapman, C. Billian, "Stimulus artifact removal in EMG from muscles adjacent to stimulated muscles," *J. Neurosci. Methods* **64** (1), 55–62 (1996).

40. Horstman, A. M., M. J. Beltman, K. H. Gerrits, P. Koppe, T. W. Janssen, P. Elich, and A. de Haan, "Intrinsic muscle strength and voluntary activation of both lower limbs and functional performance after stroke," *Clin. Physiol. Funct. Imaging* **28**, 251–261 (2008).

41. Hultman, E., H. Sjoholm, I. Jaderholm-Ek, and J. Krynicki, "Evaluation of methods for electrical stimulation of human skeletal muscle in situ," *Pflugers Arch.* **398**, 139–141 (1983).

42. Jubeau, M., J. Gondin, A. Martin, A. Sartorio, and N. A. Maffiuletti, "Random motor unit activation by electrostimulation," *Int. J. Sports Med.* **28**, 901–904 (2007).

43. Karaca, P., A. Hadzic and J. D. Vloka, "Specific nerve blocks: an update," *Curr. Opin. Anaesthesiol.* **13**, 549–555 (2000).

44. Kern, H., S. Salmons, W. Mayr, K. Rossini, and U. Carraro, "Recovery of long-term denervated human muscles induced by electrical stimulation," *Muscle Nerve* **31**, 98–101 (2005).

45. Knaflitz, M., and R. Merletti, "Suppression of stimulation artifacts from myoelectric evoked potential recordings," *IEEE Trans. Biomed. Eng.* **35**, 758–763 (1988).

46. Knaflitz, M., R. Merletti, and C. J. De Luca, "Inference of motor unit recruitment order in voluntary and electrically elicited contractions," *J. Appl. Physiol.* **68**, 1657–1667 (1990).

47. Kralj, A., and T. Bajd, *Functional Electrical Stimulation: Standing and Walking After Spinal Cord Injury*, 1st ed., CRC Press, Boca Raton, FL, 1989.

48. Lexell, J., and C. C. Taylor, "Variability in muscle fibre areas in whole human quadriceps muscle. How much and why?," *Acta Physiol. Scand.* **136**, 561–568 (1989).

49. Maffiuletti, N. A., "Physiological and methodological considerations for the use of neuromuscular electrical stimulation," *Eur. J. Appl. Physiol.* **110**, 223–234 (2010).

50. Mandrile, F., D. Farina, M. Pozzo, and R. Merletti, "Stimulation artifact in surface EMG signal: effect of the stimulation waveform, detection system, and current amplitude using hybrid stimulation technique," *IEEE Trans. Neural Syst. Rehabil. Eng.* **11**, 407–415 (2003).

51. Mang, C. S., O. Lagerquist, and D. F. Collins, "Changes in corticospinal excitability evoked by common peroneal nerve stimulation depend on stimulation frequency," *Exp. Brain Res.* **203**, 11–20 (2010).

52. Martin, V., G. Y. Millet, A. Martin, G. Deley, and G. Lattier, "Assessment of low-frequency fatigue with two methods of electrical stimulation," *J. Appl. Physiol.* **97**, 1923–1929 (2004).

53. McGill, K. C., K. L. Cummins, L. J. Dorfman, B. B. Berlizot, K. Leutkemeyer, D. G. Nishimura, and B. Widrow, "On the nature and elimination of stimulus artifact in nerve signals evoked and recorded using surface electrodes," *IEEE Trans. Biomed. Eng.* **29**, 129–137 (1982).

54. McComas, A. J., P. R. Fawcett, M. J. Campbell, and R. E. Sica, "Electrophysiological estimation of the number of motor units within a human muscle," *J. Neurol. Neurosurg. Psychiatry* **34**, 121–131 (1971).

55. McComas, A. J., "Invited review: motor unit estimation: methods, results, and present status," *Muscle Nerve* **14**, 585–597 (1991).

56. Merletti, R., A. Botter, F. Lanfranco, and M. A. Minetto, "Spinal involvement and muscle cramps in electrically elicited muscle contractions," *Artif. Organs* **35**, 221–225 (2011).

57. Merletti, R., A. Botter, A. Troiano, E. Merlo, and M. A. Minetto, "Technology and instrumentation for detection and conditioning of the surface electromyographic signal: state of the art," *Clin. Biomech.* **24**, 122–134 (2009).

58. Merletti, R., A. Rainoldi, and D. Farina, "Myoelectric manifestations of muscle fatigue," in R. Merletti, and P. A. Parker, *Electromyography. Physiology, Engineering, and Non-invasive Applications*, John Wiley–IEEE Press, Hoboken, NJ, 2004.

59. Merletti, R., "Electrical stimulation of the periferal nervous system" in A. Rainoldi, M. A. Minetto, and R. Merletti, eds., *Biomedical engineering in exercise and Sports*, Edizioni Minerva Medica, Torino, Italy, 2006.

60. Merletti, R., M. Knaflitz, and C. J. De Luca, "Myoelectric manifestations of fatigue in voluntary and electrically elicited contractions," *J. Appl. Physiol.* **69**, 1810–1820 (1990).

61. Merletti, R., M. Knaflitz, and C. J. De Luca, "Electrically evoked myoelectric signals," *Crit. Rev. Biomed. Eng.* **19**, 293–340 (1992).

62. Merletti, R., and L. Lo Conte, "Advances in processing of surface myoelectric signals, Part I and II," *Med. Biol. Eng. Comp* **33**, 362–384 (1995).

63. Mesin, L., and D. Farina, " Estimation of M-wave scale factor during sustained contractions at high stimulation rate," *IEEE Trans. Biomed. Eng.* **52**, 869–877 (2005).

64. Mesin, L., and D. Cocito, "A new method for the estimation of motor nerve conduction block," *IEEE Trans. Neural Syst. Rehabil. Eng.* **11**, 407–415 (2003).

65. Minetto, M.A., A. Holobar, A. Botter, and D. Farina, "Origin and development of muscle cramps," *Exerc. Sport Sci. Rev.* **41**, 3–10 (2013).

66. Newsam, C. J., and L. L. Baker, "Effect of an electric stimulation facilitation program on quadriceps motor unit recruitment after stroke," *Arch. Phys. Med. Rehabil.* **85**, 2040–2045 (2004).

67. O'Keeffe, D. T., G. M. Lyons, A. E. Donnelly, and C. A. Byrne, " Stimulus artifact removal using a software-based two-stage peak detection algorithm," *J. Neurosci. Methods* **109**, 137–145 (2001).

68. Piitulainen, H., A. Botter, R. Merletti, and J. Avela, "Muscle fiber conduction velocity is more affected after eccentric than concentric exercise," *Eur. J. Appl. Physiol.* **111**, 261–273 (2011).

69. Shefner, J., M. Cudkowicz, M. Zhang, D. Schoenfeld, and D. Jillapalli, "The use of statistical MUNE in a multicenter clinical trial," *Muscle Nerve* **30**, 463–469 (2004).

70. Smith, G. V., G. Alon, S. R. Roys, and R. P. Gullapalli, "Functional MRI determination of a dose–response relationship to lower extremity neuromuscular electrical stimulation in healthy subjects," *Exp. Brain Res.* **150**, 33–39 (2003).

71. Snyder-Mackler, L., A. Delitto, S. L. Bailey, and S. W. Stralka, "Strength of the quadriceps femoris muscle and functional recovery after reconstruction of the anterior cruciate ligament. A prospective, randomized clinical trial of electrical stimulation," *J. Bone Jt Surg. Am.* **77**, 1166–1173 (1995).

72. Solomonow, M., "External control of the neuromuscular system," *IEEE Trans. Biomed. Eng.* **31**, 752–763 (1984).

73. Stashuk, D. W., T. J. Doherty, A. Kassam, and W. F. Brown, "Motor unit number estimates based on the automated analysis of F-responses," *Muscle Nerve* **17**, 881–890 (1994).

74. Stein, R. B., S. L. Chong, K. B. James, A. Kido, G. J. Bell, L. A. Tubman, and M. Belanger "Electrical stimulation for therapy and mobility after spinal cord injury," *Prog. Brain Res.* **137**, 27–34 (2002).

75. Stein, R. B., and J. F. Yang, "Methods for estimating the number of motor units in human muscles," *Ann. Neurol.* **28** (4),**487–495** (1990).

76. Stephens, J. A., R. Garnett, and N. P. Buller, "Reversal of recruitment order of single motor units produced by cutaneous stimulation during voluntary muscle contraction in man," *Nature* **272**, 362–364 (1978).

77. Trimble, M. H., and R. M. Enoka, "Mechanisms underlying the training effects associated with neuromuscular electrical stimulation," *Phys. Ther.* **71**, 273–280 (1991).

78. Vanderthommen., M., and J. Duchateau, "Electrical stimulation as a modality to improve performance of the neuromuscular system," *Exerc. Sport Sci. Rev.* **35**, 180–185 (2007).

79. van Dijk, J. P., J. H. Blok, B. G. Lapatki, I. N. van Schaik, M. J. Zwarts, and D. F. Stegeman, "Motor unit number estimation using high-density surface electromyography," *Clin. Neurophysiol.* **119**, 33–42 (2008).

80. van Dijk, J. P., H. J. Schelhaas, I. N. Van Schaik, H. M. Janssen, D. F. Stegeman, and M. J. Zwarts, "Monitoring disease progression using high-density motor unit number estimation in amyotrophic lateral sclerosis," *Muscle Nerve* **42**, 239–244 (2010).

81. Vieira, T. M. M., A. Botter, C. Itiki, and R. Merletti, "Study of the compartmentalization of the human medial gastrocnemius with incremental M-wave: preliminary results," in *Proceedings of the XVIII Congress of the International Society of Electrophysiology and Kinesiology*, Dario Farina and Debora Falla, eds., Aalborg, Denmark, 2010.

82. Vodovnik, L., and S. Rebersek,"Improvements in voluntary control of paretic muscles due to electrical stimulation. Neural organization and its relevance to prosthetics," in *Lojze*

Vodovnik Collected Works, D. Miklavcic, T. Kotnik, and G. Sersa, University of Ljubljana, Faculty of Electrical Engineering, Ljubljana, Slovenia, 2003.

83. Wang, F. C., and P. J. Delwaide, "Number and relative size of thenar motor units estimated by an adapted multiple point stimulation method," *Muscle Nerve* **18**, 969–979 (1995).

84. Wust, R. C., C. I. Morse, A. de Haan, D. A. Jones, and H. Degens, "Sex differences in contractile properties and fatigue resistance of human skeletal muscle," *Exp. Physiol.* **93**, 843–850 (2008).

12

SURFACE EMG APPLICATIONS IN NEUROPHYSIOLOGY

S. Baudry,[1] M. A. Minetto,[2] and J. Duchateau[1]

[1]Laboratory of Applied Biology and Neurophysiology, ULB Neuroscience Institute (UNI), Université Libre de Bruxelles (ULB), Brussels, Belgium
[2]Division of Endocrinology, Diabetology and Metabolism, Department of Medical Sciences, University of Turin, Turin, Italy

12.1 INTRODUCTION

Human motion is the consequence of highly complex interactions between systems (nervous, muscular, joint) that have multiple degrees of freedom. Even a simple task such as upright standing is surprisingly complex and requires the integration of sensory inputs from muscles, eyes, and vestibular system to control the signals sent to the muscles [49]. Understanding the processes that underlie the sensorimotor integration is challenging as the recruitment of muscles and the modulation of their activity vary with the task being performed. Moreover, the efficiency of muscle action depends on its anatomical arrangement and the properties of its motor units, as well as on the current state of the system (muscle length, joint angular position, type of contraction, speed, etc.). Last but not least, several strategies can be used to perform a task, making it more difficult to identify relevant and nonrelevant control signals. With regard to this complexity and the diversity of conditions (upright stance, walking, lifting loads, etc.) that form human motor behaviors, research in motor control most often requires the use of noninvasive methods that provide indirect pieces of information. In this context, the surface electromyogram (EMG) technique can be considered as a relevant tool for the assessment of muscle

Surface Electromyography: Physiology, Engineering, and Applications, First Edition.
Edited by Roberto Merletti and Dario Farina.
© 2016 by The Institute of Electrical and Electronics Engineers, Inc. Published 2016 by John Wiley & Sons, Inc.

activation and, when interpreted with caution, to study movement strategies defined as the neural activation patterns that are associated with achieving different movement goals [29]. Moreover, an EMG can be combined with other electrophysiological techniques, opening the door to the study of the neural circuitries (located at either the supraspinal or spinal level) that are directly involved in the motor command or contribute to its adjustment.

In this chapter, we provide a basic description of electrophysiological methods and their applications to different topics related to acute adjustments and chronic adaptations, with the goal to emphasize the relevance of the electrophysiological tools based on EMG to investigate movement strategies in humans.

12.2 SURFACE EMG ACTIVITY

As detailed in Chapters 1 and 2, the sEMG signal is the summation of the electrical potential fields generated by the depolarization of the sarcolemma of the muscle fibers detected by the electrodes. The recorded sEMG signal represents the electrical activity of numerous motor units. In fact, the signal results from the algebraic summation of multiple motor unit action potentials (MUAP) detected at various distances from the source, each MUAP being distorted by the spatial low-pass filtering effect of the volume conductor and possibly superimposed with MUAPs of other motor units. Nonetheless, when intramuscular and sEMG are compared in the biceps brachii, they are highly correlated for dynamic and isometric contractions [17], suggesting that sEMG may reflect muscle activity if caution is taken to avoid pitfalls in the interpretation of this complex signal (see below).

12.2.1 Surface EMG Variables

Before considering the sEMG variables, it is important to mention appropriate placement of the surface electrodes. As reviewed in Chapter 2, Chapter 3, and elsewhere [38], general guidelines should be followed to allow proper sEMG recordings: Sensors (electrode pairs) should be placed with the electrodes oriented in parallel with muscle fiber orientation. For fusiform muscles, the location of the electrodes should be halfway between the end plates (innervation zone) and a muscle tendon, and away from the edge with others muscles or subdivisions of the targeted muscle (see Chapter 3). The situation is different for pinnate muscles (see Chapter 5).

From the sEMG recordings, two parameters are commonly calculated to quantify the amplitude of the sEMG signal, providing an estimate of the magnitude of muscle activity that can be used to assess indirectly the neural drive to the muscle: (1) The average value of the rectified sEMG signal, referred to as ARV (average rectified value, aEMG, is also often used), that corresponds to the mean value of the absolute amplitude over a given period of time (moving average) or to the low-pass filtered rectified value (linear envelope) (for details, see Chapter 3); (2) the root mean square (RMS) value, whose square represents the variance, that is the power of the signal.

The unit of both ARV value and RMS is mV or μV. Although ARV is proportional to RMS amplitude, for a given and constant probability density function, the latter provides a better measure of muscle activity (see Chapters 3 and 5). Nonetheless, in a number of applications, the choice of ARV or RMS is of minor importance as both measurements provide quite similar information, especially during isometric muscle contractions [30].

12.2.2 Normalization Procedures

Regardless of the parameter chosen (ARV or RMS), sEMG signals should be normalized because absolute values (mV or μV) are inadequate to infer the level of muscular activation relative to the maximal activation capacity of an individual or to compare individuals. Indeed, the raw EMG amplitude depends on numerous factors (location and size of the electrodes, distance between electrodes, and thickness of the volume conductor between source and detection points; see Chapter 2) that do not have any neurophysiological meaning but can confound the interpretation and the comparison of the signal parameters between conditions or individuals (see Chapter 3). Therefore, one of the most common procedures is to express the sEMG (ARV or RMS) as a percentage of the value recorded during a maximal voluntary contraction (MVC), whose value is defined slightly differently among researchers (peak value of one contraction, average of the peak of N contractions, etc.). This may help, for example, to compare individuals performing the same motor task, as for example when comparing young and elderly adults (see Section 12.4.4, dedicated to Aging). Nonetheless, care should be taken when using such normalization procedure as the amount of muscle activity for a given force produced during an isometric contraction varies with the angular joint position reflecting muscle length and the rate at which force is developed (see below). Moreover, the association between sEMG and force is not always linear [36], depending on numerous factors such as the mechanical model (single- or multi-joint system), the presence of synergistic muscles, and the range of MU recruitment threshold.

In order to reduce the error related to these latter factors, it is possible to assess the relation between sEMG amplitude and torque from low to maximal force output [40]. This consists of transposing sEMG (neurophysiological parameter) as an estimate of the force generated from an activated muscle or muscle group, leading to consider the biomechanical resultant of the muscle activity. It is important to construct the sEMG–torque relation taking into account various factors such as the joint angle (muscle tension–length relation). For example, the sEMG–torque relations for the gastrocnemii (medialis and lateralis) muscles differ regarding the knee angle joint whereas such differences are not observed for the soleus muscle [21], illustrating the relevance to consider muscle characteristics (mono- or bi-articular) in relation with the posture of the subject, in order to construct a representative sEMG–torque relation. Nonetheless, one should be aware that the sEMG recorded from the muscle underneath the electrodes may not represent the activity of the other muscles (synergists) contributing to the torque produced (see Chapter 10). Therefore, considering only one muscle may flaw the sEMG–torque relation. This effect (referred to as "load sharing") can be reduced if several muscles acting on the join of interest are monitored, as such procedure allows to assess changes in their respective contribution, as described in Chapter 10.

When considering nonisometric contractions, additional factors should be taken into account before using the sEMG–torque relation. First, as the electrodes are fastened to the skin, a large range of motion induces changes in the recording site at the muscle level (because of the changing relative position of the electrodes with respect to the innervation zone, among other factors) and in the number of muscle fibers in the recording volume as well. Therefore, sEMG should be measured within the same, quite reduced range of motion for building the sEMG–torque relation and in conditions as close as possible to the target motor task. In dynamic conditions, the sEMG is also influenced by movement velocity with the sEMG increasing with movement velocity [18], and contraction type with a greater movement-velocity/ sEMG slope for shortening compared with lengthening contractions [12]. A greater sEMG amplitude has been observed during shortening than lengthening contractions [12], although contrasting results have been reported [7]. As for the isometric contractions, it is important when constructing the sEMG–torque relation to control the posture adopted by the subject, especially for multi-joint movements. For example, the peak torque and median frequency of sEMG in the gastrocnemius medialis are significantly reduced during isokinetic knee flexion when ankle joint is in a plantarflexion position compared with dorsiflexion position [22]. Based on the time-dependent characteristic of these factors, it is important to keep in mind that the duration of the time epoch used for estimation of sEMG features should be the same when comparing two contraction types.

Finally, another means to normalize the raw sEMG activity is to express its value relative to the maximal amplitude of the compound muscle action potential (M wave) induced by electrical stimulation of the motor nerve (see below and Chapters 10 and 11). Although this procedure has some limitations [43], it controls for changes in recording conditions, allowing an accurate comparison between subjects.

Even if all these factors are taken into account, sEMG should always be interpreted with caution, especially when trying to deduce neural drive from its recording. Nonetheless, sEMG remains a major method in motor control studies to understand the interactions between the nervous and muscular systems in a variety of human movements and tasks.

12.2.3 Muscle Fiber Conduction Velocity

The global characteristics of the sEMG, such as its amplitude distribution and power spectral density, depend on the shape, conduction velocity, number and timing of the motor unit action potentials, as well as on the membrane properties of the muscle fibers. Muscle fiber conduction velocity (MFCV or CV), which can be noninvasively determined using sEMG (see Chapter 5), mainly reflects the diameter of muscle fibers (since diameter determines the cytoplasmic resistance of a fiber) and the sarcolemmal properties of muscle fibers. In fact, MFCV is related to the concentration of essential ions on both sides of the membrane and the properties of numerous ion channels, mainly the sodium and potassium channels and the Na^+–K^+ pump. However, MFCV is also influenced by muscle temperature, contraction type and duration [31], and motor unit firing rate (MFCV increases by ~1% for each increase of 1 pps in discharge rate) due to the velocity recovery function of muscle fibers [32].

Several studies have tried to establish a relation between MFCV (or its rate of change during a fatiguing contraction) and fiber type distribution. This association is indirect and should follow the relation between fiber diameter and fiber type. In other words, as MFCV and fiber type are related to the same variable (i.e., fiber diameter: type II fibers usually present greater values of MFCV than type I fibers due to their greater diameter), they should correlate to each other. So far, this indirect relation has been postulated on the basis of few animal and human studies. Several studies, for example, have tried to link MFCV (or its rate of change during a fatiguing contraction) with fiber type composition on the basis of the comparison of MFCV between different groups of subjects that were presumed to be distinguishable with respect to their fiber type composition, such as sprinters and endurance athletes [70] or young and elderly subjects [55]. Contrary to the above approach, Farina et al. [33] investigated the relation between MFCV and myosin heavy chain (MHC) isoforms (contractile proteins that influence the velocity properties of muscle contraction) and found a significant positive correlation between MFCV and the proportion of MHC-I (mainly composing type I muscle fiber) in vastus lateralis samples from healthy subjects involved in submaximal cycling (when mostly type I fibers were possibly active) and concluded with concomitant increases in MFCV and the proportion and diameter of type I fibers. In a recent paper, however, Minetto et al. [57] found no correlation either between vastus lateralis MFCV obtained during isometric quadriceps contractions at 70% MVC and MHC-II relative content or between vastus lateralis MFCV obtained at 10–20–40% MVC (when mostly type I fibers were presumably active) and MHC-I relative content.

These results emphasize the problem when attempting to indirectly estimate the distribution (and the training-induced changes) of muscle fiber types from MFCV or other sEMG variables as well as to physiological and anatomical factors. In fact, a number of factors may mask the association between fiber type and MFCV: (1) the overlap of MFCV values for the different fiber types and (2) the within-muscle variation of muscle fiber properties: A nonrandom arrangement of fiber types has been observed in the quadriceps, with the superficial portions having more type II fibers than the deeper portions [50]. Nevertheless, fibers in the superficial muscle parts (those mainly contributing to sEMG) are generally smaller (and, therefore, conduct more slowly) than fibers in the deeper parts, although the distribution may depend on the differences in the functional adaptation of various muscle portions to physical demand.

In conclusion, MFCV is a basic physiological parameter that mainly reflects the diameter and the sarcolemmal properties of muscle fibers. It increases with contraction level and decreases with fatigue, whereas the proportion of different fiber types cannot be estimated from this electrophysiological variable (see also Chapter 10). Values of MFCV depend on the conditions in which conduction velocity is measured due to the velocity recovery function of muscle fibers and to modulation effects of muscle temperature, and contraction type and duration.

12.3   EVOKED POTENTIAL

Another useful approach to investigate the neuromuscular aspects of human performance is to evoke potentials by electrical stimulation of motor and sensory nerve
fibers at the periphery or from magnetic stimulation of motor cortical areas and spinal
cord at the cervicomedullary junction (Fig. 12.1). Individually, these evoked potentials can provide valuable information on the modulation of the neural drive to the
muscle or on changes at the neuromuscular junction and beyond. When combined,
these techniques can be useful to infer the loci of changes within the neural circuitry
while performing a motor task, or in response to chronic adaptations, such as those
induced by physical training or aging. As mentioned above, evoked potentials can
result from different sites and types of stimuli, and it is therefore essential to master the
basic technical background for each of them to fully understand their meaning and
limitations. The following paragraphs will provide this basic knowledge, distinguishing electrical and magnetic stimulation methods. The potentials evoked by mechanical stimuli will not be addressed in this chapter. Nonetheless, the readers can refer to
comprehensive monography on this topic [68].

12.3.1   Electrical Stimulation

The Compound Muscle Action Potential—M Wave.   A compound muscle action
potential is the electrophysiological response recorded from sEMG when stimulating
motor axons with a current intensity sufficient to activate some or all the motor units of a
muscle. This response is commonly called motor (M) wave in neurophysiology (see
Chapter 11). The M wave is the summation of action potentials of the synchronously
activated muscle fibers in a muscle that results from the direct depolarization of the
α-motor axons. As all evoked potential, its shape reflects the extracellular recording of
muscle fiber action potentials that are filtered by physiological tissues (skin, fat,
aponeuroses, etc.) located between the muscles fibers and the surface electrodes.

Electrode Configuration The M wave is generally evoked by stimulating peripheral
nerve with short (~0.2-ms duration, close to the axon's chronaxy) rectangular
monophasic electrical pulse with the two electrodes placed close to or far from
each other. The cathode can be placed over the nerve trunk and the anode, usually
much larger, is often fastened on the opposite side of the limb (see Chapter 10). The
M wave in the soleus muscle, for example, is evoked by placing the cathode in the
popliteal fossa and the anode above the patella [75]. This configuration permits us to
stimulate a deep nerve trunk and ensure quite consistent recording even during
movements. However, one pitfall is the risk to stimulate other nerves in the vicinity of
the targeted nerve. In contrast, placing both electrodes close to each other limits such
risk as both cathode and anode are placed over the nerve trunk at short interelectrode
distance (2–3 cm). For example, contraction of some intrinsic muscles of the hand can
be induced by stimulation of the median nerve at the wrist level by placing both
electrodes over the nerve trunk, in the medial part of the forearm, with the cathode
distally located relative to the anode [23].

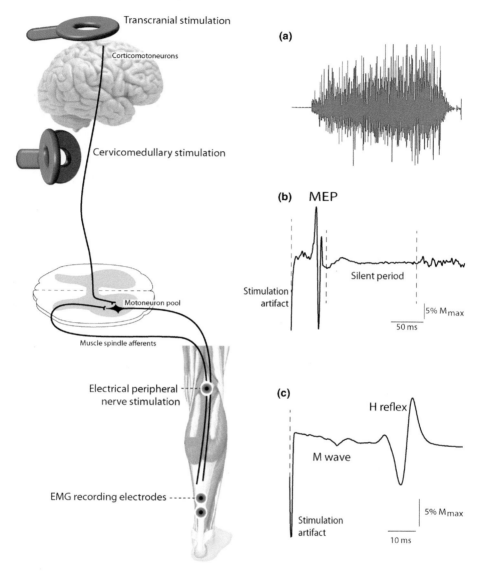

FIGURE 12.1   Left panel: Schematic representation of the corticospinal pathway and muscle spindle afferent pathway, along with the different locations of stimulation to assess changes in the neural adjustments occurring during various motor tasks or with aging. (**a**) sEMG recorded on the soleus during voluntary isometric contraction of the ankle plantar flexor muscles. (**b**) Motor evoked potential (MEP) induced by transcranial magnetic stimulation (TMS; averaged over five sweeps) in the soleus during voluntary isometric contraction of the ankle plantar flexor muscles. The right dashed line determines the end of silent period. (**c**) Hoffmann (H) reflex (averaged over 10 responses), preceded by a small M wave, evoked in the soleus muscle by an electrical stimulation applied to the tibial nerve.

The distal position of the cathode is assumed to avoid anodal block that results from the hyperpolarization of the nerve cell membrane at the anode, although such phenomenon, observed in vitro and predicted by models, does not seem to occur during neurophysiological studies.

The M wave can also be evoked by stimulating the muscle at the motor point, defined as the location where the branches of the motor axons enter the muscle belly (also called motor entry point). The motor point is usually described as the site of stimulation producing the most effective muscle contraction in response to a given small stimulus [23]. Nonetheless, the smaller distance between stimulating and recording electrodes (both types being placed over the muscle) can induce an overlapping between the stimulus artifact and the onset, compared wth nerve stimulation, of the M wave (see Chapter 11), reducing the ability to measure accurately some variables (duration, area) from the latter. Moreover, stimulating over the motor point does not allow to evoke an H reflex (see below), and therefore this electrode configuration has limited applications in neurophysiology, although it is relevant for neuromuscular electrical stimulation in rehabilitation and sport training (see Chapter 19).

The positioning of the stimulating and recording electrodes is crucial to correctly use sEMG in neurophysiology, and consistent placement of both types of electrodes is required to compare subjects or to investigate the long-term effects of an intervention in an adequate way.

Electrical Nerve Stimulation and Motor Unit Recruitment Order In contrast to the physiological recruitment of motor units determined by mechanisms related to the motor neuron size, smaller motor neurons being recruited first (size principle) [37], axonal activation by electrical stimulation is thought to reverse the recruitment order. This peculiar feature of peripheral nerve stimulation results from the inverse relation between axonal diameter and axial resistance, increasing the probability of a current flow in large axons [77]. Accordingly, the use of M wave to infer specific changes between low- and high-threshold motor units should be done with caution (see Chapters 10 and 11). However, the reversal in the recruitment order in response to peripheral motor axon stimulation has been debated [34], and it has been found that motor unit recruitment during percutaneous stimulation also depends on the morphological organization of the motor axons within the nerve trunk (or in the muscle, for motor point stimulation).

The Maximal Amplitude of the M Wave—M_{max} The main variable measured from the M wave is its amplitude and, more specifically, its maximal amplitude (M_{max}). The M_{max} represents the maximal sEMG response to an electrical stimulus sufficient to activate all the motor axons in the nerve trunk. In order to ensure that the current intensity is sufficient to depolarize all the motor axons that innervate a muscle, the current intensity should be progressively increased until the amplitude of the M wave reaches a plateau. From this point, it is recommended to set the current intensity at least 20% above the M_{max} intensity to preclude the influence of small changes in stimulation conditions or decreased axonal excitability during sustained activation. It is to note that the M_{max} depends also on the number and size of the muscle fibers that are being recorded.

The M_{max} is often used as a variable to normalize other evoked potentials or background sEMG because it is not modulated by neural inputs converging onto the motor neuron pool and represents the sEMG response to the full synchronous activation of the muscle motor units. Such normalization procedure allows, therefore, an estimate of the proportion of the motor neuron pool activated by an evoked potential and provides information on the integrity of the neuromuscular junction and muscle fiber action potential propagation that influence any kind of evoked potentials. Care must be taken, however, to maintain muscle temperature constant because M_{max} may change during long-lasting experiments that involve long period of rest and possible temperature changes. A decrease in conduction velocity may decrease the amplitude of the M_{max} but increase its duration. Indeed, decreased conduction velocity can reduce synchronization of individual action potentials and increase their duration. In this context, the M_{max} area is sometimes a good index to avoid misinterpretation of a change in M_{max} amplitude [14].

The Hoffmann Reflex. The Hoffmann (H) reflex was originally described in human by Paul Hoffmann in 1910. It consists of a sEMG response evoked by a submaximal electrical stimulation applied to the homonymous peripheral nerve trunk, and has been coined "Hoffmann reflex" by Magladery and McDougal [52], in 1950, in reference to the pioneering work of Hoffmann. In the first half of the twentieth century, experiments on animals and humans have provided priceless information on the pathway conveying the H reflex, as summarized by Paillard [66]. Indeed, the similarity in the characteristics of the H reflex and the stretch reflex (reflex induced by rapid muscle stretch such as a tendon tap), the involvement of large sensory nerves and the short central delay (<1 ms) emphasized a monosynaptic reflex pathway originating from muscle spindle afferents (Ia afferents). The H reflex has been considered for a long time as a tool to investigate the excitability of the motor neuron pool. However, the existence of presynaptic mechanisms that can depress the amount of neurotransmitter release by Ia afferents forced to reconsider the interpretation of H-reflex data with more caution [68]. Indeed, a modulation in the level of Ia presynaptic inhibition influences the synaptic efficacy of Ia afferents to discharge in the absence of change in the excitability of motor neurons. Moreover, the poor time resolution of the technique and the possible involvement of disynaptic spinal pathway influencing the size of the H reflex provided further arguments to consider the H reflex as a tool to investigate changes in transmission in spinal neural pathways rather than to study motor neuron excitability per se. As already highlighted almost 60 years ago by Paillard [66], the use of H reflex requires to take into account numerous factors to conduct appropriate experiments and avoid misinterpretation of the data. The following sections will describe the basic methodology to record H reflex and the relevant factors to consider when assessing its modulations. Further considerations on recording H reflex can be found in Chapter 11 and elsewhere [68].

The H-Reflex and M-Wave Recruitment Curve Percutaneous stimulation of a nerve trunk implies that both afferent and motor axons are activated. Accounting for the greater excitability of large axons (see above), Ia afferents fibers are recruited at a lower current intensity compared with α-motor axons. Therefore, increasing progressively the current intensity applied to a peripheral nerve depolarizes first the large afferent axons and then the α-motor axons (Fig. 12.2). The increased H-reflex amplitude with the increase in stimulus intensity results from the growing group I afferent volley that recruits progressively more motor neurons according to the Henneman's size principle, up to a maximum commonly referred to as H_{max}. However, increasing the

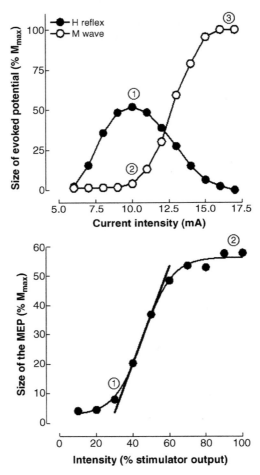

FIGURE 12.2   Top panel: Recruitment curve for H reflex and M wave (averaged over five responses) in the soleus muscle of an individual during upright standing. The measured parameters are: the maximal amplitude of the H reflex (H_{max}; 1), the motor threshold (2), and the maximal compound muscle action potential (3). **Bottom panel:** Input–output curve of the motor evoked potential (MEP; averaged over five responses) to transcranial magnetic stimulation in the soleus muscle of an individual during upright standing. The measured parameters are: the motor threshold (1) and the MEP_{max} (2). The straight line represents the MEP_{slope} calculated by differentiating the central portion of the input–output relation. The data points are fitted with a Boltzmann function.

intensity of the electrical stimulus also depolarizes progressively α-motor axons. The intensity at which this recruitment begins is named "motor threshold" and is characterized by a M wave preceding the H-reflex response on the sEMG signal. Further increase in current intensity evokes greater direct motor volley (increased M-wave amplitude) and progressive decrease in H-reflex size. This inverse relation between the H reflex and the M wave after the H_{max} is mainly due to the increase in motor volley (from stimulation of the motor axons) propagating both orthodromically (toward muscle producing the M wave) and antidromically (toward the soma of the motor neuron). The antidromic motor volley collides with the H-reflex volley and progressively eliminates the latter, meaning that when M_{max} is reached, no H reflex remains visible on the signal.

An example of H–M recruitment curves is illustrated in Fig. 12.2, from which the ascending and descending limbs of the H-reflex recruitment curve can be seen, as well as the motor threshold and the sigmoid-like shape of the recruitment curve of the M wave. It is important to mention that only H reflexes in the ascending limb of the recruitment curve are sensitive to changes in both excitatory and inhibitory inputs received by the motor neuron pool.

Stimulation Parameters The ability to record consistent H reflex is related to the characteristics of the electrical stimulus, especially the duration, the electrode configuration and the frequency of stimulation.

As highlighted by Paillard in 1955 [66], a low, long-duration, rectangular pulse (1 ms) is more suitable than a shorter one to evoke a H reflex. This is because of the different strength–duration curves of motor and sensitive nerve fibers. The main features of the electrode configuration discussed for evoking an M wave are similar for the H reflex, distinguishing between electrodes being far from each other, suitable for femoral, tibial and fibular nerve, and electrodes close to each other, more suitable for radial and median nerve at the arm level.

Because the H-reflex amplitude can be variable in some conditions, it is required to collect and average several responses, raising the question of the number of responses needed and the frequency at which H reflexes can be evoked. When assessing the intra-session reliability of H-reflex measurements in supine and standing position in the soleus muscle, intraclass correlation coefficients (ICC) above 0.8 calculated for 4 to 10 responses have been reported [62]. Nonetheless, depending on the task being performed and to reduce the risk of statistical errors and misinterpretation of the data, a minimum of 10 responses is likely suitable, especially for experiments involving elderly adults [62]. One potential concern with evoking successive H reflexes is related to "homosynaptic postactivation depression", a reduction in neurotransmitter release from Ia terminals (presynaptic level) due to repeated activation [79], that may reduce progressively the amplitude of the H reflexes within the sequence. In relaxed muscle, this presynaptic mechanism can depress the H-reflex amplitude for intervals below 10 s. However, the influence of this mechanism is minimized, during voluntary contractions, for stimulation rates up to 4 Hz [79]. Therefore, in order to combine the need to collect several responses but to avoid long-lasting recording periods, stimulations should be mostly evoked using at least 10-s interval when assessed at rest, but a shorter interval can be used when H reflexes are evoked during voluntary contractions.

Factors Influencing the H Reflex Several factors need to be considered when recording H reflexes. First, a H reflex cannot be recorded in all muscles at rest, and most motor neurons need to be facilitated by a weak voluntary contraction that places their resting membrane potential closer to the exitation threshold. In most individuals, H reflexes can be evoked at rest in muscles such as the soleus, quadriceps, hamstring, flexor carpi radialis and abductor pollicis brevis. The amplitude of the H reflex increases with the muscle torque developed during a voluntary contraction up to ~60% MVC [6,45] due to the "automatic gain compensation" mechanism. Therefore, when comparing H reflexes between subjects or conditions, it is important to ensure similar background sEMG activity of the target muscle, or to take into account differences in the neural drive received by the motor neuron pool when interpreting modulation in H-reflex amplitude. Furthermore, some authors have suggested that H reflexes must be recorded during voluntary contractions when trying to understand mechanisms involved in motor control. Another relevant factor when investigating neural modulation through H-reflex recordings is the posture adopted and the type of movement performed by the subjects. Despite the limitations and factors to consider when recording H reflexes, it remains a main tool to investigate the neural control of movement in humans. In combination with other methods, H-reflex recording is a useful probe to obtain detailed understanding on the neural adjustments underlying motor performance as well as those that accompanied physical training, disease or ageing process.

12.3.2 Magnetic Stimulation

Principle. Transcranial magnetic stimulation (TMS) has been introduced by Barker et al. in 1985 [1]. They provided evidence that the corticospinal tract can be activated by a brief, fast changing magnetic field pulse applied over the scalp. The principle of TMS is based on short-duration (<1 ms), high-intensity electrical current conveyed through a coil that creates a short-lasting and fast-varying magnetic field whose derivative induces an electrical current in the conductive and excitable tissues. The induced current is directed perpendicular to the coil surface, and its intensity is proportional to the original current intensity of the stimulator [73]. When applied to the scalp, the induced current depolarizes cortical neurons and initiates action potentials in neurons or changes the level of neural excitability, although the exact mechanisms that drive the active stimulation are still under debate [86]. When applied to the motor cortex, TMS seems to preferentially activate pyramidal neurons by a transsynaptic mechanism. In response to TMS applied over the motor cortex, a mechanical twitch can be obtained along with a motor evoked potential (MEP), recorded by sEMG, in the targeted muscle of the contralateral limb [73]. Because TMS may not be focused on only one muscle, the muscle twitch can be difficult to interpret. In contrast, MEP provides relevant information on the excitability of the corticospinal tract during various motor tasks [67].

Magnetic stimulation of the spinal cord can also be induced at the cervicomedullary junction (mastoids level) to evoke the cervicomedullary MEP (CMEP) that reflects the responsiveness of spinal motor neurons to the stimulation [83]. Therefore, combining MEP and CMEP can provide more details on the exact loci of changes in the corticospinal pathway (supraspinal vs. spinal) [81]. As for the MEP and H reflex, magnetic stimulation of the cervicomedullary junction is more successful to evoke CMEP during voluntary contraction that makes the spinal motor neurons more excitable. The cervicomedullary junction can also be activated electrically, but the pain induced by such stimulation overcomes the disadvantage of magnetic stimulation due to unmatched shape of the coil with some subject morphology.

Type of Coil.   The distribution, depth and focus point of the magnetic field are highly dependent on the size and shape of the stimulating coil. The circular coil induces deep stimulation but have a low spatial resolution. The figure-of-eight-shaped coils have a more focal field compared with circular coils. A figure-of-eight coil with the two components at an angle (double-cone coil) increases the power at the intersection of the two magnetic field and allows deep stimulation with a good spatial resolution. The double-cone coil is well designed to stimulate cortical areas of thigh and leg muscles.

Type of Stimulation.   Several types of stimulation can be used with TMS ranging from single to high-frequency repetitive stimulations. Repetitive TMS (rTMS) can be useful to change the physiological activity of specific cortical areas [53], whereas single- or paired-pulse stimulations are more suitable to investigate change in excitability of cortical neurons. Accordingly, the next section will focus on these last two types of stimulation.

As already mentioned, single-pulse TMS creates a short-lasting magnetic field that induces a brief current pulse depolarizing cortical neurons. Single TMS pulse is commonly used to investigate alterations in the corticospinal tract, changes in cortical excitability, to assess the effects of practice and physical training, or to study the ageing process [68]. It is also important to keep in mind that MEP recordings do not allow to distinguish the sites of neural adjustment in the corticospinal tract (supraspinal or spinal). Nonetheless, when combined with other methods (H reflex or CMEP), the loci responsible for the changes in corticospinal pathway excitability can be addressed more accurately. Furthermore, spatial facilitation methods have been developed to better understand the neural adjustments underlying motor performance (more information can be found in Petersen et al. [67]).

The paired-TMS pulses method consists of a test stimulation preceded by a conditioning stimulation, separated by a determined interval, and allows the investigation of changes in intracortical interneurons with facilitatory or inhibitory effects [24]. Short interpulse intervals (<5 ms) involve inhibitory interneurons whereas longer interpulse intervals produced intracortical facilitation, although such effects depends also on the intensity of the two stimulations. It is to note that these pathways have been mainly studied for the motor cortex.

MEP Variables. Several complementary variables can be measured from the MEP. The MEP threshold corresponds classically to the TMS intensity at which three out of four evoked responses are discerned above background sEMG levels [73]. This threshold provides information on the state of excitability of the targeted cortical area. The amplitude of the MEP, measured as peak to peak, is used to assess changes in the excitability of the corticospinal pathway. For a given intensity of stimulation, the MEP amplitude increases with the force developed during a voluntary contraction up to the maximal force in ankle plantar flexor muscles [65], whereas it decreases for contraction intensities greater than 50% MVC force for elbow flexors and first dorsal interosseous. Such decrease for high contraction intensity mainly results from a reduced responsiveness of spinal motor neurons due to the increase in their firing rate with contraction force. As for the M wave, the input–output curve can be generated from MEP responses recorded during increased TMS intensity until the MEP amplitude reaches a plateau or up to the maximal capacity of the stimulator. Because of the inherent fluctuation of the MEP size, the input/output curve can be analyzed more carefully by fitting the experimental data with a sigmoidal Boltzmann function. From the above procedure, the MEP_{slope} parameter can be calculated by differentiating the input–output equation to inform on a change in the input–output gain (greater or lower excitability of the corticospinal tract). The duration of the period of sEMG silence that follows each MEP, known as the silent period (SP) [64], is measured from onset of the MEP to the return of background sEMG, although it can be easier to define the onset of SP as the TMS artifact. As the size of the MEP also influence the SP duration, the ratio between SP and MEP amplitude (SP/MEP ratio) is more accurate.

12.4 APPLICATIONS

In this section we will illustrate how the methods described above can provide relevant information on the neural adjustments occurring in different physiological conditions. The topics have been chosen to relate both acute adjustments (cramps, fatigue) and chronic adaptations (training, aging) that involve different mechanisms.

12.4.1 Muscle Cramps

Fasciculations and cramps represent involuntary muscle phenomena associated with repetitive discharges of motor unit action potentials that have been respectively referred to as *fasciculation potentials* and *cramp discharge.*

The clinical manifestation of a fasciculation is the random and spontaneous twitching of a group of muscle fibers belonging to a single motor unit. The twitch may produce movement of the overlying skin (if in limb or trunk muscles) or mucous membrane (if in the tongue). When the number of muscle fibers ongoing fasciculation is large enough, an associated joint movement may be observed. The clinical manifestation of a cramp is a sudden, involuntary, painful contraction of a muscle or part of it, self-extinguishing within seconds to minutes, often accompanied by a palpable knotting of the muscle.

Although fasciculations and cramps may occur in patients with neurological disorders, they also often occur in healthy subjects with no history of neuromuscular disorders (such as during sleep and strenuous physical exercise) and have been respectively defined as "benign fasciculation potentials" and "benign or idiopathic cramps" [56]. Even though it is generally accepted that fasciculations and cramps have a neurogenic nature, their origin has been long discussed [56]. Surface EMG signals detected during fasciculations and cramps have been analyzed in many studies to unravel their neural aspects/origin, and recent electrophysiological findings providing new insights into the physiopathology of fasciculations and cramps are presented in the following paragraphs.

The physiological mechanism of fasciculations and their site(s) of origin have been controversial topics. Fasciculations were first thought to originate peripherally, proximally to the terminal branches of motor axons, with 80% of fasciculation potentials arising in the distal axonal branches of the spinal motor neuron. However, fasciculation potentials could sometimes be recorded synchronously in different muscles innervated by different motor nerves but belonging to the same myotome and suggested that these fasciculation potentials had a central, perhaps cortical, origin.

In recent years, high-density surface EMG (see Chapter 5) studies were performed to investigate the spatiotemporal aspects of fasciculation potentials in neurological patients and healthy subjects [47,59]. High-density sEMG enables to identify fasciculation potentials, to recognize their origin, progression, and discharge rate, and to construct interspike interval histograms. In a recent paper, Kleine et al. [47] showed that fasciculation potentials recorded at rest in the medial head of the gastrocnemius muscle were followed by F-waves in 71% of healthy subjects and 57% of amyotrophic lateral sclerosis patients. Nonetheless, interspike intervals measured in this study indicated that some fasciculation potentials originated from more proximal generators in the nerve trunk. These observations reinforced the need to define differences, if any, between fasciculation potentials arising from different sites of the nerve trunk.

Cramp pathophysiology has been poorly understood partly because of the unpredictable occurrence of cramps that makes them relatively difficult to be studied in classic experimental settings. One hypothesis is that cramps result from spontaneous discharges of the motor nerves or abnormal excitation of the terminal branches of motor axons (peripheral or axonal origin hypothesis) [71]. In fact, it has been shown that involuntary contractions could be induced by electrical stimulation distal to a peripheral nerve block and that voluntarily elicited cramps may spread over a large muscle area during their development, as a possible result of activation of adjacent motor nerve terminals by ephaptic transmission.

Another hypothesis is that cramps result from the hyperexcitability of motor neurons (central or spinal origin hypothesis). In fact, action potentials of similar shape repeat consistently over time during cramps, so that they are presumably motor unit action potentials [58]. Moreover, motor unit discharge rates decrease over time during a cramp [58], which is consistent with the decrease in motor unit discharge rate during fatiguing voluntary contractions. Recent findings have proved unambiguously the relevance of central mechanisms in the generation and development of muscle

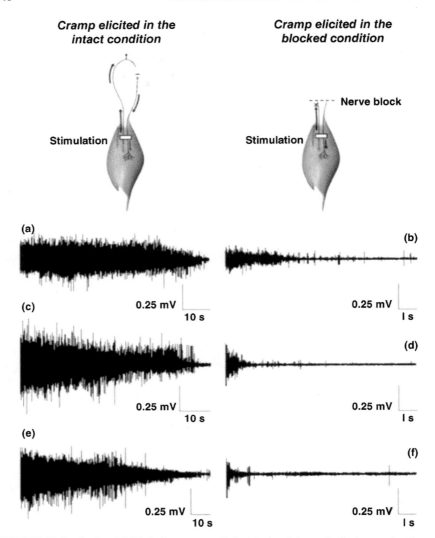

FIGURE 12.3   Surface EMG during cramps elicited in the abductor hallucis muscle. Cramps were elicited in absence (intact condition: (**a, c, e**) and presence (blocked condition: (**b, d, f**)) of an anaesthetic block of the posterior tibial nerve. Note different horizontal scales. Each row (**a–b; c–d; e–f**)) represents one subject. Adapted from Minetto et al. [60], with permission.

cramps [60]. In fact, involuntary contractions elicited by electrical stimulation distal to a peripheral nerve block (blocked condition) were compared with cramps elicited with intact spinal loop (intact condition). Figure 12.3 shows examples of sEMG activity during such contractions. Duration and intensity of the contractions elicited in the intact condition were substantially greater than the duration and intensity of those elicited in the blocked condition. Moreover, intramuscular EMG signal analysis showed that the motor unit activity detected during the intact condition presented the typical characteristics of the motor neuron discharges, including range of discharge rates and decline in discharge over time, while the motor unit activity in the blocked condition presented the characteristics of the spontaneous discharge of motor nerves, including short interval of activity, high discharge rates, and high discharge variability [60].

Despite recent progress, several unresolved issues in the physiopathology of muscle cramps still remain and require further investigation.

12.4.2 Neuromuscular Fatigue

Neuromuscular fatigue has been a research topic for more than one century and the subject of numerous reviews [28]. This section, therefore, will not consist of a review of muscle fatigue but rather on how sEMG and evoked potentials have helped to highlight some of the most relevant mechanisms underlying fatigue. The myoelectric manifestations of muscle fatigue are discussed in Chapter 10. As emphasized elsewhere [28], fatigue manifestations and mechanisms are highly dependent on the characteristics of the task being performed. In this context, the intensity of the effort (often expressed as percentage of the maximal force or joint torque) is the most influential parameters and will thereby serve to base our analysis.

The sEMG (ARV or RMS amplitude) of the main agonist muscles increases progressively during sustained submaximal isometric contractions [26] while it decreases for sustained maximal or near maximal contractions [13]. The extent of the sEMG increases during submaximal contractions, which mainly reflects the recruitment of additional motor units [19], differs depending on the target force, with a greater increase for low-force contractions compared with high-force contraction [16]. Based on previous work that indicated an upper limit of motor unit recruitment of about 80–90% MVC torque for most limb muscles, the lower increase in sEMG when exerting force above 50% of maximal [26] is consistent with a reduced available range of motor unit recruitment. Similar observations were made during submaximal dynamic contractions [10]. However, at task failure, it is commonly observed that sEMG does not reach the maximal values that are obtained during an MVC of the same muscle group for sustained isometric contractions and repeated dynamic contractions. These results indicate limiting factors in the ability to increase the neural drive during submaximal contractions [28] that may differ depending on contraction intensities.

Before considering physiological factors, it is important to point out a methodological limitation in sEMG to describe changes in the neural drive. The sEMG can be biased during muscle contractions by the algebraic addition (and partial cancellation) of the motor unit action potentials. This modulation of sEMG amplitude that is not related to change in physiological mechanisms can be substantial and reach a value greater than 40% for submaximal contractions. Furthermore, when sEMG and a single motor unit were recorded simultaneously, sEMG appears insensitive to modest changes in motor unit activity (discharge rate) during sustained submaximal contractions [6,61]. Therefore, it is recommended to combine sEMG amplitude with other variables and methods to investigate physiological mechanisms of muscle fatigue.

The mechanisms responsible for neuromuscular fatigue have often been subdivided as central or peripheral, depending on their location. Central fatigue refers to alterations located in the central nervous system (CNS) comprising sites involved in the initiation of motor tasks, whereas peripheral alterations correspond to those occurring from the motor axon to cross-bridge interaction within the muscle fibers.

As mentioned previously, the M_{max} represents the summation of individual muscle action potentials in response to the complete activation of the motor axons in the nerve trunk. Accordingly, a change in M_{max} may indicate alteration in the transmission at the neuromuscular junction and conduction along the nerve and muscle fibers. Although there is a safety margin, such changes can lead to decrease muscle activation and thereby to a reduction in the force produced by the muscle (peripheral fatigue). These alterations, however, may vary with contraction intensity. Moreover, M_{max} duration may provide information on changes in sarcolemmal (muscle membrane) conduction velocity and action potential propagation [41]. In this vein, repeated shortening maximal contractions induced a greater increase in M_{max} duration, compared with lengthening contractions [7]. These results suggest that shortening contractions involved a higher level of muscular activation that could have altered more rapidly processes involved in muscle contraction and consequently its tension output. Moreover, the decline in mean power frequency of the sEMG observed during submaximal contractions performed with plantar flexor muscles also suggests a decrease in conduction velocity.

To investigate neural adjustments during fatiguing contractions (central fatigue), a change in the H-reflex size has often been used. For example, Duchateau et al. [26] reported a decrease in H-reflex amplitude during sustained contractions performed at 25% and 50% MVC with the abductor pollicis brevis associated with an increase in aEMG. These results indicate recruitment of additional motor units despite a decrease in synaptic efficacy of Ia afferents, which could involve presynaptic inhibitory interneurons. Presynaptic inhibition of Ia afferents can be modulated either by the corticospinal tract or by feedback transmitted by group III and IV afferents. The increase in mean arterial pressure reported during sustained isometric submaximal contractions is likely accompanied by an increase in the feedback transmitted by group III and IV afferents [42] that could increase presynaptic inhibition of the Ia afferents during fatiguing contractions. This is also consistent with the rather slow time course of the decrease in H-reflex amplitude and the absence of recovery during ischemic conditions [26]. Finally, by combining H-reflex and TMS methods, Klass et al. [46] provided additional arguments on a possible increase in Ia presynaptic inhibition during sustained submaximal contractions. Indeed, these authors observed a concomitant decrease in the amplitude of the H reflex with an increase of the MEP amplitude recorded in the biceps brachii, partly ruling out a decrease in spinal motor neuron excitability (Fig. 12.4) during a sustained contraction of the elbow flexor muscles. Nonetheless, the H-reflex amplitude during sustained fatiguing contractions has also been shown to increase [51] or remain unchanged [4]. These changes in H-reflex amplitude are likely the consequence of adjustments due to varying experimental conditions (muscle studied, amplitude of the initial H reflex, and intensity of the contraction, resting muscle versus contraction).

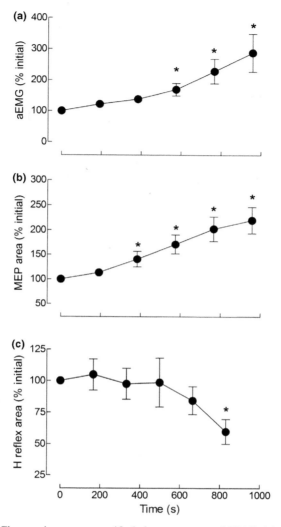

FIGURE 12.4 Changes in average rectified electromyogram (aEMG) (**a**), area of the TMS-induced MEP (**b**), and area of the H reflex (**c**) for the biceps brachii during isometric contraction in elbow flexion direction sustained until failure at 20% of the maximal force. Data are expressed as percentage of initial, and each data point represents mean ± SEM of 11 (**a, b**) and six subjects (**c**). *Significant difference from the initial (*P* < 0.05).

During submaximal sustained contractions, changes at the motor cortical level can be inferred from MEP and SP parameters obtained in response to TMS. For both elbow flexor and ankle plantar flexor muscles, submaximal sustained contractions induced an increase in MEP amplitude, indicating an enhanced corticospinal excitability, likely reflecting the increase in voluntary drive sent to the muscle [46].

Interestingly, these changes appear to be focused to the muscle involved in the fatiguing contraction, suggesting that neural adjustments can be specific to the fatigued muscle. Moreover, the SP induced by TMS increased during the sustained contraction [46], suggesting an increased inhibition of cortical output. In contrast, paired TMS pulses indicated that short-interval cortical inhibition decreased with repeated isometric MVC [11]. These apparent contradictory results are likely related to the fact that short-interval (paired TMS pulses) and long-interval cortical inhibition (SP) are mediated by different neural networks, underlining the complexity of the neural adjustments during sustained contractions [54].

Collectively, the studies discussed in this section emphasize that sEMG activity can provide relevant information (increased neural drive) when confirmed with the recordings of evoked potential by electrical and magnetic stimulations. Therefore, methods based on sEMG recordings (M_{max}, H reflex, MEP, and CMEP) are suitable to investigate noninvasively the complexity of the neural adjustments that occur during sustained fatiguing contractions. Other methods based on TMS and electrical peripheral nerve stimulation can be used to investigate neuromuscular fatigue, but these methods are beyond the scope of this chapter. The reader can find detailed description of these methods elsewhere [68].

12.4.3 Strength Training

Beside investigating acute changes in response to physical activity (fatigue) or transient altered functioning of the neuromuscular system (cramps), sEMG and the associated techniques described in this chapter can be used to explore the neural adaptations induced by physical training (see also Chapter 19). It is well documented that strength training involves both neural and muscular factors that have a different time course throughout the training process [25]. Indeed, at the early part of a training program, strength increases before any detectable changes in muscle cross-sectional area (hypertrophy), suggesting that the first gains in strength rely on adaptations within the nervous system, the so-called "neural factors". The aim of this section is to highlight the relevance of sEMG and associated methods to investigate plasticity in the nervous system following chronic exercise.

Changes in voluntary muscle activation in response to a training program can be assessed by recording sEMG before and after the training period (see Chapter 19). The classic approach consists of recording changes in sEMG activity during an MVC. Accordingly, the amplitude of sEMG has been shown to increase after a strength training program [35]. As sEMG is sensitive to changes in the properties of the sarcolemma and the size of the muscle fibers, its value is sometimes normalized to the M_{max} in order to assess more adequately the training effects on muscle activation. In addition to an increase in voluntary activation of the agonist muscle, the sEMG approach also points towards other neural adaptations involving antagonist muscles. As the force developed by the antagonist muscle counteracts that produced by agonist muscle, activation of the antagonist muscle decreases the net measured torque output, although an adequate level of antagonist activation is assumed to contribute to joint stability [78]. Following strength training, the agonist/antagonist ratio, measured with sEMG, decreases following strength training of the knee extensor muscles [35] contributing to increase the net torque (see also Chapter 19).

Furthermore, training programs using fast (ballistic) contractions have been shown to induce briefer muscle activation and greater sEMG at the onset of the contraction [84]. These results indicate that voluntary sEMG can be considered as a relevant tool to investigate changes in muscle activation, and to provide information on training-related adaptation in the neural strategy of co-activation. Here also, voluntary sEMG cannot provide further information on the location of the neural adaptations occurring during training. For this purpose, additional methods, based on recordings of elicited sEMG, must be used.

As described previously, the H-reflex method can be used to investigate changes located at the spinal level, involving either presynaptic (changes in the synaptic efficacy of Ia afferents) or postsynaptic mechanisms (changes in motor neuron excitability). Several studies have shown that strength training increases the amplitude of the H reflex when tested during muscle contraction, changes accompanied by an increase in muscle force. The concomitant increase in H-reflex amplitude and strength suggests that spinal mechanisms may contribute to changes in strength, due to an increase in motoneuronal excitability or a reduced presynaptic inhibition. Interestingly, the H reflex did not change when investigated in the untrained contralateral limb which also showed an increase in force (cross-education)[48], indicating that cross-education effect of strength training should rely more on supraspinal than spinal mechanisms. Furthermore, the H-reflex shows no significant changes in amplitude after training involving contractions exercises designed to improve sensorimotor integration for balance control, whereas it increases after strength training, suggesting that spinal adaptations can be very specific to the training modalities [80]. Similarly, a strength training program of eccentric contractions of the ankle plantar flexor muscles increased the H-reflex amplitude of the soleus only during eccentric MVC [27]. In contrast, the increased H reflex observed in the bi-articular medial gastrocnemius did not depend on the contraction type (isometric, concentric, and eccentric contractions), suggesting that the specificity of training effects can vary with the muscle involved, regarding for example the joint position [27]. These results underscore that the H-reflex method can provide relevant information on the neural adjustments induced by strength training, particularly the adaptations at the spinal level, and the specificity of those adaptations.

Results on supraspinal adaptations assessed by TMS are more controversial. For example, Carroll et al. [20] found that an increase in maximal isometric force of a hand muscle after strength training was not accompanied by changes in MEP amplitude when measured at rest but MEP was reduced when elicited during voluntary contractions [20]. Similarly, MEP evoked by transcranial electrical stimulation (TES), a method assumed to activate the corticospinal fibers at the axon hillock, did not change after training when assessed at rest but decreased during a similar absolute torque level. These results suggest that strength training involves changes in the functional properties of the spinal cord circuitry, but not in the output from the motor cortex [28,80].

12.4.4 Aging

With advancing age, motor control declines, and old adults execute gross motor tasks more slowly and perform fine motor skills with less accuracy. The decline in motor control is primarily ascribed to reductions in muscle strength via the sarcopenia process (age-related loss of muscle mass) in addition to changes in cognitive, endocrine, metabolic, and cardiovascular functions. Of course, changes occurring in the central and peripheral nervous systems also contribute to the age-related alteration of the sensorimotor abilities.

The aging process is associated with numerous changes in the muscular and nervous systems, some of them being susceptible to be investigated by sEMG and associated techniques. For example, the level of leg muscle activation during functional tasks, such as upright standing, has been reported to be greater in elderly compared with young adults, even after normalization of the sEMG of upright standing to sEMG recorded during MVC of the ankle plantar flexor muscles [2,8,15]. This increase in sEMG during upright standing is related to the well-documented decrease in force capacity with aging that could explain the decrease in balance stability in elderly adults [15]. In contrast, the normalized sEMG during ballistic contractions of the tibialis anterior is lower for elderly compared with young adults [44], in agreement with the reduced motor unit discharge rate observed in elderly adults during these contractions [44], suggesting a reduction in the ability to produce fast voluntary contractions, likely related to the loss of type II motor units.

In addition, the level of agonist/antagonist ratio is likely among the most evident age-related changes in motor control. In comparison with young adults, elderly adults activate to a greater extent the antagonist muscle during isometric [3,82] and dynamic contractions [82], when keeping upright stance [2], or during locomotion [39]. These results indicate a change in the strategy adopted to produce the required net torque at a joint.

In aging leg muscles (ankle plantar and dorsi-flexors), the amplitude of the M_{max} is significantly reduced and its duration increased [85]. In absence of difference in the latency of the M_{max} (the time interval between the onset of the stimulus artifact up to the first deflection of M wave), these changes in M_{max} parameters should be due to alterations at the muscle level. Although the amplitude of the H reflex also decreases with aging in leg [2,5,45,74] and forearm muscles [3], the H-reflex duration did not always show changes with aging [3,45,74]. In contrast, the latency of the H reflex is increased in elderly adults. These findings imply a decrease in conduction velocity of Ia afferents and/or a prolonged synaptic transmission. Furthermore, the reduced modulation of the H-reflex amplitude with contractions intensity and during upright standing in different balance conditions indicates a change in the neural control of these tasks.

Together, these findings indicate age-related alterations at the muscle level due to sarcopenia (depressed M_{max}) and in the proprioceptive pathway (increased latency and decreased amplitude of the H reflex). Furthermore, the reduced modulation of the H reflex indicates that elderly adults may rely more on central than peripheral mechanisms to regulate motor output. As task-specific modulation of the excitatory afferent feedback is important for motor control, these age-related adaptations may contribute to alter motor skills and postural control in elderly adults.

Only a few studies have investigated the effect of advancing age on motor corticospinal pathway. A pioneering work by Rossini et al. [72] reported that the MEP duration and amplitude recorded from a foot muscle are increased and reduced, respectively, in elderly compared with young adults, whereas only the MEP duration is influenced by age for a hand muscle. However, MEP amplitude was not normalized to the corresponding M_{max}, leaving possible differences in sEMG recording conditions as changes in muscle tissue influencing at a peripheral level the evoked potentials. More recent studies assessing the effect of age on the excitability of the corticospinal pathway with TMS revealed contrasting results, as some reported a decrease in MEP amplitude [63] whereas other did not find any age-related difference [63,76]. This may result from the muscle tested [72] and its state (rest vs. active) [63,76]. For example, when the amplitude of the MEP was tested at rest, no age-related difference was observed for the first dorsal interosseus [63,76], whereas in the same muscle the MEP was lower in elderly adults when assessed during voluntary contractions [63]. Furthermore, MEP amplitude (normalized to M_{max}) in the soleus was greater in elderly adults during upright standing compared with young adults, a difference that relies in part on balance control [9], suggesting, in combination with the decreased H_{max}, that aging is accompanied by an increase in corticospinal excitability during upright standing.

The inconsistent results reported above likely reflect the influence of muscle activation on corticospinal excitability as well as the influence of the task being performed [67].

In conclusion, sEMG and evoked potentials provide relevant information on age-related changes within the muscle and the nervous system, as well as on the neural adjustments required to perform various motor tasks. Here again, the consistent and/or complementary results obtained from the different methods confirm that either voluntary or electrically induced sEMG can be a valuable noninvasive method to assess changes related to aging.

12.5 CONCLUSIONS

In this chapter we focused on the basics of sEMG and related methods for the study of human motor control and its adaptations. Recordings of voluntary sEMG provide partial information on the mechanisms involved in muscle activity. However, the combination of sEMG and methods based on electrically and magnetically evoked potentials allows stepping further in a comprehensive approach of movement strategies during tasks, and the influence of various factors (etiology of cramps, fatigue, training, aging) on such strategies. Nonetheless, one must keep in mind that electrophysiological methods only provide indirect evidence on physiological mechanisms and should therefore always be interpreted as such.

REFERENCES

1. Barker, T., R. Jalinous, and I. L. Freeston, "Non-invasive magnetic stimulation of human motor cortex," *Lancet* **1**(8437), 1106–1107 (1985).

2. Baudry, S., F. Penzer, and J. Duchateau, "Input–output characteristics of soleus homonymous Ia afferents and corticospinal pathways during upright standing differ between young and elderly adults," *Acta Physiol (Oxf)* **210**, 667–677 (2014).

3. Baudry, S., A. H. Maerz, and R. M. Enoka, "Presynaptic modulation of Ia afferents in young and old adults when performing force and position control," *J. Neurophysiol.* **103**, 623–631 (2010).

4. Baudry, S., A. H. Maerz, J. R. Gould, and R. M. Enoka, "Task- and time-dependent modulation of Ia presynaptic inhibition during fatiguing contractions performed by humans," *J. Neurophysiol.* **106**, 265–273 (2011).

5. Baudry, S., and J. Duchateau, "Age-related influence of vision and proprioception on Ia presynaptic inhibition in soleus muscle during upright stance," *J. Physiol.* **590**(21), 5541–5554 (2012).

6. Baudry, S., G. Lecoeuvre, and J. Duchateau, "Age-related changes in the behavior of the muscle-tendon unit of the gastrocnemius medialis during upright stance," *J. Appl. Physiol.* **112**, 296–304 (2012).

7. Baudry, S., K. Jordan, and R. M. Enoka, "Heteronymous reflex responses in a hand muscle when maintaining constant finger force or position at different contraction intensities," *Clin. Neurophysiol.* **120**, 210–217 (2009).

8. Baudry, S., M. Klass, B. Pasquet, and J. Duchateau, "Age-related fatigability of the ankle dorsiflexor muscles during concentric and eccentric contractions," *Eur. J. Appl. Physiol.* **100**, 515–525 (2007).

9. Baudry, S., S. Sarrazin, and J. Duchateau, "Effects of load magnitude on muscular activity and tissue oxygenation during repeated elbow flexions until failure," *Eur. J. Appl. Physiol.* **113**, 1895–1904 (2013).

10. Baudry, S., T. Rudroff, L. A. Pierpoint, and R. M. Enoka, "Load type influences motor unit recruitment in biceps brachii during a sustained contraction," *J. Neurophysiol.* **102**, 1725–1735 (2009).

11. Benwell, N. M., P. Sacco, G. R. Hammond, M. L. Byrnes, F. L. Mastaglia, and G. W. Thickbroom, "Short-interval cortical inhibition and corticomotor excitability with fatiguing hand exercise: a central adaptation to fatigue?," *Exp. Brain Res.* **170**, 191–198 (2006).

12. Bigland, B., and O. C. Lippold, "The relation between force, velocity and integrated electrical activity in human muscles," *J. Physiol.* **123**, 214–224 (1954).

13. Bigland-Ritchie, B., C. G. Kukulka, O. C. Lippold, and J. J. Woods, "The absence of neuromuscular transmission failure in sustained maximal voluntary contractions," *J. Physiol. (Lond.)* **330**, 265–278 (1982).

14. Bigland-Ritchie, B., R. Johansson, O. C. Lippold, S. Smith, and J. J. Woods, "Changes in motoneurone firing rates during sustained maximal voluntary contractions," *J. Physiol.* **340**, 335–346 (1983).

15. Billot, M., E. M. Simoneau, J. Van Hoecke, and A. Martin, "Age-related relative increases in electromyography activity and torque according to the maximal capacity during upright standing," *Eur. J. Appl. Physiol.* **109**, 669–680 (2010).

16. Booghs, C., S. Baudry, R. Enoka, and J. Duchateau, "Influence of neural adjustments and muscle oxygenation on task failure during sustained isometric contractions with elbow flexor muscles," *Exp. Physiol.* **97**, 918–929 (2012).

17. Bouisset, S., and B. Maton, "Quantitative relationship between surface EMG and intramuscular electromyographic activity in voluntary movement," *Am. J. Phys. Med.* **51**, 285–295 (1972).

18. Carpentier, A., J. Duchateau, and K. Hainaut, "Velocity-dependent muscle strategy during plantarflexion in humans," *J. Electromyogr. Kinesiol.* **6**, 225–233 (1996).

19. Carpentier, A., J. Duchateau, and K. Hainaut, "Motor unit behaviour and contractile changes during fatigue in the human first dorsal interosseus," *J. Physiol.* **534**(3), 903–912(2001).

20. Carroll, T. J., S. Riek, and R. G. Carson, "Corticospinal responses to motor training revealed by transcranial magnetic stimulation," *Exerc. Sport Sci. Rev.* **29**, 54–59 (2001).

21. Cresswell, G., W. N. Löscher, and A. Thorstensson, "Influence of gastrocnemius muscle length on triceps surae torque development and electromyographic activity in man," *Exp. Brain Res.* **105**, 283–290 (1995).

22. Croce, R. V., J. P. Miller, and P. St. Pierre, "Effect of ankle position fixation on peak torque and electromyographic activity of the knee flexors and extensors," *Electromyogr. Clin. Neurophysiol.* **40**, 365–373 (2000).

23. Desmedt, J. E., "Methods of studying neuromuscular function in humans: isometric myogram, excitation electromyogram, and topography of terminal innervation," *Acta Neurol Psychiatr. Belg.* **58**, 977–1017 (1958).

24. Di Lazzaro, V., D. Restuccia, A. Oliviero, P. Profice, L. Ferrara, A. Insola, P. Mazzone, P. Tonali, and J. C. Rothwell, "Magnetic transcranial stimulation at intensities below active motor threshold activates intracortical inhibitory circuits," *Exp. Brain Res.* **119**, 265–268 (1998).

25. Duchateau, J., and S. Baudry, "Training adaptation of the neuromuscular system," in *The Encyclopaedia of Sports Medicine: An IOC Medical Commission Publication, Neuromuscular Aspects of Sports Performance*, P. V. Komi, Wiley-Blackwell, Oxford, UK, 2010.

26. Duchateau, J., C. Balestra, A. Carpentier, and K. Hainaut, "Reflex regulation during sustained and intermittent submaximal contractions in humans," *J. Physiol.* **541**(3), 959–967(2002).

27. Duclay, J., A. Martin, A. Robbe, and M. Pousson, "Spinal reflex plasticity during maximal dynamic contractions after eccentric training," *Med. Sci. Sports Exerc.* **40**, 722–734 (2008).

28. Enoka, R. M., *Neuromechanics of Human Movement*, Human Kinetics, Champaign, IL, 2008.

29. Enoka, R. M., S. Baudry, T. Rudroff, D. Farina, M. Klass, and J. Duchateau, "Unraveling the neurophysiology of muscle fatigue," *J. Electromyogr. Kinesiol.* **21**, 208–219 (2011).

30. Farfan, D., J. C. Politti, and C. J. Felice, "Evaluation of EMG processing techniques using Information Theory," *Biomed. Eng. Online* **9**, 72 (2010).

31. Farina, D., and D. Falla, "Effect of muscle-fiber velocity recovery function on motor unit action potential properties in voluntary contractions," *Muscle Nerve* **37**, 650–658 (2008).

32. Farina, D., L. Arendt-Nielsen, and T. Graven-Nielsen, "Effect of temperature on spike-triggered average torque and electrophysiological properties of low-threshold motor units," *J. Appl. Physiol.* **99**, 197–203 (2005).

33. Farina, D., R. A. Ferguson, A. Macaluso, and G. De Vito, "Correlation of average muscle fiber conduction velocity measured during cycling exercise with myosin heavy chain composition, lactate threshold, and VO2max," *J. Electromyogr. Kinesiol.* **17**, 393–400 (2007).

34. Feiereisen, P., J. Duchateau, and K. Hainaut, "Motor unit recruitment order during voluntary and electrically induced contractions in the tibialis anterior," *Exp. Brain Res.* **114**, 117–123 (1997).

35. Häkkinen, K., M. Kallinen, M. Izquierdo, K. Jokelainen, H. Lassila, E. Mälkiä, W. J. Kraemer, R. U. Newton, and M. Alen, "Changes in agonist-antagonist EMG, muscle CSA, and force during strength training in middle-aged and older people," *J. Appl. Physiol.* **84**, 1341–1349 (1998).

36. Hasan, Z., and R. M. Enoka, "Isometric torque-angle relationship and movement-related activity of human elbow flexors: implications for the equilibrium-point hypothesis," *Exp. Brain Res.* **59**, 441–450 (1985).

37. Henneman, E., G. Somjen, and D. O. Carpenter, "Functional significance of cell size in spinal motoneurons," *J. Neurophysiol.* **28**, 560–580 (1965).

38. Hermens, J., B. Freriks, C. Disselhorst-Klug, and G. Rau, "Development of recommendations for SEMG sensors and sensor placement procedures," *J. Electromyogr. Kinesiol.* **10**, 361–374 (2000).

39. Hortobágyi, T. and P. DeVita, "Muscle pre- and coactivity during downward stepping are associated with leg stiffness in aging," *J. Electromyogr. Kinesiol.* **10**, 117–126 (2000).

40. Hunter, S. K., D. L. Ryan, J. D. Ortega, and R. M. Enoka, "Task differences with the same load torque alter the endurance time of submaximal fatiguing contractions in humans," *J. Neurophysiol.* **88**, 3087–3096 (2002).

41. Jones, D. A., "High- and low-frequency fatigue revisited," *Acta Physiol. Scand.* **156**, 265–270 (1996).

42. Kaufman, M. P., "The exercise pressor reflex in animals," *Exp. Physiol.* **97**, 51–58 (2012).

43. Keenan, K. G., D. Farina, R. Merletti, and R. M. Enoka, "Influence of motor unit properties on the size of the simulated evoked surface EMG potential," *Exp. Brain Res.* **169**, 37–49 (2006).

44. Klass, M., M. Lévénez, R. M. Enoka, and J. Duchateau, "Spinal mechanisms contribute to differences in the time to failure of submaximal fatiguing contractions performed with different loads," *J. Neurophysiol.* **99**, 1096–1104 (2008).

45. Klass, M., S. Baudry, and J. Duchateau, "Age-related decline in rate of torque development is accompanied by lower maximal motor unit discharge frequency during fast contractions," *J. Appl. Physiol.* **104**, 739–746 (2008).

46. Klass, M., S. Baudry, and J. Duchateau, "Modulation of reflex responses in activated ankle dorsiflexors differs in healthy young and elderly subjects," *Eur. J. Appl. Physiol.* **111**, 1909–1916 (2011).

47. Kleine, U., W. A. Boekestein, I. M. P. Arts, M. J. Zwarts, H. J. Schelhaas, and D. F. Stegeman, "Fasciculations and their F-response revisited: high-density surface EMG in ALS and benign fasciculations," *Clin. Neurophysiol.* **123**, 399–405 (2012).

48. Lagerquist, O., E. P. Zehr, and D. Docherty, "Increased spinal reflex excitability is not associated with neural plasticity underlying the cross-education effect," *J. Appl. Physiol.* **100**, 83–90 (2006).

49. Lestienne, F. G., and V. S. Gurfinkel, "Posture as an organizational structure based on a dual process: a formal basis to interpret changes of posture in weightlessness," *Prog. Brain Res.* **76**, 307–313 (1988).

50. Lexell, J., K. Henriksson-Larsén, and M. Sjöström, "Distribution of different fibre types in human skeletal muscles. 2. A study of cross-sections of whole m. vastus lateralis," *Acta Physiol. Scand.* **117**, 115–122 (1983).

51. Loscher, W. N., A. G. Cresswell, and A. Thorstensson, "Excitatory drive to the alpha-motoneuron pool during a fatiguing submaximal contraction in man.," *J. Physiol.* **491** (1), 271–280(1996).

52. Maglaredy, J. W., and D. B. McDougal, Jr., "Electrophysiological studies of nerve and reflex activity in normal man. I. Identification of certain reflexes in the electromyogram and the conduction velocity of peripheral nerve fibers," *Bull. Johns Hopkins Hosp.* **86**, 265–290 (1950).

53. Matsunaga, K., A. Maruyama, T. Fujiwara, R. Nakanishi, S. Tsuji, and J. C. Rothwell, "Increased corticospinal excitability after 5 Hz rTMS over the human supplementary motor area," *J. Physiol. (Lond.)* **562**(1), 295–306(2005).

54. McNeil, J., S. Giesebrecht, S. C. Gandevia, and J. L. Taylor, "Behaviour of the moto-neurone pool in a fatiguing submaximal contraction," *J. Physiol.* **589**(14), 3533–3544 (2011).

55. Merletti, R., L. R. Lo Conte, C. Cisari, and M. V. Actis, "Age related changes in surface myoelectric signals," *Scand. J. Rehabil. Med.* **24**, 25–36 (1992).

56. Miller, T. M., and R. B. Layzer, "Muscle cramps," *Muscle Nerve* **32**(4), 431–442(2005).

57. Minetto, M. A., A. Botter, O. Bottinelli, D. Miotti, R. Bottinelli, and G. D'Antona, "Variability in muscle adaptation to electrical stimulation," *Int. J. Sports Med.* **34**, 544–553 (2013).

58. Minetto, M. A., A. Botter, R. Ravenni, R. Merletti, and D. De Grandis, "Reliability of a novel neurostimulation method to study involuntary muscle phenomena," *Muscle Nerve* **37**, 90–100 (2008).

59. Minetto, M. A., A. Holobar, A. Botter, and D. Farina, "Discharge properties of motor units of the abductor hallucis muscle during cramp contractions," *J. Neurophysiol.* **102**, 1890–1901 (2009).

60. Minetto, M. A., A. Holobar, A. Botter, R. Ravenni, and D. Farina, "Mechanisms of cramp contractions: peripheral or central generation?," *J. Physiol. (Lond.)* **589**(23), 5759–5773(2011).

61. Mottram, J., J. M. Jakobi, J. G. Semmler, and R. M. Enoka, "Motor-unit activity differs with load type during a fatiguing contraction," *J. Neurophysiol.* **93**, 1381–1392 (2005).

62. Mynark, R. G., "Reliability of the soleus H-reflex from supine to standing in young and elderly," *Clin. Neurophysiol.* **116**, 1400–1404 (2005).

63. Oliviero, A., P. Profice, P. A. Tonali, F. Pilato, E. Saturno, M. Dileone, F. Ranieri, and V. Di Lazzaro, "Effects of aging on motor cortex excitability," *Neurosci. Res.* **55**, 74–77 (2006).

64. Orth, M., and J. C. Rothwell, "The cortical silent period: intrinsic variability and relation to the waveform of the transcranial magnetic stimulation pulse," *Clin. Neurophysiol.* **115**, 1076–1082 (2004).

65. Oya, T., B. W. Hoffman, and A. G. Cresswell, "Corticospinal-evoked responses in lower limb muscles during voluntary contractions at varying strengths," *J. Appl. Physiol.* **105**, 1527–1532 (2008).

66. Paillard, J., "Electrophysiologic analysis and comparison in man of Hoffmann's reflex and myotatic reflex," *Pflugers Arch.* **260**, 448–479 (1955).

67. Petersen, N. T., H. S. Pyndt, and J. B. Nielsen, "Investigating human motor control by transcranial magnetic stimulation," *Exp. Brain Res.* **152**, 1–16 (2003).

68. Pierrot-Deseilligny, E., and D. Burke, *The Circuitry of the Human Spinal Cord: Its Role in Motor Control and Movement Disorders.* Cambridge University Press, Cambridge, UK, 2005.

69. Pitcher, J. B., K. M. Ogston, and T. S. Miles, "Age and sex differences in human motor cortex input–output characteristics," *J. Physiol.* **546**(2), 605–613(2003).

70. Rainoldi, A., M. Gazzoni, and G. Melchiorri, "Differences in myoelectric manifestations of fatigue in sprinters and long distance runners," *Physiol. Meas.* **29**, 331–340 (2008).

71. Roeleveld, K., B. G. van Engelen, and D. F. Stegeman, "Possible mechanisms of muscle cramp from temporal and spatial surface EMG characteristics," *J. Appl. Physiol.* **88**, 1698–1706 (2000).

72. Rossini, P. M., M. T. Desiato, and M. D. Caramia, "Age-related changes of motor evoked potentials in healthy humans: non-invasive evaluation of central and peripheral motor tracts excitability and conductivity," *Brain Res.* **593**, 14–19 (1992).

73. Rothwell, J. C., P. D. Thompson, B. L. Day, S. Boyd, and C. D. Marsden, "Stimulation of the human motor cortex through the scalp," *Exp. Physiol.* **76**, 159–200 (1991).

74. Scaglioni, G., M. V. Narici, N. A. Maffiuletti, M. Pensini, and A. Martin, "Effect of ageing on the electrical and mechanical properties of human soleus motor units activated by the H reflex and M wave," *J. Physiol.* **548**(2), 649–661(2003).

75. Schieppati, M., "The Hoffmann reflex: a means of assessing spinal reflex excitability and its descending control in man," *Prog. Neurobiol.* **28**, 345–376 (1987).

76. Smith, E., M. V. Sale, R. D. Higgins, G. A. Wittert, and J. B. Pitcher, "Male human motor cortex stimulus-response characteristics are not altered by aging," *J. Appl. Physiol.* **110**, 206–212 (2011).

77. Solomonow, M., "External control of the neuromuscular system," *IEEE Trans. Biomed. Eng.* **31**, 752–763 (1984).

78. Solomonow, M., and M. Krogsgaard, "Sensorimotor control of knee stability. A review," *Scand. J. Med. Sci. Sports* **11**, 64–80 (2001).

79. Stein, R. B., K. L. Estabrooks, S. McGie, M. J. Roth, and K. E. Jones, "Quantifying the effects of voluntary contraction and inter-stimulus interval on the human soleus H-reflex," *Exp. Brain Res.* **182**, 309–319 (2007).

80. Taube, W., N. Kullmann, C. Leukel, O. Kurz, F. Amtage, and A. Gollhofer, "Differential reflex adaptations following sensorimotor and strength training in young elite athletes," *Int. J. Sports Med.* **28**, 999–1005 (2007).

81. Taylor, J. L., and S. C. Gandevia, "Noninvasive stimulation of the human corticospinal tract," *J. Appl. Physiol.* **96**, 1496–1503 (2004).

82. Tracy, B. L., and R. M. Enoka, "Older adults are less steady during submaximal isometric contractions with the knee extensor muscles," *J. Appl. Physiol.* **92**, 1004–1012 (2002).

83. Ugawa, Y., J. C. Rothwell, B. L. Day, P. D. Thompson, and C. D. Marsden, "Percutaneous electrical stimulation of corticospinal pathways at the level of the pyramidal decussation in humans," *Ann. Neurol.* **29**, 418–427 (1991).

84. Van Cutsem, M., J. Duchateau, and K. Hainaut, "Changes in single motor unit behaviour contribute to the increase in contraction speed after dynamic training in humans," *J. Physiol.* **513**(1), 295–305(1998).

85. Vandervoort, A. and A. J. McComas, "Contractile changes in opposing muscles of the human ankle joint with aging," *J. Appl. Physiol.* **61**, 361–367 (1986).

86. Wagner, T., J. Rushmore, U. Eden, and A. Valero-Cabre, "Biophysical foundations underlying TMS: setting the stage for an effective use of neurostimulation in the cognitive neurosciences," *Cortex* **45**, 1025–1034 (2009).

13

SURFACE EMG IN ERGONOMICS AND OCCUPATIONAL MEDICINE

M. Gazzoni, B. Afsharipour, and R. Merletti

Laboratory for Engineering of the Neuromuscular System, Politecnico di Torino, Torino, Italy

13.1 INTRODUCTION

The term "musculoskeletal disorders (MSDs)" is used to describe a variety of conditions that affect the muscles, bones, joints, and connective tissue and include repetitive strain injuries (RSI) and cumulative trauma disorders (CTD). Work-related musculoskeletal disorders are reported for different occupations, ranging from factory assembling lines to performing arts. They can be caused by repetitive or sustained work, monotonous job tasks, and localized muscular loadings and fatigue and have a high social cost and impact on quality of life [18,21,70,73].

Prevention of musculoskeletal disorders is one of the main goals in ergonomics. The development of methods to quantify muscle force, fatigue, and muscle involvement in a work task is important for this purpose. Furthermore, information about muscular fatigue, recovery, and rest can support the design of ergonomic workstations and the planning of postures and proper work–rest patterns [121,122].

MSDs caused by work affect more than four million workers in the European Union (EU) and account for about half of all work-related disorders in EU countries [44], The European Working Conditions Survey (EWCS) published by the European Foundation has shown that 24.7% of workers across the EU report suffering frequently from backache and 22.8% from muscular pain [97,107].

MSDs are also one of the most important causes of long-term sickness absence. The Labour Force Survey (LFS) ad hoc module 2007 showed that in 61% of persons

Surface Electromyography: Physiology, Engineering, and Applications, First Edition.
Edited by Roberto Merletti and Dario Farina.
© 2016 by The Institute of Electrical and Electronics Engineers, Inc. Published 2016 by John Wiley & Sons, Inc.

with a work-related health problem in the previous 12 months, musculoskeletal disorders (bone, joint, or muscle) were the main problems. The European Commission estimates that MSDs account for 49.9% of all absences from work lasting 3 days or longer and for 60% of permanent work incapacity.

Work-related MSDs also rank among the most serious health problems within the aging workforce [109]. Figures from Eurostat illustrate that as workers get older, they are more likely to experience a long-term health problem [32]. The incidence of chronic illness or disability continues to rise as a result of an aging population in the EU [1].

Not included in these figures are the individuals who had already quit working because of severe musculoskeletal conditions, which are, next to mental disorders, the most common reason for early poor health retirement [93]. Part of these problems are because, despite declining physical capacities [94], exposure to certain physical risks, such as handling heavy loads, working in painful or tiring postures, or repetitive hand/ arm movements, remains considerably high in elderly workers [98].

On the basis of this data, MSDs represent an estimated cost to society of between 0.5% and 2% of the gross domestic product (GDP) while the burden of illness within an aging workforce has serious economic implications for businesses and social security systems in European countries [12].

If the negative effects of MSDs, on both quality of life and work disability, are to be minimized, early diagnosis and treatment are critical. Although a considerable amount of knowledge about the physiopathology and epidemiology of MSDs has been accumulated over the past two decades, they remain, as shown before, the most common work-related diseases (WRDs) in the EU and workers of all ages and in all sectors and occupations can be affected [18,70].

Health promotion can be carried out in various locations: for example, the community, the medical schools, health care facilities, and workplaces [114]. The workplace, as a setting for health promotion, deserves special attention, because adults spend more time at the workplace than in any other location. The World Health Organization (WHO) states that the workplace "has been established as one of the priority settings for health promotion into the 21st century" because it influences "physical, mental, economic and social wellbeing" and it "offers an ideal setting and infrastructure to support the promotion of health of a large audience" [123].

The literature and the institutional reports mentioned above indicate that it is crucial to (a) design workplaces in such a way that the manifestation (or, at least, the aggravation) of these illnesses can be prevented and (b) detect such illnesses as early as possible, by mean of suitable programs for monitoring exposed workers, so that actions can be taken to enable employees to work up to the regular retirement age [124]. This would also result in considerable economic benefits [1].

13.2 SURFACE EMG IN ERGONOMICS AND OCCUPATIONAL MEDICINE

Surface electromyography (sEMG) and kinematic measurements are the main tools for the quantitative evaluation of risks related to work activity. First of all, sEMG measurements provide objective information about the timing and relative intensity of muscle activity, not appreciable otherwise. In addition, sEMG techniques can be used to quantify and validate existing subjective/qualitative methods for the ergonomic evaluation mostly based on scales and descriptions. For example, the OCRA (Occupational Repetitive Actions) methodology for the evaluation of risks related to the repetitiveness of upper limb movements contains a section about the evaluation of the force needed to perform particular actions in a work activity. Such evaluation is simply based on a questionnaire compiled by the worker, in which he/she describes the kind of effort he/she has been performing, by choosing among some descriptions provided by the ergonomists. These methodologies, such as the Borg scale [15], obviously introduce a marked degree of subjectivity, depending on age, gender, general health conditions, and mood/attitude of the worker in that particular moment. EMG-based techniques provide a more direct and objective assessment, reducing the subjectivity introduced by the declarations of the worker. Furthermore, the activity of agonist/antagonist muscular groups involved in the execution of a particular task can be measured through (a) the activation intervals, (b) the degree of co-contraction, and (c) myoelectric manifestations of muscle fatigue (see Chapters 10 and 12) and/or lack of (or duration of) complete relaxation. This data leads to the definition of the risk level associated to a particular workstation or work activity and to fatigue. The US Department of Health and Human Services addressed this issue in 1992 in the Report on "Selected topics in surface electromyography for use in occupational settings" [110].

The development of sEMG imaging techniques in the last 10 years, providing the spatial distribution (maps) of sEMG parameters, such as amplitude and fatigue indicators, over the surface of one or more muscles, opens important new perspectives in the field of ergonomics and provides new tools for the analysis of the neuro-muscular system in the work environment, as described in Chapters 5 and 10 and Section 13.5. An example is provided by Fig. 13.1. which depicts the distribution of amplitude of the sEMG signal (ARV) in the medial portion of the upper, middle, and lower trapezius muscles (left and right) during typing with and without forearm support. The very different level of upper–middle trapezius activation is not perceived by the subject but is detected by the monitoring equipment and can be used to activate a warning system indicating bad posture and lack of resting time (see "gap analysis" in Section 13.5.3) or a continuous biofeedback system for user training or re-training as indicated in Chapter 18.

13.3 BASIC WORKLOAD CONCEPTS AND TECHNICAL ISSUES

The literature regarding the potential of sEMG techniques in ergonomic studies is extensive and concerns many applications, some of which are listed below:

1. work at display terminals and office workstations [27,28,76,86,113,117],
2. production line workstations [26,50,59],
3. work with surgical workstations [31,75,82,84,103,116],

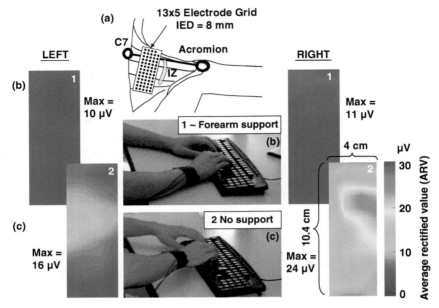

FIGURE 13.1 (a) An electrode grid (13 rows by 5 columns with IED = 8 mm) is placed medially with respect to the innervation zone (IZ) of the upper and middle trapezius muscle, on the right and left sides of a computer operator. The interpolated images show the distribution of the EMG average rectified values (ARV) below the array. (b) Condition 1: ARV distribution while typing with the forearms resting on the desk. (c) Condition 2: ARV distribution while typing with the forearms not resting on the desk. The maps show a remarkable difference in EMG amplitude and distribution.

4. workspace optimization [61,68,88,92],
5. lifting tasks [14,19,39,40,67,91],
6. performing art, use of musical instruments [3,10,11,41–43,51,62,66,104]
7. and others [83,108].

The concept of workload is often incorrectly associated to high forces and fatigue. This is not necessarily the case. A number of factors have to be taken into account in this concept, such as: (a) physical effort and the way it is produced (some strategies are better than others; consider, for example, the possible ways of lifting a heavy object from the floor); (b) incorrect load sharing among agonist muscles and unnecessary co-contraction of the antagonists; (c) awkward posture; (d) high repetitivity; (e) duration of effort; (f) lack of recovery periods. In most of the above cases, sEMG has been proposed as a monitoring technique or as a source of feedback information during training [80,99,119]. In particular, low loads are not without risk, if sustained for a long time. In 1991 G. Hägg proposed the "Cinderella Hypothesis" [52], which assumes that there are low-threshold motor units which are recruited as soon as the muscle is activated and stay active until total muscle relaxation. These motor units would face overloading, which may lead to mitochondrial disturbances in type I fibers, degenerative processes, inflammation, and pain [53].

As is the case for other bio-signals (e.g., ECG, EEG), the use of sEMG requires considerable expertise and training in the proper detection and interpretation of the signals. The ergonomist or the occupational doctor or therapist must be aware of the many limitations and conditions under which sEMG signals are reliable. Only superficial muscles may be monitored today. Unstable electrode–skin contacts must be detected and corrected. If an electrode pair is used, it must be placed according to specific rules in relation to the muscle innervation zone (IZ) as indicated in Barbero et al. [9] and in Chapter 2. Power-line interference is a very common problem in an industrial environment and must be recognized and reduced with suitable interventions such as the "Driven Right Leg" system, or other measures such as identifying disturbing equipment and moving away from them or reducing their influence. Offline software reduction of power-line interference is the last resort. In many applications, electrode arrays or grids are preferable with respect to an electrode pair but more complex to apply. Surface EMG variables are affected by multiple physiological factors (e.g., increase of the EMG amplitude of a muscle may reflect either increasing force, antagonist co-contraction, different load sharing, or fatigue) and many confounding factors must be considered in the interpretation [36].

In the following, two main issues that can be approached via sEMG in ergonomics will be discussed in some detail: (1) the analysis of exerted forces and torques and (2) the analysis of muscle fatigue. Both are presented in Chapters 4, 5, and 10 of this book, from different points of view.

13.4 EMG–FORCE RELATIONSHIP

13.4.1 A Word of Caution

High muscle contraction forces imply rapid muscle fatigue, associated with high sEMG amplitude and fast changes of sEMG features (myoelectric manifestations of muscle fatigue such as changes of mean and median spectral frequencies, muscle fiber conduction velocity, etc.), as mentioned in Chapters 4 and 10. In dynamic conditions the muscle slides under the skin and the lever arm of its force changes as the joint angle changes in time. These geometrical changes modify the relationship between force and torque acting on the joint as well as the relationship between muscle force and sEMG, resulting in (i) a system that is very complex to model and (ii) data very difficult to interpret and highly misleading if incorrectly interpreted. For this reason, muscles are studied mostly in isometric conditions, often during constant force contractions. In many occupational activities, forces are relatively low and some muscles (e.g., of the back and shoulder) indeed operate in almost isometric conditions, making the analysis relatively simple. Since sEMG is affected by many anatomical, physiological, and geometric factors [36], a clear relationship between sEMG and force cannot be defined except in particular conditions when one factor is changing and most, possibly all, of the other affecting factors are constant.

In addition, the force or torque measured at a joint during voluntary contractions is the result of the contribution of many muscles while the sEMG is mostly due to the one above which the electrodes are placed. If force produced by synergic or antagonist muscles is attributed to the muscles on which sEMG is read, the sEMG–force relation is obviously incorrect (see Chapter 10). Finally, even in conditions of constant neural drive, the sEMG amplitude is continuously fluctuating because of the random constructive and destructive algebraic summation of MUAPs and occasional synchronization of motor units [65].

Figure 13.2a shows a schematic example, explaining some of the issues mentioned above, relative to a simple nonphysiological two-muscle system which becomes much more complex in situations like that depicted in Fig. 13.2b when more muscles are simultaneously active. Figure 13.2 recalls the concepts described in Figs. 10.1 and 10.2 and shows that, in the same muscle, recruitment of identical MUs, either from superficial to deep ones or from deep to superficial ones, produces different sEMG–force profiles. Such profiles are further modified by the fact that the MUs in a muscle

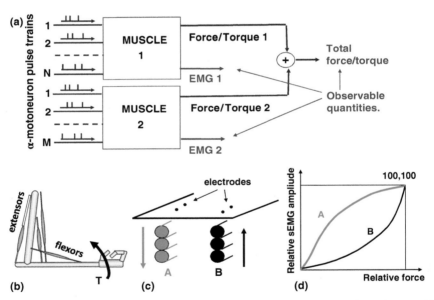

FIGURE 13.2 (a) Example of two muscles acting on the same skeletal joint . The N and M motor units (MU) of muscle 1 and muscle 2 are driven by the respective N and M α-motorneuron action potential trains and produce force/torque and sEMG. The sEMG signals and the total force/torque can be measured. Individual muscle force contributions cannot be measured directly. (b) Example of a joint with a number of flexors and extensors acting on it. (c) Example of extreme and opposite motor unit recruitment modalities, from superficial to deep MUs in A and from deep to superficial MUs in B. (d) Qualitative example of the relative sEMG versus relative force for each muscle of the example in part (c). Depending on the recruitment modality, size and depth of the MUs, the curve sEMG =f(force) may have upward or downward concavity or be more or less linear. (see also Chapter 10).

are not identical at all and are recruited, according to the Henneman's Principle [54], beginning with the smaller ones. The global load may be shared by different muscles (with agonists and antagonists possibly co-contracting) in a time-varying mode so that the sEMG of a specific muscle might even decrease, while the measured force is constant or increasing, or vice versa, because of a changing load sharing strategy among muscles. The issue of changing "load sharing" is a topic of current research [4,7,16] and is addressed in Chapter 10. Despite the above considerations, which suggest great caution in the estimation of force based on sEMG, many articles have reported very questionable sEMG–force associations.

13.4.2 Techniques for Estimating Muscle Force or Joint Torque from sEMG

The force or torque estimation techniques from sEMG reported in the literature are mostly related to isometric conditions and are based on sEMG amplitude estimation (see Chapter 10). Very different approaches have been reported. EMG-based force or torque estimation in dynamic conditions is much more difficult because the muscles move under the electrodes, lever arms change, and inertial components are not easy to measure. An overview of the force estimation techniques is provided below as a summary of the material presented in Chapters 4, 9, and 10. A review of sEMG in dynamic contractions is presented in Farina et al. [35].

Average Rectified Value (ARV) and Root Mean Square (RMS). These simple amplitude estimators still dominate most applied studies involving sEMG. They are calculated over a specific time interval (epoch), usually $0.25\,\text{s}$ or $0.5\,\text{s}$ or $1.0\,\text{s}$. Even for a stationary signal, they show fluctuations from epoch to epoch (see Chapter 10). Interferential sEMG is a zero mean stochastic signal and has a probability density function (pdf) which depends on the contraction level [8]. For any given pdf, ARV and RMS are related by a specific equation. At relatively high contraction levels (very interferential signal) the sEMG has a zero mean Gaussian pdf and the "true" (estimated on an infinitely long epoch) ARV and RMS are related by the equation $\text{ARV} = \text{RMS}\sqrt{2/\pi} \cong 0.8\,\text{RMS}$, where $\text{RMS} = \sigma$ that is the standard deviation of the Gaussian pdf. For weaker contractions the sEMG pdf is better approximated by a decaying bilateral exponential ($y = \frac{b}{2}e^{-|bx|}$, where $b = \sqrt{2}/\sigma$) [25].

In turn, for a given epoch length, ARV and RMS are random variables and have their own probability density function (pdf) whose discrete version (probability histogram) indicates the likelihood of finding values in a particular bin of the histogram.

Single or Multichannel Adaptive Whitening. Since the amplitude estimation of a random signal shows a smaller variance when the samples are uncorrelated (Chapter 4), ensuing investigations found that it is appropriate to decorrelate sEMG samples. This process is referred to as "whitening" because the autocorrelation function of the signal approaches the Dirac function in zero, $\delta(t)$, and the power spectrum (which is the Fourier transform of the autocorrelation function) approaches a constant (all the harmonics have the same power and the power spectrum is "white").

Combining multiple EMG channels into a single amplitude estimate has also been shown to further reduce the estimator's variance [24]. Because of the shape of the sEMG power spectrum, "whitening" the signal implies a particular type of high-pass filter (described in Chapter 4).

High-Pass Filtering of Surface EMG Signals. Potvin and Brown [102] proposed that high-pass filtering of the sEMG signals with cutoff above 300 Hz prior to amplitude estimation reduces errors in force estimation. The idea at the origin of the technique is that sEMG reflects not only the MUAPs, but also the tissue filtering that occurs between the fibers and the electrodes. This factor attenuates the high-frequency content and it is reasonable to expect that the sEMG power spectrum will not necessarily represent the true electrophysiological processes occurring at the fibers that are at the origin of both sEMG and of muscle force production [38].

Bayesian Filtering. Sanger [106] introduced a nonlinear recursive filter based on Bayesian estimation for the purpose of joint torque estimation. All estimators imply a presumed model of the statistics of the driving signal, and results depend on assumptions about the bandwidth of the driving signal. For example, estimation of muscle force by low-pass filtering the rectified EMG signal corresponds to the assumption that force does not change more rapidly than a specific frequency. Comparison of the Bayesian algorithm with linear methods showed that the Bayesian algorithm has more rapid response to EMG signal onset and offset maintaining the ability to provide smooth output during slowly varying force contractions.

Principal Component Analysis (PCA) and Independent Component Analysis (ICA). Grids of electrodes collect many sEMG signals over a relatively small surface. As described in Chapter 5, the signals can be considered as a multidimensional dataset that contains redundant information. Principal component analysis (PCA) can be used to detect this type of redundancy in multivariate data by dimensional reduction. A method using PCA for force/torque estimation was propsed by Staudenmann et al. [112] and applied to monopolar recordings. In Staudenman's work, PCA was used to remove the common information from a set of signals and therefore perform a spatial filtering operation. The advantage of this method with respect to classic spatial filtering is that it does not involve prior assumptions on fiber direction and, when monopolar sEMG is collected from a grid, the method is independent of the alignment between electrodes and fiber orientations.

MUAP Rate (MR). This technique, proposed by Kallenberg and Hermens [64], is based on the global firing rate of the detected MUs of a muscle (total number of firings per second, often referred to as "muscle activity"). Assuming an average force contribution per firing, the MR should reflect muscle force.

Activity Index (AI). The use of activity index as muscle force estimator was introduced by Istenič [60]. It is based on the Convolution Kernel Compensation (CKC) described in Holobar et al. [48,56]. The method decomposes multichannel sEMG into trains of neural spikes of the individual motor neurons and provides information about the global firing rate of the detected MUs of a muscle (like MR).

13.5 DOSE AND EXPOSURE IN ERGONOMICS

13.5.1 The Concept of Exposure

The concept of "dose" of a harmful agent (e.g., radiation, pollutant) during a time interval is the integral, over such interval, of the intensity of the agent. "Exposure" is the result of this integral divided by the integration time—that is, the average intensity of the agent.

In ergonomics, the "agent" is an external load, a particular posture, or the rate of repetition of a movement. The term "external exposure" is associated with conditions related to the work task or to the external environment, while "internal exposure" is associated to conditions internal to the body, such as joint angles or muscle forces. The second is the cause of fatigue, discomfort, and, eventually, disorders and pain. Despite efforts carried out mostly in the last decades of the twentieth century, as well as more recently [2,23,28,105], an efficient and accepted indication of "exposure" does not exist; all indicators have drawbacks, and consensus on their use is lacking. The methods providing the indicators of exposure most frequently mentioned in the literature are exposure variation analysis (EVA), gap analysis, and joint analysis of spectrum and amplitude (JASA), all with remarkable limitations.

13.5.2 Exposure Variation Analysis (EVA)

Exposure variation analysis (EVA) provides a joint histogram of muscle activity and its timing and is used to assess the physiological and biomechanical effects in ergonomics. The EVA allows analysis of activity bursts and short pauses in work and comparison of individual differences at performing a task [41].

EVA is based on a two-dimensional histogram whose X axis reports the EMG amplitude bins, the Y axis reports the durations of time intervals during which the sEMG amplitude is within an amplitude bin. Figure 13.3 provides a didactic example by showing the RMS amplitude computed over each time unit (epoch) for 25 time units. The marginal distribution of the plot, obtained by integrating (or summing occurrences) along the Y axis, is the sEMG amplitude histogram. The amplitude histograms obtained from RMS distributions in time shown in Figs. 13.3a and 13.3c are identical. The marginal distribution of the plot, obtained by integrating (or summing occurrences) along the X axis, is the cumulative number of times the muscle is continuously active for an interval of given duration (regardless of the level).

The sEMG amplitude histograms corresponding to two different tasks or workloads can be equal while the timing of activities is different as depicted in Figs. 13.3a and 13.3c. In ergonomics, the two cases should be considered as being different and distinguishable and they are interpreted differently. Adding another dimension that is

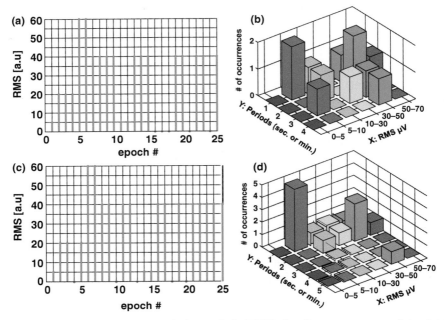

FIGURE 13.3 **(a, c)** Exposure variation analysis (EVA) plots for two sequences of simulated EMG RMS values estimated every time unit (e.g., minute). The RMS histogram is the same in the two cases, but the time sequence of RMS values is not. **(b, d)** This difference is outlined and is relevant from the point of view of exposure. In the first row, along Y, for $0 < X < 5\,\mu V$ the EMG is at the noise level and the corresponding time intervals (Periods) are called "gaps". They are relevant from the point of view of exposure and the object of "gap analysis". Each bar of the 2D histogram shows how many times the sEMG has amplitude X for a duration Y during the observation time.

called "periods" to the histogram enables us to distinguish between two different load or sEMG profiles or individuals.

13.5.3   Gap Analysis

If the lowest amplitude bin is the noise RMS level, at which the sEMG amplitude is considered to be zero, such as in Figs. 13.3a and 13.3c, then the duration of each time interval during which this is true is called a "gap". Based on evidence that more frequent brief periods of rest in muscles during work are associated with fewer symptoms and complaints, sEMG gap analysis was developed to quantify the number and duration of periods of very low sEMG activity [118]. Surface EMG gap analysis provides two important summary measures: the *gap frequency*, which is the average number of gaps per minute over the analysis time period, and the *percent time of muscular rest*, which is the summed duration of all gaps divided by the duration of the sEMG recording [118]. Increases in the frequency of EMG gaps and the total gap duration have been associated with reduced risk of muscle myalgia.

13.5.4 Joint Analysis of Spectrum and Amplitude (JASA)

A signal is characterized (nonunivocally!) by its amplitude probability density function (pdf) and related features (such as RMS, ARV), as well as by its power spectral density (PSD) and related features (such as MNF, MDF) as described in Chapter 4. Amplitude and spectral features change as functions of time, and their variations may be in the same or in opposite directions. The joint analysis of these variations is referred to as JASA, as proposed by Luttmann et al. [78,79], and is described in Fig. 13.4. A rather basic analysis of the sEMG features indicates that the progressive increase of force implies the recruitment of motor units with progressively larger fibers, having progressively higher conduction velocity and therefore higher MNF or MDF (Fig. 13.4a).

During a sustained, isometric, and constant force contraction, *with a stable number of recruited motor units*, the progressive decrease of average conduction velocity (CV) results in an increase of RMS and a decrease of MNF (Fig. 13.4b). When the contraction level decreases, larger motor units drop out, RMS decreases, and average CV decreases, causing a decrease of MNF (Fig. 13.4c). If a low contraction level

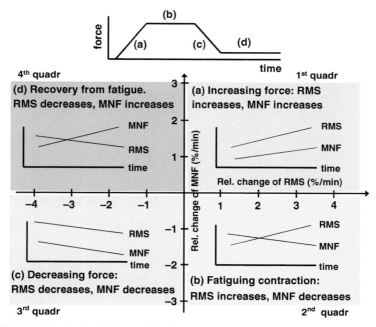

FIGURE 13.4 Example of joint amplitude and spectral analysis (JASA). Variations of sEMG amplitude (e.g., RMS) and spectral feature (e.g., MNF) may be grouped in the four classes associated to the four quadrants of the plot showing relative changes of MNF versus relative changes of RMS. However. a number of factors may alter the behavior of the two variables. Among these factors, the rotation (dropout and recruitment) of motor units may be very important, especially at low contraction levels (see Fig. 10.7), and may substantially alter the expected pattern. This approach has drawbacks of which the user must be aware (see text).

follows a fatiguing contraction (recovery), RMS decreases, while CV and MNF increase, recovering their normal values (Fig. 13.4d). These four behaviors are represented in the four quadrants (a, b, c, and d) of Fig. 13.4. These are theoretically expected behaviors that, in general, are applicable to short isometric contraction performed in well-controlled laboratory conditions.

In general, if the motor unit pool is not stable or the intracellular action potential changes shape, the observed behavior may be different from the expected one (see Fig. 10.7). Other factors may modify the expected standard behavior and make this classification not always reliable or correct. In addition, the extrapolation of this classification to dynamic or field situations is extremely questionable.

13.6　NORMALIZATION MODELS

The same muscle control strategy may generate signals with very different amplitudes, depending on the subcutaneous tissue thickness and locations of the active motor units within the muscle [37]. Because of the many factors affecting sEMG amplitude and spectral variables (see Chapters 4, 5, and 10), a specific sEMG amplitude–force relation cannot be valid for different muscles or subjects and must be identified on a subject-by-subject and muscle-by-muscle basis. Normalization with respect to a force or to a specific sEMG value is crucial in order to estimate the force exerted by a muscle during a voluntary contraction. This problem was largely studied in the past in order to compare the activity of different muscles or different subjects. Several studies were focused on calibration of sEMG with respect to muscle force or joint moments [6,13,29,30,55,69,71,72,81].

The normalization of the sEMG signal was made with respect to its ARV or RMS value at the MVC level or with respect to a submaximal value. As an example, Fjellman-Wiklund et al. [42] used 30% MVC as the reference voluntary contraction (RVC) in musician studies.

The normalization problem is not yet solved in a satisfactory way, even for isometric contractions, and is very serious in dynamic contractions, even when great care is applied in positioning the electrodes. This problem represents a major limitation to the widespread use of EMG in ergonomics and is a field of active research. Some steps forward may be expected in the near future with the more widespread use of EMG imaging techniques and, possibly, with the solution of the "load sharing" problem.

13.7　HIGH-DENSITY EMG RECORDINGS IN ERGONOMICS

In the previous sections of this chapter, the terms "sEMG signal", "sEMG channel", "RMS", or "ARV" values were referred to a single signal detected in monopolar or single differential mode.

The two-dimensional (2-D) electrode arrays, described in Chapter 5 and in Fig. 13.1, provide "high-density EMG recordings"—that is, images that describe the monopolar or single differential intensity of the signal over a surface above a muscle or a group of muscles. Each electrode (or electrode pair) provides a pixel of an image whose intensity is evolving in time and is associated to a local EMG amplitude or a spectral variable computed on a defined epoch whose duration is typically 0.25 s, 0.5 s, or 1.0 s, as indicated in Figs. 5.5 and 5.6. In addition, a 2-D electrode array allows surface EMG decomposition, as described in Chapter 7, and the monitoring of discharge rate, recruitment, and derecruitment of the identified motor units [48,56,87] provides a powerful tool for the identifications of the level of activity and of the control strategies of the muscles, and of the muscle compartments, under the array. Adhesive arrays and elastic sleeves, carrying 32, 64, or 128 electrodes, are available for this application to limb and back muscles (see Figs. 5.8, 10.4, and 10.8). As discussed in Chapter 5, if the interelectrode distance is in the 5 to 10 mm range, the image can be interpolated, without introducing artefacts due to spatial aliasing, to produce smooth high-resolution images such as those depicted in Figs. 13.1, 13.5, and 13.7 [22].

Recent advances concerning sEMG electrode arrays that have the same acoustic impedance of the skin, and are therefore transparent to ultrasound, provide simultaneous echographic images of the muscle and of its EMG surface distribution [17].

13.8   EXAMPLES OF APPLICATIONS

13.8.1   sEMG and Force Perception

Figure 13.5 shows a laboratory application of the HDsEMG technique to the study of the upper trapezius muscle and examples of HDsEMG maps obtained from 5×13 electrode arrays with IED $= 8$ mm [115]. Panels (a) to (c) show the experimental arrangement. Panel (d) shows the interpolated maps of RMS values detected on the upper trapezius of three healthy subjects with different activity distributions; the white dashed lines connect the acromion to C7. The different distribution topographies may reflect different individual anatomies or different individual strategies for producing the required force. If a strategy is not optimal, the online presentation of the sEMG image, as a form of biofeedback, may help to improve it and reduce the likelihood of strain and possible future pain.

Figure 13.6a shows the Borg subjective index of perceived exertion (CR10) reported by 14 subjects performing isometric efforts of shoulder lifting at levels of 10% to 80% MVC as described in Fig. 13.5. Figure 13.6b reports the corresponding distributions of RMS values. Figure 13.7c shows excellent correlation ($R = 0.99$) between the subjective and objective mean values of these indices. The RMS values are averaged in space over a portion of the grid on one side of the innervation zone (see Chapter 5). These results indicate that properly detected RMS of the surface EMG may be good objective indicators of the perceived effort.

Perceived exertion ratings have been used to study the subjective feeling of the effect of an exercise. Different scales were compared in the literature [49]. Borg scale has been applied to both resistance [46] and endurance [101] exercises, as

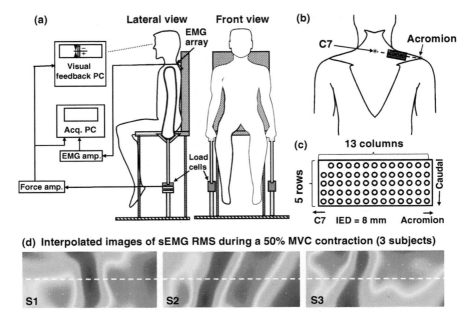

FIGURE 13.5 System for measurement and display of force generated by selective and isometric contraction of the upper trapezius muscle. **(a)** Position of the subject on the chair: lateral and front view. Subject's feet are off the floor, arms are kept straight along the body, and the back is positioned against the chair back. **(b)** Position of the two-dimensional electrode array: The array's columns are parallel to the line connecting C7 to the acromion, and the center of the array is located halfway. **(c)** Two-dimensional array with interelectrode distance of 8 mm. The subject was asked to raise both shoulders while keeping the elbows fully extended, as well as to match the target force level displayed on the screen. **(d)** Maps of sEMG amplitude detected by the array from three healthy subjects (S1, S2, S3) showing different distribution of sEMG intensity on their right upper trapezius muscles. Parts a and b are reproduced from Troiano et al. [115] with permission.

well as to both isometric [90] and dynamic contractions [15]. Perceived exertion ratings are considered an acceptable approach in work related studies [90] to measure the perceived response at different loads [47] or to test the effect of rest periods on recovering perceived efficiency [126]. The association of the Borg index with sEMG, demonstrated by the work of Troiano et al. [115] for the upper trapezius muscle, should be assessed by further research work on other muscles.

It is well known that the interpretation of sEMG of muscles in dynamic field conditions is particularly challenging [35]. However, the study of postural or stabilizing muscles, which operate in almost isometric conditions during a working task, can be attempted to understand neuromuscular control strategies in ergonomically relevant situations. The trapezius and the erector spinae are among these muscles and have been extensively investigated [33,34,58,63,85].

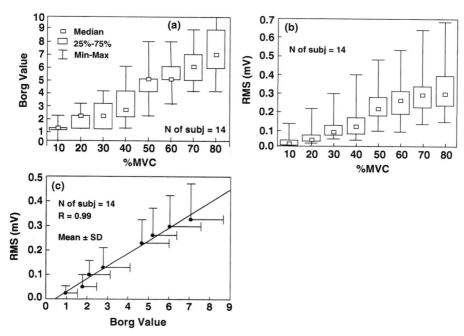

FIGURE 13.6   Data obtained from the setup depicted in Fig. 13.5. (**a**) Borg scale of subjective force values provided by 14 subjects in response to contractions at target values (%MVC) that were randomized and unknown to the subject, (**b**) Global RMS values obtained by averaging, in time and space, the RMS of N channels visually selected by the operator on one side of the innervation zone; results are reported as median, quartile range, and min–max values. Note that the Borg scale has only integer values, and therefore the median value may overlap with the quartile edge or the quartile edge may overlap with the min value. (**c**) Relation between Borg ratings and RMS estimates at different force levels (mean ± standard deviation). Reproduced from Troiano et al. [115] with permission.

13.8.2   Surface EMG Topographical Changes and Fatigue

Variations of the distribution of surface sEMG intensity among muscles or muscle regions suggest a time-varying load sharing strategy among muscles or their compartments. Such changes seem to prevent the overload of muscle portions or groups of motor units during a fatiguing task and may also be triggered by a momentary change of force or a variation of task. Variations in activation of different regions in the same muscle, described and quantified by high-density sEMG recordings, can be used to assess muscle overload as a risk factor. This hypothesis comes from basic physiological studies showing motor unit substitution when a brief variation in force is introduced during a static task [34,120]. Farina et al. [37] observed that the surface distribution of sEMG intensity (RMS) over the trapezius changes during a contraction sustained to the endurance time, and its centroid moves upward, indicating the progressive involvement of motor units in the upper part of the muscle, as indicated in Fig. 13.7. These authors also observed that the subjects

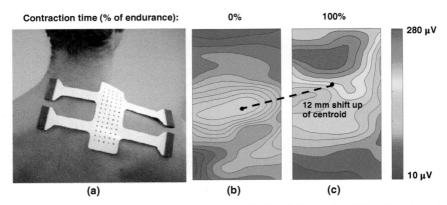

FIGURE 13.7 (a) Placement of the electrode grid (13×5 electrodes, IED = 8 mm) on the upper trapezius of a subject. The subject was asked to abduct both arms at 90° with elbows extended and palms facing down and to hold this position until endurance (about 6–7 min), (b, c) Topographical maps of sEMG RMS at the beginning of the exercise and at the endurance time. The map evolves in time showing a progressive shift of its centroid in the cranial direction. Reproduced from Farina et al. [37] with permission.

showing the largest shifts of the centroid of the sEMG activity distribution were those showing longer endurance times.

It is interesting to observe that people can learn to voluntarily redistribute muscle activity using biofeedback techniques without compromising task performance. Palmerud et al. [95] showed that subjects are able to reduce trapezius muscle activity (using sEMG feedback) while maintaining the arm in an isometric, abducted position. The target posture was maintained by redistributing activity to the rhomboids and transverse part of trapezius [96]. Similarly, other studies investigated whether it is possible to selectively activate different neuromuscular compartments within the trapezius muscle. Eleven out of 15 subjects learned, after one hour of training based on feedback of the sEMG amplitude of four anatomical subdivisions within the trapezius muscle, to selectively activate at least one of the four muscle subdivisions [57,58].

13.8.3 Simulation of Precision Industrial Work

A simulation of an industrial work task, with kinematic and sEMG measurements, is provided in Gazzoni and co-workers [20,45]. In these studies the subject was sitting at a bench and was requested to perform 60 repetitions of the lifting task of a shell (1.5 kg) placing and removing it in front of him on a reference at shoulder height, as indicated in Fig. 13.8a–c. The shell had three cylindrical prongs to be inserted in three holes, as indicated in Fig. 13.8d. Two different levels of complexity were tested: (1) sharply cut prongs that had to match the holes exactly and (2) prongs with rounded terminations providing a guide that would facilitate the task by avoiding the need for exact alignment. The sEMG signals were acquired from biceps brachii (long and short head), triceps brachii, anterior, lateral, and posterior deltoid, upper trapezius, and

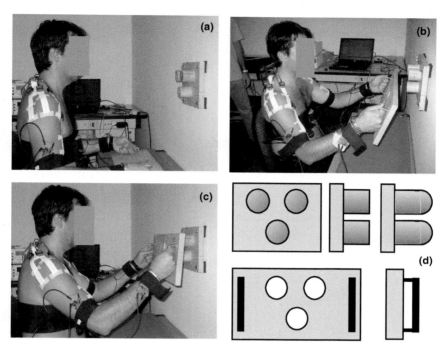

FIGURE 13.8 Simulation of an industrial task. The subject was asked to perform 60 repetitions of the lifting task of a shell (1.5 kg), placing and removing it on a reference in front of him. The task cadence was provided by a metronome. **(a)** The subject sits in front of a reference frame at a distance equal to the upper limb length. The height of the chair is adjusted to align the C7 vertebra with the center of the support. **(b, c)** Kinematic and sEMG data acquisition. Xsens sensors were fixed with Velcro straps on the right and left arm and forearm. A reference sensor was fixed on the sternum. sEMG signals were detected from biceps brachii (long head), triceps brachii (lateral head), anterior deltoid, lateral deltoid, posterior deltoid, and upper trapezius. Linear electrode arrays (eight electrodes, 5-mm interelectrodes distance) were used to detect sEMG signals. **(d)** Front and lateral view of load and reference.

longissimus dorsi muscles (arrays of eight electrodes, 5 mm interelectrodes distance, on each muscle) together with the kinematics of the arms using Xsens sensors (www.xsens.com) on the forearms, arms, and trunk. The sEMG amplitude (RMS), spectral frequency (MNF) variables and conduction velocity (CV) were estimated for muscles listed above in correspondence of 20% of the cycle length. Results are shown in Fig. 13.9. While no differences can be detected in the kinematics signals, sEMG envelopes showed a higher muscle activation when sharply cut prongs were used (Fig. 13.9a). This highlights the importance of a combined analysis of kinematics and EMG in workstation characterization/optimization. The observed decrease of estimated CV is associated with a decrease of sEMG amplitude (Fig. 13.9b). This trend may reflect a fatiguing muscle but, more likely, a central motor control optimization: The subject learns how to perform the task with progressively lower muscle activation and lower muscle forces. The sEMG amplitude decreases because of the recruitment

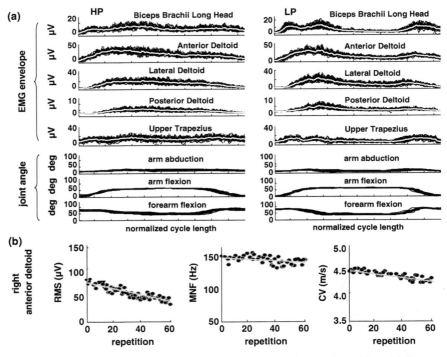

FIGURE 13.9 Results from the setup of Fig. 13.8 for a representative subject. **(a)** Superposition of sEMG envelopes and joint angles estimated for each cycle of the lifting task. The profiles are normalized in time with respect to each cycle duration. The mean waveform is plotted in white. HP: High-precision task (sharp prongs). LP: low-precision task (rounded prongs). **(b)** Time course of RMS, MNF, and CV estimated from the right anterior deltoid in correspondence of 20% of the maximum height reached by the wrist during the cycle. Reproduced from Gazzoni [45] with permission.

of fewer motor units. CV decreases because the MUs with higher CV, recruited in the previous cycles, are released. This example highlights the sEMG differences associated with different tools and the learning process by the worker.

13.8.4 Playing-Related Musculoskeletal Disorders

Over the past 20 years, there has been a growing interest in the medical problems of performing artists, as well as in musician's physiology and medicine. Due to the high physical and psychological demands of their work, musicians have a high risk of developing a range of health problems. The main causes of their musculoskeletal disorders are overuse, nerve compression/inflammation, and focal dystonia. The musician's intrinsic motivation to practice and to repeat motor patterns to perfection posits the exposure to repetitive strain injuries (RSI) and cumulative trauma disorders (CTD). Pianists and bow instrument musicians are the most affected. The well-known

Teatro alla Scala in Milano, Italy, started a rehabilitation service in 1994 and during its first 14 years of activity treated 311 of the 415 musicians who have been orchestra members, for tendinopathy (15%), overuse (30%), arthrosis (5%), nerve compression (5%), dystonia (2%), and back pain (43%) [111].

Among the 1353 instrumentalists (60% females and 40% men) evaluated and reported in the work of Lederman [74], musculoskeletal disorders were diagnosed in 64% of the cases, peripheral nerve problems in 20%, and focal dystonia in 8%. The average age at the time of evaluation was 37 years for men and 30 years for women. Regional muscle pain syndromes, particularly of the upper limb, neck, and shoulders, are most common together with (a) ulnar neuropathy at the elbow and (b) carpal tunnel syndrome. Each instrument group shows a characteristic distribution of symptoms and signs that are associated with the static and dynamic stresses inherent in the playing of the instrument, as well as with its misuse, overuse, and repetition.

Medical and medicolegal controversies exist about the definition of "misuse", "overuse", and "repetition" [125]. "Misuse" generally means incorrect technical handling of the instrument or posture of the body, "overuse" generally means excessive use, and "repetition" generally means execution of the same movement over and over. While many musicians complete a full career without suffering from any complaints, others do not, indicating that subjective characteristics are relevant. Microruptures, at the cell level, and inflammation may take place when a muscle is subjected to repeated submaximal loading as in the playing of string instruments. This fact may concern either the muscles directly involved in the task or stabilizing and postural muscles, such as the trapezius and the back muscles, which may be active for a long time without breaks (see Section 13.5.3).

Fatigue and exhaustion are reversible protective phenomena preventing injury. If the resulting discomfort is ignored, it may generate delayed-onset muscle soreness (DOMS), well known in sports, starting 2–48 hours after the performance and lasting 1–2 days. As soreness develops, the musician will keep playing and adjust posture and technique to reduce the pain, possibly worsening the situation.

Muscle dystonia is not yet well understood and involves unintentional movement of the fingers and painful cramping: intrinsic hand muscle and wrist and finger flexors contract uncontrollably and force the musician to interrupt the activity and rest.

The recent literature review of de Souza Moraes [89], on violinists and violists, includes 50 articles (24 of which of good to high level) and provides a state of the art. Some investigators, quoted in this review, found that violists and violinists have a higher incidence of pain in the neck, shoulder, elbow, and forearm than pianists. Others observed that violists report more pain in the shoulder and arms due to the greater weight and size of the viola compared to the violin. Another study observed prevalence of pain in the neck and shoulder of 35.3% of the violists and violinists who showed predominant disorders in the left upper limb such as tendinitis and compressive syndromes. The left hand of violinists presents around twice as many problems as the right hand; this may derive from the more awkward position of the wrist and from the action of the fingers. Flexors and extensors of the right forearm and wrist are used to control the bow whose movement requires shoulder and back stabilization through the action of the trapezius and erector spinae muscles. The maintenance of the technical posture adopted by the violinist and violist represents a considerable factor of discomfort for the performance of the activity, due to the increase in muscle effort required to maintain such posture on chairs that often do not have adjustment devices. One study reported postural issues as associated with discomfort in at least 90% of the musicians. Another study reported the most common problems during the seated posture as poor positioning of the feet, pelvic rotation, increase of lumbar lordosis and rotation to the right, elevation of shoulders and elbows, head bent and turned to the left. Most of the quoted works deal with clinical and biomechanical issues.

Both central and peripheral adaptations of the neuromuscular system following musical training result in subject-specific strategies of movement. Variability of sEMG activity in the trapezius muscle during string performance, for example, was assessed by Fjellman-Wiklund et al. in 2004 [41,42]. Twelve string players (9 violins, 2 violas, 1 cello) performed a 7-minute-long excerpt from a classical piece of music, at a given time and 12 weeks later. Exposure variation analysis and principal components analysis were used to process sEMG, collected with a pair of electrodes placed at two-thirds of the line acromion-C7.

Generally, each musician showed the same sEMG pattern in the two sessions (no intraindividual differences), but great differences were pointed out among individual subjects (large interindividual differences). The authors referred these differences as due to different styles adopted by the players, as observed in references [11] and [100]. Moreover, these authors [41,42] reported that intersubjects differences were greater when difficult pieces of music were played and stressed the different role of the left–right trapezius during the musical performance: Left shoulder, in fact, is mainly involved in static tasks, while dynamic contractions are performed with the right upper limb. Since different "schools" provide different recommendations, the question "Are intersubject differences due to individual optimization of muscle activity or to the lack of proper uniform training and teaching?" remains to be answered.

Despite its recent developments and its potential, surface electromyography has not been used often; when it is used, however, it is, mostly in its traditional form based on single electrode pairs (see Chapter 4). Surface EMG biofeedback has been used as a form of prevention and treatment, in the attempt to minimize unnecessary stress, especially in violinists [77] as mentioned in Chapter 18.

Detecting and interpreting the EMG in dynamic conditions is a difficult task. A first step, in a dynamic situation, is the study of muscles operating in almost isometric conditions such as the trapezius and the erector spinae of a sitting player. Linear and 2-D electrode arrays covering the trapezius and the lumbar part of the erector spinae provide means to investigate the spatial distribution of activity and myoelectric manifestations of muscle fatigue (see Chapter 5). Figure 13.10 depicts a study situation where the distributions of sEMG activities of the trapezius and back muscles were monitored during playing different strings of a violin. Figure 13.10 shows a violin player repetitively playing the first and fourth string of the violin in 10 large bowings lasting one second each. A 2-D adhesive array (16 rows by 4 columns) was placed on the right trapezius of the bowing arm, and two 1-D (16 electrodes) arrays were placed on the right and left erector spinae muscles as indicated in the figure. The RMS value was computed from single differential signals over 10 bowings (10 s). The

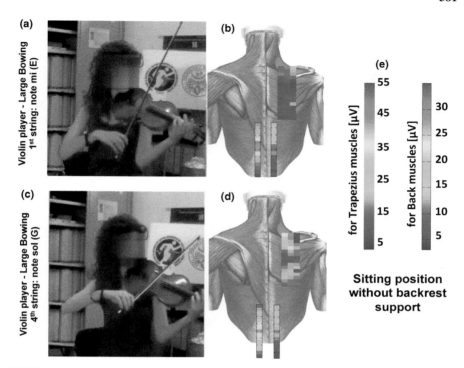

FIGURE 13.10   Surface EMG activity of the trapezius and erector spinae muscles in a violin student player during playing **(a)** the first string (most medial string; note: mi/E) and **(c)** the fourth string (most lateral string; note: sol/G) in large bowing (the entire length of the bow slides on the string in one second). RMS maps (over 10-s recording) from single differential signals, recorded from the trapezius muscle of right arm and the left and right erector spinae muscles, are presented in **(b)** for the first string and in **(d)** for the fourth string. Detection grids for the trapezius muscle (middle and lower trapezius) included 16×4 electrodes. Two 16×1 electrode arrays had been applied to the right and left erector spinae muscles. **(e)** Color scales for the trapezius and back muscles.

sensitivity of the trapezius and erector spinae EMG amplitude maps to different positions of the arm is very clear. It is also evident that a single electrode pair might have detected either significant amplitude differences, or no difference at all, between playing the two strings, depending on its location on the muscles. The technique has considerable potential for the identification of optimal and nonoptimal postures of string instrument players (e.g., using or not using a back rest, holding the instrument in a particular way, etc.).

Exposure variation analysis (EVA) has been used in musician studies [41,42]. Previous studies in this field have been done using conventional bipolar electrodes. Recent technologies, based on high-density sEMG (HDsEMG), combine EVA with HDsEMG and provide results related to a muscle region (or muscle group) rather than to a specific electrode position.

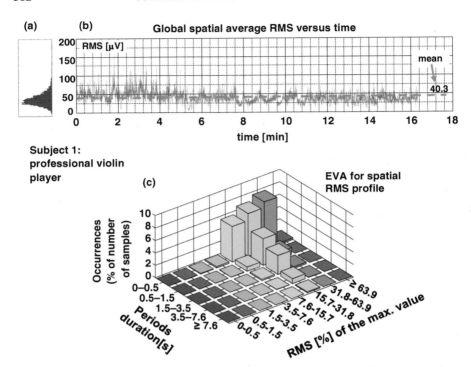

FIGURE 13.11 Example of distribution of the right (bowing arm) trapezius activity during a 17-min nonstop performance of a "difficult" piece of music chosen by the subject (professional player) based on the level of his proficiency. **(a)** The probability distribution of global spatial average of RMS values (16×3 single differential channels) computed over 250-ms epoch duration for each channel along 17-min recording in monopolar configuration (16×4 channels). **(b)** Average in space (over 16×3 channels) RMS values versus time. The mean RMS value along the time axis is 40.3 μV. **(c)** The exposure variation analysis (EVA) plot from the data shown in panel b.

Figures 13.11 and 13.12 show two examples of the combined EVA and HDsEMG techniques. Figures 13.11a and 13.11b show the distribution of the right (bowing arm) trapezius muscle activity during a 17-minute nonstop performance. The subject is a professional violin player and played different/difficult pieces of music (pieces of music were chosen by the subject based on the level of proficiency). The global spatial average (panel b) was computed as the average over the entire grid (16×3 single differential channels) of the RMS values computed over 250-ms epoch duration for each channel. The EVA plot concerning the global spatial average presented in panel b is shown in panel c. The same concept for the trapezius muscle of another violin player (student) is shown in Fig. 13.12.

The EVA plot analysis requires further research. Many additional considerations should be taken into account and specific protocols should be prepared for a comparison between subjects or conditions. The two plots are reported here only to demonstrate a method whose outcome remains to be investigated [5].

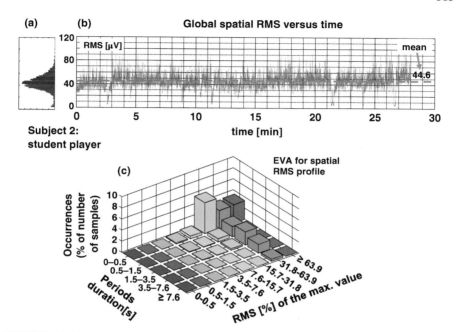

FIGURE 13.12 Example of distribution of the right (bowing arm) trapezius activity during a 28-min nonstop performance of a "difficult" piece of music chosen by the subject (student player) based on the level of her proficiency. **(a)** The probability distribution of global spatial average of RMS values (16×3 single differential channels) computed over 250-ms epoch duration for each channel along 28 min of recording. **(b)** Average in space (over 16×3 channels) RMS values versus time. The mean RMS value along the time axis is 44.6 µV. **(c)** Exposure variation analysis (EVA) plot from the data shown in panel b. For very similar mean RMS value, the EVA plots in Fig. 13.11c and Fig. 13.12c, are different.

13.9 CONCLUSIONS

Surface EMG has a remarkable potential role in the prevention, monitoring, documentation, and possibly treatment of work related disorders. Insufficient research funding and standardization efforts are still limiting its applications. Much greater research efforts are required to transform its potential into a practical technique with clinical and legal relevance and economic impact on work health.

REFERENCES

1. *Active Ageing and Solidarity Between Generations—A Statistical Portrait of the European Union*: Publication Office of the European Union, Luxembourg, 2012.
2. Abbiss, C. R., L. Straker, M. J. Quod, D. T. Martin, and P. B. Laursen, "Examining pacing profiles in elite female road cyclists using exposure variation analysis," *Br. J. Sport Med.* **44**, 437–442 (2010).

3. Ackermann, B., R. Adams, and E. Marshall, "The effect of scapula taping on electromyographic activity and musical performance in professional violinists," *Aust. J. Physiother.* **48**, 197–203 (2002).

4. Afsharipour, B.,"Estimation of load sharing among muscles acting on the same joint and applications of surface electromyography," PhD dissertation, Department of Electronics and Telecommunications, Politecnico di Torino, Italy, 2014.

5. Afsharipour, B., T. M. Vieira, and R. Merletti, "Comparing three segmentation algorithms applied to simulated monopolar EMG maps," in XX Congress of the International Society of Electrophysiology and Kinesiology (ISEK). Rome, Italy (2014).

6. Anders, C., S. Bretschneider, A. Bernsdorf, K. Erler, and W. Schneider, "Activation of shoulder muscles in healthy men and women under isometric conditions," *J. Electromyogr. Kinesiol.* **14**, 699–707 (2004).

7. Aventaggiato, M.,"Torque estimation from surface EMG signals during isometric elbow flexion–extension," M.Sc. thesis, Politecnico di Torino, Italy, 2008.

8. Ayachi, F. S., S. Boudaoud, and C. Marque, "Evaluation of muscle force classification using shape analysis of the sEMG probability density function: a simulation study," *Med. Biol. Eng. Comp.* **52**, 1–12 (2014).

9. Barbero, M., R. Merletti, and A. Rainoldi, *Atlas of Muscle Innervation Zones*, Springer Verlag, Italy, 2012.

10. Berque, P., "Influence of neck–shoulder pain on trapezius muscle activity among professional violin and viola players," *Physiotherapy* **89**, 126 (2003).

11. Berque, P., and H. Gray, "The influence of neck–shoulder pain on trapezius muscle activity among professional violin and viola players: an electromyographic study," *Med. Probl. Performing Arts* **17**, 68–75 (2002).

12. Bevan, S., T. Quadrello, R. McGee, M. Mahdon, A. Vavrovsky, and L. Barham,"Fit for work? Musculoskeletal disorders in the European workforce," The Work Foundation, London, 290003, 30 September 2009.

13. Bogey, R. A., J. Perry, and A. J. Gitter, "An EMG-to-force processing approach for determining ankle muscle forces during normal human gait," *IEEE Trans. Neural Syst. Rehabil. Eng.* **13**, 302–310 (2005).

14. Bonato, P., G. R. Ebenbichler, S. H. Roy, S. Lehr, M. Posch, J. Kollmitzer, et al., "Muscle fatigue and fatigue-related biomechanical changes during a cyclic lifting task," *Spine (Philadelphia 1976)* **28**, 1810–1820 (2003).

15. Borg, G., "Psychophysical scaling with applications in physical work and the perception of exertion," *Scand J. Work Environ. Health* **16 Suppl. 1**, 55–58 (1990).

16. Botter, A., H. R. Marateb, B. Afsharipour, and R. Merletti, "Solving EMG–force relationship using particle swarm optimization," *Conf. Proc. IEEE Eng. Med. Biol. Soc.* **2011**, 3861–3864 (2011).

17. Botter, A., T. M. Vieira, I. D. Loram, R. Merletti, and E. F. Hodson-Tole, "A novel system of electrodes transparent to ultrasound for simultaneous detection of myoelectric activity and B-mode ultrasound images of skeletal muscles," *J. Appl. Physiol. (1985)* **115**, 1203–1214 (2013).

18. Buckle, P. W., and J. J. Devereux, "The nature of work-related neck and upper limb musculoskeletal disorders," *Appl. Ergonom.* **33**, 207–217 (2002).

19. Caboor, D. E., M. O. Verlinden, E. Zinzen, P. Van Roy, M. P. Van Riel, and J. P. Clarys, "Implications of an adjustable bed height during standard nursing tasks on spinal motion, perceived exertion and muscular activity," *Ergonomics* **43**, 1771–1780 (2000).

20. Cescon, C., M. Gazzoni, E. Guasco, F. Mastrangelo, and R. Merletti, "Repetitive task evaluation by means of electromyographic and kinematic signals acquired from upper limbs," in *VII Congresso Nazionale della Società Italiana di Analisi del Movimento in Clinica*, Cuneo, Italy, p. 12, 2007.

21. Chandran, A., C. Schaefer, K. Ryan, R. Baik, M. McNett, and G. Zlateva, "The comparative economic burden of mild, moderate, and severe fibromyalgia: results from a retrospective chart review and cross-sectional survey of working-age U.S. adults," *J. Manag. Care Pharm.* **18**, 415–426 (2012).

22. Chatterjee, P., S. Mukherjee, S. Chaudhuri, and G. Seetharaman, "Application of Papoulis–Gerchberg Method in image super-resolution and inpainting," *Comput. J.* **52**, 80–89 (2009).

23. Ciccarelli, M., L. Straker, S. E. Mathiassen, and C. Pollock, "Variation in muscle activity among office workers when using different information technologies at work and away from work," *Hum. Factors* **55**, 911–923 (2013).

24. Clancy, E. A., and N. Hogan, "Multiple site electromyograph amplitude estimation," *IEEE Trans. Bio-med. Eng.* **42**, 203–211 (1995).

25. Clancy, E. A., and N. Hogan, "Probability density of the surface electromyogram and its relation to amplitude detectors," *IEEE Trans. Bio-med. Eng.* **46**, 730–739 (1999).

26. Clasby, R. G., D. J. Derro, L. Snelling, and S. Donaldson, "The use of surface electro-myographic techniques in assessing musculoskeletal disorders in production operations," *Appl. Psychophysiol. Biofeedback* **28**, 161–165 (2003).

27. Cook, C., R. Burgess-Limerick, and S. Papalia, "The effect of upper extremity support on upper extremity posture and muscle activity during keyboard use," *Appl. Ergonom.* **35**, 285–292 (2004).

28. Dennerlein, J. T., and P. W. Johnson, "Different computer tasks affect the exposure of the upper extremity to biomechanical risk factors," *Ergonomics* **49**, 45–61 (2006).

29. Doorenbosch, C. A. M., and J. Harlaar, "Accuracy of a practicable EMG to force model for knee muscles," *Neurosci. Lett.* **368**, 78–81 (2004).

30. Doorenbosch, C. A. M., A. Joosten, and J. Harlaar, "Calibration of EMG to force for knee muscles is applicable with submaximal voluntary contractions," *J. Electromyogr. Kinesiol.* **15**, 429–435 (2005).

31. Emam, T. A., G. Hanna, and A. Cuschieri, "Ergonomic principles of task alignment, visual display, and direction of execution of laparoscopic bowel suturing," *Surg. Endosc. Other Intervent. Tech.* **16**, 267–271 (2002).

32. *European Commission: Major and chronic diseases: Policy.* (2012). Available from: http://ec.europa.eu/health/major_chronic_diseases/policy/index_en.htm

33. Falla, D., H. Andersen, B. Danneskiold-Samsøe, L. Arendt-Nielsen, and D. Farina, "Adaptations of upper trapezius muscle activity during sustained contractions in women with fibromyalgia," *J. Electromyogr. Kinesiol.* **20**, 457–464 (2010).

34. Falla, D., and D. Farina, "Non-uniform adaptation of motor unit discharge rates during sustained static contraction of the upper trapezius muscle," *Exp. Brain Res.* **191**, 363–370 (2008).

35. Farina, D., "Interpretation of the surface electromyogram in dynamic contractions," *Exerc. Sport Sci. Rev.* **34**, 121–127 (2006).

36. Farina, D., C. Cescon, and R. Merletti, "Influence of anatomical, physical, and detection-system parameters on surface EMG," *Biol. Cybern.* **86**, 445–456 (2002).

37. Farina, D., F. Leclerc, L. Arendt-Nielsen, O. Buttelli, and P. Madeleine, "The change in spatial distribution of upper trapezius muscle activity is correlated to contraction duration," *J. Electromyogr. Kinesiol.* **18**, 16–25 (2008).

38. Farina, D., and A. Rainoldi, "Compensation of the effect of sub-cutaneous tissue layers on surface EMG: a simulation study," *Med. Eng. Phys.* **21**, 487–497 (1999).

39. Ferguson, S. A., W. S. Marras, and D. Burr, "Workplace design guidelines for asymptomatic vs. low-back-injured workers," *Appl. Ergonom.* **36**, 85–95 (2005).

40. Ferguson, S. A., W. S. Marras, D. L. Burr, K. G. Davis, and P. Gupta, "Differences in motor recruitment and resulting kinematics between low back pain patients and asymptomatic participants during lifting exertions," *Clin. Biomech.* **19**, 992–999 (2004).

41. Fjellman-Wiklund, A., H. Grip, H. Andersson, J. Stefan Karlsson, and G. Sundelin, "EMG trapezius muscle activity pattern in string players: part II—influences of basic body awareness therapy on the violin playing technique," *Int. J. Ind. Ergonom.* **33**, 357–367 (2004).

42. Fjellman-Wiklund, A., H. Grip, J. S. Karlsson, and G. Sundelin, "EMG trapezius muscle activity pattern in string players: part I—is there variability in the playing technique?," *Int. J. Ind. Ergonom.* **33**, 347–356 (2004).

43. Furuya, S., R. Osu, and H. Kinoshita, "Effective utilization of gravity during arm downswing in keystrokes by expert pianists," *Neuroscience* **164**, 822–831 (2009).

44. Gauthy, R., *Musculoskeletal Disorders: An Ill-Understood Pandemic*. European Trade Union Institute, Brussels, 2007.

45. Gazzoni, M., "Multichannel surface electromyography in ergonomics: potentialities and limits," *Hum. Factors Ergonom. Manufact. Service Indust.* **20**, 255–271 (2010).

46. Gearhart, R. E., F. L. Goss, K. M. Lagally, J. M. Jakicic, J. Gallagher, and R. J. Robertson, "Standardized scaling procedures for rating perceived exertion during resistance exercise," *J. Strength Cond. Res.* **15**, 320–325 (2001).

47. Gearhart, R. F., Jr., F. L. Goss, K. M. Lagally, J. M. Jakicic, J. Gallagher, K. I. Gallagher, et al., "Ratings of perceived exertion in active muscle during high-intensity and low-intensity resistance exercise," *J. Strength Cond. Res.* **16**, 87–91 (2002).

48. Glaser, V., A. Holobar, and D. Zazula, "Real-time motor unit identification from high-density surface EMG," *Neural Syst. Rehabil. Eng., IEEE Trans.* **21**, 949–958 (2013).

49. Grant, S., T. Aitchison, E. Henderson, J. Christie, S. Zare, J. McMurray, et al., "A comparison of the reproducibility and the sensitivity to change of visual analogue scales, borg scales, and likert scales in normal subjects during submaximal exercise∗," *CHEST J.* **116**, 1208–1217 (1999).

50. Grieshaber, D. C., and T. J. Armstrong, "Insertion loads and forearm muscle activity during flexible hose insertion tasks," *Hum. Factors* **49**, 786–796 (2007).

51. Guettler, K., H. Jahren, and K. Hartviksen, "On the muscular activity of the performing violinist," in *British Performing Arts Medicine Trust: Health and the Musician*, York, Chico, CA, 1997.

52. Hägg, G. M., "Static work load and occupational myalgia—A new explanation model," in P. Anderson, D. Hobart, and J. Danoff, eds., *Electromyographical Kinesiology*, Elsevier Science Publishers, Amsterdam, 1991.

53. Hägg, G. M., "Human muscle fibre abnormalities related to occupational load," *Eur. J. Appl. Physiol.* **83**, 159–165 (2000).

54. Henneman, E., "Relation between size of neurons and their susceptibility to discharge," *Science* **126**, 1345–1347 (1957).

55. Hof, A. L., C. N. Pronk, and J. A. van Best, "Comparison between EMG to force processing and kinetic analysis for the calf muscle moment in walking and stepping," *J. Biomech.* **20**, 167–178 (1987).

56. Holobar, A., and D. Zazula, "Multichannel blind source separation using convolution kernel compensation," *IEEE Trans. Signal Processing* **55**, 4487–4496 (2007).

57. Holtermann, A., C. Grönlund, J. Stefan Karlsson, and K. Roeleveld, "Spatial distribution of active muscle fibre characteristics in the upper trapezius muscle and its dependency on contraction level and duration," *J. Electromyogr. Kinesiol.* **18**, 372–381 (2008).

58. Holtermann, A., K. Roeleveld, P. J. Mork, C. Grönlund, J. S. Karlsson, L. L. Andersen, et al., "Selective activation of neuromuscular compartments within the human trapezius muscle," *J. Electromyogr. Kinesiol.* **19**, 896–902 (2009).

59. Iritani, T., I. Koide, and Y. Sugimoto, "Strategy for health and safety management at an automobile company—from the prevention of low back pain to Toyota's verification of assembly line (TVAL)," *Indust. Health* **35**, 249–258 (1997).

60. Istenič, R.,"EMG-based muscle force estimation by convolution kernel compensation: research report," Faculty of Electrical Engineering and Computer Science, System Software Laboratory, University of Maribor, Slovenia (2006).

61. Iwakiri, K., S. Yamauchi, and A. Yasukouchi, "Effects of a standing aid on loads on low back and legs during dishwashing," *Indust. Health* **40**, 198–206 (2002).

62. Jennifer Wales, B. K.,"3D movement and muscle activity patterns in a violin bowing task," Master of Science, Applied Health Sciences, Brock University, 2007.

63. Jensen, C., O. Vasseljen, and R. H. Westgaard, "The influence of electrode position on bipolar surface electromyogram recordings of the upper trapezius muscle," *Eur. J. Appl. Physiol. Occup. Physiol.* **67**, 266–273 (1993).

64. Kallenberg, L. A., and H. J. Hermens, "Behaviour of motor unit action potential rate, estimated from surface EMG, as a measure of muscle activation level," *J. NeuroEng. Rehabil.* **3**, 15 (2006).

65. Keenan, K. G., D. Farina, K. S. Maluf, R. Merletti, and R. M. Enoka, "Influence of amplitude cancellation on the simulated surface electromyogram," *J. Appl. Physiol.* **98**, 120–131 (2005).

66. Kenny, D. T., and B. Ackermann, *Optimising Physical and Psychological Health in Performing Musicians*, Oxford University Press, Oxford, UK, pp. 390–400, 2008.

67. Kingma, I., T. Bosch, L. Bruins, and J. H. van Dieën, "Foot positioning instruction, initial vertical load position and lifting technique: effects on low back loading," *Ergonomics* **47**, 1365–1385 (2004).

68. Kofler, M., A. Kreczy, and A. Gschwendtner, "Occupational backache–surface electromyography demonstrates the advantage of an ergonomic versus a standard microscope workstation," *Eur. J. Appl. Physiol.* **86**, 492–497 (2002).

69. Koo, T. K. K., and A. F. T. Mak, "Feasibility of using EMG driven neuromusculoskeletal model for prediction of dynamic movement of the elbow," *J. Electromyogr. Kinesiol.* **15**, 12–26 (2005).

70. Kumar, S., and A. Mital, *Electromyography in Ergonomics*, CRC Press, Boca Raton, FL, 1996.

71. Laursen, B., B. R. Jensen, G. Németh, and G. Sjøgaard, "A model predicting individual shoulder muscle forces based on relationship between electromyographic and 3D external forces in static position," *J. Biomech.* **31**, 731–739 (1998).

72. Laursen, B., K. Søgaard, and G. Sjøgaard, "Biomechanical model predicting electromyographic activity in three shoulder muscles from 3D kinematics and external forces during cleaning work," *Clin. Biomech.* **18**, 287–295 (2003).

73. Lazaro, P., E. Parody, R. Garcia-Vicuna, G. Gabriele, J. A. Jover, and J. Sevilla, "Cost of temporary work disability due to musculoskeletal diseases in Spain," *Reumatol. Clin.* **10**, 109–112 (2014).

74. Lederman, R., "Neuromuscular and musculoskeletal problems in instrumental musicians," *Muscle & Nerve* **27**, 549–561 (2003).

75. Lee, G., T. Lee, D. Dexter, R. Klein, and A. Park, "Methodological infrastructure in surgical ergonomics: a review of tasks, models, and measurement systems," *Surg. Innov.* **14**, 153–167 (2007).

76. Lee, T.-H., "Ergonomic comparison of operating a built-in touch-pad pointing device and a trackball mouse on posture and muscle activity," *Perceptual and Motor Skills* **101**, 730–736 (2005).

77. LeVine, W. R., and J. K. Irvine, "In vivo EMG biofeedback in violin and viola pedagogy," *Biofeedback Self Regul.* **9**, 161–168 (1984).

78. Luttmann, A., M. Jäger, and W. Laurig, "Electromyographical indication of muscular fatigue in occupational field studies," *Int. J. Ind. Ergonom.* **25**, 645–660 (2000).

79. Luttmann, A., M. Jäger, J. Sökeland, and W. Laurig, "Electromyographical study on surgeons in urology. II. Determination of muscular fatigue," *Ergonomics* **39**, 298–313 (1996).

80. Madeleine, P., P. Vedsted, A. K. Blangsted, G. Sjøgaard, and K. Søgaard, "Effects of electromyographic and mechanomyographic biofeedback on upper trapezius muscle activity during standardized computer work," *Ergonomics* **49**, 921–933 (2006).

81. Manal, K., and T. S. Buchanan, "A one-parameter neural activation to muscle activation model: estimating isometric joint moments from electromyograms," *J. Biomech.* **36**, 1197–1202 (2003).

82. Manukyan, G. A., M. Waseda, N. Inaki, J. R. Torres Bermudez, I. A. Gacek, A. Rudinski, et al., "Ergonomics with the use of curved versus straight laparoscopic graspers during rectosigmoid resection: results of a multiprofile comparative study," *Surg. Endosc.* **21**, 1079–1089 (2007).

83. Matern, U., C. Giebmeyer, R. Bergmann, P. Waller, and M. Faist, "Ergonomic aspects of four different types of laparoscopic instrument handles with respect to elbow angle," *Surg. Endosc. Other Intervent. Tech.* **16**, 1528–1532 (2002).

84. Matern, U., G. Kuttler, C. Giebmeyer, P. Waller, and M. Faist, "Ergonomic aspects of five different types of laparoscopic instrument handles under dynamic conditions with respect to specific laparoscopic tasks: An electromyographic-based study," *Surg. Endosc. Other Intervent. Tech.* **18**, 1231–1241 (2004).

85. Mathiassen, S. E., and T. Aminoff, "Motor control and cardiovascular responses during isoelectric contractions of the upper trapezius muscle: evidence for individual adaptation strategies," *Eur. J. Appl. Physiol. Occup. Physiol.* **76**, 434–444 (1997).

86. McLean, L., M. Tingley, R. N. Scott, and J. Rickards, "Computer terminal work and the benefit of microbreaks," *Appl. Ergonom.* **32**, 225–237 (2001).

87. Merletti, R., M. Aventaggiato, A. Botter, A. Holobar, H. Marateb, and T. M. Vieira, "Advances in surface EMG: recent progress in detection and processing techniques," *Crit. Rev. Biomed. Eng.* **38**, 305–345 (2010).

88. Möller, T., S. E. Mathiassen, H. Franzon, and S. Kihlberg, "Job enlargement and mechanical exposure variability in cyclic assembly work," *Ergonomics* **47**, 19–40 (2004).

89. Moraes, G. F., and A. P. Antunes, "Musculoskeletal disorders in professional violinists and violists. Systematic review," *Acta Ortop. Brasil.* **20**, 43–47 (2012).

90. Neely, G., G. Ljunggren, C. Sylven, and G. Borg, "Comparison between the Visual Analogue Scale (VAS) and the Category Ratio Scale (CR-10) for the evaluation of leg exertion," *Int. J. Sports Med.* **13**, 133–136 (1992).

91. Nevala, N., J. Holopainen, O. Kinnunen, and O. Hänninen, "Reducing the physical work load and strain of personal helpers through clothing redesign," *Appl. Ergonom.* **34**, 557–563 (2003).

92. Odell, D., A. Barr, R. Goldberg, J. Chung, and D. Rempel, "Evaluation of a dynamic arm support for seated and standing tasks: a laboratory study of electromyography and subjective feedback," *Ergonomics* **50**, 520–535 (2007).

93. OECD, *Sickness, Disability and Work: Breaking the Barriers—A Synthesis of Findings Across OECD Countries*, Organisation for Economic Co-operation and Development (OECD) Publishing, Paris, 2010.

94. Okunribido, O., and T. Wynn, "Ageing and work-related musculoskeletal disorders. A review of the recent literature," Health and Safety Laboratory for the Health and Safety Executive, Buxton Research Report RR799, 2010.

95. Palmerud, G., R. Kadefors, H. Sporrong, U. L. F. Järvholm, P. Herberts, C. Högfors, et al., "Voluntary redistribution of muscle activity in human shoulder muscles," *Ergonomics* **38**, 806–815 (1995).

96. Palmerud, G., H. Sporrong, P. Herberts, and R. Kadefors, "Consequences of trapezius relaxation on the distribution of shoulder muscle forces: an electromyographic study," *J. Electromyogr. Kinesiol.* **8**, 185–193 (1998).

97. Parent-Thirion, A., E. F. Macías, J. Hurley, and G. Vermeylen, "Fourth European Working Conditions Survey" European Foundation for the Improvement of Living Standards, Dublin, Report ef0698, 22 February 2007.

98. Parent-Thirion, A., G. Vermeylen, G. v. Housten, M. Lyly-Yrjanainen, I. Biletta, and J. Cabrita, "Fifth European working conditions survey," Eurofound, Luxembourg ef1182, 12 April 2012.

99. Peper, E., V. S. Wilson, K. H. Gibney, K. Huber, R. Harvey, and D. M. Shumay, "The integration of electromyography (SEMG) at the workstation: assessment, treatment, and prevention of repetitive strain injury (RSI)," *Appl. Psychophysiol. Biofeedback* **28**, 167–182 (2003).

100. Philipson, L., R. Sorbye, P. Larsson, and S. Kaladjev, "Muscular load levels in performing musicians as monitored by quantitative electromyography," *Med. Problems of Performing Artists* **5**, 79–82 (1990).

101. Pincivero, D. M., A. Coelho, and R. M. Campy, "Gender differences in perceived exertion during fatiguing knee extensions," *Med. Sci. Sports Exerc.* **36**, 109–117 (2004).

102. Potvin, J. R., and S. H. M. Brown, "Less is more: high pass filtering, to remove up to 99% of the surface EMG signal power, improves EMG-based biceps brachii muscle force estimates," *J. Electromyogr. Kinesiol.* **14**, 389–399 (2004).

103. Quick, N. E., J. C. Gillette, R. Shapiro, G. L. Adrales, D. Gerlach, and A. E. Park, "The effect of using laparoscopic instruments on muscle activation patterns during minimally invasive surgical training procedures," *Surg. Endosc. Other Intervent. Tech.* **17**, 462–465 (2003).

104. Rickert, D. L., M. Halaki, K. A. Ginn, M. S. Barrett, and B. J. Ackermann, "The use of fine-wire EMG to investigate shoulder muscle recruitment patterns during cello bowing: The results of a pilot study," *J. Electromyogr. Kinesiol.* **23**, 1261–1268 (2013).

105. Samani, A., and P. Madeleine, "A comparison of cluster-based exposure variation and exposure variation analysis to detect muscular adaptation in the shoulder joint to subsequent sessions of eccentric exercise during computer work," *J. Electromyogr. Kinesiol.* **24**, 192–199 (2014).

106. Sanger, T. D., "Bayesian filtering of myoelectric signals," *J. Neurophysiol.* **97**, 1839–1845 (2007).

107. Schneider, E., and X. Irastorza,"OSH in figures: Work-related musculoskeletal disorders in the EU—Facts and figures," European Agency for Safety and Health at Work, Luxembourg, May 4, 2010.

108. Sillanpää, J., M. Nyberg, and P. Laippala, "A new table for work with a microscope, a solution to ergonomic problems," *Appl. Ergonom.* **34**, 621–628 (2003).

109. Silva, A., and I. Steinbuka,"Health and safety at work in Europe (1999–2007): A statistical portrait," Eurostat, Belgium, 20 July 2010.

110. Soderberg, G. L., W. Marras, R. Lamb, D. Hobart, D. G. Gertleman, T. Cook, et al., *Selected Topics in Surface Electromyography for Use in Occupational Setting: Expert Perspective*, DHHS (NIOSH) Publication No. 91-100, NIOSH Publications and Products, 1992.

111. Spotti, C., L. Tamborlani, and R. Converti,"A rehabilitation service in the theater for orchestra musicians: 14 years of experience," in *12th European Congress and 3rd International Congress on Musician's Medicine*, Milan, pp. 137–138, 2008.

112. Staudenmann, D., K. Idsart, A. Daffertshofer, D. F. Stegeman, and J. H. van Dieen, "Improving EMG-based muscle force estimation by using a high-density EMG grid and principal component analysis," *IEEE Trans. Biomed. Eng.* **53**, 712–719 (2006).

113. Tepper, M., M. M. R. Vollenbroek-Hutten, H. J. Hermens, and C. T. M. Baten, "The effect of an ergonomic computer device on muscle activity of the upper trapezius muscle during typing," *Appl. Ergonom.* **34**, 125–130 (2003).

114. Tones, K., and S. Tilford, *Health Promotion: Effectiveness, Efficiency and Equity*, 3rd ed. Nelson Thornes Ltd., Cheltenham, UK, 2001.

115. Troiano, A., F. Naddeo, E. Sosso, G. Camarota, R. Merletti, and L. Mesin, "Assessment of force and fatigue in isometric contractions of the upper trapezius muscle by surface EMG signal and perceived exertion scale," *Gait & Posture* **28**, 179–186 (2008).

116. Uhrich, M. L., R. A. Underwood, J. W. Standeven, N. J. Soper, and J. R. Engsberg, "Assessment of fatigue, monitor placement, and surgical experience during simulated laparoscopic surgery," *Surg. Endosc. Other Intervent. Tech.* **16**, 635–639 (2002).

117. van Dieen, J. H., M. P. De Looze, and V. Hermans, "Effects of dynamic office chairs on trunk kinematics, trunk extensor EMG and spinal shrinkage," *Ergonomics* **44**, 739–750 (2001).

118. Veiersted, K. B., R. H. Westgaard, and P. Andersen, "Pattern of muscle activity during stereotyped work and its relation to muscle pain," *Int. Arch. Occup. Environ. Health* **62**, 31–41 (1990).

119. Voerman, G., L. Sandsjö, M. R. Vollenbroek-Hutten, P. Larsman, R. Kadefors, and H. Hermens, "Effects of ambulant myofeedback training and ergonomic counselling in female computer workers with work-related neck-shoulder complaints: a randomized controlled trial," *J. Occup. Rehabil.* **17**, 137–152 (2007).

120. Westad, C., R. H. Westgaard, and C. J. De Luca, "Motor unit recruitment and derecruitment induced by brief increase in contraction amplitude of the human trapezius muscle," *J. Physiol.* **552**, 645–656 (2003).

121. Westgaard, R. H., "Work-related musculoskeletal complaints: some ergonomics challenges upon the start of a new century," *Appl. Ergonom.* **31**, 569–580 (2000).

122. Westgaard, R. H., and J. Winkel, "Guidelines for occupational musculoskeletal load as a basis for intervention: a critical review," *Appl. Ergonom.* **27**, 79–88 (1996).

123. WHO, *Workplace health promotion—Benefits.* (2009). Available from http://www.who.int/occupational_health/topics/workplace/en/index1.html.

124. Winkemann-Gleed, A.,"Demographic change and implications for workforce ageing in Europe: raising awareness and improving practice," Working Lives Research Institute, London Metropolitan University, London, 2011.

125. Winspur, I., "Controversies surrounding "misuse," "overuse," and "repetition" in musicians," *Hand Clin.* **19**, 325–329 (2003).

126. Woods, S., T. Bridge, D. Nelson, K. Risse, and D. M. Pincivero, "The effects of rest interval length on ratings of perceived exertion during dynamic knee extension exercise," *J. Strength Cond. Res./Nat. Strength Cond. Assoc.* **18**, 540–545 (2004).

14

APPLICATIONS IN PROCTOLOGY AND OBSTETRICS

R. Merletti

Laboratory for Engineering of the Neuromuscular System (LISiN), Politecnico di Torino, Torino, Italy

14.1 INTRODUCTION

The pelvic floor is a thick muscular diaphragm that holds the urogenital organs like a hammock. Its neurocontrol has been described by Vodusek [50]. Muscular disorders of this diaphragm often lead to some degree of fecal or urinary incontinence, a symptom that strongly influences the quality of life of the affected subjects and that is still underreported due to social stigma and embarrassment. For this reason, the social extent and cost of incontinence is underestimated. Although urinary incontinence is quite diffuse, this chapter will focus on the anal sphincter, on the puborectalis muscle, and on fecal incontinence resulting from their iatrogenic lesions.

Faecal incontinence is 9–10 times more common in females than in males, presumably because of child delivery. It affects about 2% of the total population, about 7% of the self-sufficient population above 65 years, and up to 20%–50% of the residents in nursing homes. A few authors estimated social and medical costs of fecal incontinence ranging from 3000 to 6000 US$/year/subject with an average total cost per subject near 17,000 US$ [13,30,35].

The puborectalis and the anal sphincter muscles provide fecal continence. The puborectalis (PR) is shaped like a sling with both sides attached to the pubic bone, as indicated in Figs. 14.1a and 14.1b. The anal sphincter comprises an internal involuntary smooth muscle layer (IAS) and an external voluntary striated muscle

Surface Electromyography: Physiology, Engineering, and Applications, First Edition.
Edited by Roberto Merletti and Dario Farina.
© 2016 by The Institute of Electrical and Electronics Engineers, Inc. Published 2016 by John Wiley & Sons, Inc.

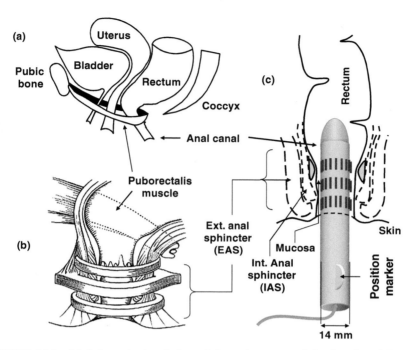

FIGURE 14.1 (**a**) Schematic description of the main organs of the female pelvic region sectioned in the sagittal plane. (**b**) The external anal sphincter (EAS) and the puborectalis muscle. (**c**) Longitudinal section of the anal canal showing the external and internal anal sphincter and a probe carrying three circumferential arrays of 16 electrodes each.

layer (EAS). The innervation of the EAS may be affected by the mode of child delivery [22,51]. The EAS muscle has been studied with the traditional coaxial needle technique [33]. This chapter deals with minimally invasive EMG detection systems based on intra-anal probes.

The motor units of the EAS are arranged circularly, and their action potentials (MUAP) can therefore be detected with a cylindrical probe with electrodes equally spaced along the circumference. Figure 14.1c shows such a probe with three circular arrays of 16 electrodes each, providing a monopolar EMG image of 48 pixels (see Chapter 5). A marker provides a position reference as well as proper and repeatable positioning of the probe in different EMG measurements.

A single-array probe is depicted in Fig. 14.2a and is inserted so that the electrodes are near the anal opening where the EAS is closest to the mucosa (Fig. 14.1c). The existence of a longitudinal anal muscle (LAM), placed in between the IAS and the EAS, is controversial [55].

Some commercially available "anal plugs" feature two solid ring electrodes. Obviously, these systems cannot read MUAPs from the EAS; however, they detect action potentials from the LAM and from a large volume including nearby muscles whose fibers are not arranged circularly. Other commercially available devices feature

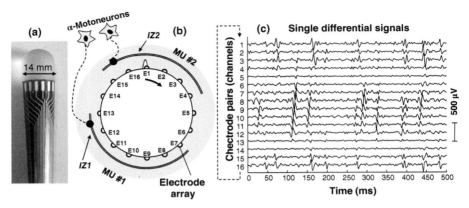

FIGURE 14.2 (**a**) Cylindrical probe carrying one printed circuit circumferential array of 16 electrodes (1 × 5 mm each). (**b**) Cross section of the array and indication of two motor units around it. (**c**) Example of half a second signal epoch of a single differential recording, showing MUAPs of motor units innervated under electrodes 12–13 (propagating counterclockwise) and under electrodes 15–16 (propagating clockwise). Modified from Merletti et al. [36] with permission.

two longitudinal strips on opposite sides of the "anal plug." They also detect signals from a large and poorly defined tissue volume. The limitations of a single bipolar detection system have been outlined in other chapters of this book and are particularly strong in the case of the EAS and puborectalis muscles (Chapters 4 and 5).

The detection distance (radial or longitudinal) depends on many parameters, such as electrode size and shape, mucosa thickness and conductivity, muscle conductivity anisotropy ratio, and so on. Assuming to have a reference circular EAS fiber just external to the mucosa and the IAS (each taken to be 1 mm thick, Fig. 14.1c), the monopolar EMG signal would drop to 50% and to 10% of that produced in the reference position (near the IAS), when the fiber is displaced radially by 1.5 mm and by 4 mm, respectively [37]. These two values cover the range of the EAS thickness. The differential signal would drop faster with distance between the detection electrodes and the source (depending on diameter of the probe and interelectrode distance).

As a consequence, the probes depicted in Figs. 14.1 and 14.2 could detect most of the EAS motor units but not necessarily all of them. Those further away radially would produce MUAPs with amplitudes at the noise level (a few μV_{RMS}).

14.2 EMG AND INNERVATION OF THE EXTERNAL ANAL SPHINCTER

The EAS is usually described as a circular muscle made by a right and a left hemimuscle, respectively innervated by the right and left pudendal nerves; however, Wunderlich and Swash [54] demonstrated overlapping of the innervation by the two pudendal nerves.

The motor units are arranged along the circumference of the anal canal, to form a contractile ring. An electrode array designed to study this muscle should therefore be cylindrical and have its contact areas distributed along a circumference, as depicted in Fig. 14.2a, in order to detect the propagation of motor unit action potentials and their staring point—that is, the innervation zone (IZ)—as indicated in the cross-section image of Fig. 14.2b. It should be observed that, differently from skeletal muscles, the 16-electrode circular array provides 16 (not 15) sequential single differential signals (channels), since the electrode pair 1–16 follows the pair 16–15 [36]. Intra-anal probes (as well as electrode arrays glued on gloved fingers) may require a lubricant for easier insertion: The lubricant can be neither a conductive gel (it would short out the electrodes), nor an oily insulator (it would prevent the contact between the electrodes and the mucosa). Ultrasound gel is electrically conductive and not suitable for the purpose. A tiny drop of glycerol on the probe has been found to be a reasonable compromise.

The recorded single differential channels depicted in Fig. 14.2c present a number of propagating patterns corresponding to MUAPs of motor units that are not innervated in the middle, as is usually the case for skeletal muscles. All the motor units whose MUAPs are depicted in Fig. 14.2c show predominantly monodirectional propagation; that is, they are innervated at one end, rather than in the center. Propagation patterns initiate at channel 13 in counterclockwise direction (for MU # 1) and at channel 15 in clockwise direction (for MU # 2). More than one motor unit is innervated at each location. Furthermore, EMG amplitude is greater in channels 7–12 and channels 15–3 than in channels 4–6 and in channels 13–14 (Fig. 14.2).

MUAPs from two additional subjects are depicted in Figs. 14.3a and 14.3b. The IZs are identified with a circle, and the propagating patterns are outlined with straight lines. In both cases, no motor unit appears to be innervated in the middle; all motor units are innervated at (or near) one end. Subject 1 shows innervation locations at channels 2–3, channel 12, and channels 15–16 while subject 2 shows innervation locations at channel 2, channel 5, and channels 16–1. In both cases, more than one motor unit is innervated at each location [36].

In 2004, within the European Project "On Asymmetry in Sphincters," Enck et al. [15–17] investigated the distribution of the innervation zone locations in 15 healthy males and 37 healthy female subjects. They found the distributions reported in Fig. 14.4. These distributions do not show a predominant location; rather, the distribution of IZ locations appears to be rather uniform, indicating a very large intersubject variability. Each of the 16 locations has about the same probability of representing an IZ location (approximately 1/16, that is 6.25%). This situation is very different from that of skeletal muscles where the IZ location(s) are quite similar, across subjects, for each given muscle [2].

Another important difference between the EAS and the skeletal muscles concerns the concept of interelectrode distance and muscle fiber conduction velocity. The circular nature of the 16-electrode array implies that each monopolar or differential channel spans an angle of $2\pi/16$ radiants (0.3926 radiants) and conduction velocity should be considered as angular (rad/s) rather than linear (m/s). As a consequence, motor units located radially at different distances from the surface of the array

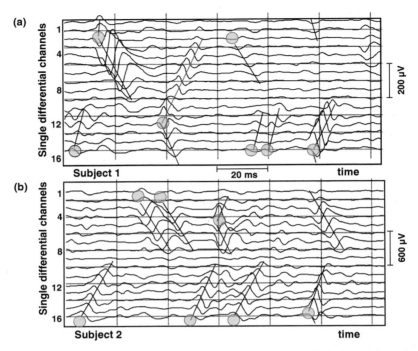

FIGURE 14.3 Examples of a differential 128-ms EMG recording from the 16 channels of a cylindrical array during a voluntary contraction of the external anal sphincter of two subjects. Propagating MUAPs and innervation zones are respectively outlined with lines and with gray circles. Reproduced from Merletti et al. [36] with permission.

(mucosa of the anal canal) and, having the same linear conduction velocity of radially closer motor units, have different angular velocities and their propagation patterns show slopes lower than those of closer motor units. As an example, consider an intra-anal probe with a radius of 7 mm (Fig. 14.2a); further consider a motor unit whose fibers have an average radial distance R_1 of 10 mm from the axis of the probe (3 mm from the surface) and a second motor unit whose fibers have an average radial distance R_2 of 15 mm from the axis of the probe (8 mm from the surface). Assuming that both motor units have the same linear conduction velocity v of 4 m/s (4 mm/ms), the first motor unit will show a MUAP propagation pattern with a slope of $\omega = v/R_1 = 0.40$ rad/ms with an interchannel time shift of $(2\pi/16)/\omega = 0.98$ ms/ch whereas the second will show a MUAP propagation pattern with a slope of $\omega = v/R_2 = 0.26$ rad/ms with an interchannel time shift of $(2\pi/16)/\omega = 1.47$ ms/ch. As a consequence, the slope of the propagation patterns (straight lines in Fig. 14.3) is inversely proportional to the radial distance of the motor unit from the center of the probe and directly proportional to the linear muscle fiber conduction velocity.

The distribution of IZs of the EAS shows a very large intersubject variability. Such distribution, as well as the EMG activity of the EAS, may be more or less symmetric. The relevance of asymmetry has been investigated within the EU project "On

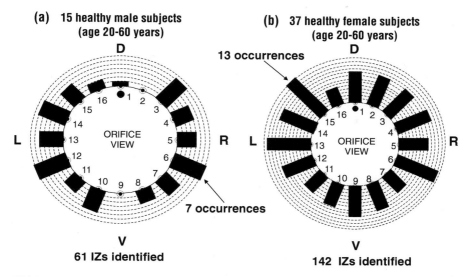

FIGURE 14.4 Circular histograms of the number of motor units with innervation zones observed in the external anal sphincter of 15 healthy males (**a**) and 37 healthy nonpregnant females (**b**). The probability distribution is approximately uniform in these two groups, suggesting that there is no preferred zone of innervation (see also Fig. 14.12). Reproduced from Enck et al. [15] with permission.

Asymmetry In Sphincters (OASIS)" and will be outlined in Section 14.5. Figure 14.4 shows the histogram of the distribution of IZs observed in 15 healthy males and 37 healthy female subjects by Enk et al. [15]. The distribution is substantially uniform (slightly bimodal), indicating that the probability of finding an IZ under a specific electrode is about 1/16.

Probes with multiple circular electrode arrays, forming a cylindrical grid, provide more information and facilitate (a) the decomposition of surface EMG into the constituent MUAP trains—for example, by means of the CKC algorithm (Chapter 7 and references [28,29])—and (b) the localization and orientation of motor units.

Figure 14.5a shows such a probe [5] with seven "columns" and 16 "rows" while Fig. 14.5b shows the superposition of 30 MUAPs (or discharges, or firings) of a single motor unit identified with the CKC decomposition algorithm; the signal "columns" of 16 channels each are presented one below the other while in Fig. 14.5c they are presented one beside to the other. The motor unit is innervated under electrode 14 and extends from ch. 7 to ch. 4–5, embracing most of the circle, and its MUAP is detected under columns 4 to 7 (Fig. 14.5a,b). Most motor units of the EAS are indeed detected near the orifice, where the mucosa is thin and the IAS is absent (Fig. 14.1c and Fig. 14.5), and a single circumferential array is usually sufficient to detect many of them. A two-dimensional array (Fig. 14.5a) provides considerable information about the arrangement of the motor units in the anal canal and can be considered a tool for noninvasive "electrophysiological anatomy" [8].

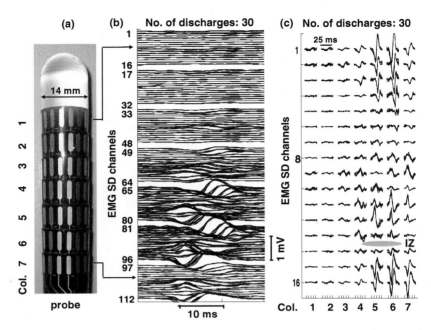

FIGURE 14.5 (a) Intra-anal probe with 7×16 two-dimensional electrode array (patent pending). (b) Single differential EMG of a single motor unit discharging 30 action potentials as seen from the seven columns. (c) Representation of the same potential in two-dimensional spatial coordinates and time.

An accurate estimation of the location of the IZ of MU of the EAS is important in relation to lesions associated to child delivery (Section 14.5). Assessing the repeatability of this estimate is, of course, equally important to prove that the outcome is not affected by the operator, possible discomfort, and other factors. Repeatability of the IZ in successive tests was observed to be within one angular interelectrode distance, that is, $2\pi/16$ rad [18].

14.3   EMG AND INNERVATION OF THE PUBORECTALIS MUSCLE

The puborectalis muscle (Figs. 14.1a and 14.1b) can be clearly felt by digital examination, above the anal sphincter. A flexible disposable screen-printed or printed-circuit electrode array (8 electrodes), glued to the glove on the index fingertip, as indicated in Fig. 14.6a, provides means to detect the EMG signals from this muscle on the right, left, or posterior side [7]. Figures 14.6b and 14.6c show these three positions and examples of signals detected from two subjects. It is clear that the innervation zones of the puborectalis are lateral (right and left) and the motor units connect, presumably to connective tissue, in the posterior region, with some overlap. The technique described in Fig. 14.6 provides information about the symmetry, the

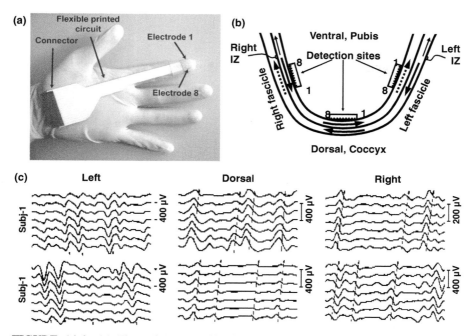

FIGURE 14.6 (**a**) Electrode array with eight contacts, placed on a gloved finger for exploration of the puborectalis muscle. (**b**) Three positions of the array: on the left fascicle, the right fascicle, and the dorsal aspect of the canal. (**c**) Examples of differential EMG signals detected in the three positions from two subjects and showing MUAPs propagating in different directions. Reproduced from Cescon et al. [7] with permission.

length, and the number of the motor units in the two fascicles. As in the case of the sphincter, lubricants must be used with caution because they might either short the electrodes together (ECG or ultrasound gels) or isolate them from the mucosa (oily lubricants).

14.4 MODELING OF THE EMG OF THE ANAL SPHINCTER

Farina et al. [19] developed a model with multilayer description of the volume conductor. This type of model allows simulation of cylindrical structures with external electrodes (like a limb) or with internal electrodes (like a sphincter). Simulations of two single fibers of a sphincter are depicted in Fig. 14.7. Figure 14.7a describes the system: letters A–R identify the electrodes (and monopolar signals) while numbers 1–16 identify the differential channels. The mucosa is 1 mm thick. The two fibers F1 and F2 have the same angular extension (but different length) and conduction velocity (2.3 m/s) and radial distance of 2 mm and 5 mm from the probe surface (including 1-mm mucosa). Figures 14.7b and 14.7c depict the monopolar signals produced by the two fibers and detected by the A–R electrodes. The

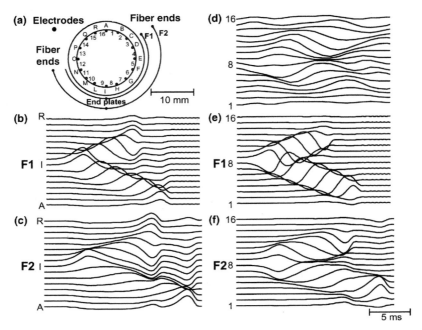

FIGURE 14.7 Examples of monopolar (**b, c**) and single differential (**e, f**) signals produced by two simulated sphincter fibers with the geometry described in (**a**). In part a the letters A–R indicate the electrodes (i.e., the monopolar channels), while the numbers 1–16 indicate single differential channels. The model assumes an insulating probe, a 1-mm-thick mucosa, and a radially infinite muscle layer with anisotropic conductivity tensor. Two fibers (F1 and F2) at a depth of 2- and 5-mm distance from the probe surface have been simulated. A low conduction velocity value 2.3 m/s has been chosen on the basis of experimental observations. The angular extension of the two fibers is 200°. To enhance shape differences, in all panels, the potentials are normalized with respect to their maximum values; thus their amplitudes cannot be compared. (**d**) A motor unit action potential extracted from an experimental recording is shown. There was no attempt to match the experimental signals with the simulated ones. The experimental signals are reported only as representative examples of the features of surface EMG signals detected from the anal sphincter. Reproduced from [19] with permission.

relative end-of-fiber effect is quite evident, particularly for the more distal fiber F2, as expected (see Chapter 2). Figures 14.7e and 14.7f display the simulated differential signals (channels 1–16) generated by the two fibers. Figure 14.7d shows an experimentally detected MUAP whose pattern resembles that of Fig. 14.7f. The end-of-fiber effect is less evident in Fig. 14.7d, presumably because of the dispersion of endings of the fibers of the motor unit being greater than in the simulated case.

The circular geometry of the motor units of the EAS also allows easier removal of common mode interference and reconstruction of monopolar signals from the differential signals [38].

14.5   CHILD-DELIVERY-RELATED LESIONS AND EAS DENERVATION

Recent data [16,17,52] suggest that either natural or iatrogenic asymmetry of innervation of the EAS may contribute to the occurrence and severity of incontinence symptoms resulting from aging or pelvic floor trauma, in particular after child delivery [51]. The association of pelvic floor trauma with child delivery modality (in particular to episiotomy) is discussed in many studies [1,3,4,11,12,14,23]. Although right EAS partial denervation consequent to right mediolateral episiotomy has been ascertained [11], its role in determining fecal incontinence at a later age is still being investigated [20,22,24–27,31,32,34,40–42].

Episiotomy is the surgical enlargement of the vaginal canal by an incision of the perineum during the last part of the second stage of labor. This procedure is done with scissors or scalpel and requires repair by suturing [48]. This intervention produces a clean surgical cut whose purpose is to reduce the likelihood of a ragged tear or laceration more difficult to suture [21]. It is believed to minimize pressure on the fetal head and shorten the second stage of labor. The surgical incision of the recto-vaginal septum is directed to the right or left, at an angle of about 40 degrees. In almost all cases it is performed to the right because most operators are right-handed. Midline episiotomy, performed in the past, is now abandoned as unsafe. Mediolateral episiotomy has claimed benefits but has many risks and may later lead to incontinence [46,47,51–53].

It is intuitive that performing medio-lateral episiotomy on the side carrying fewer axonal branches innervating the EAS would reduce the risk of partial denervation of this muscle. This concept is illustrated in Fig. 14.8. Identification of the location of the

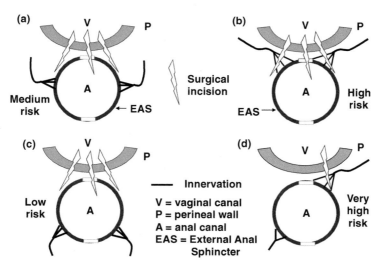

FIGURE 14.8   The surgical incision (episiotomy) performed medially or mediolaterally (predominantly on the right because operators are mostly right-handed) may imply different levels of risk of EAS denervation, depending on the individual innervation pattern. Four innervation patterns are depicted. Knowledge of the individual pattern is expected to help the operator to choose the right or left side, or opt for no episiotomy.

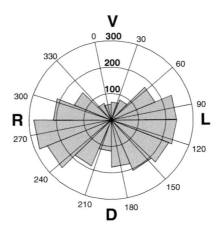

N = 2748 motor units

FIGURE 14.9 Innervation zone distribution in the detected motor units of the external anal sphincter of 500 pregnant women. (See also Figs. 14.4 and 14.12). Reprinted from Cescon et al. [9] with permission.

innervation zones of motor units of the EAS is therefore a meaningful task and a research challenge that is being actively pursued [6,49].

Considering that innervation is about equally probable in any location around the EAS (Fig. 14.4b) and that the length of the episiotomy incision might extend for about 2–3 channels of the 16-channel detection system, the estimated probability of damaging nerve fibers or axonal branches, and causing some degree of EAS denervation consequent to episiotomy, is in the range of 12.5% to 18.7%. However, some motor units may be more important than others because of their size or because of their higher discharge rate. Their IZs should therefore be "weighted" to attribute to them greater importance than to others. This leads to the concept of "weighted" distribution of IZ [39].

Reinnervation might take place after lesion, but this phenomenon has not been investigated. If reinnervation takes place, at the growth rate of axons, or of axonal branches, of the order of 1–2 mm/day [45] it should be completed within 2–3 weeks.

A recent multicenter clinical study [9] investigated the distribution of IZ of EAS motor units in 500 pregnant women and found it to be less uniform than in the previous work by Enck et al. [15] and slightly bimodal (right and left peaks) as indicated in Fig. 14.9. Figure 14.10 shows the distribution of IZ for one subject. About 350 of the 500 women participating to this study [9,10] were tested again six weeks after delivery, when any possible spontaneous reinnervation, following laceration or episiotomy, was very likely completed. Eighty-six of the subjects underwent episiotomy, and many of these presented post-partum changes of the EAS IZs and of the EMG amplitude distribution around the probe circumference. Figure 14.11 shows one case while Fig. 14.12 shows the global distribution of IZs before and after delivery with episiotomy [10].

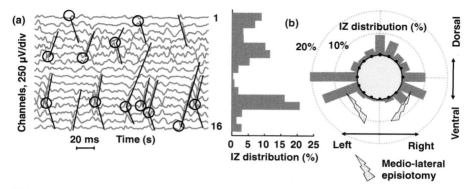

FIGURE 14.10 Innervation zone distribution of the detected motor units of the external anal sphincter of a single subject. (**a**) Raw single differential EMG signals and outline of motor units and of their innervation zones (short epoch of 0.2 s). (**b**) Linear and circular histograms of the innervation zone distribution for the subject in part a, obtained over a 10-s epoch. Reprinted from Enck et al. [18] with permission.

Anal sphincter motor units and innervation zones

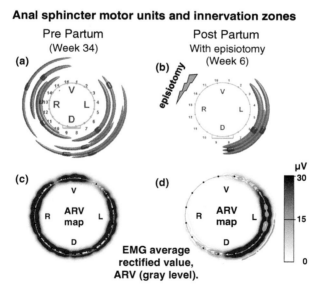

FIGURE 14.11 (**a**) Distribution of the innervation zones of the motor units detected from the external anal sphincter of a woman five weeks prepartum and (**b**) six weeks postpartum with right mediolateral episiotomy. Eight motor units were detected prepartum and three postpartum. Each gray circle arc represents a motor units and the black sleeve on it represents the innervation zone. (**c**) Distribution of EMG intensity (average rectified value, ARV) of each channel, represented as level of gray, prepartum and (**d**) postpartum. Despite the significant loss of motor units and muscle activity, no incontinence was reported six weeks postpartum. Reprinted from [10] with permission.

Loss of innervation zone in RV quadrant of women after mediolateral right episiotomy

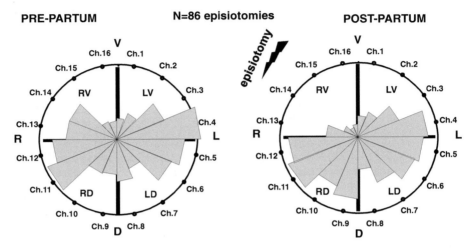

FIGURE 14.12 Circular histogram of the distribution of innervation zones of the external anal sphincter in 86 women who delivered a child with right mediolateral episiotomy. The loss of innervation in the right-ventral quadrant of the muscle is statistically significant ($p < 0.05$).

Prepartum screening of the EAS innervation and the recording of this information on the proposed "perineal card" could "guide" episiotomy (when strictly required) and reduce the prevalence of fecal incontinence in women [23,41,44]. In addition, it would also reduce the likelihood of litigations for cases of medical malpractice [43].

REFERENCES

1. Andrews, V., A. H. Sultan, R. Thakar, and P. W. Jones, "Risk factors for obstetric anal sphincter injury: a prospective study," *Birth* **33**, 117–122 (2006).
2. Barbero M., R. Merletti, and A. Rainoldi, *Atlas of Muscle Innervation Zones: Understanding Surface Electromyography and Its Applications*, Springer, Berlin, 2012.
3. Borello-France, D., K. L. Burgio, H. E. Richter, H. Zyczynski, M. P. Fitzgerald, W. Whitehead, P. Fine, I. Nygaard, V. L. Handa, A. G. Visco, A. M. Weber, and M. B. Brown, "Fecal and urinary incontinence in primiparous women," *Obstet. Gynecol.* **108**, 863–872 (2006).
4. Carroli, G., and J. Belizan, "Episiotomy for vaginal birth," Cochrane Database System Review CD000081 (2000).
5. Cescon, C., and R. Merletti, "Conic Disposable Rectal Probe," International patent N. TO2009A000814 (2010).
6. Cescon, C., "Automatic location of muscle innervation zones from multi-channel surface EMG signals," Proceedings MeMeA2006, Benevento, Italy, 2006.

7. Cescon, C., A. Bottin, F. Azpiroz, and R. Merletti, "Detection of individual motor units of the puborectalis muscle by non-invasive EMG electrode arrays" *J. Electromyogr. Kinesiol.* **28**, 382–389 (2008).

8. Cescon, C., L. Mesin, M. Nowakowski, and R. Merletti, "Geometry assessment of anal sphincter muscle based on monopolar multichannel surface EMG signals," *J. Electromyogr. Kinesiol.* **21**, 394–401 (2011).

9. Cescon C., E. E. Raimondi, V. Začesta, K. Drusany-Starič, K. Martsidis, and R. Merletti "Characterization of the motor units of the external anal sphincter in pregnant women with multichannel surface EMG," *Int. Urogynecol. J.* **25**, 1097–1103 (2014).

10. Cescon, C., D. Riva, V. Začesta, K. Drusany-Starič, K. Martsidis, O. Protsepko, K. Baessler, and R. Merletti, "Effect of vaginal delivery on external anal sphincter muscle innervation pattern evaluated with multichannel surface EMG (Results of the multicenter study TASI-2)," *Int. Urogynecol. J.* **25**, 1491–1499 (2014).

11. Christianson, L. M., V. E. Bovbjerg, E. C. McDavitt, and K. L. Hullfish, "Risk factors for perineal injury during delivery," *Am. J. Obstet. Gynecol.* **189**, 255–260 (2003).

12. De Leeuw, J. W., M. E. Vierhout, P. C. Struijk, W. C. Hop, and H. C. Wallenburg, "Anal sphincter damage after vaginal delivery: functional outcome and risk factors for fecal incontinence," *Acta Obstet. Gynecol. Scand.* **80**, 830–834 (2001).

13. Deutekom, M., A. C. Dobben, M. G. Dijkgraaf, M. P. Terra, J. Stoker, and P. M. Bossuyt, "Costs of outpatients with fecal incontinence." *Scand. J. Gastroenterol.* **40**, 552–558 (2005).

14. Dudding, T. C., C. J. Vaizey, and M. A. Kamm, "Obstetric anal sphincter injury: incidence, risk factors, and management," *Ann Surg.* **247**, 224–237 (2008).

15. Enck, P., H. Franz, F. Azpiroz, X. Fernandez-Fraga, H. Hinninghofen, K. Kaske-Bretag, A. Bottin, S. Martina, and R. Merletti, "Innervation zones of the external anal sphincter in healthy male and female subjects," *Digestion* **69**, 123–130 (2004a).

16. Enck, P., H. Hinninghofen, H. Wietek, and H. D. Becker, "Functional asymmetry of pelvic floor innervation and its role in the pathogenesis of fecal incontinence," *Digestion* **69**, 102–111 (2004b).

17. Enck, P., H. Hinninghofen, R. Merletti, and F. Azpiroz, "The external anal sphincter and the role of surface electromyography," *Neurogastroenterol. Motil.* **17**(Suppl.1), 60–67 (2005).

18. Enck, P., H. Franz, E. Davico, F. Mastrangelo, L. Mesin, and R. Merletti, "Repeatability of innervation zone identification in the external anal sphincter muscle," *Neurourol. Urodyn.* **29**, 449–457 (2010).

19. Farina, D., L. Mesin, S. Martina, and R. Merletti, "A surface EMG generation model with multilayer cylindrical description of the volume conductor," *IEEE Trans. Biomed. Eng.* **51**, 415–426 (2004).

20. Fenner, D., "Anal incontinence: relationship to pregnancy, vaginal delivery, and cesarean section," *Semin. Perinatol.* **30**, 261–266 (2006).

21. Fitzgerald, M. P., A. M. Weber, N. Howden, G. W. Cundiff, and M. B. Brown, "Risk factors for anal sphincter tear during vaginal delivery," *Obstet. Gynecol.* **109**, 29–34 (2007).

22. Franz H., H. Hinninghofen, A. Kowalski, R. Merletti, and P. Enck "Mode of delivery affects anal sphincter innervation." *Gastroenterology* **130**(Suppl.2) S724 (2006).

23. Fritel, X., J. Schaal, A. Fauconnier, V. Bertrand, C. Levet, and A. Pigné, "Pelvic floor disorders four years after first delivery: a comparative study of restrictive versus systematic episiotomy," *BJOG* **115**, 247–252 (2008).

24. Gee, A. S., R. S. Jones, and P. Durdey, "On line quantitative analysis of surface electromyography of the pelvic floor in patients with faecal incontinence," *Br. J. Surg.* **87**, 814–818 (2000).

25. Gregory, W. T., J. S. Lou, K. Simmons, and A. L. Clark, "Quantitative anal sphincter electromyography in primiparous women with anal incontinence," *Am. J. Obstet. Gynecol.* **198**, 550.e1–6 (2008).

26. Guise, J. M., C. Morris, P. Osterwell, H. Li, D. Rosenberg, and M. Greenlick, "Incidence of fecal incontinence after childbirth," *Obstet. Gynecol.* **109**, 281–288 (2007).

27. Handa, V. L., J. L. Blomquist, K. C. McDermott, S. Friedman, and A. Muñoz, "Pelvic floor disorders after vaginal birth: effect of episiotomy, perineal laceration, and operative birth," *Obstet Gynecol.* **119**(2Pt 1), 233–239 (2012).

28. Holobar A,. and D. Zazula "Multichannel blind source separation using convolution kernel compensation," *IEEE Trans. Signal Process.* **55**, 4487–4496 (2007).

29. Holobar, A."On repeatability of motor unit identification in multi-channel surface EMG of the external sphincter muscle," *16th International Electrotechnical and Computer Science Conference, Portorož, Slovenia*, pp. 24–26, 2007.

30. Kamm, M. A. "Faecal incontinence," *BMJ.* **327**, 1299–1300 (2003).

31. Kudish, B., S. Blackwell, S. G. McNeeley, E. Bujold, M. Kruger, S. L. Hendrix, and R. Sokol, "Operative vaginal delivery and midline episiotomy: a bad combination for the perineum," *Am. J. Obstet. Gynecol.* **195**, 749–754 (2006).

32. Lewicky-Gaupp, C., Q. Hamilton, J. Ashton-Miller, M. Huebner, J. Q. DeLancey, and D. E. Fenner, "Anal sphincter structure and function relationships in aging and fecal incontinence," *Am. J. Obstet. Gynecol.* **200**, 559.e1–5 (2009).

33. López, A., Nilsson, B. Y., Mellgren, A., Zetterström, J. and Holmström, B. "Electromyography of the external anal sphincter: comparison between needle and surface electrodes," *Dis. Colon Rectum* **42**, 482–485 (1999).

34. McKinnie, V., S. E. Swift, W. Wang, P. Woodman, A. O'Boyle, M. Kahn, M. Valley, D. Bland, and J. Schaffer, "The effect of pregnancy and mode of delivery on the prevalence of urinary and fecal incontinence," *Am. J. Obstet. Gynecol.* **193**, 512–517 (2005).

35. Melgren, A., L. L. Jensen, Zetterstrom, W. D. Wong, J. H. Hofmeister, and A. C. Lowry, "Long-term cost of fecal incontinence secondary to obstetric injuries," *Dis. Colon Rectum* **42**, 857–867 (1999).

36. Merletti, R., A. Bottin, C. Cescon, D. Farina, L. Mesin, M. Gazzoni, S. Martina, M. Pozzo, A. Rainoldi, and P. Enck, "Multichannel surface EMG for the non-invasive assessment of the anal sphincter muscle," *Digestion* **69**, 112–122 (2004).

37. Mesin L., and R. Gervasio, "Detection volume of simulated electrode systems for recording sphincter muscle electromyogram," *Med. Eng. Phys.*, **30**, 896–904 (2008).

38. Mesin L., "Estimation of monopolar signals from sphincter muscles and removal of common mode interference," *Biomed. Signal Proc. Control*, **4**, 37–48 (2008).

39. Mesin, L., M. Gazzoni, and R. Merletti, "Automatic localisation of innervation zones: a simulation study of the external anal sphincter," *J. Electromyogr. Kinesiol.* **19**, 413–421 (2009).

40. Murphy, D., M. Macleod, R. Bahl, K. Goyder, L. Howarth, and B. Strachan, "A randomized controlled trial of routine versus restrictive use of episiotomy at operative vaginal delivery: a multicenter pilot study," *BJOG.* **115**, 1695–1703 (2008).

41. Nelson, R. L., "Epidemiology of faecal incontinence," *Gastroenterology* **126**, 3–7 (2004).

42. Oberwalder, M., A. Dinnewitzer, M. K. Baig, K. Thaler, K. Cotman, J. J. Nogueras, E. G. Weiss, J. Efron, A. M. Vernava, and S. D. Wexner, "The association between late-onset fecal incontinence and obstetric anal sphincter defects," *Arch. Surg.* **139**, 429–432 (2004).

43. Paternoster, M., P. Di Lorenzo, M. Niola, and V. Graziano, "Episiotomia: esigenze cliniche ed implicazioni medico-legali," Arch. Med. Legale, *Sociale Criminol. "Zacchia"* **31**, 143–171 (2013) (in Italian).

44. Pirro, N., B. Sastre, and I. J. Sielezneff, "What are the risk factors of anal incontinence after vaginal delivery?," *J. Chir. (Paris)* **144**, 197–202 (review in French) (2007).

45. Recknor, J. B., and S. K. Mallapragada, "Nerve regeneration: tissue engineering strategies," in *The Biomedical Eng. Handbook: Tissue Engineering and Artificial Organs*, J. D. Bronzino, ed., Taylor & Francis, New York, 2006.

46. Sartore, A., F. De Seta, G. Maso, R. Pregazzi, E. Grimaldi, and S. Guaschino, "The effects of mediolateral episiotomy on pelvic floor function after vaginal delivery," *Obstet. Gynecol.* **103**, 669–673 (2004).

47. South, M. M., S. S. Stinnet, D. B. Sanders, and A. C. Weidner, "Levator ani denervation and reinnervation six months after childbirth," *Am. J. Obstet. Gynecol.* **200**, 519.e1–7 (2009).

48. Thacker, S. B., and H. D. Banta, "Benefits and risks of episiotomy: an interpretative review of the English language literature: 860–1980," *Obstet. Gynecol. Surv.* **38**, 322–338 (1983).

49. Ullah K., C. Cescon, B. Afsharipour, and R. Merletti "Automatic detection of motor unit innervation zones of the external anal sphincter by multichannel surface EMG," *J. Electromyogr. Kinesiol.* **24**, 860–867 (2014).

50. Vodušek, D., "Anatomy and neurocontrol of the pelvic floor," *Digestion* **69**, 87–92 (2004).

51. Wheeler, T. L., and H. E. Richter, "Delivery method, anal sphincter tears and fecal incontinence: new information on a persistent problem," *Curr. Opin. Obstet. Gynecol.* **19**, 474–479 (2007).

52. Wietek, B. M., H. Hinninghofen, E. C. Jehle, P. Enck, and H. B. Franz, "Asymmetric sphincter innervation is associated with fecal incontinence after anal sphincter trauma during childbirth," *Neurourol Urodyn.* **26**, 134–139 (2007).

53. Woolley, R. J., "Benefits and risks of episiotomy: a review of the English-language literature since 1980. Part I and Part II," *Obstet. Gynecol. Surv.* **50**, 806–835 (1995).

54. Wunderlich, M., and M. Swash, "The overlapping innervation of the two sides of the external anal sphincter by the pudendal nerves," *J. Neurol. Sci.* **59**, 97–109 (1993).

55. Zbar A., M. Guo, and M. Pescatori, "Anorectal morphology and function analysis: the Shafik legacy," *Tech. Coloproctol.* **12**, 191–200 (2008).

15

EMG AND POSTURE IN ITS NARROWEST SENSE

T. M. Vieira,[1] D. Farina,[2] and I. D. Loram[3]

[1]*Laboratory for Engineering of the Neuromuscular System, Politecnico di Torino, Torino, Italy, and Escola de Educação Física e Desportos, Universidade Federal do Rio de Janeiro, Brasil*
[2]*Department of Neurorehabilitation Engineering, University Medical Center Göttingen, Göttingen, Germany*
[3]*Cognitive Motor Function Research Group, School of Healthcare Science, Manchester Metropolitan University, United Kingdom*

15.1 INTRODUCTION

In a broad sense, there is a repertoire of postures the human body might assume. Sustaining a specific configuration of the body and its segments is an inherent demand of every single motor task. Any bodily configurations (i.e., postures) could not be maintained without muscle activation. In anticipation of and during movements, the nervous system coordinates the timing and the degree of activity of a specific set of skeletal muscles. When timely activated, these muscles ensure appropriate opposition to the continuous downward pull of gravity; posture is maintained steadily.

Surface electromyography has been used to identify which muscles could possibly have a significant postural role. Typically, such postural role is inspected by recording surface electromyograms (EMGs) from different muscles when subjects sustain their body in different postures. A classic, elegant example relating EMG to posture is provided in the observational study of Portnoy and Morin [72]. With pairs of brass suction surface electrodes, these authors visually inspected the activation of leg, thigh and trunk muscles in different, postural circumstances. From the normal standing

Surface Electromyography: Physiology, Engineering, and Applications, First Edition.
Edited by Roberto Merletti and Dario Farina.
© 2016 by The Institute of Electrical and Electronics Engineers, Inc. Published 2016 by John Wiley & Sons, Inc.

posture, when subjects leaned forward over their ankle, distal and proximal dorsal muscles were activated collectively. When subjects maximally flexed their trunk, however, the sacrospinalis muscle was surprisingly silenced while the other dorsal muscles in the thigh and leg actively sustained the forward body lean (cf. Fig. 2 in Portnoy and Moring [72]). Crucially, these results indicate that (i) movements occurring over a single joint elicit the activation of muscles spanning a set of body joints and (ii) the relative contribution of certain muscles to posture maintenance changes in certain conditions. More generally, these results indicate how complex it might be to study muscle activation and posture in its broad sense (see also Chapter 6).

The activation of different skeletal muscles, however, can be well characterized for a specific posture. Here, we deal with electromyographic studies of the human standing posture. Basmajian and De Luca [3], in their book *Muscles Alive*, defined the bipedal standing as the narrowest sense of posture: "In the narrowest sense, posture may be considered to be the upright, well balanced stance of the human subject in a 'normal' position" [3, p. 252]. In this chapter, specifically, we discuss how surface electromyography has been used early and recently to study posture in its narrowest sense. Focus is given to lower limb muscles, in particular the gastrocnemius muscle.

Surface EMGs have been recorded to answer different questions related to the understanding of human standing control. Some questions were successfully answered while others still remain open. In Sections 15.2 and 15.3, we summarize how, specifically, electromyography has been used to investigate (i) the mechanisms possibly involved in the control of standing posture, (ii) which and how muscles are activated in standing, (iii) the sources of perturbation to the standing posture, (iv) the characteristics and the goal of the feedback system involved in standing control, and (v) the effect of dementias on the activation of postural muscles. Not all of these questions can be investigated with the subjects standing naturally. This is why this chapter is sectioned in "EMG and natural standing" and "EMG and postural perturbations."

In Section 15.4 we further narrow our argument of study to the activation of the human gastrocnemius muscle in standing. Our knowledge on how motor units are represented in the surface EMGs collected from pinnate muscles advanced substantially in the last years. Of greatest relevance to the study of standing posture is the representativeness of gastrocnemius activity in conventional surface EMGs. The issue of representativeness is extensively covered in Section 15.4.

15.2   EMG AND NATURAL STANDING

Since the seminal studies of Hellebrandt [30], the orthostatic posture has been associated with some form of movement. Spontaneous and small postural sways of the body are observed during quiet standing [22,30,87], even when one stands as still as possible [48]. These small oscillations continuously threaten the maintenance of the inherently mechanically unstable, quiet standing posture. If the projection of body center of gravity (CoG), in the transverse anatomical plane, deviates from the ankle axis of rotation by any arbitrarily small distance and in any direction, the gravitational force causes the body to move away from its equilibrium position. Consequently, if not compensated for, the spontaneous postural sways would lead to a fall. Compensation of body sways is aggravated by intersegmental movements, requiring the control of multiple degrees of freedom within a short period [96], and internal sources of disturbances (e.g., respiration and heart beats [34]. While some people exhibit outstanding balancing skills (dancers and gymnasts), others show severe impairments (subjects with neurologic and muscle–skeletal dysfunctions). Efforts have been driven in an attempt to understand how humans succeed in the task of balancing their body in quiet standing posture. However, the control mechanisms have not been completely unraveled, and a widely accepted paradigm of postural control is not currently available.

Remote propositions have suggested that upright stance could be safely achieved without muscular effort. These were mostly based on empirical reports on minimal or absent muscle activity represented in surface EMGs, in particular for the trunk muscles [24,72]. According to this line of reasoning, passive tension in the ligamentous tissues would ensure sufficient resistance to gravity. Two decades ago, the effortless view of standing posture has been biomechanically revisited. While verifying the validity of the mechanical relationship between body CoG and center of pressure (CoP), Winter et al. [96] suggested that intrinsic ankle stiffness could suffice for balancing the body in the natural, upright stance. Direct measurements of ankle stiffness, however, indicate a partial, though modest (from ~45% to ~90%), passive contribution of muscle and tendon tissues to the stabilization of standing posture [7,45,49,55]. The remnant torque necessary for the avoidance of falls must then be driven actively by the nervous system.

15.2.1   What Control Mechanisms Have Been Inferred from Studying EMG in Standing?

Muscle activation during standing has long been observed. First attempts to identify the muscles predominantly active during the standing posture were made in the early fifties. Joseph and Nightingale [39], for example, recorded surface EMGs from different leg muscles while 12 subjects stood quietly. Pairs of circular surface electrodes (1-cm diameter; 2-cm center-to-center distance) were positioned on muscles crossing the ankle ventrally (tibialis anterior) and dorsally (medial and lateral gastrocnemius and soleus). Tibialis anterior was consistently silent across subjects. Surface potentials from the gastrocnemius muscle were recorded for half of the 12 subjects tested. The soleus muscle of all 12 participants, instead, showed surface potentials with high (70 μV) peak-to-peak amplitude. This amplitude was twice as large as the noise amplitude reported by the authors. A few years later, these authors further investigated how greater EMGs recorded in quiet standing are with respect to those obtained when subjects relaxed their muscle as much as possible [40]. Even though the surface potentials recorded at rest were three times greater than the noise level ($2 \mu V_{pp}$), during standing these potentials reached amplitudes markedly higher than during rest. As in their previous studies, Joseph et al. [40] recorded largest potentials for the calf muscles. Tibialis anterior and thigh muscles (i.e., quadriceps and hamstrings) showed potentials with negligible amplitude. Such a predominant activation of the calf muscles in quiet standing has also been reported in different, recent studies [14,60]. On one hand, compelling evidence suggests that active plantar flexion torque is necessary for the arrest of postural sways. On the other hand, the mechanism governing the activation of the calf muscles in standing is controversial.

There is a long established tradition of physiological investigation into mechanisms of postural control. Since Sherrington [82,83], the concept of the reflex arc, in which motor output reflects sensory input, has been well established. More importantly, the neurophysiological mechanisms of spinal and transcortical stretch reflexes have been well established [73]. Through these mechanisms, rotation of the joints, such as the ankle, causes continuous modulation of muscle activity up to frequencies well beyond the voluntary bandwidth (1–2 Hz) and over 10 Hz [8,18,74]. From the calf muscles, the soleus is particularly well endowed with muscle spindles and fatigue resistant fibers relevant to the postural role of tonic and dynamic stretch reflexes [14,51]. Coupled with the common experience that postural control appears automated and involuntary, and coupled with the early demonstration of the postural capability of spinal and decerebrate preparations, the reflex paradigm has been richly used for explaining postural control [20,21,23,78]. The reflex explanation of postural control has strong theoretical underpinning in the servo mechanism of control engineering. This classic, continuous, negative feedback controller is the archetype of position regulation and thus is naturally the first explanation for biological control of posture [27,59,61,62].

More recently, there has been growing interest in the role of more sophistsicated mechanisms regulating muscle activity even in the simplified paradigm of control of a single segment via the ankle musculature [48,50]. Postural control requires capabilities beyond local regulation of joint angle in relation to a commanded set point. The postural activity required in any muscle, including the calf, depends on a wider context including current task goals, planned bodily movements, envirnomental contraints such as objects that one is in contact with, and the current multi-segmental configuration of the body. While the short latency stretch reflex mechanism is local to the spinal cord and possesses limited flexibility via supraspinal input [52], the so-called long latency functional stretch reflex crosses the cortex and receives input generally from multiple brain regions. Consequently, this transcortical reflex system, which has been extensively researched over many years, has been shown to be modulated by all the general factors listed above while still providing rapid response at 100 ms latency [73]. Nonetheless, there are limitations to this more sophisticated system. Flexibility is limited to modulating the threshold and strength of response: Modulating the direction of response requires more intentional mechanisms of greater latency, and this is relevant to the control of posture [11,52]. Moreover, while the long latency system provides an initial response, alone, it is indequate in reliability and efficacy for position control [58]. Effective, adequate position regulation requires the complete reflex-voluntary system [58], and this provides reasons why postural control is limited to the lower-frequency bandwidth of the voluntary system (1–2 Hz) rather than the high-frequency bandwidth (>10 Hz) of the reflexive systems [47,52]. This perspective underlies the hypothesis and accumulating evidence that the main feedback loop regulating postural control contains an intermittent, serial ballistic mechanism [50,54,89]. Additional rationale supporting intermittent rather than continuous control of posture concerns the possibilty of increased robustness and adaptability of intermittent controllers [2,5,47,89].

A prediction of this intermittent control hypothesis is that postural EMG will show evidence of control actions related to the voluntary rather than reflex bandwidth (1–2 Hz rather than >10 Hz) [47,52]. Evidence from EMG in line with this prediction comes from traditional observation of sporadic postural activity for example in the gastrocnemius muscle (Joseph's papers [39,40]). Analysis of the minimization of postural sway size by accurate torque modulation described as "drop and catch" pattern is also consistent with the hypothesis [48–50]. EMG and ultrasonography revealed two to three sluggish and impulsive muscle adjustments per second during standing, associated with the process of maintaining balance [53,54]. More precisely, intramuscular recording within the gastrocnemius during quiet standing shows recruitment of individual motor units at a modal rate of two to three per second. This rate is consistent with the rate of serial ballistic events in intermittent manual control which optimizes position and velocity regulation [47]. Alternatively, the view that continuous feedback mechanisms dominate preservation of balance is well supported [92]. While evidence is now available discriminating against continuous control of manual tracking [91] and postural movement [89], evidence discriminating against continuous control of balance remains to be published.

15.3   EMG AND POSTURAL PERTURBATIONS

Identifying the system involved in postural control or its characteristics from EMGs is not possible without opening the standing loop. As opening the loop is not physiologically possible, challenging the control of standing with the application of systematic perturbations constitutes an attractive paradigm to investigate (i) the generation of reflexes responses, (ii) responses time delay, (iii) formation of postural synergies, (iv) the source of postural perturbations, (v) alterations in muscle activation with dementias, and (vi) feedback pathways involved in standing control.

15.3.1   Which and How Muscles Are Activated in Response to Postural Perturbations?

Steady maintenance of any posture is accounted for by the collective rather than the isolated activation of groups of muscles performing specific functions. The reason for a collective postural action relies on the gravitational torque acting on different body joints. Consider, for example, the upper limb posture when someone holds a small book at the eye level; arms relaxed alongside the trunk and elbows flexed at ~100 degrees and pronated at 15 degrees with respect to its anatomical reference position. The book weight tends to passively extend and supinate the elbow as well as to move the wrist and fingers in extension. Isolated, sustained activation of elbow flexors (i.e., biceps brachii, brachioradialis and brachialis) would ensure the forearm facing-eye position, otherwise the book would fall due to forearm supination and wrist extension. The activation of elbow, wrist, forearm, and finger muscles must thus be coordinated to ensure such a reading posture. This same coordination issue is present in the perturbed standing posture.

Postural sways induced at the ankle joint are compensated for by the coordinated activation of ankle, hip, and trunk extensors and flexors. With the use of surface electromyography, Nashner [70] observed that coordination was a consequence of preprogrammed and fixed response patterns rather than of a sequence of reflexive, mechanically coupled responses. Nashner recorded surface EMGs from ventral (tibialis anterior and quadriceps) and dorsal (gastrocnemius, hamstrings, and sacrospinalis) muscles while the standing posture was perturbed in two different manners: surface translation back and forward and surface rotation in dorsal and plantar flexion directions. As shown in Figure 15.1A, the temporal sequence of tibialis anterior and quadriceps activation was observed regardless of whether the ankle was directly rotated in plantar flexion or backward sways were induced with surface forward translations. This same sequence of distal to proximal activation was observed for the dorsal muscles when reversing the direction of surface translation and rotation (Figure 15.1B). Furthermore, Nashner [70] observed that once triggered, subjects could not modify the set of postural responses. Such unaltered pattern of muscle activation with perturbation mode indicates that muscle coordination is programmed prior to the response itself. Functionally, the sequential activation of distal–proximal muscles, usually termed ankle strategy, ensures the body moves as an inverted pendulum over the ankle joint. Under certain circumstances, as for example when subjects respond to perturbations while standing over a narrow surface, the order of activation sequence changes [36]. In this case, proximal muscles tend to be activated prior to distal muscles. CoG position is then controlled predominantly by movements about the hip joint, thus characterizing the hip strategy [36]. These early findings suggest that balancing reactions are organized on the level of muscle synergies; each being the coordinated activation of a functional set of muscles (see Chapter 6).

The degree of muscle activation and the number of muscles participating in a given postural synergy depend on the perturbation direction. Henry et al. [31] measured the response of several muscles in the leg, thigh, and trunk to force-plate translations in 12 directions. The integrated envelope of surface EMGs was considered to quantify how much a given muscle contributed to responses elicited by a specific direction. The tensor fascia latae muscle in the left thigh, for example, showed markedly larger EMG envelopes when the surface translation was directed towards the opposite side (postural sway induced to the left side). Interestingly, their results suggest a gradual change in the amplitude of EMG envelopes with perturbation direction (cf. EMG envelopes shown in Fig. 1 in Henry et al. [31]). The onset of muscle activation, instead, changed somewhat discretely with perturbation direction. While other leg and thigh muscles showed a consistent onset of activation across perturbation directions, the tensor latae, rectus abdominis, and erector spinae were activated either early or late. Such discrete change in the latency of activation of proximal muscles led the authors to consider the possibility that central mechanisms control the timing of muscle activation, with priority given to trunk stabilization. By applying

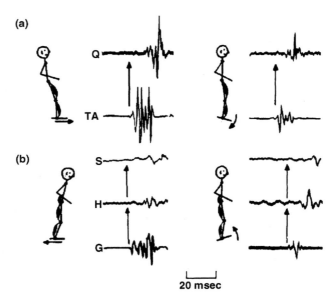

FIGURE 15.1 Examples of the raw EMGs from which the latency of gastrocnemius (G), hamstrings (H), tibialis anterior (TA), quadriceps (Q), and sacrospinalis (S) contractions were determined in relation to perturbation onset. (**A**) Examples of sequence of TA-Q contraction during backward swaying and during direct plantarflexion of the ankles. (**B**) Examples of the sequence of G–H–S contraction during forward swaying and during direct dorsiflexion of the ankles. Reproduced from Nashner [70] with permission.

multidirectional surface translations, Torres-Oviedo and Ting [88] investigated which muscle synergies are predominantly elicited in response to specific perturbation directions. From surface EMGs, six postural synergies were identified, involving a combination of leg, thigh, and/or trunk muscles. Most interestingly, different combinations among these six synergies were sometimes observed in response to a single perturbation direction. It is uncertain whether this intertrial variability is a consequence of (i) environmental aspects, (ii) attention, (iii) flexible combination of different synergies to achieve a fixed postural goal (e.g., control of body CoG, minimization of muscular effort), or (iv) low representativeness of muscle activity in surface EMGs (see Section 15.4). In any case, evidence from studies combining electromyography and multidirectional perturbations revealed the existence of stabilization mechanism of marked computational efficiency. Postural responses seem not to be organized at the level of individual muscles but instead at the level of a few, functional muscle synergies.

15.3.2 Alterations in EMG Responses with Dementias

Translations of the supporting surface might be useful to distinguish different neurological impairments in surface EMGs. As shown in Fig. 15.1, a fixed pattern of muscle activation is triggered in response to surface translations. Firstly, distal muscles are activated prior to the proximal muscles. Secondly, sways induced in backward direction elicit activation of only ventral muscles whereas forwardly induced sways trigger activation of dorsal muscles exclusively. It is reasonable then to expect a change of these response patterns in neurological patients. With surface electrodes positioned on the leg, thigh, and trunk muscles, the reaction of patients with diabetic peripheral neuropathy and of parkinsonian patients to surface translations has been investigated. Likely in virtue of their somatosensory loss (i.e., increased excitation thresholds), neuropathic patients showed a similar though systematically delayed pattern of distal–proximal muscle response when compared to healthy individuals [38]. In parkinsonians, conversely, Horak et al. [37] observed a hindered ability to coordinate muscle activation. Specifically, while young healthy subjects showed patterns similar to those reported above in the left panels of Fig. 15.1 (see Fig. 15.2A), parkinsonians showed a marked co-activation of both ventral and dorsal muscles (Fig. 15.2B). These results indicate how differently neuropathy and

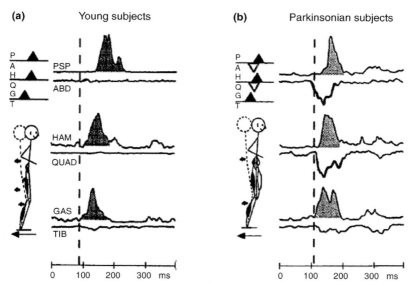

FIGURE 15.2 Typical patterns of muscle activation (rectified, low-pass filtered EMGs) in a representative young (**A**) and parkinsonian (**B**) subjects in response to backward surface translations. Dorsal sEMG bursts associated with an ankle strategy are shaded and ventral bursts associated with a hip strategy are thickened in both the sEMG traces and their schematic, triangular representations. Surface EMGs are averaged from the last 5 of 10 consecutive trials. EMG gains are constant with burst peaks ranging from 0.5 mV to 1 mV. Time zero indicates onset of surface translation and dashed lines show onset of first EMG responses. Stick figures show muscles activated and small arrows indicate direction of active correction for forward sway displacement. Adapted from Horak et al. [37] with permission. PSP, lumbar paraspinalis; ABD, rectus abdominis; HAM, biceps femoris; QUAD, rectus femoris; GAS, medial gastrocnemius; TIB, tibialis anterior.

parkinson syndrome might be represented in surface EMGs recoded during surface translations. Patients suffering from the former pathology tend to respond with significant delays to postural perturbations while the parkinsonians show marked difficulty in adapting to the postural stimulus. It should be noted though that these interpretations are based on the assumption that conventional surface EMGs sample from a representative muscle volume; the issue of representativeness will be covered in Section 15.4.

15.3.3 Internal Sources of Postural Perturbations

In addition to the externally imposed postural challenges, when stabilizing the standing posture the nervous system must account for perturbations with internal origins. These might result from individuals' volition and body dynamics. Any segmental movement, for instance, is accompanied by shifts in the whole-body CoG. Anticipatory activation of a set of muscle is necessary for counteracting these CoG shifts inherent to voluntary body movements. If anticipatory postural adjustments are not issued by the nervous system, the motor goal might not be achieved or its achievement might be impaired. Such orchestration of anticipatory activation of postural muscles is often reported from surface EMGs. One typical example is related to protocols based on intentional arms raising [1,4,67]. By asking subjects to bilaterally raise their arms, Aruin and Latash [1] observed activation of lower limb and trunk muscles in advance of activation of the shoulder movers. Overtly, throughout a given movement, continued activation of postural muscles is necessary to compensate for further shifts in the location of body CoG and, thus, to ensure a steady posture. Postural compensation is not only triggered during movements; as evidenced from surface EMGs, it is also an incessant demand imposed by vital functions—that is, breathing and heart beating. Indeed, it is not surprising that, regardless of how effortful subjects breathe, there is a moderately high peak observed in the coherence function evaluated between movements of the chest wall and envelopes of surface EMGs recorded from trunk and hip muscles [34]. Very likely, postural adjustments occurring during or in advance of movements reflect preprogrammed responses. Whether such programming takes place centrally or peripherally in the nervous system remains an attractive subject of study.

15.3.4 Are the Mechanisms Involved in the Control of Perturbed and Natural Standing the Same?

In several aspects, the control of quiet and perturbed standing might differ. Firstly, during perturbed standing the characteristics of the disturbance are unpredictable. Secondly, for perturbation protocols typically considered in the literature, sensory receptors operate far over their threshold [21]. As a consequence, postural responses are primarily driven by reflex loops. During quiet standing, on the other hand, the sources of disturbance are internal and the velocity and amplitude of body movements are close to the perceptual threshold. Finally, calf muscle stiffness is inversely related to the amplitude of ankle dorsal flexion. With a sophisticated apparatus, Loram et al. [55,56] applied rotational stimuli to the supporting surface while standing subjects were looking at their EMGs to suppress calf muscle activation. Rotations ranged from −1.5 to 5.5 degrees of dorsal flexion. The values of chordal ankle stiffness decreased logarithmically with the rotation amplitude [55] and such a reduction was accounted for by the muscle contractile components [56]. These findings indicate that the calf muscles are extremely stiff for the amplitude of body sways occurring in quiet standing (few tenths of degree), but not for ankle movements higher than ~0.5 degrees. Consequently, (i) calf muscle spindles seem to lose their ability to sense muscle lengthening in quiet standing [51], (ii) state variables relating to body sways are estimated without ambiguity in response to surface perturbations, which are automatically triggered with 60- to 130-ms delay [31,68]. For these reasons, any inference on the mechanisms involved in the control of quiet standing posture should be made circumspectly from perturbation results.

Some perturbations, however, might be designed to reveal what happens in quiet standing. These perturbations have magnitude and bandwidth comparable to that of the spontaneous sways occurring in standing [20,46]. In the next section, we highlight some standing-like-perturbation protocols that have been considered to investigate key aspects of the system controlling the human standing posture.

15.3.5 What Are the Characteristics and What Is the Goal of the Feedback System Involved in Standing Control?

In general, the human upright posture is unstable, and thus the preservation of balance and posture requires sustained feedback [7,45,49,53]. The human postural control system contains much redundancy and equivalent possibilities at many levels. There are many kinematic possibilities for keeping the center of mass over the base of support. There are many equivalent muscle activation patterns capable of maintaining appropriate tonic and dynamic joint moments. There is choice of variables to regulate, including single-joint and multisegmental configuration, degree of sway and joint rotation, and the amount of muscle activity in maintaining balance [13,46]. In general the task of inferring the control strategy requires perturbations, measurement of muscle activity and kinematic response, and comparison with models representing alternative choices in control strategy.

A seminal analysis investigated the feedback loop regulating quiet standing through mechanical force perturbations applied at the waist and sensory perturbations applied through galvanic vestibular stimulation [20]. These perturbations allowed analysis of the open-loop transfer functions relating sway to soleus EMG, and soleus EMG to sway, allowing calculation of the feedback loop gain. This investigation showed that the gain of the feedback loop was relatively low around unity, which distinguished biological, postural control from engineering servo control, and reflex conceptions in which feedback gain would be expected to be high to maintain positional stability [20]. This result is typical of biological systems which include relatively long feedback loop delays, low-frequency bandwidth, and low feedback gain which makes them unsuitable for simple feedback control of this kind [41,43]. Prediction is inherent to the functionality of the nervous system and is the normal engineering solution to the "problem" of time delays [28,29]. Intermittent control including flexible response selection is an appropriate solution to the problem of the flexible control of redundant, low bandwidth systems [57,77,89,91]. The recent demonstration that continuous feedback is unnecessary for postural control and that intermittent control is more robust [47] is increasing awareness of the intermittent control paradigm in relation to motor and postural control [10,41,42,57,85,89,90].

In an approach building on that of Fitzpatrick [20], the human postural control strategy has been identified using mechanical perturbations at the waist and shoulder and sensory perturbations using movement of the visual scene. Using EMG from multiple muscles to construct the control signal for the ankle and hip joints generated by the nervous system, the open loop mapping from leg and trunk segment angles to EMG and from EMG to leg and trunk angles were both identified allowing analysis of the complete feedback pathway. By comparison with optimal control models representing a variety of cost functions, analysis showed the feedback strategy was more similar to one which minimizes muscle activity than one which minimizes sway movements. The postural system ensures FBOS (finite base of support) stability with minimal muscle activation [46]. More recently, the utilization of segmental degrees of freedom has been tested more strongly by applying gentle asymmetric perturbations to the knee joint. While minimally disturbing balance or center-of-mass position, these perturbations test the extent to which, and manner in which, the ankle, knee, and hip joints are regulated during posture. Large variation between participants revealed two main postural goals. Most common was generalized restriction of segmental motion. Less common was higher allowance of segmental motion, indicating a prior goal to minimize muscle activation [13]. These results indicate the normal population lies some distance from the baseline for minimizing muscle activation and allowing segmental movement.

In summary, much has been learned from surface EMGs collected from different skeletal muscles during natural and perturbed standing. All this evidence, however, prompts from conventional electromyography individual couples of surface electrodes placed on different muscles. A crucial aspect related to surface EMGs recorded by conventional means is the assumption that the degree and timing of muscle activity is unequivocally represented in the EMGs. The use of multiple electrodes to sample the surface distribution of activity in individual muscle volumes, however, has shown that this might not be case. In the next section, the reader is provided with recent insights into the physiological interpretation of surface EMGs recorded from the pinnate gastrocnemius muscle during quiet standing.

15.4   NEW PHYSIOLOGICAL AND POSTURAL INSIGHTS GAINED FROM GASTROCNEMIUS, HIGH-DENSITY SURFACE ELECTROMYOGRAMS

In this section, we revisit some of the issues outlined above with high-density surface EMG (HDsEMG). Even though much has been done lately regarding the investigation of muscle activation with high-density systems of electrodes, very little has been done in terms of understanding posture. Specific studies relying on the use of such systems of electrodes to characterize the distribution of activity within postural muscles are incipient; these have been mostly focused on the activation of the calf muscles in natural standing. We, therefore, further narrow the concept of EMG and posture as discussed in the previous sections to the activation of the gastrocnemius muscle in quiet standing.

Initially, a technical note on the definition of HDsEMG is warranted (see also Chapter 5). Readers familiar with advanced electromyography are certainly aware that "multichannel" and "high-density" terminologies have been used indistinguishably in the literature, with both referring to the use of multiple electrodes for the detection of surface EMGs. "Multichannel" is a general terminology that might indicate multiple electrodes sampling from the same or from different muscles. In the perturbation paradigms illustrated above (Figs. 15.1 and 15.2), postural strategies have been identified with different pairs of electrodes positioned on skin surfaces covering different muscles. These studies used "multichannel" systems of surface electrodes. On the other hand, in an attempt to investigate where on the skin the quality and significance of surface EMGs are the highest, Sacco et al. [79] positioned multiple pairs of relatively large electrodes over the same muscle. In this study, different regions of a same muscle volume were also sampled with a "multichannel" system. "High-density" then seems more appropriate than "multichannel" to denote the use of multiple electrodes to sample EMGs from a single muscle volume. What are, however, the number of and the distance between electrodes characterizing a "high-density" system? While there is no specific criterion classifying a system of electrodes as "high-density," we refer to "high-density" surface EMGs as those recorded with any type of array of electrodes positioned over a single muscle volume (see Chapter 5 for technical details).

15.4.1   Interpreting Surface EMGs from Gastrocnemius Muscles

The possibility that surface EMGs collected with a single pair of electrodes do not genuinely reflect the activation of the gastrocnemius muscles has been of concern in the classic literature. When studying the latency of muscle response to surface translations and rotations, Nashner [70] questioned whether estimated response delays were representative of the actual activation delay: "Because each EMG was recorded from a very limited surface of a large muscle, errors in the measure of activation latency were possible" (page 20 in Nashner [70]). A similar statement was made 25 years earlier by Joseph and Nightingale [39]. From surface EMGs recorded with pairs of surface electrodes in standing, these authors observed surface potentials in the soleus muscle of all subjects whereas surface potentials in gastrocnemius were observed inconsistently across subjects. Besides questioning "whether it is correct to assume that in some subjects soleus is the only muscle which is active and in others both soleus and gastrocnemius are active," these authors further considered "that the failure of the gastrocnemius electrodes to record potentials may be due not to absence of activity in that muscle but to the arrangements of its fibers or a layer of fat between the muscle and the electrodes or to the inadequate sensitivity of the recording apparatus" (page 489 in Joseph and Nightingale [39]). Sadly, these early and relevant concerns are frequently not considered nor acknowledged in contemporary and recent studies on EMG and posture.

Postural studies are typically focused on identifying when and how strongly the gastrocnemius muscles are activated [31,80,95,98]. If the timing and the intensity of gastrocnemius activation is investigated with a couple of surface electrodes, then it is necessarily presumed that sampled EMGs reflect the degree of activation of a whole muscle volume. Should this assumption be true, redundant information would be collected with a high-density system of electrodes (i.e., timing and degree of muscle activation would be represented equally well in all electrodes). However, crucial differences exist in surface EMGs collected from different regions of a single muscle, in particular for muscles with pinnate architecture.

The degree of representation of motor unit action potentials in surface EMGs depends on the position of electrodes in relation to the gastrocnemius fibers. Currently available recommendations suggest to position surface electrodes over the belly of gastrocnemius muscles [32]. These recommendations are based on the obsolete view that when located over the gastrocnemius belly, surface electrodes detect genuine activity from gastrocnemius motor units with little interference from the other nearby muscles. There is a problem here, however, related to the number of gastrocnemius motor units residing within the detection volume defined by a couple of electrodes. Empirical and theoretical evidence suggests that, because of its pinnate geometry, the activity of a markedly small portion of the gastrocnemius large volume is sampled by surface electrodes [66,84,93,95]. In the next paragraphs, we explain to readers that (i) surface EMGs from the gastrocnemius muscle sample the activity of a local muscle portion and (ii) the local representation of action potentials of gastrocnemius motor unit in surface EMGs is a consequence of the muscle pinnate architecture.

The physiological information conveyed in surface EMGs recorded from the gastrocnemius muscle is not similar to that obtained from EMGs in skin-parallel-fibered muscles. Consider, for example, the single-differential EMGs collected by 16 electrodes (15 differential channels) covering the biceps brachii muscles and shown in Fig. 15.3. From these signals, action potentials of the same motor units are represented in several different channels, from channel 6 to 15. Signals recorded in the more proximal channels, up to channel 5, show potentials of negligible amplitude. This difference between groups of channels is related to their positioning with respect to the muscle and tendon tissues. Potentially useful anatomical and physiological information might be obtained from the EMGs shown in Fig. 15.3, such as the location of the innervation zone (between channels 10 and 11; see Garcia and Vieira [26]) and the conduction velocity of muscle fibers (i.e., by estimating the delay between EMGs in consecutive channels; see Farina and Merletti [19]). As consecutive electrodes are parallel to the muscle fibers then, in virtue of the propagation of action potentials, it could be argued that a pair of electrodes positioned within channels 6–14 would likely suffice to study the timing and the degree of biceps activation. For this particular subject, EMGs representative of biceps activity could be detected over a skin region as large as ∼8 cm. Equally large regions for the pickup of representative gastrocnemius activity in surface EMGs are not viable. Because of the

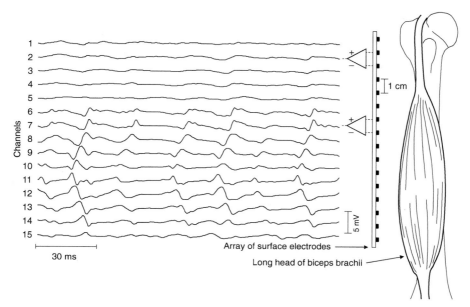

FIGURE 15.3   Single differential surface EMGs recorded with an array of 16 electrodes from the biceps brachii muscle are shown. A schematic representation of the position of the array of electrodes with respect to the muscle is shown in the right panel. For convenience, only the long head of the biceps brachii is shown. Potentials from different motor units are observable from channels 6 to 15. Phase opposition between surface potentials is evident between channels 10 and 11, that is, location of the innervation zone. Because of propagation and of the parallel arrangement between fibers and electrodes, a temporal delay between successive potentials is clear in the figure. At the tendon regions, negligible potentials are seen.

pinnate arrangement of gastrocnemius fibers, consecutive surface electrodes are not parallel to muscle fibers. Only at the most distal region might electrodes and gastrocnemius fibers be arranged in parallel planes (Fig. 15.4A).

Considering the example shown in Fig. 15.4, it is clear that gastrocnemius motor units are represented locally in the surface EMGs. Representation of the most distal units though tends to be less localized than those of proximal units (compare distal and proximal surface potentials in Figure 15.4B). Indeed, different aspects of most distal and proximal gastrocnemius units are represented differently in the surface EMGs. As observed for the signals in Fig. 15.4, the surface potentials detected from channels 11 to 15 in Fig. 15.4B clearly show phase opposition and temporal delay. Apart from the differences in fiber length, surface EMGs recorded from biceps brachii and from the very distal gastrocnemius region show similar features. The surface potentials recorded from the proximal gastrocnemius region, however, do not show propagation or phase opposition (see the potentials in channels 3–4 shown in Fig. 15.4B). What physiological information might then be obtained from these proximal surface potentials?

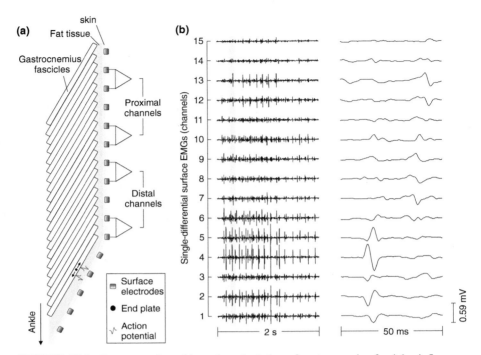

FIGURE 15.4 Representation of how the orientation of gastrocnemius fascicles influences the patterns of myoelectric activity recorded at different proximal–distal regions. **(A)** Relative positioning of surface electrodes with respect to gastrocnemius fascicles. Close to the muscle–tendon junction, fascicles are orientated more parallel to the skin; electrode channels now lie along fascicles and propagation of action potentials might be visible in electrode channels located over this region. **(B)** Surface EMGs from one subject recorded during quiet standing (2-s period; left panel), with signals from a 50-ms epoch (gray block) time zoomed in the right panel. Adapted from Hodson-Tole et al. [35].

Surface EMGs recorded by electrodes over the gastrocnemius, superficial apo-neurosis reflect the number and the location of active fibers beneath electrodes. Such an association between surface EMGs and the distribution of active fibers beneath the recording electrodes is a consequence of the muscle pinnate geometry. Because of pinnation, consecutive electrodes over the superficial aponeurosis do not sample from the same group of fibers. Each electrode is instead positioned over the distal extremity of a different group of fibers (see Fig. 15.4A). Therefore, it is reasonable to expect a predominant contribution of the end-of-fiber effect, resulting from the extinction of the action potentials at the fibers' distal tendons, to the surface EMG collected over gastrocnemius superficial aponeurosis. Whether a given surface electrode detects a large potential depends on the number of action potentials reaching the distal tendon (i.e., superficial aponeurosis) of fibers residing within the electrode detection volume. This concept is illustrated by simulation results reported in Fig. 15.5. Surface potentials simulated for skin-parallel fibers are recorded by largely spaced electrodes aligned along the fibers (Fig. 15.5A). For a pinnate fiber, instead, surface potentials

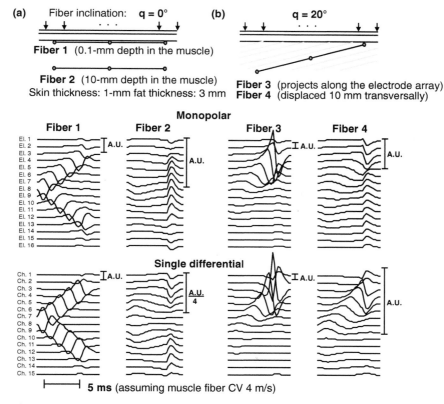

FIGURE 15.5 Simulation of surface EMGs detected with an array of 16 electrodes (5-mm interelectrode distance) located over the longitudinal projection of a single muscle fiber on the skin surface. Surface potentials were simulated for a skin-parallel (**A**) and a pinnate fiber (**B**). Monopolar and single-differential signals are shown for different fibers, corresponding to different positions. Note the different scales representing the surface potentials. Adapted from Mesin et al. [66] with permission.

are represented only in the EMGs recorded nearby the fiber superficial ending (Fig. 15.5B). It should be noted that EMGs shown in Fig. 15.5 were all simulated for a single fiber. Given that the action potential of a single motor unit corresponds to the algebraic summation of action potentials from each of its fibers, the localized representation of individual motor unit action potentials in surface EMGs depends on the size of motor unit territories and on the distribution of fibers within the territory [93]. Therefore, HDsEMGs posit a valuable means to study the spatial organization of postural activity in pinnate muscles.

Regional modulation in the activity of gastrocnemius muscles during standing has been, indeed, observed with the use of a high-density system of electrodes. Specifically, Vieira et al. [95] positioned a grid of 128 electrodes (8 rows × 16 columns; 1-cm interelectrode distance) on the right calf of standing subjects. This grid was positioned in such a way that medial and lateral gastrocnemius was each covered by half of the

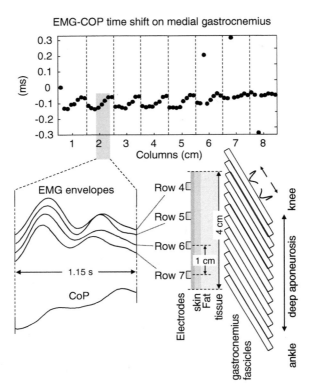

FIGURE 15.6 Schematic of how the progressive delay in modulations of EMG amplitude might occur within the same muscle during standing. **Top:** Solid circles illustrate the time shifts between EMG envelopes and displacements in the body center of pressure (CoP) for the medial gastrocnemius of a single subject. EMG envelopes and CoP are shown in the bottom left. The position of electrodes detecting such envelopes is illustrated in the bottom right. Rectangles represent the gastrocnemius fascicles, while arrows indicate the direction of propagation of action potentials. Adapted from Vieira et al. [95].

128 electrodes. Ultrasound scanning was used for guiding the grid positioning (see supplemental material in Vieira et al. [94]). A key result prompted from this study was the variable temporal delay between consecutive modulations in EMG amplitude. Figure 15.6 schematically shows how the amplitude of surface EMGs (i.e., EMG envelopes) changed equally, though with a progressive delay, across consecutive channels in the grid. It is tempting to attribute the delay between envelopes in Fig. 15.6 to the propagation of action potentials. However, as discussed above (see Figs. 15.4 and 15.5), consecutive electrodes positioned over the superficial aponeurosis in the gastrocnemius muscle are not sensitive to the propagation of action potentials. Furthermore, if the temporal delay between EMG envelopes were due to potential propagation, then (i) the amplitude of consecutive envelopes should be markedly different (see Fig. 15.5B) and (ii) the direction of propagation should not change. Occasionally, temporal delays were observed in both distal-to-proximal and

proximal-to-distal sequences. (iii) Given the physiological range of values of muscle fiber conduction velocity ($3–6\,\text{ms}^{-1}$) and considering the 1-cm distance between consecutive electrodes, the temporal delay between envelopes should be within the range ~1.5–3.5 ms. Delays between consecutive envelopes were often far greater than the upper limit of this range (see top panel in Fig. 15.6). This evidence led us to consider the possibility that progressive delays across consecutive electrodes were reflecting the sequential activation of motor units in different gastrocnemius regions during standing. Indeed, different motor unit action potentials have been observed in different regions of the same gastrocnemius muscle (Fig. 15.4; see Figs. 6 and 7 in Merletti et al. [64].

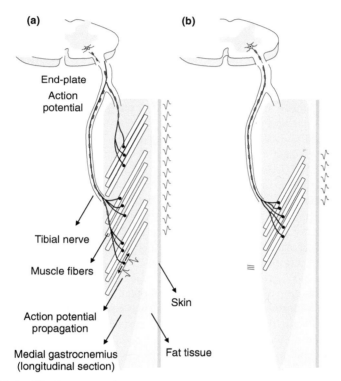

FIGURE 15.7 (**A**) Illustration of the hypothesis of muscle units (i.e., fibers supplied by individual motor neurons) extending widely along the medial gastrocnemius muscle. The body of one motor neuron is located at the spinal cord ventral root, while its axon runs through the tibial nerve and branches to supply muscle fibers scattered along the whole muscle. (**B**) Scheme showing the hypothesis of spatially localized muscle units. Axonal branching occurs close to the muscle and supplies muscle fibers distributed locally. Due to the gastrocnemius pinnate arrangement, and since action potentials propagate along muscle fibers and not along the muscle, intramuscular potentials might appear locally or widely on the surface of the skin, depending on how large the territories of gastrocnemius muscle units are (compare the schematic representation of action potentials on the surface for both cases). Modified from Vieira et al. [93] with permission.

The localized organization of postural activity in the gastrocnemius muscle has crucial methodological and physiological implications. Methodologically, a few closely spaced surface electrodes sample the activity of a small gastrocnemius region, unrepresentative of the large muscle volume. The expert user of surface electromyography must then be aware that conventional bipolar recordings hardly provide EMGs representing the actual timing and degree of gastrocnemius activation. Physiologically, surface potentials recorded by only a few consecutive electrodes suggest that muscle fibers belonging to individual motor units occupy a discrete gastrocnemius region. Otherwise, if the territory of gastrocnemius motor unit was large with respect to the muscle length, localized activation would not be evident in high-density surface EMGs. The next section gives a detailed description on the quantification of the territory of postural motor units in the gastrocnemius muscle from high-density surface EMG.

15.4.2   Estimating the Territory Size of Postural Motor Units from High-Density Surface EMGs

By combining surface and intramuscular EMGs, our group was able to estimate how extensively the fibers of postural motor units distribute along the longitudinal axis of the human medial gastrocnemius muscle. Bearing in mind that surface EMGs collected from pinnate muscles sample from the fibers ending nearby the recording electrodes (Fig. 15.5), one might promptly realize that the spatial distribution of surface EMGs reflects the distribution of active fibers beneath an array of electrodes (Fig. 15.7). Therefore, to investigate how localized postural motor units are in the human gastrocnemius muscle, we triggered surface EMGs detected by an array of electrodes with the firings of motor units identified from intramuscular EMGs while subjects stood at ease. Surface potentials detected by several electrodes would indicate a large territory whereas motor unit potentials represented in a few electrodes would suggest small territory. These two hypotheses are illustrated in Fig. 15.7. The amplitude distribution of individual motor unit potentials across the surface electrodes was fitted with a Gaussian curve and its spread was compared in relation to the typical values reported for the gastrocnemius muscle length at rest (~25 cm; Narici et al. [69]). The following paragraphs provide a detailed description on the validation of the method and on the experimental results.

The proposed association between the spread of Gaussian curves fitting the spatial distribution of surface action potential amplitude and the size of motor unit territories was validated using an advanced biophysical model of surface EMG generation in pinnate muscles. The model is an extension of that developed by Mesin and Farina [65], to further include the adipose and the skin layers separating the muscle tissue from the recording electrodes. The geometry of the model is shown in Fig. 15.8. Because of the muscle pinnation, the volume conductor between the source and the skin surface is not space invariant [65]. As a consequence, the shape of the surface potential changes as the intramuscular action potential propagates. For this reason, the surface representation of an action potential propagating along a single, pinnate fiber was simulated by computing the impulse response for sources positioned at different

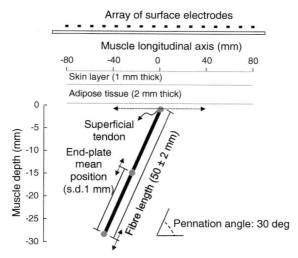

FIGURE 15.8 A schematic representation of the model used for the simulation of action potentials in pinnate fibers is shown. A single fiber is illustrated in the figure. All muscle fibers were simulated 30 degrees inclined with respect to the skin (1 mm thick) and adipose tissue (2 mm thick). The distal tendon of each fiber was located 0.1 mm deeper than the fat tissue and was allowed to vary by 160 mm along the muscle longitudinal axis. Values of fiber length were extracted from a Gaussian distribution with 50 mm mean and 2-mm standard deviation. The mean position of end plates coincided with the center of each muscle fiber, and its standard deviation was set at 1 mm. For each of the 3200 muscle fibers simulated, their surface potential was sampled with 16 electrodes, equally spaced by 1 cm along the skin.

locations along the fiber longitudinal axis. Compound action potentials of individual motor units were obtained, finally, as a summation of the action potentials of single fibers.

Single fibers action potentials were simulated with the pinnate fibers being distributed within a longitudinal muscle dimension of 8 cm (Fig. 15.8). This size was chosen to ensure that some of the 16 surface electrodes simulated were not located at skin regions above the superficial endings of the pinnate fibers (i.e., the center of the array of electrodes coincided with the center of the muscle region simulated). Muscle tissue was considered as anisotropic while the other layers were regarded as isotropic (Mesin et al. [66]; conductivities: skin $\sigma_S = 4.3 \times 10^{-4}$ S/m; fat $\sigma_F = 4 \times 10^{-2}$ S/m; muscle longitudinal conductivity $\sigma_L = 40 \times 10^{-2}$ S/m; transversal conductivity $\sigma_T = 9 \times 10^{-2}$ S/m). The thickness of the adipose tissue and skin, the pinnation angle, the length of the pinnate fibers, and the location of end plates were all chosen according to values reported in the literature [44,54,69], or estimated with ultrasound scanning, for the human gastrocnemius muscle during standing.

Two libraries of single fiber action potentials were created. Each library comprised the surface potentials of 3200 fibers distributed normally or uniformly (40 fiber/mm) along the muscle region simulated. The longitudinal position of each simulated fiber was set by extracting a random number from a Gaussian (standard deviation: 2 cm) or a uniform (range: 8 cm) distribution. For each library, a population of 300 motor units was simulated. The longitudinal projection of the territory of these units along the skin varied from 1 cm to 6 cm, randomly. As the number of muscle fibers composing the small, postural motor units in the human gastrocnemius muscle is uncertain, territories were simulated with three densities. Motor unit territories included 200, 300, and 400 fibers (about 400 fibers compose a single, large muscle unit in the cat gastrocnemius muscle [6]). For each motor unit, the average rectified amplitude of its simulated surface action potential was computed and, then, fitted with a Gaussian curve. The standard deviation of the optimal Gaussian was compared with the longitudinal size of the territories simulated to test for the validity of our method (i.e., to test for a systematic association between the standard deviation of the Gaussian fitting curve and the size of the motor unit territory).

Regardless of whether the pinnate fibers were distributed normally or uniformly, the surface representation of motor unit action potentials reflected the distribution of fibers within the motor unit territory (Fig. 15.9A–H). The more fibers were located beneath skin regions where the surface electrodes were simulated, the higher the amplitude of the surface potentials (Fig. 15.9A,B,D,E). For skin regions not covering motor unit fibers, the amplitude of surface potentials was dramatically small. Therefore, it is highly possible that surface EMGs detected from muscles whose fibers are oblique with respect to the skin convey anatomical information of individual motor units. Namely, the amplitude distribution of surface potentials reflects both the longitudinal location of the superficial endings of single, gastrocnemius muscle units, and the proportion of fibers within their territory.

On average, the Gaussian curves represented accurately the distribution of amplitude of the surface potentials (Fig. 15.9A,D). Less than 10% of the 1800 populations of motor units simulated yielded unrepresentative Gaussian fittings (Fig. 15.9G). This less frequent observation of inaccurate fittings resulted from the occasional concentration of fibers within different regions along the simulated territory. For example, two dense clusters of fibers where located at about 0 cm and 4 cm, for the muscle unit shown in Fig. 15.9H. These clusters led to large surface potentials at two separated regions (see potentials in Fig. 15.9H). Consequently, the amplitude distribution of surface potentials was bi-modal and its optimal Gaussian represented, accurately, the highest peak (Fig. 15.9G). Motor unit territories, however, do not seem to comprise clusters of fibers in specific regions. Indeed, we did not observe multi-modal amplitude distributions for the motor units detected experimentally [93].

The validity of our method prompts from the ability of the Gaussian curves to capture the dispersion of surface potentials rather than from the quality of the fitting. Expectedly, Gaussian fittings represented more accurately the surface potentials for motor units whose fibers were distributed normally, rather than uniformly, within their territory (Fig. 15.9I; ANOVA, $P < 0.0001$, $N = 1800$ units), independently of the density of fibers simulated (ANOVA, $P = 0.3$). Notwithstanding this difference in fitting quality, both distributions of fibers resulted in Gaussian curves significantly correlated with the size of the motor unit territories (Pearson $R > 0.9$, $P < 0.01$ for both plots shown in Fig. 15.10; $N = 900$ units). This result demonstrates a consistent

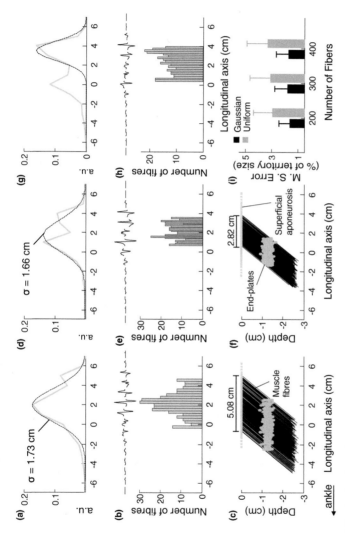

FIGURE 15.9 (A) Distribution of the average rectified amplitude (solid line), and its optimal Gaussian curve (dashed line), for the surface action potentials of a single, simulated motor unit. (B) Histogram created with the position, on the superficial aponeuroses, of all muscle fibers for the motor unit whose amplitude distribution is shown in part A. The raw, surface potentials of this unit are shown on top. Note how closely the amplitude and the location of surface potentials relate to the distribution of muscle fibers. (C) All muscle fibers simulated for this motor unit, including the variations in the fiber length and in the location of end plates along the fibers. The location of the superficial aponeuroses was distributed normally along the simulated territory (5.08 cm). (D–F) As shown in parts A–C, for another simulated unit. Muscle fibers were distributed uniformly along the motor unit territory (2.82 cm long, as shown in part F). (G) Bimodal amplitude distribution, resulting from the bimodal distribution of muscle fibers within the motor unit territory, shown in part H. (H) In this case, the Gaussian curve fits, optimally, only one peak. (I) Mean error values (whiskers: standard deviation; $N = 300$ units) obtained by fitting Gaussian curves to amplitude distributions, when the motor unit territories were simulated with 200, 300, and 400 fibers and when the fibers were distributed normally (■) or uniformly (■) in the territory.

429

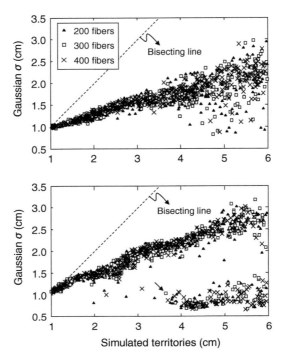

FIGURE 15.10 The standard deviations of optimal Gaussian curves are plotted against the actual longitudinal size of the territories of simulated motor units. Gaussians were fitted to the amplitude distributions of simulated surface potentials. Motor unit territories were simulated with 200 (▲), 300 (□), and 400 (×) muscle fibers ($N = 300$ motor units). **Top and bottom:** the distribution of muscle fibers along the territory of each unit was Gaussian and uniform, respectively. Less frequently, large motor unit territories resulted in the appearance of either a narrow peak or multiple peaks in the amplitude distribution (The arrow in the bottom panel indicates the motor unit whose fibers distribution and surface potentials are shown in Fig. 15.11G,H). This explains the higher variability in the standard deviation of optimal Gaussian curves for larger motor unit territories. Then, the cluster formed by low values of σ and large territories in the bottom panel was not included in the regression and correlation analysis. These cases (i.e., multiple peaks in amplitude distributions), reported here for completeness, were not observed experimentally [93] and depend on the very large number of simulation conditions, of which not all have a correspondence with experimental observations.

association between size of territories and spread of the Gaussian fitting. Regression analysis between the standard deviation (σ) of the optimal Gaussians and the territory longitudinal size (Fig. 15.10) showed a strong association between these two variables and yielded slopes equal to 0.26 and 0.32, when muscle fibers were distributed normally and uniformly within the motor unit territory, respectively. These values for the slope of the linear regression indicate that the size of motor unit territory corresponds to approximately 4 times the standard deviation of the Gaussian fittings. Therefore, simulations strongly support the statement that most, if not all, of

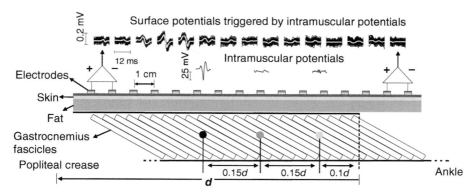

FIGURE 15.11   Schematic representation of electrodes positioning (**bottom**) and the surface potentials (**top**) triggered with the firing pattern of the motor units identified in the proximal portion of the medial gastrocnemius. Note how locally the action potential identified for a single motor unit in the most proximal intramuscular EMG (*black circle*) is represented in the surface EMGs collected along the gastrocnemius muscle. Average surface potentials (*gray traces*) are observable mostly from the third to the sixth channel starting from left.

the muscle fibers of gastrocnemius muscle units are confined to a longitudinal region defining 95% of the Gaussian area (i.e., 4σ).

From the theoretical indications above, the longitudinal size of the territory of postural motor units was found to be small in relation to the gastrocnemius length. Specifically, action potentials of individual motor units identified in intramuscular EMGs detected from three locations along the gastrocnemius muscle of standing subjects were represented locally in the surface EMGs [93]. The upper panel in Fig. 15.11 shows one example for a motor unit identified in the most proximal intramuscular EMG. Regardless of whether motor units were identified from one of the three specific locations where wire electrodes were inserted, about 95% of the area of all Gaussian curves was distributed within 4 cm. It should be noted, however, that changes in the amplitude of surface potentials across electrodes (i.e., changes in the standard deviation of Gaussian curves) depend on the distribution of fibers within the territory of gastrocnemius units and also on how far these surface potentials are detected from their source [76]. Consequently, the spread value defining an area below the Gaussians of 95% is likely an overestimation of the actual territory sizes. It is very likely then that postural motor neurons in the medial gastrocnemius muscle supply fibers confined to a longitudinal region smaller than 4 cm. This figure indicates a territory size fairly smaller (about six times smaller) than the gastrocnemius length.

Small territory, however, does not necessarily indicate independent activation of muscle units residing in specific gastrocnemius regions. Independent activation of gastrocnemius subvolumes or compartments demands access to motor neurones supplying gastrocnemius fibers in discrete muscle regions. This issue and the potential relevance of gastrocnemius compartmentalization in standing is discussed in the next section.

15.4.3   Is the Human Gastrocnemius Muscle Compartmentalized? What Is the Relevance of Compartments in Standing?

Originally, neuromuscular compartments were defined as groups of muscle fibers supplied by the primary branches of the nerve serving a muscle ([15]; see English et al. [16] for a review). This definition, however, does not imply that muscle units are organized about the primary nerve branches. For example, if the axon of motor neurons ramifies through different main nerve branches, single muscle units would be supplied by at least two primary nerve branches (Fig. 15.7A). Activation of individual motor neurones would therefore lead to activation of fibers in different muscle locations; fibers supplied by different primary nerve branches would be activated collectively. The elegant studies of English and Weeks [15,17] on the cat lateral gastrocnemius showed compelling electrophysiological and histochemical evidences positing the organization of single muscle units about the primary nerve branches. On this view, neuromuscular compartmentalization should be defined in terms of grouping of single muscle units rather than of muscle fibers supplied by single primary nerve branches.

Evidence associating the localization of single muscle units with the neuromuscular compartmentalization of the gastrocnemius muscles exists. With the experimental denervation followed by reinnervation of the cat gastrocnemius nerve, Rafuse and Gordon [75] observed that the territory of muscle units was related to how proximally the axonal branching occurs. When the cut nerve was sutured directly onto the gastrocnemius fascia, the territory of regenerated motor units was smaller than that for the normal motor units. The smaller territory of regenerated motor units was likely due to the restricted number of pathways along which axonal branching occurred [75]. Figure 15.7 illustrates how the axonal branching might affect the territory of gastrocnemius muscle units. This figure was created based on results from human gastrocnemius muscles showing that the tibial nerve splits, initially, into main branches subserving distinct muscle portions and, finally, into terminal branches innervating the muscle fibers [81,97]. Two main branches of the tibial nerve are shown in Fig. 15.7, supplying the central and distal gastrocnemius portions. If the axonal branching occurs proximally before the nerve splitting, muscle fibers in the central and distal locations might be innervated by the same motorneuron (muscle unit with large territory; Fig. 15.7A). Conversely, if the axons branch distally, after the nerve splitting, only the muscle fibers supplied by one of the main nerve branches would be innervated by the same motor neuron (muscle unit with small territory; Fig. 15.7B). Therefore, according to the definition of neuromuscular compartmentalization, recent results positing the spatial localization of muscle units [93] suggest that the human gastrocnemius muscle is, indeed, organized in neuromuscular compartments.

On one hand, motor units in the gastrocnemius muscle have small territories, thus allowing for the local occupation of muscle fibers within the total muscle volume. On the other hand, small territories do not ensure independent activation of individual compartments (i.e., of individual gastrocnemius subvolumes). For the local muscle activation to occur, the nervous system must be provided with a mechanism of selective activation of motor neurons serving muscle units in specific compartments. Empirical evidence showing regional variations in the amplitude of surface EMGs collected from the calf muscles suggests that such a mechanism likely exists [25,63,84,94]. One physiologically plausible mechanism accounting for the local muscle activation might be the unequal spatial distribution of excitatory synaptic inputs across gastrocnemius motor neurons. If, for example, motor neurons of similar excitability (similar rheobase and cronaxie) receive different amounts of excitatory inputs, then the likelihood of being activated would be overtly higher for those receiving more excitatory inputs. Recent circumstantial evidence from intramuscular EMG is consistent with such uneven distribution of excitatory inputs to gastrocnemius motor neurons in quiet standing. Postural activation therefore might occur at the level of populations of motor units, possibly belonging to specific gastrocnemius compartments, more likely than at the global muscle level.

The functional relevance of neuromuscular compartmentalization in the gastrocnemius muscle for the control of standing posture remains to be elucidated. What could be the advantage of activating individual gastrocnemius compartments differently during standing? From intramuscular EMGs, different researchers observed that specific motor units tend to be more easily recruited when forces over the body joints are exerted at specific directions [12,33,86]. These findings suggest that motor unit recruitment might be shaped by the kinematic efficiency of specific motor unit populations. Indirect evidence relating the regional activation within and between calf muscles to the direction of ankle torque prompts high-density surface EMGs [84]. Similar results reported in standing are incipient. We, however, observed that while subjects stand at ease, their medial gastrocnemius muscle is activated more frequently when bodily sways are deviated laterally [95]. It is unknown, however, whether the direction of ankle torque in standing might depend on the uneven activation of different gastrocnemius compartments.

The notion of local, postural activation of the gastrocnemius muscle is not in opposition with the theory of posture control in terms of functional modules. It is also not associated with greater complexity in central processing by the nervous system. Actually, the fact that activation may be organized at the level of populations of motor units possibly explains why some muscles may take part in different postural synergies (see Fig. 4 in Torres-Oviedo and Ting [88]). It is reasonable to admit that the issue of limited central resources might be solved by mapping muscle activation in terms of task-dependent synergies [71]. Task-dependent synergies, however, might not be coded in terms of muscles but of populations of motor units with greatest task-specific efficiency. On this view, therefore, much could be gained by revisiting postural synergies with HDsEMG.

REFERENCES

1. Aruin, A. S., and M. L. Latash, "Directional specificity of postural muscles in feed-forward postural reactions during fast voluntary arm movements," *Exp. Brain Res.* **103**, 323–332 (1995).

2. Asai, Y., Y. Tasaka, K. Nomura, T. Nomura, M. Casadio, and P. Morasso, "A model of postural control in quiet standing: robust compensation of delay-induced instability using intermittent activation of feedback control," *PLoS One* **4**, e6169 (2009).

3. Basmajian, J., and C. J. De Luca, *Muscles Alive: Their Functioning Revealed by Electromyography*, Wiliams & Wilkins, Baltimore, 1985.

4. Bleuse, S., F. Cassim, J. L. Blatt, E. Labyt, P. Derambure, J. D. Guieu, and L. Defebvre, "Effect of age on anticipatory postural adjustments in unilateral arm movement," *Gait & Posture* **24**, 203–210 (2006).

5. Bottaro, A., Y. Yasutake, T. Nomura, M. Casadio, and P. Morasso, "Bounded stability of the quiet standing posture: An intermittent control model," *Hum. Mov Sci.* **27**, 473–495 (2008).

6. Burke, R. E., and P. Tsairis, "Anatomy and innervation ratios in motor units of cat gastrocnemius," *J. Physiol.* **234**, 749–765 (1973).

7. Casadio, M., P. G. Morasso, and V. Sanguineti, "Direct measurement of ankle stiffness during quiet standing: implications for control modelling and clinical application," *Gait & Posture* **21**, 410–424 (2005).

8. Cochrane, D. J., I. D. Loram, S. R. Stannard, and J. Rittweger, "Changes in joint angle, muscle–tendon complex length, muscle contractile tissue displacement, and modulation of EMG activity during acute whole-body vibration," *Muscle Nerve* **40**, 420–429 (2009).

9. Creed, R. S., D. Denny-Brown, J. C. Eccles, E. G. T. Liddell, and C. S. Sherrington, *Reflex Activity of the Spinal Cord*, Clarendon Press, Oxford, 1932.

10. D'avella, A., and F. Lacquaniti, "Control of reaching movements by muscle synergy combinations," *Front Comput. Neurosci.* 2013 Apr 19;7:42. doi:10.3389/fncom.2013.00042.

11. Day, B. L., and I. N. Lyon, "Voluntary modification of automatic arm movements evoked by motion of a visual target," *Exp. Brain Res.* **130**, 159 (2000).

12. Desmedt, H. E., and E. Godaux, "Spinal motoneuron recruitment in man: rank deordering with direction but not with speed of voluntary movement," *Science* **214**, 933–936 (1981).

13. Di Giulio, I., V. Baltzopoulos, C. N. Managanaris, and I. D. Loram, "Human standing: does the control strategy pre-program a rigid knee?," *J. Appl. Physiol.* **114**, 1717–1729 (2013).

14. Di Giulio, I., C. N. Maganaris, V. Baltzopoulos, and I. D. Loram, "The proprioceptive and agonist roles of gastrocnemius, soleus and tibialis anterior muscles in maintaining human upright posture," *J. Physiol.* **587**, 2399–2416 (2009).

15. English, A. W., and O. I. Weeks, "Compartmentalization of single muscle units in cat lateral gastrocnemius," *Exp. Brain Res.* **56**, 361–368 (1984).

16. English, A. W., S. L. Wolf, and R. L. Segal, "Compartmentalization of muscles and their motor nuclei: the partitioning hypothesis," *Phys. Ther.* **73**, 857–867 (1993).

17. English, A. W., and O. I. Weeks, "Electromyographic cross-talk within a compartmentalized muscle of the cat," *J. Physiol.* **416**, 327–336 (1989).

18. Evans, C. M., S. J. Fellows, P. M. Rack, H. F. Ross, and D. K. Walters, "Response of the normal human ankle joint to imposed sinusoidal movements," *J. Physiol.* **344**, 483–502 (1983).

19. Farina, D., and R. Merletti, "Estimation of average muscle fiber conduction velocity from two-dimensional surface EMG recordings," *J. Neurosci. Met.* **134**, 199–208 (2004).

20. Fitzpatrick, R., D. Burke, and S. C. Gandevia, "Loop gain of reflexes controlling human standing measured with the use of postural and vestibular disturbances," *J. Neurophysiol.* **76**, 3994–4008 (1996).

21. Fitzpatrick, R., and D. I. McCloskey, "Proprioceptive, visual and vestibular thresholds for the perception of sway during standing in humans," *J. Physiol.* **478**(Pt1), 173–186 (1994).

22. Fitzpatrick, R. C., R. B. Gorman, D. Burke, and S. C. Gandevia, "Postural proprioceptive reflexes in standing human subjects: bandwidth of response and transmission character-istics," *J. Physiol.* **458**, 69–83 (1992).

23. Fitzpatrick, R. C., J. L. Taylor, and D. I. McCloskey, "Ankle stiffness of standing humans in response to imperceptible perturbation: reflex and task-dependent components," *J. Physiol.* **454**, 533–547 (1992).

24. Floyd, W. F., and P. H. Silver, "The function of the erectores spinae muscles in certain movements and postures in man," *J. Physiol.* **129**, 184–203 (1955).

25. Gallina, A., R. Merletti, and T. M. M. Vieira, "Are the myoelectric manifestations of fatigue distributed regionally in the human medial gastrocnemius muscle?," *J. Electro-myogr. Kinesiol.* **21**, 929–938 (2011).

26. Garcia, M. A. C., and T. M. M. Vieira, "Surface electromyography: why, when and how to use it," *Rev. Andal. Med. Deporte* **4**, 17–28 (2011).

27. Gatev, P., S. Thomas, T. Kepple, and M. Hallett, "Feedforward ankle strategy of balance during quiet stance in adults," *J. Physiol.* **514**(Pt3), 915–928 (1999).

28. Gawthrop, P., I. Loram, and M. Lakie, "Predictive feedback in human simulated pendulum balancing," *Biol. Cybern.* **101**, 131–146 (2009).

29. Gollee, H., A. Mamma, I. Loram, and P. Gawthrop, "Frequency-domain identification of the human controller," *Biol. Cybern.* **106**, 359–372 (2012).

30. Hellebrandt, F. A., "Standing as a geotropic reflex: the mechanism of the asynchronous rotation of motor units," *Am. J. Physiol.* **121**, 471–474 (1938).

31. Henry, S. M., J. Fung, and F. B. Horak, "EMG responses to maintain stance during multidirectional surface translations," *J. Neurophysiol.* **80**, 1939–1950 (1998).

32. Hermens, H. J., B. Freriks, C. Disselhorst-Klug, and G. Rau, "Development of recom-mendations for SEMG sensors and sensor placement procedures," *J. Electromyogr. Kinesiol.* **10**, 361–374 (2000).

33. Herrmann, U., and M. Flanders, "Directional tuning of single motor units," *J. Neurosci.* **18**, 8402–8416 (1998).

34. Hodges, P. W., V. S. Gurfinkel, S. Brumagne, T. C. Smith, and P. C. Cordo, "Coexistence of stability and mobility in postural control: evidence from postural compensation for respiration," *Exp. Brain Res.* **144**, 293–302 (2002).

35. Hodson-Tole, E. F., I. D. Loram, T. M. M. Vieira, "Myoelectric activity along human gastrocnemius medialis: different spatial distributions of postural and electrically elicited surface potentials," *J. Electromyogr. Kinesiol.* **23** (1),43–50 (2013).

36. Horak, F. B., and L. M. Nashner, "Central programming of postural movements: adapta-tion to altered support-surface configurations," *J. Neurophysiol.* **55**, 1369–1381 (1986).

37. Horak, F. B., J. G. Nutt, and L. M. Nashner, "Postural inflexibility in parkinsonian subjects," *J. Neurol. Sci.* **111**, 46–58 (1992).

38. Inglis, J. T., F. B. Horak, C. L. Shupert, and C. Jones-Rycewicz, "The importance of somatosensory information in triggering and scaling automatic postural responses in humans," *Exp. Brain Res.* **101**, 159–164 (1994).

39. Joseph, J., and A. Nightingale, "Electromyography of muscles of posture: leg muscles in males," *J. Physiol.* **117**, 484–491 (1952).

40. Joseph, J., A. Nightingale, and P. L. Williams, "A detailed study of the electric potentials recorded over some postural muscles while relaxed and standing," *J. Physiol.* **127**, 617–625 (1955).

41. Karniel, A., "Open questions in computational motor control," *J. Integr. Neurosci.* **10**, 385–411 (2011).

42. Karniel, A., "The minimum transition hypothesis for intermittent hierarchical motor control," *Front. Comput. Neurosci.* 7–12 (2013).

43. Karniel, A., and G. F. Inbar, "Human motor control: Learning to control a time-varying, nonlinear, many-to-one system," *IEEE Trans. Syst. Man Cyber. Part C—App. Rev.* **30**, 1–11 (2000).

44. Kawakami, Y., Y. Ichinose, and T. Fukunaga, "Architectural and functional features of human triceps surae muscles during contraction," *J. Appl. Physiol.* **85**, 398–404 (1998).

45. Kiemel, T., A. J. Elahi, and J. J. Jeka, "Identification of the plant for upright stance in humans: multiple movement patterns from a single neural strategy," *J. Neurophysiol.* **100**, 3394–3406 (2008).

46. Kiemel, T., Y. Zhang, and J. J. Jeka, "Identification of neural feedback for upright stance in humans: stabilization rather than sway minimization," *J. Neurosci.* **31**, 15144–15153 (2011).

47. Loram, I. D., H. Gollee, M. Lakie, and P. J. Gawthrop, "Human control of an inverted pendulum: is continuous control necessary? Is intermittent control effective? Is intermittent control physiological?," *J. Physiol.* **589**, 307–324 (2011).

48. Loram, I. D., S. M. Kelly, and M. Lakie, "Human balancing of an inverted pendulum: is sway size controlled by ankle impedance?," *J. Physiol.* **532**, 879–891 (2001).

49. Loram, I. D., and M. Lakie, "Direct measurement of human ankle stiffness during quiet standing: the intrinsic mechanical stiffness is insufficient for stability," *J. Physiol.* **545**, 1041–1053 (2002).

50. Loram, I. D., and M. Lakie, "Human balancing of an inverted pendulum: position control by small, ballistic-like, throw and catch movements," *J. Physiol.* **540**, 1111–1124 (2002).

51. Loram, I. D., M. Lakie, I. Di Giulio, and C. N. Maganaris, "The consequences of short-range stiffness and fluctuating muscle activity for proprioception of postural joint rotations: the relevance to human standing," *J. Neurophysiol.* **102**, 460–474 (2009).

52. Loram, I. D., M. Lakie, and P. J. Gawthrop, "Visual control of stable and unstable loads: what is the feedback delay and extent of linear time-invariant control?," *J. Physiol.* **587**, 1343–1365 (2009).

53. Loram, I. D., C. N. Maganaris, and M. Lakie, "Active, non-spring-like muscle movements in human postural sway: how might paradoxical changes in muscle length be produced?," *J. Physiol.* **564**, 281–293 (2005).

54. Loram, I. D., C. N. Maganaris, and M. Lakie, "Human postural sway results from frequent, ballistic bias impulses by soleus and gastrocnemius," *J. Physiol.* **564**, 295–311 (2005).

55. Loram, I. D., C. N. Maganaris, and M. Lakie, "The passive, human calf muscles in relation to standing: the non-linear decrease from short range to long range stiffness," *J. Physiol.* **584**, 661–675 (2007).

56. Loram, I. D., C. N. Maganaris, and M. Lakie, "The passive, human calf muscles in relation to standing: the short range stiffness lies in the contractile component," *J. Physiol.* **584**, 677–692 (2007).

57. Loram, I. D., C. van de Kamp, H. Gollee, and P. J. Gawthrop, "Identification of intermittent control in man and machine," *J. R. Soc. Interface* **9**, 2070–2084 (2012).

58. Marsden, C. D., P. A. Merton, H. B. Morton, J. C. Rothwell, and M. M. Traub, "Reliability and efficacy of the long-latency stretch reflex in the human thumb," *J. Physiol.* **316**, 47–60 (1981).

59. Masani, K., M. R. Popovic, K. Nakazawa, M. Kouzaki, and D. Nozaki, "Importance of body sway velocity information in controlling ankle extensor activities during quiet stance. *J. Neurophysiol.* **90**, 3774–3782 (2003).

60. Masani, K., A. H. Vette, N. Kawashima, and M. R. Popovic, "Neuromusculoskeletal torque-generation process has a large destabilizing effect on the control mechanism of quiet standing," *J. Neurophysiol.* **100**, 1465–1475 (2008).

61. Masani, K., A. H. Vette, and M. R. Popovic, "Controlling balance during quiet standing: proportional and derivative controller generates preceding motor command to body sway position observed in experiments," *Gait Posture* **23**, 164–172 (2006).

62. Maurer, C., and R. J. Peterka, "A new interpretation of spontaneous sway measures based on a simple model of human postural control," *J. Neurophysiol.* **93**, 189–200 (2005).

63. McLean, L., and N. Goudy, "Neuromuscular response to sustained low-level muscle activation: within- and between-synergist substitution in the triceps surae muscles," *Eur. J. Appl. Physiol.* **91**, 204–216 (2004).

64. Merletti, R., A. Botter, C. Cescon, M. A. Minetto, and T. M. M. Vieira, "Advances in surface EMG: recent progress in clinical research applications," *Crit. Rev. Biomed. Eng.* **38**, 347–379 (2010).

65. Mesin, L., and D. Farina, "Simulation of surface EMG signals generated by muscle tissues with inhomogeneity due to fiber pinnation," *IEEE Trans. Biomed. Eng.* **51**, 1521–1529 (2004).

66. Mesin, L., R. Merletti, and T. M. M. Vieira, "Insights gained into the interpretation of surface electromyograms from the gastrocnemius muscles: A simulation study," *J. Biomech.* **44**, 1096–1103 (2011).

67. Mochizuki, G., T. D. Ivanova, and S. J. Garland, "Postural muscle activity during bilateral and unilateral arm movements at different speeds," *Exp. Brain Res.* **155**, 352–361 (2004).

68. Nardone, A., and M. Schieppati, "Group II spindle fibres and afferent control of stance. Clues from diabetic neuropathy," *Clin. Neurophysiol.* **115**, 779–789 (2004).

69. Narici, M. V., T. Binzoni, E. Hiltbrand, J. Fasel, F. Terrier, and P. Cerretelli, "In vivo human gastrocnemius architecture with changing joint angle at rest and during graded isometric contraction," *J. Physiol.* **496**(Pt1), 287–297 (1996).

70. Nashner, L. M., "Fixed patterns of rapid postural responses among leg muscles during stance," *Exp. Brain Res.* **30**, 13–24 (1977).

71. Neilson, P. D., and M. D. Neilson, "An overview of adaptive model theory: solving the problems of redundancy, resources, and nonlinear interactions in human movement control," *J. Neural. Eng.* **2**, S279–S312 (2005).

72. Portnoy, H., and F. Morin, "Electromyographic study of postural muscles in various positions and movements," *Am. J. Physiol.* **186**, 122–126 (1956).

73. Pruszynski, J. A., and S. H. Scott, "Optimal feedback control and the long-latency stretch response," *Exp. Brain Res.* **218**, 341–359 (2012).

74. Rack, P. M., H. F. Ross, A. F. Thilmann, and D. K. Walters, "Reflex responses at the human ankle: the importance of tendon compliance," *J. Physiol.* **344**, 503 (1983).

75. Rafuse, V. F., and T. Gordon, "Self-reinnervated cat medial gastrocnemius muscles. II. analysis of the mechanisms and significance of fiber type grouping in reinnervated muscles," *J. Neurophysiol.* **75**, 282–297 (1996).

76. Roeleveld, K., D. F. Stegeman, H. M. Vingerhoets, and A. Van Oosterom, "Motor unit potential contribution to surface electromyography," *Acta Physiol. Scand.* **160**, 175–183 (1997).

77. Ronco, E., T. Arsan, and P. J. Gawthrop, "Open-loop intermittent feedback control: Practical continuous- time GPC," *IEEE Proc. D-Control Theor. Appl.* **146**, 426–434 (1999).

78. Rothwell, J., *Control of Human Voluntary Movement*, Chapman & Hall, London, 1994.

79. Sacco, I. C., A. A. Gomes, M. E. Otuzi, D. Pripas, and A. N. Onodera, "A method for better positioning bipolar electrodes for lower limb EMG recordings during dynamic contractions," *J. Neurosci. Meth.* **180**, 133–137 (2009).

80. Schieppati, M., A. Nardone, S. Corna, and M. Bove, "The complex role of spindle afferent input, as evidenced by the study of posture control in normal subjects and patients," *Neurol. Sci.* **22**, S15–S20 (2001).

81. Segal, R. L., S. L. Wolf, M. J. DeCamp, M. T. Chopp, and A. W. English, "Anatomical partitioning of three multiarticular human muscles," *Acta Anat.* **142**, 261–266 (1991).

82. Sherrington, C. S., *Integrative Action of the Nervous System*, Constable, London, 1906.

83. Sherrington, C. S., *Integrative Action of the Nervous System*, Cambridge University Press, Cambridge, 1947.

84. Staudenmann, D., I. Kingma, A. Daffertshofer, D. F. Stegeman, and J. H. van Dieen, "Heterogeneity of muscle activation in relation to force direction: a multichannel surface electromyography study on the triceps surae muscle," *J. Electromyogr. Kinesiol.* **19**, 882–895 (2009).

85. Suzuki, Y., T. Nomura, M. Casadio, and P. Morasso, "Intermittent control with ankle, hip, and mixed strategies during quiet standing: a theoretical proposal based on a double inverted pendulum model," *J. Theor. Biol.* **310**, 55–79 (2012).

86. ter Haar Romeny, B. M., J. J. van der Gon, and C. C. Gielen, "Relation between location of a motor unit in the human biceps brachii and its critical firing levels for different tasks," *Exp. Neurol.* **85**, 631–650 (1984).

87. Thomas, D. P., and R. J. Whitney, "Postural movements during normal standing in man," *J. Anat.* **93**, 524–539 (1959).

88. Torres-Oviedo, G., and L. H. Ting, "Muscle synergies characterizing human postural responses," *J. Neurophysiol.* **98**, 2144–2156 (2007).

89. Van De Kamp, C., P. Gawthrop, H. Gollee, M. Lakie, and I. D. Loram, "Interfacing sensory input with motor output: does the control architecture converge to a serial process along a single channel?," *Front. Comput. Neurosci.* **7**, (2013).

90. Van de Kamp, C., P. Gawthrop, H. Gollee, and I. Loram, "Refractoriness in a whole-body human balance task," in *Proceedings ISPGR Congress*, Trondheim, 2012.

91. van de Kamp, C., P. J. Gawthrop, H. Gollee, and I. D. Loram, "Refractoriness in sustained visuo-manual control: is the refractory duration intrinsic or does It depend on external system properties?," *PLoS Comput. Biol.* **9**, e1002843 (2013).

92. van der Kooij, H., and E. de Vlugt, "Postural responses evoked by platform pertubations are dominated by continuous feedback," *J. Neurophysiol.* **98**, 730–743 (2007).

93. Vieira, T. M. M., I. D. Loram, S. Muceli, R. Merletti, and D. Farina, "Postural activation of the human medial gastrocnemius muscle: are the muscle units spatially localised?," *J. Physiol.* **589**, 431–443 (2011).

94. Vieira, T. M. M., R. Merletti, and L. Mesin, "Automatic segmentation of surface EMG images: improving the estimation of neuromuscular activity," *J. Biomech.* **43**, 2149–2158 (2010).

95. Vieira, T. M. M., U. Windhorst, and R. Merletti, "Is the stabilization of quiet upright stance in humans driven by synchronized modulations of the activity of medial and lateral gastrocnemius muscles?," *J. Appl. Physiol.* **108**, 85–97 (2010).

96. Winter, D. A., A. E. Patla, F. Prince, M. Ishac, and K. Gielo-Perczak, "Stiffness control of balance in quiet standing," *J. Neurophysiol.* **80**, 1211–1221 (1998).

97. Wolf, S. L., and J. H. Kim, "Morphological analysis of the human tibialis anterior and medial gastrocnemius muscles," *Acta Anat.* **158**, 287–295 (1997).

98. Zajac, F. E., "Muscle and tendon: properties, models, scaling, and application to biomechanics and motor control," *Crit. Rev. Biomed. Eng.* **17**, 359–411 (1989).

16

APPLICATIONS IN MOVEMENT AND GAIT ANALYSIS

A. Merlo and I. Campanini

LAM—Motion Analysis Laboratory, Department of Rehabilitation, AUSL of Reggio Emilia, Correggio, Italy

16.1 THE RELEVANCE OF ELECTROMYOGRAPHY IN KINESIOLOGY

Motion analysis is accomplished by measuring trajectories, rotations of joints, torque values, power values, ground reaction forces, temporal events, and muscle activities [35]. The simultaneous acquisition of all signals allows multifactorial analysis of the investigated task.

Trajectories and joint rotations are obtained with motion capture systems that reconstruct the position in space (trajectories) of markers placed on a subject's skin. These systems also use computational models to compute both the location (pose) of bone segments and joint rotations [14,54]. Some joint angles can be measured with electrogoniometers, as both ankle and knee flexion/extension angles shown in Fig. 16.1. Gyroscopes and accelerometers are suitable for assessing voluntary as well as involuntary movements such as tremor, chorea, dystonia, and dyskinesia in clinical settings [9]. The development of inertial sensor units (IMUs) and dedicated biomechanical protocols enables the kinematics of joints to be assessed, both indoors and outdoors [18].

Force plates embedded in the floors of motion analysis laboratories measure force transfer between foot and ground during gait. Numerical indices can be obtained from force plate data regarding an individual's amount of vertical loading, pushing, and progression during stance [11,44,66]. These indices can be used together with

Surface Electromyography: Physiology, Engineering, and Applications, First Edition.
Edited by Roberto Merletti and Dario Farina.
© 2016 by The Institute of Electrical and Electronics Engineers, Inc. Published 2016 by John Wiley & Sons, Inc.

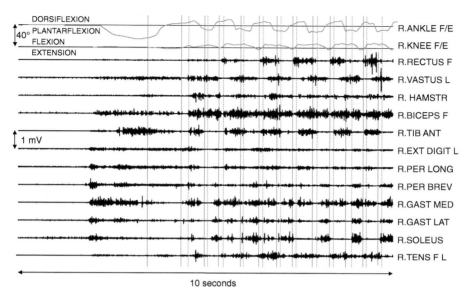

FIGURE 16.1 Ankle and knee sagittal kinematic surface EMG, foot contacts (*continuous vertical lines*), and foot-offs (*dashed vertical lines*) of a neurological patient at the beginning of a walking session. The green lines indicate right-limb events, and the red lines indicate left-limb events. The period between two equal (e.g., green) continuous lines corresponds to the stride or gait cycle. Within a cycle, homolateral temporal events permit determination of the stance phase (from foot contact to foot-off) and the swing phase (from foot-off to next foot contact). Within stance, contralateral limb events permit recognition of the first double-support, single-support, and second double-support phases. The following muscles are presented, from the top: *rectus femoris, vastus lateralis, hamstrings, biceps femoris, tibialis anterior, extensor digitorum longus, peroneus longus, peroneus brevis, gastrocnemius medialis, gastrocnemius lateralis, soleus,* and *tensor fasciae latae.* Miniaturized electrodes (area = 5 mm²) were placed on the right limb according to Andersen et al. [3].

walking speed to monitor the progress of a patient during rehabilitation. A calibration procedure enables both the coordinates of joint centers and data from force plates to be expressed in the same reference system. Subsequently, the external joint moments can be computed, and joint powers can ultimately be obtained by combining the angular velocities and the joint torques [12,13,65]. Cameras for video acquisition and a surface electromyography (sEMG) system complete the typical equipment of a motion analysis laboratory.

Kinematics, kinetics, and sEMG are used clinically to identify the primary causes of motion impairment and the resulting compensatory movements, thus enhancing clinical assessment and supporting decisions such as choice of surgery, local muscle inhibition by drugs, orthoses, rehabilitation protocols, and so on [26,44,63]. Further-more, instrumental motion analysis can be used to verify and quantify the effective-ness of a delivered treatment [20,62].

Several sensors and technologies can be used to determine joint kinematics, but only sEMG can provide information on the origin of motion produced by muscle activation. Since several signals from muscles are simultaneously acquired in motion analysis, multiple terms such as polyelectromyography, multichannel electromyography, dynamic sEMG and kinesiological sEMG, have been used to indicate this one examination [24]. A comprehensive history of the study of skeletal muscle function over the centuries, with emphasis on kinesiological electromyography, can be found in Blanc and Dimanico [4].

The number of channels acquired (that is, the number of muscles being investigated) in motion analysis applications can range from 5 to 8 in the assessment of simple tasks, to 12–16 in the evaluation of complex exercises. For example, to completely assess a reaching task, the data on the muscles of the hand, forearm, arm, shoulder girdle, and trunk (contralateral to movement) should be acquired. As a rule of thumb, the greater the number of muscles involved, the easier the device should be to use. Thus, electromyographs used in laboratories typically perform simple single differential detection.

Interestingly enough, electromyography during task execution enables muscle activities to be assessed rather than inferred from a combination of observation of movements and clinical evaluation performed under static conditions (for example, with the subject on the examination table). Inferring muscle activity based on observation may be unreliable and misleading in a number of clinical applications. During observation, lack of movement may be interpreted as deficient recruitment of the agonist muscles, although it may also be due to either the presence of passive constraints or co-activation of agonist and antagonist muscles. For example, in patients with hemiplegia subsequent to stroke, shortening of the calf muscles can limit ankle dorsiflexion even though the ankle dorsiflexors (*tibialis anterior* and *extensor digitorum longus*) are properly activated. Also, the involuntary elbow flexion that may occur in stroke patients during walking may be caused by any one of the three elbow flexors (*biceps brachii*, *brachialis*, and *brachioradialis*) or by the combined activity of two or three of these muscles, a situation which cannot be identified by visual inspection but requires sEMG assessment to be understood and properly treated.

To properly interpret sEMG traces, temporal markers that link muscle activity to recorded motion are required. Signals from switches (foot switches in gait analysis) can be acquired to obtain information on the temporal evolution of the task. An sEMG-synchronized video with an adequate frame rate, or kinematic data when available, can be also used to define temporal markers. During gait analysis, for example, both the foot strike (continuous vertical lines in Figs. 16.1 and 16.2) and the foot off (dashed vertical lines in Figs. 16.1 and 16.2) events of both legs should be recorded in order to distinguish stance from swing and, within stance, to differentiate between double support phases and the single support phase (Fig. 16.2). Temporal events are essential for interpreting patients' muscle activity properly, as the timing of muscle activity may be different from normal reference values, as indicated by the variation in the duration of the gait phases. For example, during the gait cycle, delayed activation of the ankle dorsiflexors with respect to normal reference onset time is not pathological if the patient's swing phase is delayed too, due to a prolonged stance phase.

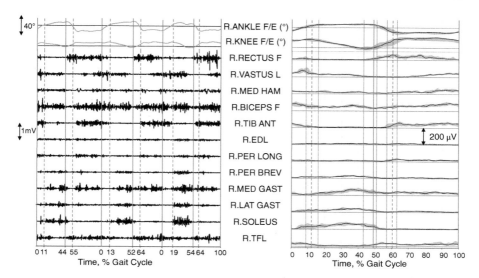

FIGURE 16.2 Three consecutive right gait cycles and the median envelope, whose computation was based on all available strides. Time axes as well as temporal events are expressed as a percentage of the gait cycle. The data refer to the same subject and trial as in Fig. 16.1.

Since the temporal events that define the beginning of movements have been recorded and are thus available, cyclic movements such as walking, cycling, and repetitive reaching are typically time-normalized over a 0–100% time base [44]. Figure 16.2 shows three consecutive cycles of the right limb, along with the median envelope of data from Fig. 16.1.

16.2 EXPERIMENTAL SETTING AND sEMG IN MOTION ANALYSIS

In motion analysis, sEMG acquisitions are usually performed to answer specific questions regarding clinical applications or for research purposes, and the acquisition must be set up to achieve the accuracy needed to answer the question. Typically, in order to answer a research question, homogeneous data are collected from one or more groups of subjects or patients, mean profiles or values are computed, and statistical methods are employed to check for group differences or for associations between sEMG-based numerical indices and previous exposition of the subjects to one or more (risk) factors. The answer to a clinical question may require a patient-specific approach, and the setup may vary from patient to patient. Studies on healthy volunteers may rely on complex setups with multiple sensors and lengthy acquisitions [33]. Conversely, in the clinical routine, the setup must comply with constraints such as the patient's functional limitations, the assessment time available (e.g., 30 minutes), the devices that are available, the operators' clinical and technical skills, and the overall cost of the examination. Thus, the simplest setup that answers the clinical question is often the best solution. The acquisition of maximum voluntary contraction (MVC) that is paradigmatic in research studies on healthy subjects [8,64] cannot be carried out on patients with pathologies affecting muscle recruitment or selective control, or on subjects with pain.

Once the question to be addressed by the sEMG assessment has been established—together with either the required accuracy or the minimum clinically significant variation—the procedure to be used can be summarized as follows:

- Choice of the muscles targeted for data acquisition
- Choice of the type of electrode to use
- Choice of the electrode placement protocol
- Verification of the selectivity of the detection system
- Choice of task
- Acquisition
- Verification of data quality
- Generation of a report (for clinical evaluation)
- Data analysis and extraction of relevant information.

The **choice of the muscles targeted for data acquisition** is crucial and requires a profound knowledge of functional anatomy, which must be supplemented with an understanding of pathology when patients are being dealt with.

The **type of the electrodes to use** must be specifically chosen to obtain the level of accuracy required by the measurement. It is essential for the detected signal to originate from the target muscle only, without containing components coming from adjacent muscles (crosstalk). An sEMG that is not crosstalk-free can dramatically affect clinical decision making. Thus, selectivity must always be a priority higher than signal amplitude, despite what is suggested in Sacco et al. [48]. The detected volume of a pair of surface electrodes mainly depends on their interelectrodic distance (IED), as described in Chapters 2 and 3. Selectivity increases as IED is reduced. Since the minimum distance between adjacent electrodes depends on their dimensions, small electrodes with a diameter of a few millimeters are preferable in applications of motion analysis to electrodes with a diameter of 1–2 cm which were originally designed for pediatric electrocardiographic (ECG) acquisition. Moreover, large electrodes may even be wider than the target muscle and directly pick up the sEMG activity of a neighboring muscle. This phenomenon is common if ECG electrodes are used on children's muscles, on the forearm muscles of adults, and, in general, when the transverse size of the target muscle is smaller than electrode diameter (e.g., 2 cm). In the adult leg, this might occur during detection of the *peroneus longus* muscle. The latter is active in stance, and its trace can be corrupted by activity of the *extensor digitorum longus*, which occurs in swing [10]. Indwelling electrodes (fine wires) must be used to measure deep muscles and whenever surface sEMG may be affected by crosstalk and also whenever surface electrodes may cover neighboring muscles. Acquisition by fine wires may be also affected by crosstalk in some cases; thus, the caveats reported in the literature should be taken into account [19,30,49]. Additional technical indications concerning electrodes and the electrode-skin interface (impedance and noise) are provided in Chapter 3.

In the clinical arena the topic of **electrode placement for surface sEMG** is still debated, and several placement protocols have been proposed. These are based on (a) empirical measurement of minimum crosstalk areas, that is, skin areas over a target muscle where the measured RMS amplitude of crosstalk was minimum during a sequence of selective contractions of all surrounding muscles or muscle groups (i.e., antagonists muscles) [1–3], (b) revised literature [29], and (c) the location of the innervation zone [42]. It is important to note that the available guidelines are reliable only under specific conditions: (a) when using electrodes equivalent to those used by the authors, in terms of dimensions and IED, (b) when the subject being examined meets the inclusion criteria used by the authors, and thus (c) in the absence of bone deformities and history of functional surgery (e.g., muscle transfers).

Test maneuvers should be performed to **verify the selectivity of the detection system** for each of the measured muscles by eliciting activation of a single muscle and

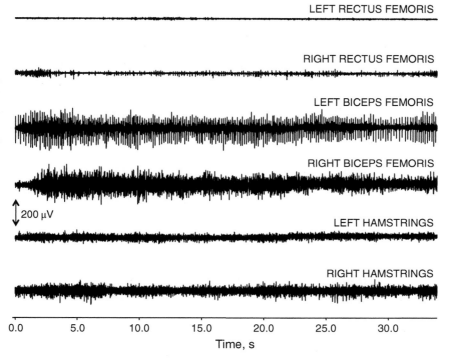

FIGURE 16.3 Bipolar thigh muscle EMG acquisition from a patient with paraparesis subsequent to spinal cord injury who is attempting to rise from the supine to a seated position on a bed. Abnormal activation of the hip extensor muscles impedes anterior pelvic-on-femoral rotation (pelvic tilt) and thus prevents the patient from verticalizing the trunk and reaching the seated position.

checking for crosstalk on all other acquired traces [2,3]. Such a check is essential when working with patients suffering from anatomical deformities in either bones or soft tissues (muscle shortening) that limit the reliability of protocols for electrode placement [3]. Once again, the relative practice requires broad competency in functional anatomy and technological issues, such as distinguishing sEMG from noise, interference, and artifacts (Chapter 3).

The **choice of the task** performed for data acquisition is directly associated with the research or clinical question. The task may be as simple as an ankle dorsiflexion or a transition from the supine to the seated position, as shown in Fig. 16.3 for a patient; or, it may be more complex, such as the chair-to-chair transition shown in Fig. 16.4 for a healthy adult. When assessing patients, it is often useful to study the changes in the sEMG pattern that result from changes in measuring conditions, such as an increase in speed from the self-selected speed. When assessing gait, the use of different shoes, gait aids and orthoses can provide important information.

In research studies, the number of repetitions performed for acquiring data on each task is defined a priori when the study is designed [25,28]. In clinical applications, the number of repetitions is a trade-off between patient fatigue and the need to obtain consistent data.

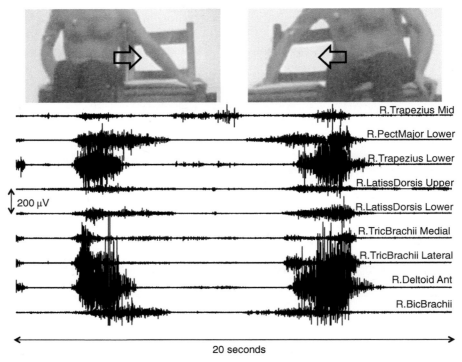

FIGURE 16.4 Transfer from one chair to another by a healthy adult, a move which simulates chair-to-wheelchair or bed-to-chair relocation. Timing and amplitude differences between different functional parts of the same muscle can be noted.

Verification of data quality is necessary to prevent misinterpretation and inappropriate clinical intervention. The sEMG signal acquired during a movement consists of both onset and offset phases, thus being nonstationary (not even in the wide sense). However, the use of power spectral density (PSD) to determine the nature of the PSD profile can be a powerful tool for checking sEMG data quality [40]. A PSD profile that lacks the typical bell shape because of motion artifacts or power line interference is a strong indicator of unreliable data. Furthermore, other unexpected artifacts can be revealed by PSD computation. These include power-line interference and periodic transmission artifacts that might occur at low battery levels, data loss (short periods at zero), and external interference produced by electric motors when training machines are used. Surface EMG filtering procedures are typically used in applications of motion analysis to remove motion artifacts (MA). High-pass filters with cutoff frequencies ranging between 10 and 30 Hz are used, as MA are often limited to the lowest part of the frequency range. However, in some cases, and frequently during gait, MA frequency content can be higher than the filter's cutoff frequency (e.g., 30 Hz), and filtered artifacts may appear as short bursts of sEMG activity. In particular, artifacts that repeat

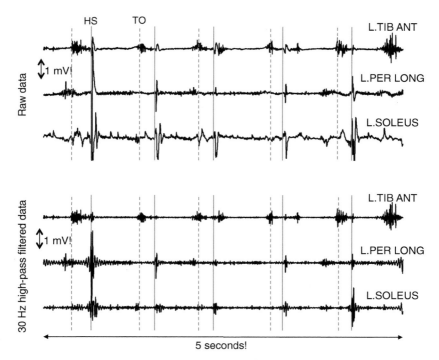

FIGURE 16.5   Raw (*above*) and filtered (*below*) data collected during the gait of a child with cerebral palsy. The motion artifacts at heel strike that are evident in the raw data resemble EMG bursts, after high-pass filtering. Motion artifacts may be a serious problem in motion analysis applications, where abrupt acceleration occurs, and the frequency content of the artifacts may partially overlap to the EMG frequency band. Motion artifacts can be avoided by proper skin preparation and preamplifier attachment.

rhythmically may lead to erroneous clinical decisions. A very important, frequent example is the motion artifact at heel strike (HS) during walking. The abrupt deceleration at HS may vibrate cables and/or preamplifiers and exert a pull on the electrodes, thus causing a large rhythmic artifact. Figure 16.5 shows sEMG data acquired during the gait of a child affected by cerebral palsy, along with the same signal filtered by a fourth-order elliptic filter with cutoff frequency of 30 Hz. Once filtered, these artifacts can no longer be recognized as such and may be interpreted as an abnormal stretch reflex (spasticity) of the *triceps surae* muscle and consequently be treated (e.g. by botulinum toxin injection). We note that motion artifacts can be minimized by adequate skin preparation and by selecting electrodes with good adhesion, avoiding tension in cables, and properly fastening preamplifiers to the skin [3,40].

In motion analysis, the activity of multiple muscles is typically displayed on a single graph, as in Fig. 16.1, together with temporal events and kinematic data, if available. In the **generation of a report** for clinical evaluation, a common *y*-axis scale must be set up before data are shared with nonexpert colleagues or physicians, in order to prevent misinterpretation that could adversely influence clinical decisions. An example of the effect on actual gait data of both autoscaling and correct full-scale setup of the *y*-axis is shown in Fig. 16.6 [40].

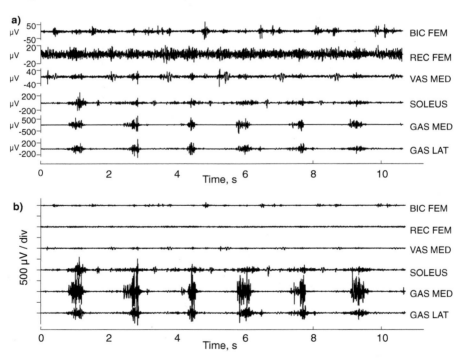

FIGURE 16.6   Surface EMG data acquired from a healthy adult during gait, presented both in autoscale (**a**) and with a fixed *y*-axis (full scale) (**b**). Because of the electrode–skin contact noise, on the autoscaled representation, *rectus femoris* seems to show continuous activity, even though it is inactive (see Chapter 3).

In conclusion, to ensure data reliability in applications of motion analysis, raw sEMG data on a fixed y-axis scale must be checked before the following are carried out: filtering, envelope computation, data normalization, clinical report creation, and statistical analysis for research purposes.

16.3   sEMG-BASED INFORMATION USED IN APPLICATIONS OF MOTION ANALYSIS

Both qualitative and quantitative sEMG-related information is used in motion analysis applications. Such information regards the timing, amplitude, and morphology (e.g., progressive or burst-like recruitment type) of muscle activation during tasks.

Normalized amplitude is widely used in research studies [8]. Under certain conditions, a linear relationship can be demonstrated between integrated or low-pass-filtered rectified myoelectric signals (measured either intramuscularly or on the surface) and certain biomechanical variables. The first set of conditions tested consisted of isometric contractions [38]. Afterward, a relationship was also found to exist for variable-force or dynamic voluntary contractions of limited intensity [7], provided that intramuscular signals from multiple recording sites were taken into account. The relationship between sEMG and force has been recently reviewed in Staudenmann et al. [51] and is discussed in Chapter 10. In clinical applications, the amplitude profile defined by surface sEMG plays a minor role. It is related only qualitatively to the force produced by muscles during dynamic contractions, since it is affected by several confounding factors, as discussed in Chapters 2, 4, and 10 [17,21]. Moreover, the absence of sEMG activity does not necessarily indicate the absence of internal moment generation in response to an applied external moment, as in the case of tensions applied to stiff, shortened muscles in some pathologies.

The amplitude profile or envelope is typically obtained through low-pass filtering of the rectified sEMG signal. MVC- and time-normalized envelopes of homogeneous healthy subjects can be merged to compute average profiles that are used as a reference when discussing the profiles of pathological subjects [6,31,44]. A smoothed profile may be easier to share with nonexperts than raw sEMG; however, it may completely mask motion artifacts (if any) and would thus be risky in clinical applications. Also, a drawback of the smoothing procedure is blurring of muscle onset and offset events. The low cutoff values (e.g., 3 Hz) suggested in Winter and Yack [60] for envelope computation by means of noncausal digital filters produce profiles that are attractive, yet are unreliable in their onset periods [24,40], as shown in Fig. 16.7

Quantitative indications of sEMG amplitude are obtained by RMS and ARV, as described in Chapters 4, 5, and 10. RMS amplitude computed over a fixed interval of a task or cycle (for example, during the stance phase of gait) allows subsequent trials of the same subject to be compared over a period of treatment, or within the same measuring session subsequent to modifications of walking conditions by means of gait aids, orthoses, concurrent cognitive tasks, and so on. The ratio between the

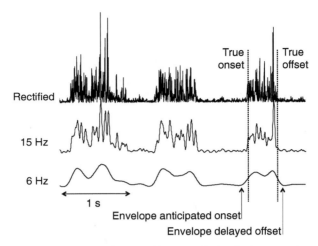

FIGURE 16.7　Envelope computation provides smooth signals, but may alter muscle activity timing, with consequent unreliable data interpretation in both clinical and research applications.

normalized RMS amplitudes of agonist and antagonist muscles has been used in the literature to assess the co-contraction level in spastic muscles of stroke patients [53].

In applications of motion analysis, estimation of the on–off timing of human skeletal muscles during movement can be either qualitative or quantitative, depending on the required level of accuracy. The change in patients' sEMG patterns can be substantial, as in the case of out-of-phase activity when electrical silence is expected. Thus, answers to clinical questions can often be obtained by visual inspection of the acquired sEMG traces. Activity patterns are classified by Perry [44] as premature, delayed, prolonged, absent, continuous, and out-of-phase. An example of the clinical use of sEMG visual inspection can be found in Keenan et al. [34]. The authors evaluated the influence of motor-control analysis with dynamic electromyography on surgical planning in patients with spastic elbow flexion deformity. The occurrences of changes of the surgical plan was 57%, and the level of agreement on surgical plan selection between two surgeons increased significantly from 68% before the study to 82% afterward [34].

Several methods are described in the literature for estimating muscle activation timing [50]. The simplest compares the amplitude of the rectified signal to a threshold set at two or three times the standard deviation of noise. An automatic procedure for threshold selection was proposed in [52]. More sophisticated methods rely on a priori knowledge of either signal amplitude distribution or the morphology of the action potentials of motor units that are summed up to produce the surface signal [5,41]. The accuracy of each of these methods depends on the signal-to-noise ratio (SNR) of the signal to be analyzed and can be as low as 10 ms when SNR exceeds 8 dB on simulated signals [41]. In applications of motion analysis, the permitted amount of error in the estimated onset time (i.e., the required accuracy) must be known before the method of onset estimation can be selected. We hypothesize that clinical interpretation of muscle timing during gait will not be affected by errors in onset estimation on the order of $\pm 1\%$ of the gait cycle, and that the subject being evaluated is a patient whose stride—from heel strike to homolateral heel strike—lasts 2 s. Thus, the method chosen must ensure that such error is lower than $\pm\,20$ ms with the SNR existing at the true moment of onset (and not throughout the entire burst of activity). If an abrupt activation takes place (e.g., local SNR > 10 dB), nearly all methods provide accurate estimates. Conversely, when mildly increasing contractions occur, the local SNR is low, and the single threshold method becomes unreliable, with a bias greater than 50 ms [41]; consequently, more accurate methods must be used. Choosing the proper approach is crucial if the estimate of onset for the muscle(s) leads to a clinical decision. For instance, two recent systematic reviews [16,61] noted a lack of homogeneity in the results of studies on delayed activation of *vastus medialis obliquus* (VMO) as compared with *vastus lateralis* (VL) in subjects with patellofemoral pain syndrome. The reported time differences in the activation of VMO relative to VL were on the order of a few tens of a millisecond, while the error in estimated onset obtained with single threshold methods used may be as large as 70 ± 140 ms. Thus, the results were unreliable, and the differences among the studies may be explained by the biases in the onset estimates [15]. Finally, it must be noted that the higher the accuracy required, the greater should be the limitation on the confounding factors that can affect muscle activity, such as task execution speed, the subject's initial body position, and so on.

A simpler, but useful, numeric indicator used to broadly quantify muscle timing is envelope peak time or time-to-peak (TTP), computed as the time distance of the peak from the beginning of the task or of the cycle. In a healthy subject, the TTP of cyclic movements is already known, so this parameter can be used as a reference when discussing patient sEMG profiles [10,17,39]. Moreover, the moments at which the MVC-normalized envelope reaches fixed amplitude levels (e.g., 5%, 10%, 20%, etc.) can be used to track the development of the sEMG profile throughout the cycle [15].

Surface sEMG morphology can be defined as the shape of the raw trace. It is a qualitative aspect of sEMG which is evaluated by visual inspection. Such morphology is affected by (1) the recruitment modality of the muscle, which is normally gradual but may be abrupt in patients; (2) the presence of separate bursts instead of a single, modulated activity; (3) the degree of EMG interference in the trace, which is due to summation of the action potentials of the motor units and may be reduced if few active sources are available [40]. Patients with neurological disease affecting recruitment may have remarkable sEMG morphologies.

Figure 16.8 shows *triceps surae* activity during walking in a subject with hemiparesis subsequent to stroke, limited range of passive ankle motion, and increased muscle stiffness. Muscle recruitment may evolve from continuous to burst-like and finally to absent over time, as degenerative changes in the intrinsic muscle properties occurred [23,27]. Numeric indicators of signal density—such as entropy [32], and time-frequency and time-scale [36,43,57,58] representations—may be useful tools for quantifying this clinical event. Although such techniques have been used primarily in research studies on healthy adults (for example, to classify sample subjects into subgroups [43,56] or to assess localized myoelectric fatigue during dynamic

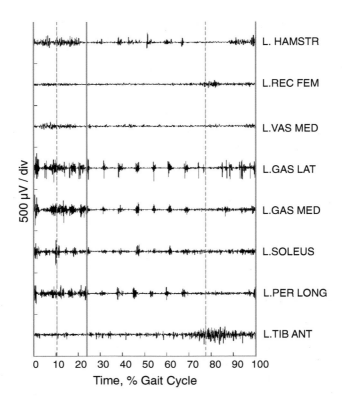

FIGURE 16.8 Typical calf muscle sEMG morphology in a chronic stroke patient with muscle shortening and range of motion (ROM) reduction.

contractions [55]), they seem to be promising in providing morphology-related indices of patient sEMG.

16.4 EXAMPLES OF APPLICATIONS IN MOTION ANALYSIS

Several examples of the use of bipolar surface sEMG in motion analysis are presented in this section.

Figures 16.1 and 16.2 show sEMG activity in the right limb of a neurological patient, along with right knee and ankle joint angles in the sagittal plane. The ankle remains plantar-flexed during swing (equinus foot deformity) and has limited dorsiflexion during stance. The foot contacts the floor with the forefoot instead of the heel. The knee extends abnormally from stance to hyperextension and does not flex properly during swing, when approximately 60° of flexion should be reached. The clinical question for this patient centered on the causes of the equinus deformity and on its possible correction to restore heel strike. sEMG data show correct activity of the dorsiflexors (*tibialis anterior* and *extensor digitorum longus*) during swing, with out-of-phase activity of the *peronei*, the lateral *gastrocnemius*, and the medial *gastrocnemius* muscles, whose action is responsible for ankle plantarflexion. The *soleus* is correctly silent in swing, so it should not be given any focal treatment. In similar patients, activity of the *tibialis anterior* in swing is a positive prognostic factor for surgical intervention to lengthen the calf muscles when restoration of the range of joint motion is required. After surgery, dorsiflexors will prevent the drop foot deformity in swing.

Figure 16.3 shows sEMG activity of the thigh muscles of a paraplegic patient while attempting to reach a seated position from the supine. The attempt was unsuccessful due to abnormal co-contraction of the hip extensors. A focal block on the hamstrings by botulinum toxin injection eliminated their interference, thus allowing the subject to seat independently once again.

The clinical assessment of spasticity after lesion to the central nervous system is performed by evaluating resistance to fast passive joint stretching on a 0–5 scale (Ashworth Scale). This procedure may be unreliable due to changes in intrinsic passive muscular properties such as length, stiffness, and viscosity [22,27]. sEMG data can be collected during stretching to assess spasticity reliably, as in Fig. 16.9.

Figure 16.4 shows a chair-to-chair transition performed by a healthy adult. The use of miniaturized surface electrodes with a conductive area of $4\,mm^2$—that is, a diameter of the conductive part of about 2.2 mm—enabled several muscles and functional muscle subparts to be detected in the trunk and arm. The situation mimics the transition that a patient performs when moving from chair to wheelchair. The

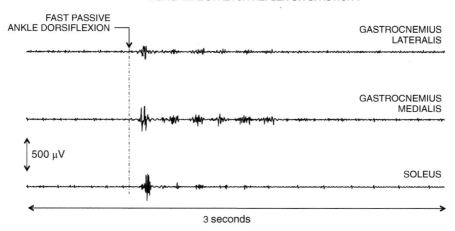

FIGURE 16.9   EMG assessment of spasticity as elicited by the Ashworth maneuver, which consists of a rapid passive stretch of the muscle with the patient at rest. The abnormal stretch reflex (spasticity) is characterized by a sequence of short bursts of activity with gradually lower amplitude levels.

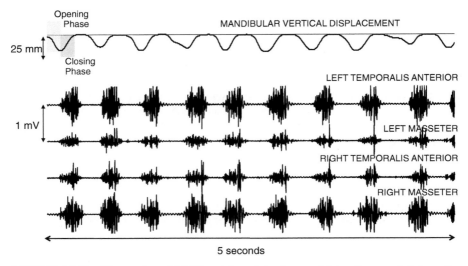

FIGURE 16.10 Kinematic and EMG activity of the jaw, as bilaterally recorded from the *masseter* and *temporalis anterior* muscles. Both EMG timing and amplitude may vary with mastication side and bolus type [47].

difference in the sEMG pattern between forward and rearward transitions can be clearly seen in the figure.

Mastication is a dynamic process involving synchronous movements of the jaws, tongue, and cheeks to position the bolus between the shearing and crushing surfaces of the teeth. Such movements can be investigated by sEMG. Mandibular movements during chewing depend on interaction of the masticatory muscles; thus, abnormal chewing kinematics are associated with altered muscle activation [46]. Abnormal chewing in children has an impact on structural growth and may lead to irreversible asymmetry of the anatomical structures involved (bones, temporomandibular joint, muscles, teeth). Signals recorded from the *masseter* and *anterior temporalis* muscles on both sides, along with studies of jaw kinematics, are used to obtain relevant clinical information about chewing function and about dental malocclusions and their treatment over time [37,45,47,59]. An example of vertical jaw displacement and sEMG activity is presented in Fig. 16.10.

16.5 CONCLUSIONS AND PERSPECTIVES

Bipolar sEMG may provide relevant information in clinical movement analysis. It may be used to complete a functional or clinical evaluation, to assist the design of clinical interventions such as rehabilitation, functional surgery and muscle inhibition by local blockage and to assess the treatment efficacy.

Several factors may alter the detected sEMG signal and may lead to misinterpretation. Most of the error sources can be avoided by treating the skin, using small electrodes with adequate adhesion, placing the electrodes according to protocols and representing raw data on a fixed y-axis scale.

It is well known (see Chapters 2, 4, and 5) that sEMG has different features (amplitude, spectral features) depending on the electrode location. This variability is due to reasons that are different in fusiform muscles, with fibers parallel to the skin, and in pinnate muscles. In fusiform muscles the innervation zone is one of the major causes of variability and should be avoided as a possible electrode location (see the book by Barbero, M., Merletti, R., and Rainoldi, A., *Atlas of Muscle Innervations Zones: Understanding Surface Electromyography and Its Applications*, Springer Verlag, 2012). In pinnate muscles, such as the gastrocnemius, where most of the sEMG is due to the fiber-end effect and action potential propagation is hard to see because of the pinnation angle, the sEMG reflects the territory of the motor units (see Chapters 5 and 15). Recent techniques are based on EMG imaging obtained by means of electrode grids with 32 to 128 contact points. These grids sample the raw signal, its ARV, or its RMS in space as well as in time, and they allow automatic segmentation of the resulting images and detection of the electrode pair (or cluster of pairs) that provide the most relevant information (such as signal strength, myoelectric manifestations of fatigue, patterns of motor unit activation in different muscle compartments, etc.). Although their use is currently limited to research laboratories, their future clinical use will solve a number of current issues eliminating the need to search for a good electrode location and providing automatic reduction of interferences and artifacts.

REFERENCES

1. Basmajian, J. V., and R. Blummenstein, *Electrode Placement in Electromyographic Biofeedback*, 5th ed., Williams & Wilkins, Baltimore, 1989.
2. Blanc, Y., "Surface electromyography (SEMG) a plea to differentiate between crosstalk and co–activation," In: Hermens, H. J., and B. Freriks, eds., *The State of the Art on Sensors and Sensor Placement Procedures*. Deliverable 5 SENIAM project. The Netherlands: Roessingh Research and Development, pp. 96–100, 1997.
3. Blanc, Y., and U. Dimanico, "Electrode placement in surface electromyography (sEMG). minimal crosstalk area (MCA)," *Open Rehabil. J.* **3**, 110–126 (2010).
4. Blanc, Y., and U. Dimanico, "History of the study of skeletal muscle function with emphasis on kinesiological electromyography," *Open Rehabil. J.* **3**, 84–93 (2010).
5. Bonato, P., T. D'Alessio, and M. Knaflitz, "A statistical method for the measurement of muscle activation intervals from surface myoelectric signal during gait," *IEEE Trans. Biomed. Eng.* **45**, 287–299 (1998).
6. Bovi, G., M. Rabuffetti, P. Mazzoleni, and M. Ferrarin, "A multiple-task gait analysis approach: kinematic, kinetic and EMG reference data for healthy young and adult subjects," *Gait Posture* **33**, 6–13 (2011).
7. Buisset, S., and B. Maton, "Quantitative relationship between surface EMG and intramuscular electromyographic activity in voluntary movement," *Am. J. Phys. Med.* **51**, 116–121 (1972).

8. Burden, A., "How should we normalize electromyograms obtained from healthy participants? What we have learned from over 25 years of research," *J. Electromyogr. Kinesiol.* **20**, 1023–1035 (2010).

9. Burkhard, P. R., H. Shale, J. W. Langston, and J. W. Tetrud, "Quantification of dyskinesia in Parkinson's disease: validation of a novel instrumental method," *Mov. Disord.* **14**, 754–763 (1999).

10. Campanini, I., A. Merlo, P. Degola, R. Merletti, G. Vezzosi, and D. Farina, "Effect of electrode location on EMG signal envelope in lower leg muscles during gait," *J. Electromyogr. Kinesiol.* **17**, 515–526 (2007).

11. Campanini, I., and A. Merlo, "Reliability, smallest real difference and concurrent validity of indices computed from GRF components in gait of stroke patients," *Gait Posture* **30**, 127–131 (2009).

12. Cappozzo A., T. Leo, and A. Pedotti, "A general computing method for the analysis of human locomotion," *J. Biomech.* **8**, 307–320 (1975).

13. Cappozzo, A., "Gait analysis methodology," *Hum. Mov. Sci.* **3**, 25–54 (1984).

14. Cappozzo, A., U. Della Croce, U., A. Leardini, and L. Chiari, "Human movement analysis using stereophotogrammetry. part 1: theoretical background," *Gait Posture* **21**, 186–196 (2005).

15. Cavazzuti, L., A. Merlo, F. Orlandi, and I. Campanini, "Delayed onset of electromyographic activity of vastus medialis obliquus relative to vastus lateralis in subjects with patellofemoral pain syndrome," *Gait Posture* **32**, 290–295 (2010).

16. Chester, R., T. O. Smith, D. Sweeting, J. Dixon, S. Wood, and F. Song, "The relative timing of VMO and VL in the aetiology of anterior knee pain: a systematic review and meta–analysis," *BMC Musculoskelet. Disord.* **9**, 64 (2008).

17. Chimera, N. J., D. L. Benoit, and K. Manal, "Influence of electrode type on neuromuscular activation patterns during walking in healthy subjects," *J. Electromyogr. Kinesiol.* **9**, 494–499 (2009).

18. Cutti, A.G., A. Ferrari, P. Garofalo, M. Raggi A. Cappello, and A. Ferrari, " 'Outwalk': a protocol for clinical gait analysis based on inertial and magnetic sensors," *Med. Biol. Eng. Comput.* **48**, 17–25 (2010).

19. English, A.W. and O. I. Weeks, "Electromyographic cross-talk within a compartmentalized muscle of the cat," *J. Physiol.* **416**, 327–336 (1989).

20. Engström, P., Å. Bartonek, K. Tedroff, C. Orefelt, Y. Haglund–Åkerlind, and E. M. Gutierrez–Farewik, "Botulinum toxin A does not improve the results of cast treatment for idiopathic toe-walking: a randomized controlled trial," *J. Bone Joint Surg. Am.* **95**, 400–407 (2013).

21. Farina, D., R. Merletti, R. M. Enoka, "The extraction of neural strategies from the surface EMG." *J. Appl. Physiol.* **96**, 1486–1495 (2004).

22. Fleuren, J. F., G. E. Voerman, C. V. Erren–Wolters, G. J. Snoek, J. S. Rietman, H. J. Hermens, and A.V. Nene, "Stop using the Ashworth Scale for the assessment of spasticity," *J. Neurol Neurosurg. Psychiatry* **81**, 46–52 (2010).

23. Foran, J. R., S. Steinman, I. Barash, H. G. Chambers, and R.L. Lieber, "Structural and mechanical alterations in spastic skeletal muscle. *Review,*" *Dev. Med. Child Neurol.* **47**, 713–717 (2005).

24. Frigo, C., and P. Crenna, "Multichannel SEMG in clinical gait analysis: a review and state–of–the–art," *Clin. Biomech.* **24**, 236–245 (2009).

25. Gabel, R. H., and R. A. Brand, "The effects of signal conditioning on the statistical analyses of gait EMG," *Electroencephalogr. Clin. Neurophysiol.* **93**, 188–201 (1994).

26. Gage, J. R., M. H. Schwartz, S. E. Koop, T. F. Novacheck, *The Identification and Treatment of Gait Problems in Cerebral Palsy*, John Wiley & Sons, Hoboken, NJ, 2009.

27. Gracies, J. M., "Pathophysiology of spastic paresis. I: Paresis and soft tissue changes. Review," *Muscle Nerve* **31**, 535–351 (2005).

28. Granata, K. P., D. A. Padua, and M. F. Abel, "Repeatability of surface EMG during gait in children," *Gait Posture* **22**, 346–350 (2005).

29. Hermens, H., B. Freriks, R. Merletti, D. Stegeman J. Blok, G. Rau C. Disselhorst–Klug and G. Hagg, *European Recommendations for Surface Electromyography*, Roessingh Research and Development, Enschede, The Netherlands, 1999.

30. Hodges, P. W., and S. C. Gandevia, "Pitfalls of intramuscular electromyographic recordings from the human," *Clin. Neurophysiol.* **111**, 1420–1424, (2000).

31. Hof, A. L., H. Elzinga, W. Grimmius, and J. P. Halbertsma, "Speed dependence of averaged EMG profiles in walking," *Gait Posture* **16**, 78–86 (2002).

32. Istenic, R., P. A. Kaplanis, C. S. Pattichis, and D. Zazula, "Multiscale entropy-based approach to automated surface EMG classification of neuromuscular disorders," *Med. Biol. Eng. Comput.* **48**, 773–781 (2010).

33. Ivanenko, Y. P., R. Grasso, M. Zago, M. Molinari, G. Scivoletto, V. Castellano, V. Macellari, and Lacquaniti F., "Temporal components of the motor patterns expressed by the human spinal cord reflect foot kinematics," *J. Neurophysiol.* **90**, 3555–3565 (2003).

34. Keenan, M. A., D. A. Fuller, J. Whyte, N. Mayer, A. Esquenazi, and R. Fidler-Sheppard, "The influence of dynamic polyelectromyography in formulating a surgical plan in treatment of spastic elbow flexion deformity," *Arch. Phys. Med. Rehabil.* **84**, 291–296 (2003).

35. Kirtley, C., *Clinical Gait Analysis: Theory and Practice*, Elsevier, Oxford, 2006.

36. Kumar, S., Y. Narayan, N. Prasad, A. Shuaib, and Z. A. Siddiqi, "Cervical electromyogram profile differences between patients of neck pain and control," *Spine* **15**, 246–253 (2007).

37. Lewin A. *Electrognathographics. An Atlas for Diagnostic Procedures and Interpretation*, Quintessence Publishing Co., Inc., Berlin, 1985.

38. Lippold, O. C. J., "The relation between integrated action potentials in a human muscle and its isometric tension," *J. Physiol.* **117**, 492–499 (1952).

39. McLoda, T. A., and A. J. Hansen, "Effects of a task failure exercise on the peroneus longus and brevis during perturbed gait," *Electromyogr. Clin. Neurophysiol.* **45**, 53–58 (2005).

40. Merlo, A., and I. Campanini, "Technical aspects of surface electromyography for clinicians," *Open Rehabil. J.* **3**, 98–109 (2010).

41. Merlo, A., D. Farina, and R. Merletti, "A fast and reliable technique for muscle activity detection from surface EMG signals," *IEEE Trans. Biomed. Eng.* **50**, 316–323 (2003).

42. Mesin, L., R. Merletti, and A. Rainoldi, "Surface EMG: the issue of electrode location," *J. Electromyogr. Kinesiol.* **19**, 719–726 (2009).

43. Nüesch, C., C. Huber, G. Pagenstert, V. von Tscharner, and V. Valderrabano, "Muscle activation of patients suffering from asymmetric ankle osteoarthritis during isometric contractions and level walking. A time–frequency analysis," *J. Electromyogr. Kinesiol.* **22**, 939–946 (2012).

44. Perry, J., *Gait Analysis, Normal and Pathological Function*, SLACK Inc., Thorofare, NJ, 1992.

45. Piancino, M. G., D. Farina, F. Talpone, A. Merlo, and P. Bracco, "Muscular activation during reverse and non-reverse chewing cycles in unilateral posterior crossbite," *Eur. J. Oral Sci.* **117**, 122–128 (2009).

46. Piancino, M. G., G. Isola, A. Merlo, D. Dalessandri, C. Debernardi, and P. Bracco, "Chewing pattern and muscular activation in open bite patients," *J. Electromyogr. Kinesiol.* **22**, 273–279 (2012).

47. Piancino, M.G., P. Bracco, T. Vallelonga, A. Merlo, and D. Farina, "Effect of bolus hardness on the chewing pattern and activation of masticatory muscles in subjects with normal dental occlusion," *J. Electromyogr. Kinesiol.* **18**, 931–937 (2008).

48. Sacco, I. C., A. A. Gomes, M. E. Otuzi, D. Pripas, and A. N. Onodera, "A method for better positioning bipolar electrodes for lower limb EMG recordings during dynamic contractions," *J. Neurosci. Methods* **180**, 133–137 (2009).

49. Solomonov, M., Baratta, R., Bernardi, M., Zhou, B., Lu, Y., Zhu, M., and Aciemo, S., Surface and wire EMG crosstalk in neighboring muscles. *J. Electromyogr. Kinesiol.* **4**, 131–142 (1994).

50. Staude, G. and W. Wolf, "Objective motor response onset detection in surface myoelectric signals," *Med. Eng. Phys.* **21**, 449–468 (1999).

51. Staudenmann, D., K. Roeleveld, D. F. Stegeman, and J. H. van Dieën, "Methodological aspects of SEMG recordings for force estimation. A tutorial and review," *J. Electromyogr. Kinesiol.* **20**, 375–387 (2010).

52. Thexton, A. J., "A randomisation method for discriminating between signal and noise recordings of rhythmic electromyographic activity," *J. Neurosci. Methods* **66**, 93–98 (1996).

53. Vinti, M., A. Couillandre, J. Hausselle, N. Bayle, A. Primerano, A. Merlo, E. Hutin and J. M. Gracies, "Influence of effort intensity and gastrocnemius stretch on co-contraction and torque production in the healthy and paretic ankle," *Clin. Neurophysiol.* **124**, 528–535 (2013).

54. Vaughan, C. L., B. L. Davis, and J. C. O'Connor, *Dynamics of Human Gait*, Kiboho Publishers, Cape Town, South Africa, 1999.

55. von Tscharner, V., "Time–frequency and principal-component methods for the analysis of EMGs recorded during a mildly fatiguing exercise on a cycle ergometer," *J. Electromyogr. Kinesiol.* **12**, 479–492 (2002).

56. von Tscharner, V., and B. Goepfert, "Gender dependent EMGs of runners resolved by time/frequency and principal pattern analysis," *J. Electromyogr. Kinesiol.* **13**, 253–272 (2003).

57. von Tscharner, V., "Intensity analysis in time–frequency space of surface myoelectric signals by wavelets of specified resolution," *J. Electromyogr. Kinesiol.* **10**, 433–445 (2000).

58. Wang, S. Y., X. Liu, J. Yianni, T. Z. Aziz, and J. F. Stein, "Extracting burst and tonic components from the surface electromyograms in dystonia using adaptive wavelet shrinkage," *J. Neurosci. Meth.* **139**, 177–184 (2004).

59. Wilding, R. J., and A. Lewin, "The determination of optimal human jaw movements based on their association with chewing performance," *Arch. Oral Biol.* **39**: 333–343 (1994).

60. Winter, D. A., and H. J. Yack, "EMG profiles during normal human walking: stride-to-stride and inter-subject variability," *Electroencephalogr. Clin. Neurophysiol.* **67**, 402–411 (1987).

61. Wong, Y.M., "Recording the vastii muscle onset timing as a diagnostic parameter for patellofemoral pain syndrome: fact or fad?," *Phys. Ther. Sport* **10**, 71–74 (2009).

62. Wren, T.A., G. E. 3rd Gorton, S. Ounpuu, and C. A. Tucker, "Efficacy of clinical gait analysis: a systematic review," *Gait Posture* **34**, 149–153 (2011).

63. Wren, T.A., K. J. Elihu, S. Mansour, S. A. Rethlefsen, D. D. Ryan, M. L. Smith, and R. M. Kay, "Differences in implementation of gait analysis recommendations based on affiliation with a gait laboratory," *Gait Posture* **37**, 206–209 (2013).

64. Yang, J., and D. Winter, "Electromyographic amplitude normalizing methods, improving their sensitivity as diagnostic tools in gait analysis," *Arch. Phys. Med. Rehabil.* **65**, 517–521 (1984).

65. Zatsiorsky, V. M., *Kinetics of Human Motion,* Champaign IL, 2002.

66. Zmitrewicz, R. J., R. R. Neptune, J. G. Walden, W. E. Rogers, and G. W. Bosker, "The effect of foot and ankle prosthetic components on braking and propulsive impulses during transtibial amputee gait," *Arch. Phys. Med. Rehabil.* **87**, 1334–1339 (2006).

17

APPLICATIONS IN MUSCULOSKELETAL PHYSICAL THERAPY

D. FALLA

Department of Institute for Neurorehabilitation Systems, Bernstein Center for Computational Neuroscience, University Medical Center Göttingen, Georg-August University, Göttingen, Germany; and Pain Clinic, Center for Anesthesiology, Emergency and Intensive Care Medicine, University Hospital Göttingen, Göttingen, Germany

17.1 INTRODUCTION

Since the discovery of surface electromyography (sEMG) in 1912 [121], this noninvasive technique for the acquisition and analysis of myoelectric signals has contributed significantly to enhancing our understanding of the function and dysfunction of the neuromuscular system and has become an essential tool in modern musculoskeletal physical therapy. Surface EMG can be considered as a summation of tissue-filtered signals generated by a number of concurrently active motor units. Detection, recording, and analysis of myoelectric signals provides a reproducible means of determining disturbances in motor control in patients with musculoskeletal disorders. Surface EMG is typically applied in physical therapy for the assessment of disturbed motor control and for monitoring change with rehabilitation. The sEMG signal can be analyzed in various ways to provide the investigator or clinician with a multitude of information about the muscle/s being studied. Examples of analyses include (1) detection of the onset or offset of muscle activity during tasks such as postural perturbations, (2) assessment of myoelectric manifestations of muscle fatigue, (3) evaluation of the magnitude of muscle activation, (4) generation of

Surface Electromyography: Physiology, Engineering, and Applications, First Edition.
Edited by Roberto Merletti and Dario Farina.
© 2016 by The Institute of Electrical and Electronics Engineers, Inc. Published 2016 by John Wiley & Sons, Inc.

tuning curves of the sEMG amplitude to relate the amplitude of the muscle response to the magnitude and direction of force, and (5) monitoring the spatial distribution of activity with high-density, two-dimensional sEMG. Throughout this chapter, these fundamental methods of EMG assessment will be explored as a means of evaluating neuromuscular impairment in patients with musculoskeletal disorders with a focus on two of the most common musculoskeletal complaints, namely, low back pain and neck pain.

17.2 TIMING OF MUSCLE ACTIVITY

17.2.1 Delayed Onset of Muscle Activity

Perturbation studies have been used extensively to explore the neural mechanisms underlying balance control. Perturbations produce accelerative forces that act on the body and head and stimulate a variety of sensory systems including visual, vestibular, and joint somatosensory inputs [85]. Internal perturbations such as a rapid movement of the arm, produces reactive forces, which are of equal magnitude and opposite direction to the forces produced by the arm movement [16,73]. These forces are transferred to body segments eliciting a series of postural adjustments to maintain equilibrium.

Detection of the onset and termination of muscle activity during activities such as postural perturbations can be used to enhance our understanding of neuromuscular control. Various methods can be employed to assess the onset and offset of EMG bursts, ranging from visual determination [94,147] to computer-based algorithms [83,112]. Evaluation of both the temporal and spatial parameters of muscle activation can provide a measure of the final outcome of all of the underlying processes occurring within the central nervous system (CNS) [69].

One CNS strategy detectable by sEMG during internal perturbations is feed-forward activation of a muscle. Extensive evidence of feed-forward activation of limb musculature in response to perturbations is available [26,62,73,95]. Feed-forward postural responses are a mechanism employed by the CNS to regulate motor control of muscles and contribute towards the maintenance of stability. This mechanism helps to maintain or regain stability in the presence of external forces such as gravity or displacements, or when a body part is required to act as a steady base for movement to occur [2]. By this means, appropriate muscles are preprogrammed to activate, which acts to provide and maintain joint stability and overall balance.

Changes in spinal posture and the intersegmental relationship of the spine are also affected during postural perturbations therefore providing a suitable means for assessing the neural control of spinal musculature. Evidence of feed-forward activation of the neck and trunk flexor muscles has been shown in healthy individuals when subjected to anterior, posterior, and rotational perturbations by standing on a hydraulically controlled platform [85]. The deep and superficial neck muscles also show feed-forward activation in healthy volunteers when subjects perform rapid movements of their arm [45,52]. Likewise, sEMG studies confirm that trunk muscles such as the lumbar erector spinae and oblique abdominal muscles are activated in advance of muscles responsible for movement of the limb in healthy volunteers [69]. These responses are direction specific with the onset of sEMG occurring earlier and with greater amplitude when the muscles oppose the direction of force on the trunk [70]. The transversus abdominis and deep fibers of the multifidus, recorded with intramuscular EMG, also display feed-forward activation; however, their activation is non-direction-specific [69,70], consistent with observations for the deep neck flexor muscles [45].

Studies evaluating the EMG responses of people with low back pain to postural perturbations have revealed delayed trunk muscle responses [24,69,70,123,124], decreased amplitude of muscle activation [102,123], as well as evidence of co-contraction [123] or higher baseline muscle activation [134]. A study examining the response to 12 directions of support surface translations in people without a history of chronic, recurrent low back pain showed (1) higher baseline sEMG amplitudes of the erector spinae muscles before perturbation onset, (2) fewer early-phase activations at the internal oblique and gastrocnemius muscles, (3) fewer late-phase activations at the erector spinae, internal and external oblique, rectus abdominae, and tibialis anterior muscles, and (4) higher sEMG amplitude of the gastrocnemius muscle following the perturbation in the patients with low back pain [74]. Delayed activation of trunk muscles in response to unexpected full body postural perturbations has also been observed during experimentally induced low back muscle pain [15].

Interestingly, individuals with low back pain during a quiescent period demonstrated earlier onset of erector spinae muscle activity in response to predictable loading, specifically on the previously painful side [102]. Similarly, healthy individuals who were anticipating induced back pain demonstrated earlier latencies in the superficial abdominal muscles following predictable loading [116]. The earlier onset latencies that have been demonstrated prior to predictable loading [102] and in anticipation of pain [116] may be due to pain related fear or fear of movement.

When a perturbation to the spine occurs, such as during a rapid arm movement, onset of the neck flexor muscles is also delayed in people with neck pain [45]. The delay in onset of these muscles exceeds the criteria for feed-forward contraction during movements which indicates a significant deficit in the automatic feed-forward control of the cervical spine. An additional observation in people with neck pain is that activation of the deep cervical flexor muscles adopts a direction-specific response which is in contrast to that observed in healthy individuals [45]. This indicates that the change is not simply a delay that could be explained by factors such as decreased motor neuron excitability, but instead is consistent with the change in the strategy used by the central nervous system to control the cervical spine. Furthermore, when patients with neck pain are exposed to full-body rapid postural perturbations, they demonstrate a reduced ability to automatically select and appropriately activate their neck muscles as seen in healthy volunteers, resulting in delayed activation of the sternocleidomatoid and splenius capitis muscles [14] (Fig. 17.1). These findings support the notion that there is a loss in predefined muscular activation patterns in individuals with chronic neck pain which is identifiable with surface EMG. The

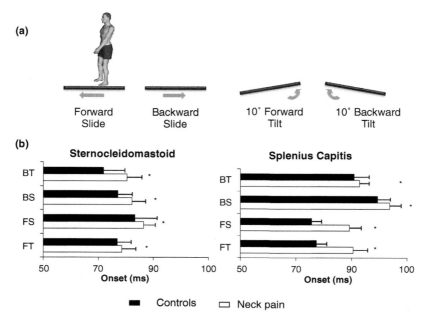

FIGURE 17.1 (a) Patients with chronic neck pain and healthy controls stood on a moveable platform and were exposed to randomized full body postural perturbations (8-cm forward slides (FS), 8-cm backward slides (BS), 10° forward tilts (FT), and 10° backward tilts (BT)). (b) Mean and SD of the onset of the sternocleidomastoid and splenius capitis muscles in response to the perturbations. Note the significantly ($* = P < 0.05$) delayed onset time of the neck muscles for the patients with neck pain regardless of the perturbation direction.

reduced ability to appropriately reestablish posture and balance following an unexpected event, such as a slip, may predispose individuals with chronic neck to further neck injury.

17.2.2 Delayed Offset

In addition to the delayed offset of muscle activity which can be seen during postural perturbations in individuals with pain [123], delayed offset of muscle activity is also seen in the form of a decreased ability of patients to relax their muscles after contraction. For example, delayed relaxation of superficial neck muscles after neck [8] and shoulder activities [61] has been demonstrated in individuals with neck pain. In these examples, the relaxation time of the muscle is simply calculated by determining the difference between (a) the onset of a light signal, indicating to the subject to relax, and (b) the time when EMG amplitude returns to the pre-test values [8]. This methodology was applied in a study conducted by Barton and Hayes [8] to determine the relaxation times of sternocleidomastoid, post maximum isometric neck flexion in subjects with and without neck pain, and associated cervicogenic headache. Although results were not significantly different between groups in this study, some patients displayed an exceedingly long relaxation time when compared to the controls and a trend was present, suggesting increased relaxation times in the symptomatic subjects. This trend towards increased muscle relaxation times in neck pain patients was later confirmed by Fredin et al. [61], who demonstrated a significant reduction in the ability to relax the upper trapezius and infraspinatus muscles between repetitive shoulder flexion contractions in patients suffering from long-term symptoms related to a whiplash injury. Furthermore, lower relative rest times were observed in female computer users with self-reported neck/ shoulder complaints during the performance of standardized short-term computer work [138]. In this study, relative rest time was calculated as the relative cumulative time with sEMG amplitude below 1% of the maximum activity, recorded during a maximum voluntary contraction (MVC), for periods of >100 ms. Additional studies have also confirmed a delayed offset of upper trapezius muscle activity in patients with neck pain following voluntary repetitive arm movements [37,117]. Likewise, some patients with neck pain demonstrate reduced ability to relax the anterior scalene and/or sternocleidomastoid muscles following voluntary activation [49].

17.3   MYOELECTRIC MANIFESTATIONS OF MUSCLE FATIGUE

Distinct modifications of the sEMG signal can be identified during sustained voluntary or electrically elicited muscle contractions, and the analysis of myoelectric manifestations of muscle fatigue can provide important information about physiological changes evolving in the muscle, as described in Chapter 10 and 11 [108,110,111]. There are several advantages associated with the use of sEMG to assess muscle fatigue in patient populations, in distinction from mechanical measures. Specifically, sEMG-related measures of muscle fatigue are less affected by subject motivation [29], require force to be maintained for a shorter duration of time [9], and are able to identify differences in synergistic muscles which may be disrupted in the presence of pain [111].

Although myoelectric manifestations of muscle fatigue can be monitored with classic bipolar recordings, the application of linear or two-dimensional (2-D) electrode arrays offer several advantages over the standard bipolar electrode configuration. Linear arrays consist of multiple equally spaced electrode contacts, which are placed along the length of a muscle. Electrode arrays provide a spatial sampling of the potential distribution along the muscle during a contraction [54] and can provide information about both peripheral and central mechanisms of control or modifications of these mechanisms in the presence of pain or muscle dysfunction [107]. As said, although both linear arrays and bipolar electrode applications can be used to assess myoelectric manifestations of muscle fatigue, the multiple channels of sEMG obtained with linear arrays provide the opportunity for additional interpretations from the sEMG signals. These include (1) estimation of the motor unit conduction velocity (CV) and CV distribution [54], (2) identification of motor units and estimation of their firing rates [54], (3) the recruitment pattern of motor units during a contraction [64,110], and (4) identification of the muscle innervation zone and tendinous regions [105,107].

The most frequently monitored sEMG variables during the assessment of myoelectric manifestations of fatigue are spectral variables, such as the mean frequency (MNF) or median frequency (MDF), amplitude variables, such as the average rectified value (ARV) or root mean square (RMS), and motor unit CV. The typical pattern observed during sustained contractions is a decrease in both the CV and spectral variables with time [9,28,88,108] and an initial increase in signal amplitude prior to the onset of mechanical fatigue [9,10,115].

The MDF of the power density spectrum can be defined as the frequency value which divides the power spectrum into two parts having equal area [9] and is in the range of 50–110 Hz [110], whereas the MNF, or centroid value, is the range of 70–130 Hz [110]. Muscle fiber CV can be defined by the equation $CV = e/d$, where e = interelectrode distance and d = delay between two (single or double differential) signals. The delay (d) is obtained by identifying the time shift required to minimize the mean square error between the Fourier transforms of two single or double differential signals [108]. Normal physiological estimates of CV are between 3-5 m/s [20]. Muscle fiber CV provides an indication of muscle membrane properties and their change under various conditions including fatigue, exercise, pain and pathology [131]. In addition, muscle fiber CV is a size principle parameter and its estimation provides information on motor unit recruitment strategies [7].

The decrement of the action potential CV over time during sustained isometric contractions [9,18] is likely due to alterations of the sarcolemma excitability [87]. The generation of action potentials induces cellular K^+ efflux and Na^+ and Cl^- influx, causing perturbations in the intracellular and extracellular K^+ and Na^+ gradient concentrations [106]. These alterations depolarize the sarcolemma and t-tubular membranes which reduce membrane excitability [25,87]. The loss of membrane excitability is partly counteracted by the Na^+-K^+ pump activity, however it is not sufficient to fully balance the K^+ efflux [25,119]. Thus during sustained contractions, the velocity of propagation of the action potential decreases, even at very low force levels [119]. Additionally, with continued muscle contraction, an increase in the concentration of H^+ ions can occur, contributing to the change in membrane excitability [63]. Subsequently, the spectral shift from high to low frequencies during sustained contraction has been attributed to changes in the average and the spread of muscle fiber CV [18,28,98,108,135], motor unit synchronization [11,86,132] and motor unit recruitment [64]. Among these factors, average muscle fiber CV is the main determinant of the spectral shift [108], as evidenced by a high correlation between relative changes in CV and spectral variables in isometric conditions [6,97,100,114].

To analyze myoelectric manifestations of muscle fatigue in isometric conditions and allow comparison between different variables, different muscles, and different subjects, a fatigue plot can be produced [109]. The fatigue plot graphs each sEMG variable across time (duration of contraction) after normalization relative to a reference value (typically the initial value or the intercept of a regression line) (Fig. 17.2). As demonstrated elsewhere [108], the fatigue plot highlights differences in myoelectric manifestations of fatigue, which might be related to different pools of activated motor units and muscle fibers (for a detailed review see references 28 and 33, Chapter 4 and 5).

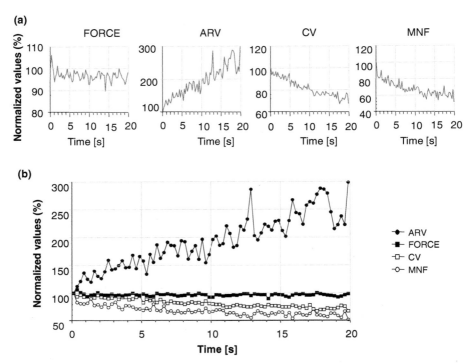

FIGURE 17.2 Fatigue plot: (a) Individual plots of the sEMG variables average rectified value (ARV), conduction velocity (CV), mean frequency (MNF), and force recorded from the anterior scalene muscle contracting 25% of the maximum voluntary contraction. Plots are obtained by normalizing each variable with respect to the initial value. (b) The time course of MNF, ARV, CV, and force are combined to produce a "fatigue plot." Note that although the force is maintained constant, the signal characteristics are modified from the onset of the contraction. Myoelectric manifestations of fatigue are identified by an increase in ARV or RMS values with time and decrease in MNF and CV values (see Chapter 10).

Alterations in the fatigue properties of the spinal muscles have been identified in patients with chronic low back pain and neck pain by investigating myoelectric manifestations of fatigue. In a study which examined fatigability of the sternocleidomastoid and anterior scalene muscles during sustained cervical flexion contractions at 25% and 50% of MVC, greater muscle fatigability was identified in chronic neck pain patients compared to control subjects [51]. This observation was characterized by a significantly greater slope of the MNF over the duration of the contraction. Similar findings have also been documented for the neck extensor muscles during sustained isometric cervical extension contractions at moderate and high loads in people with osteoarthritis of the cervical spine [65]. Furthermore, an increase of the MNF at the beginning of a sustained contraction has been observed for the neck muscles in the patients with neck pain [51]. However, absolute values of the MNF are poorly correlated with muscle fiber membrane properties (e.g., muscle fiber CV) or motor unit recruitment [42,53,58,104,113]. Thus, intersubject or intermuscle differences in absolute values of MNF cannot be attributed to specific physiological mechanisms.

Greater myoelectric manifestations of fatigue have frequently been documented for the erector spinae muscles in individuals with low back pain [30,32,82,125] with some indication of greater fatigability of the thoracic portion of the erector spinae muscles in patients with low back pain compared to the lumbar region [136]. The use of EMG-related fatigue indexes combined with force values have also been used to classify individuals with low back pain. In one study, accurate classification of 89.5% of individuals with low back pain was obtained by analyzing spectral indexes and force values during a fatigue test, which consisted of maintaining a submaximal level of 80% of MVC for 35 s. Such accuracies were achieved only when the force values obtained from the MVC were associated with the spectral variables [21].

In addition to sustained isometric contractions, fatigability of a muscle can be monitored with sEMG during dynamic contractions. Despite the difficulties due to the nonstationary signals and to the movement of the muscle under the skin, methods have been proposed to monitor muscle fiber CV during dynamic exercise (see Chapter 5). For example, the method proposed by Farina et al. [58] is based on the estimation of the delay between the signals detected by a linear array of surface electrodes. The delay is assessed by minimizing the mean square error between the aligned signals multiplied by a Gaussian window in time weighting the contribution of the action potentials. In this way, it is possible to obtain an estimate of CV local in time, representative of the average velocity of propagation of the action potentials of the motor units active in the selected time window during a movement. Instantaneous mean power spectral frequency (iMNF) of the EMG signal can also be estimated from the Choi–Williams time–frequency representation [22], which provides the most accurate estimates of instantaneous frequency from sEMG over a number of other Cohen's class time-frequency distributions [13]. Several authors have utilized spectral analysis as an indirect measure of muscle fiber CV (e.g. [145],) since a relative change in CV of an action potential is reflected by a frequency scaling of the same relative amount of its power spectrum [97]. If the average CV of a number of motor units changes over time, the characteristic spectral frequencies of the EMG signal will change proportionally. However, this relationship is disrupted if the number of active motor units fluctuates. Consequently, the comparison of spectral variables between signals generated by different populations of motor units does not reflect relative differences in CV [55]. This has been observed in several experimental conditions— for example, in high load dynamic contractions [42,104,113,122].

By monitoring changes in upper trapezius muscle fiber CV during a dynamic repetitive upper limb task, a greater decrease in upper trapezius CV over time was observed in patients with chronic neck pain indicative of greater fatigue of the upper trapezius muscle [39] (Fig. 17.3). This finding may be associated with the histological and morphological changes which have previously been identified in people with pain over the trapezius muscle such as mitochondrial disturbances in type I fibers [81,96]. In line with the observation of selective damage to the type I fibers, there is additional evidence of disruption to the microcirculation in the upper trapezius muscle in people

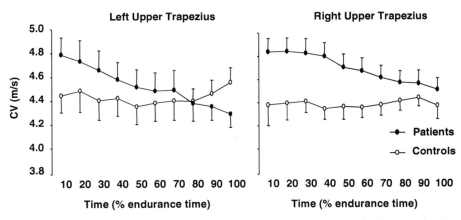

FIGURE 17.3 Mean and standard error of the upper trapezius muscle fiber conduction velocity (CV) calculated across the duration of a repetitive upper limb task in healthy controls and people with chronic neck pain. Note the higher initial values and greater decrease of CV estimates across the duration of the task for the patient group. Reprinted from Falla and Farina [39] with permission.

with trapezius myalgia [92,93]. Specifically, reduced capillarization per normal fiber cross-sectional area has been shown in people with trapezius myalgia. The presence of impaired microcirculation within the upper trapezius muscle may result in a greater buildup of metabolic byproducts such as lactic acid during muscle contractions, which could explain the significantly greater decrease in CV over the duration of the task in the patient group [18].

17.4 AMPLITUDE OF MUSCLE SIGNALS

The amplitude of the surface EMG is frequently used as a measure of muscular effort and has also been investigated as an indicator of muscle force. The amplitude of the surface EMG can be estimated by a scheme of demodulation, smoothing, and relinearization [57]. The most frequently applied estimates of sEMG amplitude are the ARV or RMS. These indices are described in detail in Chapters 4, 5, and 10 and in Farina et al. [57].

Although the amplitude of the sEMG is related to the number of recruited motor units and to their discharge rates, the EMG amplitude is influenced by many other physiological and nonphysiological factors such as location of the electrodes, thickness of the subcutaneous tissue, and the detection system used to obtain the recording [57]. For these reasons, normalization of the sEMG amplitude estimation is considered critical when comparing data across subjects or different muscles. Normalization of the EMG amplitude is typically performed by expressing the value obtained during a submaximal task as a percent relative to the activity detected during a MVC or a reference voluntary contraction (RVC). When patient populations are investigated, often a RVC is selected since a discrepancy in strength likely exists between patients and controls due to many factors including pain, deconditioning and fear of movement.

Evidence from EMG studies suggests that pain may alter the task-related modulation of muscle activity so that motor control of the spine is solved by alternative, presumably less efficient, combinations of muscle synergistic activities. For example, using a novel nasopharyngeal electrode [43,47] (Figs. 17.4a and 17.4b), reduced activation of the deep cervical flexors muscles has been observed when people with neck pain perform a clinical test of craniocervical flexion [46] (Figs. 17.4c and 17.4d). Reduced activation of the deep cervical flexor muscles during performance of this task is concomitant with increased activation of the superficial muscles, the sternocleidomastoid, and anterior scalene, indicating a reorganization of the motor strategy to perform the task [46] and possibly a measurable compensation for poor passive or active segmental support [23]. Heightened EMG activity of the superficial cervical flexor muscles in patients with neck pain compared to healthy individuals during this craniocervical flexion task has been reported in several sEMG studies and in several neck pain populations including cervicogenic headache [80], idiopathic neck pain [46], whiplash-induced neck pain [133], and work-related neck pain [75], suggesting that it is a generic finding in patients with neck pain disorders.

Surface EMG studies have also commonly shown higher activity of the superficial extensors of patients with neck pain during several tasks including a unilateral repetitive upper limb task [75], a one-hour computer typing task [137], isometric neck extension and lateral flexion [90], and isometric circular contractions in the horizontal plane [99]. However, consistent with the observation for the deep cervical flexor muscles, lower activation of the deep extensor, the semispinalis cervicis, has been observed in patients with chronic neck pain compared to healthy controls at the levels C3 [128], C2 and C5 [127]. In this example, intramuscular EMG is utilized in order to access this deep extensor muscle. Wire electrodes made of Teflon-coated stainless steel are inserted into a hypodermic needle and ~3–4 mm of insulation is removed from the tip of the wires to obtain an interference EMG signal so that classic EMG amplitude estimates can be obtained.

Similar observations of reduced EMG amplitude have been observed for the deep transversus abdominis in patients with low back pain. For example, the tonic activity of the transversus abdominis is reduced during walking [126] and during repetitive arm movements [68] in the presence of low back pain. Again, intramuscular EMG is applied in order for it accurately detect the activity of these deep muscles. On the contrary, increased erector spinae activity has been observed in patients with low back pain during the stride [1,142,144] and swing [5,91] phase of gait. Other studies have demonstrated increased co-activation of the flexor and extensor muscles during sudden unloading of the spine [123] and increased co-activation of the trunk musculature during unexpected, multidirectional translation perturbations [67,76]. During an active episode, individuals with low back pain also demonstrated inadequate or inappropriate muscle activation (co-activation) in response to unexpected perturbations [134].

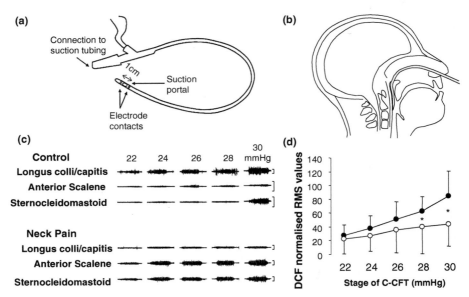

FIGURE 17.4 (a) Electrode for the detection of EMG activity from the deep longus colli and capitis muscles. The apparatus consists of custom-made, bipolar, silver wire electrode contacts (dimensions: 2×0.6 mm, interelectrode distance: 10 mm) which are attached to a suction catheter (size 10FG) with a heat-sealed distal end. The electrodes are fixed to the mucosa with a suction pressure of 30 mm Hg through a portal between the 2 electrode contacts. Reprinted from [43] with permission. (b) Using a nasopharyngeal application, surface electrodes attached to a suction catheter are positioned over the posterior oropharyngeal wall. The deep cervical flexor muscles lie directly posterior to the oropharyngeal wall, allowing myoelectric signals to be detected from these muscles. Reprinted from Falla et al. [43] with permission. (c) Representative raw EMG data are shown for a control subject and person with neck pain during performance of the cranio-cervical flexion. The task consisted of five incremental movements of increasing cranio-cervical flexion range of motion. Performance was guided by visual feedback from an air-filled pressure sensor (Stabilizer™, Chattanooga Group Inc. USA) which was placed sub-occipitally behind the subject's neck and inflated to a baseline pressure of 20 mmHg. During the test, subjects were required to perform the gentle nodding motions of cranio-cervical flexion that progressed in range to increase the pressure by five incremental levels, with each increment representing 2 mm Hg (22–30 mmHg; increments of 2 mmHg). Data are shown for the deep cervical flexors, longus colli and longus capitis, anterior scalene and sternocleidomastoid muscles. Note the incremental increase in EMG activity for all muscles with increasing cranio-cervical flexion (from stages 22 to 30 mmHg) but with lesser activity in the deep cervical flexors and greater activity in the superficial muscles for the neck pain patient suggesting a reorganization of muscle activity to perform the task. EMG calibration: 0.5 mV. (d) Normalized root mean square (RMS) values (mean and standard deviation) for the deep cervical flexor muscles for each stage of the cranio-cervical flexion test (C-CFT). * indicates significant difference between control subjects and neck pain patients ($p < 0.05$). Reprinted from Falla et al. [46] with permission.

17.5 SURFACE EMG TUNING CURVES

Another approach to monitoring the amplitude of activity of muscles with sEMG is to measure tuning curves. Surface EMG tuning curves, which depict muscle activity over a range of force or moment directions, have been used to study activation strategies of arm and neck muscles [19,49,60,84,141,143]. When tuning curves are consistent among subjects, analyzing the orientation and focus (mean direction and spread of EMG activity, respectively, defined below) of EMG tuning curves in relation to musculoskeletal mechanics has provided insight into CNS control [143].

For example, sEMG tuning curves of neck muscles can be recorded by having a subject perform contractions at a predefined force (e.g., 15 N of force) with continuous change in force direction in the range 0–360° in the horizontal plane [49]. During these circular contractions, the amplitude of the surface EMG can be estimated. The ARV of the EMG as a function of the angle of force direction can be referred to as directional activation curves. The directional activation curves represent the modulation in intensity of muscle activity with the direction of force exertion and represent a closed area when expressed in polar coordinates. The line connecting the origin with the central point of this area defines a directional vector, whose length is expressed as a percent of the mean ARV during the entire task. This normalized vector length represents the specificity of muscle activation: It is equal to zero if the muscle is active in the same way in all directions and, conversely, it corresponds to 100% if the muscle is active in exclusively one direction.

In healthy subjects, neck muscles show well-defined preferred directions of activation which are in accordance with their anatomical position relative to the spine [12,49,143] (Fig. 17.5a). These observations suggest that the CNS copes with the anatomical complexity and redundancy of the neck muscles by developing consistent muscle synergies to generate multidirectional patterns of force [12,49,143]. However, recent studies have shown that patients with either whiplash-induced neck pain or idiopathic neck pain have reduced specificity of sternocleidomastoid, splenius capitis, and semi-spinalis cervics muscle activity with respect to asymptomatic individuals [49,99,128] (Fig. 17.5b). The reduced specificity of neck muscle activity in patients with neck pain is due to a reduced modulation in discharge rate of individual motor units with force direction indicating a potential change in motor neuron excitability [49].

Polar plots or EMG tuning curves are also useful to display and compare the sEMG amplitude of a muscle in response to multidirectional perturbations. For example, Fig. 17.6 displays polar plots of the normalized sEMG amplitude of the left internal oblique, left erector spinae, tibialis anterior, and gastrocnemius muscles in response to unexpected balance perturbations performed in 12 directions in a group of individuals with and without low back pain [77]. Note the increased activity in the gastrocnemius during backward perturbations (that is, when acting as a prime mover) in the control group and increased tibialis anterior activation following perturbation directions in which the muscle would also act as a prime mover, namely perturbation directions with a forward component. In addition, the individuals with low back pain showed reduced activation of the left internal oblique in directions with either a leftward or leftward/backward component and increased left internal oblique activity during perturbations in which the left internal oblique could contribute to a hip/trunk strategy.

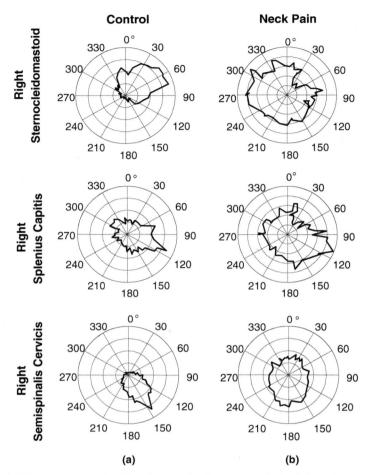

FIGURE 17.5   Representative directional activation curves obtained from the sternocleido-mastoid, splenius capitis, and semispinalis cervicis muscles during a circular contraction performed at 15 N and presented for a control subject and a patient with chronic neck pain. Note the defined activation of the sternocleidomastoid, splenius capitis, and semispinalis cervicis for the control subject (a) with minimal activity during the antagonist phase of the task. Conversely, the directional activation curves for the patient (b) indicate more even activation levels of each muscle for all directions. Reprinted from Falla and co-workers [49,99,128] with permission.

17.6   DISTRIBUTION OF MUSCLE ACTIVITY

High density, two-dimensional sEMG provides a measure of the electric potential distribution over a skin surface area during muscle contraction [56,148] (See Chapter 5). Unlike classic bipolar sEMG applications, this method provides a topographical representation of sEMG amplitude and can identify relative adaptations in the intensity of activity within regions of the muscle [148] as described in Chapter 5.

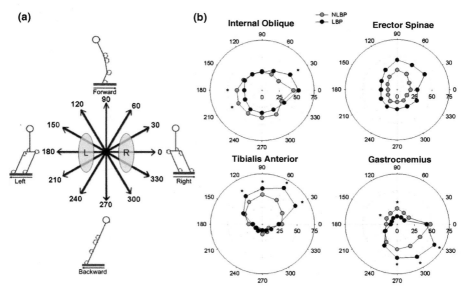

FIGURE 17.6 (a) Experimental setup for support surface translations which shows the directions of platform perturbations with the induced body sway resulting from perturbations in the cardinal directions (i.e., left, forward, right, and backward perturbations). Schematic stick figures are depicted with the subject facing to the right for the sagittal plane views and are viewed from the back for the frontal plane views. (b) Polar plots of the of the normalized EMG amplitude of the left internal oblique muscles, left erector spinae, tibialis anterior, and gastrocnemius muscles in response to unexpected balance perturbations performed in 12 direction translations in a group of individuals with and without low back pain. Reprinted from Jones et al. [77] with permission.

Spatial heterogeneity in muscle activity has been observed from multichannel sEMG recordings during sustained constant-force contractions [56,71], during contractions of increasing load [71], and during dynamic contractions [41], which suggests a nonuniform distribution of motor units or spatial dependency in the control of motor units [72].

To characterize the spatial distribution of muscle activity, two coordinates of the centroid (center of activity) of the RMS map (x- and y-axis coordinates for the medial–lateral and cranial–caudal direction, respectively) are typically calculated [56]. Studies in asymptomatic individuals show a change in the distribution of activity over the lumbar erector spinae muscle during a fatiguing sustained lumbar flexion contraction as reflected by a shift of the centroid towards the caudal region of the lumbar spine [140] (Fig. 17.7). Furthermore, a shift of activity towards the cranial region of the upper trapezius muscle is observed in healthy individuals during sustained shoulder abduction [35,40,56] as described in Fig. 17.8b. This response reflects a greater progressive recruitment of motor units within the cranial region of the upper trapezius muscle [40].

Redistribution of activity within the same muscle has been shown to be functionally important to maintain the motor output in the presence of altered afferent feedback (e.g., pain or fatigue) [56]. This variation in activation within regions of

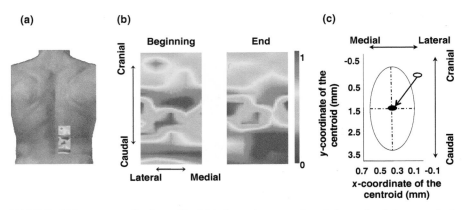

FIGURE 17.7   (a) An adhesive grid of 64 electrodes were placed above the right paraspinal muscles between the level of the L5 and L2 spinal processes as subjects performed a 6-min sustained contraction in standing with 20° forward flexion holding a weighted bar (7.5-kg load). (b) Topographical maps (interpolation by a factor 8) of the sEMG root mean square (RMS) value obtained at the beginning and end of the 6-min sustained contraction. RMS values are normalized with respect to the maximum value of the map. Dark red corresponds to maximum EMG amplitude. Note that the spatial distribution of EMG activity changed over time during the sustained contraction with a shift toward the caudal direction of the lumbar region. (c) The vector of the shift of centroid of the RMS map with respect to initial position averaged across all subjects during the sustained contraction. Note the shift in the distribution of sEMG amplitude towards the caudal region of the lumbar erector spinae. The ellipses have semi-axes equal to the standard deviations in the two directions. Modified from Tucker et al. [140] with permission.

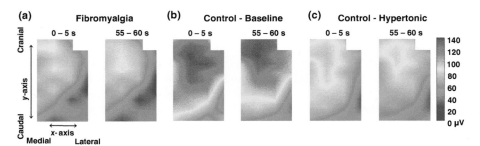

FIGURE 17.8   Representative topographical maps (interpolation by a factor 8) of the sEMG root mean square (RMS) value from the upper trapezius muscle for (a) a patient with fibromyalgia and a control subject at (b) baseline and (c) following injection of hypertonic saline into the cranial region of the upper trapezius muscle. Maps are shown for the first and last 5 s of a 60-s sustained shoulder abduction contraction. Colors are scaled between the minimum and maximum RMS values. Areas of dark blue correspond to low EMG amplitude, and areas of dark red correspond to high EMG amplitude. Reprinted from Falla et al. [34] with permission.

the same muscle is potentially relevant to avoid overload of the same muscle fibers during prolonged activation and is of particular relevance for muscles commonly exposed to repetitive or sustained activation, such as the upper trapezius muscle [146] and the lumbar erector spinae [4].

On the contrary, in the presence of either experimentally induced muscle pain [35,36,103] or clinical pain (e.g., fibromyalgia) [34] (Fig. 17.8a,c), the redistribution of activity to different regions of the muscle during sustained contractions is impaired. These findings suggest that muscle pain prevents the adaptation of muscle activity during sustained contractions as observed in nonpainful conditions, which may induce overuse of certain muscle compartments with fatigue.

17.7 MONITORING CHANGE WITH REHABILITATION

In addition to utilizing sEMG to detect changes in neuromuscular control in patients with musculoskeletal pain, as outlined above, sEMG is frequently used as a tool to document change in neuromuscular function during or after rehabilitation. High levels of repeatability of normalized sEMG estimates have been shown for both the neck and back muscles [27,38] confirming the suitability of using sEMG between sessions to monitor change.

Surface EMG has been used to monitor the physiological efficacy of multiple physical therapy interventions including exercise [78], mobilization/manipulation [59,89], and taping [66,118]. In relation to exercise, sEMG has been used to determine the most appropriate exercise to prescribe to patients [3,17,120,129], to monitor/document improvement in muscle activation post-intervention [44,79,130,139], and as a tool for biofeedback with training [31,101]. For example, sEMG studies show that low-load exercise interventions increase the activation of the deep cervical flexor muscles during a task of cranio-cervical flexion [79], enhance the speed of their activation when challenged by a postural perturbation [50,79], and enhance the degree of directional specificity of neck muscle activity during multidirectional isometric contractions of the neck [48]. Four weeks of repeated isolated training of the transversus abdominis muscle performed twice daily in patients with low back pain also results in earlier onset of the transversus abdominis muscle during rapid arm movements, indicating an enhancement of the automatic postural control strategy post-training. Such improvement was retained at a six-month follow-up [139]. Furthermore, six weeks of higher load exercise for the neck has been shown to reduce myoelectric manifestations of fatigue of the neck muscles as evidenced by a significant reduction in the and rate of change of the MNF for both the sternocleidomastoid and anterior scalene muscles during submaximal isometric cervical flexion contractions [44].

17.8 CONCLUSIONS

Surface EMG is a fundamental tool in physical therapy and has been successfully used for the assessment of disturbed motor control and to monitor the physiological efficacy of multiple physical therapy interventions. A number of analyses of the sEMG signal can be performed to gain insight into the peripheral status of the muscle and the strategies used by the CNS to control movement and posture.

This chapter has reviewed some of the most frequently applied analyses in musculoskeletal physical therapy research and presented an overview of knowledge gained from sEMG studies conducted in patients with neck and low back pain. This knowledge has provided a greater understanding of the effect of pain on neuromuscular control and places the physical therapist in a better position to provide appropriate treatment to manage patients with pain.

REFERENCES

1. Ahern, D., M. Follick, J. Council, and N. Laser-Wolston, "Reliability of lumbar paravertebral EMG assessment in chronic low back pain," *Arch. Phys. Med. Rehabil.* 762–765 (1986).

2. Alaranta, H., M. Moffroid, L. Elmqvist, J. Held, M. Pope, and P. Renstrom, "Postural control of adults with musculoskeletal impairment," *Crit. Rev. Phys. Rehabil. Med.* **4**, 337–370 (1994).

3. Andersen, L., C. Andersen, O. Mortensen, O. Poulsen, I. Bjørnlund, and M. Zebis, "Muscle activation and perceived loading during rehabilitation exercises: comparison of dumbbells and elastic resistance," *Phys. Ther.* **90**, 538–549 (2010).

4. Andersson, G., and R. Ortengren, "Assessment of back load in assembly line work using electromyography," *Ergonomics* **27**, 1157–1168 (1984).

5. Arendt-Nielsen, L., T. Graven-Nielsen, H. Svarrer, and P. Svensson, "The influence of low back pain on muscle activity and coordination during gait: a clinical and experimental study," *Pain* **64**, 231–240 (1996).

6. Arendt-Nielsen, L., and K. R. Mills, "The relationship between mean power frequency of the EMG spectrum and muscle fibre conduction velocity," *Electroencephalogr. Clin. Neurophysiol.* **60**, 130–134 (1985).

7. Arendt Nielsen, L., and K. R. Mills, "Muscle fiber conduction velocity, mean power frequency, mean EMG voltage and force during submaximal fatigueing contractions of human quadriceps," *Eur. J. Appl. Physiol. Occup. Physiol.* **58**, 20–25 (1988).

8. Barton, P. M., and K. C. Hayes, "Neck flexor muscle strength, efficiency, and relaxation times in normal subjects and subjects with unilateral neck pain and headache," *Arch. Phys. Med. Rehabil.* **77**, 680–687 (1996).

9. Basmajian, J. V., and C. J. DeLuca, *Muscles Alive: Their functions revealed by electromyography*, Williams and Wilkins, Baltimore, 1985.

10. Bigland-Ritchie, B., F. Cafarelli, and N. Vollestad, "Fatigue of submaximal static contractions," *Acta Physiol. Scand.* **128**, 137–148 (1986).

11. Bigland-Ritchie, B,. E. F. Donovan, and C. S. Roussos, "Conduction velocity and EMG power spectrum changes in fatigue of sustained maximal efforts," *J. Appl. Physiol.* **51**, 1300–1305 (1981).

12. Blouin, J. S., G. P. Siegmund, M. G. Carpenter, and J. T. Inglis, "Neural control of superficial and deep neck muscles in humans," *J. Neurophysiol.* **98**, 920–928 (2007).

13. Bonato, P., G. Gagliati, and M. Knaflitz, "Analysis of surface myoelectric signals recorded during dynamic contractions," *IEEE Eng. Med. Biol. Mag.* **15**, 102–111 (1996).

14. Boudreau, S., and D. Falla,"Individuals with chronic neck pain show alternative automatic motor control strategies," in *Proceedings of 14th World Congress on Pain,* Milan, 2012.

15. Boudreau, S., D. Farina, L. Kongstad, D. Buus, J. Redder, E. Sverrisdóttir, and D. Falla, "The relative timing of trunk muscle activation is retained in response to unanticipated postural-perturbations during acute low back pain," *Exp. Brain Res.* **210,** 259–267 (2011).

16. Bouisset, S., and M. Zattara, "Biomechanical study of the programming of anticipatory postural adjustments associated with voluntary movement," *J. Biomech.* **20,** 735–742 (1987).

17. Bressel, E., D. Dolny, and M. Gibbons, "Trunk muscle activity during exercises performed on land and in water," *Med. Sci. Sports Exerc.* **43,** 1927–1932 (2011).

18. Brody, L. R., M. T. Pollock, S. H. Roy, C. J. De Luca, and B. Celli, "pH-induced effects on median frequency and conduction velocity of the myoelectric signal," *J. Appl. Physiol.* **71,** 1878–1885 (1991).

19. Buchanan, T., G. Rovai, and W. Rymer, "Strategies for muscle activation during isometric torque generation at the human elbow," *J. Neurophysiol.* **62,** 1201–1212 (1989).

20. Buchthal, F., C. Guld, and P. Rosenfalck, "Propogation velocity in electrically activated fibres in man," *Acta Physiol. Scand.* **34,** 75–89 (1955).

21. Candotti, C., J. Loss, A. Pressi, F. Castro, M. La Torre, O. Melo Mde, L. Araújo, and M. Pasini, "Electromyography for assessment of pain in low back muscles," *Phys. Ther.* **88,** 1061–1067 (2008).

22. Choi, H. I., and W. J. Williams, "Improved time-frequency representation of multi-component signals using exponential kernels," *IEEE Trans. Acoust. Speech Sig. Proc.* **37,** 862–871 (1989).

23. Cholewicki, J., M. M. Panjabi, and A. Khachatryan, "Stabilizing function of trunk flexor–extensor muscles around a neutral spine posture," *Spine* **22,** 2207–2212 (1997).

24. Cholewicki, J., S. P. Silfies, R. A. Shah, H. S. Greene, N. P. Reeves, K. Alvi, and B. Goldberg, "Delayed trunk muscle reflex responses increase the risk of low back injuries," *Spine* **30,** 2614–2620 (2005).

25. Clausen, T., "Na+−K+ pump regulation and skeletal muscle contractility," *Physiol. Rev.* **83,** 1269–1324 (2003).

26. Cordo, P. J., and L. M. Nashner, "Properties of postural adjustments associated with rapid arm movements," *J. Neurophysiol.* **47,** 287–302 (1982).

27. Dankaerts, W., P. O'Sullivan, A. Burnett, L. Straker, and L. Danneels, "Reliability of EMG measurements for trunk muscles during maximal and sub-maximal voluntary isometric contractions in healthy controls and CLBP patients," *J. Electromyogr. Kinesiol.* **14,** 333–342 (2004).

28. De Luca, C. J., "Myoelectrical manifestations of localized muscular fatigue in humans," *Crit. Rev. Biomed. Eng.* **11,** 251–279 (1984).

29. De Luca, C. J., "The use of the surface EMG signal for performance evaluation of back muscles," *Muscle Nerve* **16,** 210–216 (1993).

30. Dehner, C., A. Schmelz, H. Völker, J. Pressmar, M. Elbel, and M. Kramer, "Intramuscular pressure, tissue oxygenation, and muscle fatigue of the multifidus during isometric extension in elite rowers with low back pain," *J. Sport Rehabil.* **18,** 572–581 (2009).

31. Dellve, L., L. Ahlstrom, A. Jonsson, L. Sandsjö, M. Forsman, A. Lindegård, C. Ahlstrand, R. Kadefors, and M. Hagberg, "Myofeedback training and intensive muscular strength training to decrease pain and improve work ability among female workers on long-term

sick leave with neck pain: a randomized controlled trial," *Int. Arch. Occup. Environ. Health*, **84**, 335–346 (2011).

32. Elfving, B., A. Dedering, and G. Nemeth, "Lumbar muscle fatigue and recovery in patients with long-term low-back trouble: electromyography and health-related factors," *Clin. Biomech.* **18**, 619–639 (2003).

33. Enoka, R. M., and A. J. Fuglevand, "Motor unit physiology: some unresolved issues," *Muscle & Nerve* **24**, 4–17 (2001).

34. Falla, D., H. Andersen, B. Danneskiold-Samsøe, L. Arendt-Nielsen, and D. Farina, "Adaptations of upper trapezius muscle activity during sustained contractions in women with fibromyalgia," *J. Electromyogr. Kinesiol.* **20**, 457–464 (2010).

35. Falla, D., L. Arendt-Nielsen, and D. Farina, "Gender-specific adaptations of upper trapezius muscle activity to acute nociceptive stimulation," *Pain* **138**, 217–225 (2008).

36. Falla, D., L. Arendt-Nielsen, and D. Farina, "The pain-induced change in relative activation of upper trapezius muscle regions is independent of the site of noxious stimulation," *Clin. Neurophysiol.* **120**, 150–157 (2009).

37. Falla, D., G. Bilenkij, and G. Jull "Patients with chronic neck pain demonstrate altered patterns of muscle activation during performance of a functional upper limb task," *Spine* **29**, 1436–1440 (2004).

38. Falla, D., P. Dall'Alba, A. Rainoldi, R. Merletti, and G. Jull, "Repeatability of surface EMG variables in the sternocleidomastoid and anterior scalene muscles," *Eur J. Appl. Physiol.* **87**, 542–549 (2002).

39. Falla, D., and D. Farina, "Muscle fiber conduction velocity of the upper trapezius muscle during dynamic contraction of the upper limb in patients with chronic neck pain," *Pain* **116**, 138–145 (2005).

40. Falla, D., and D. Farina, "Non-uniform adaptation of motor unit discharge rates during sustained static contraction of the upper trapezius muscle," *Exp. Brain Res.* **191**, 363–370 (2008).

41. Falla, D., D. Farina, and T. Graven Nielsen, "Spatial dependency of trapezius muscle activity during repetitive shoulder flexion," *J. Electromyogr. Kinesiol.* **17**, 299–306 (2007).

42. Falla, D., T. Graven-Nielsen, and D. Farina, "Spatial and temporal changes of upper trapezius muscle fiber conduction velocity are not predicted by surface EMG spectral analysis during a dynamic upper limb task," *J. Neurosci. Meth.* **156**, 236–241 (2006).

43. Falla, D., G. Jull, P. Dall'Alba, A. Rainoldi and R. Merletti, "An electromyographic analysis of the deep cervical flexor muscles in performance of craniocervical flexion," *Phys. Ther.* **83**, 899–906 (2003).

44. Falla, D., G. Jull, P. W. Hodges, and B. Vicenzino, "An endurance–strength training regime is effective in reducing myoelectric manifestations of cervical flexor muscle fatigue in females with chronic neck pain," *Clin. Neurophysiol.* **117**, 828–837 (2006).

45. Falla, D., G. Jull, and P. W. "Hodges Feedforward activity of the cervical flexor muscles during voluntary arm movements is delayed in chronic neck pain," *Exp. Brain Res.* **157**, 43–48 (2004).

46. Falla, D., G. Jull, and P. W. Hodges, "Patients with neck pain demonstrate reduced electromyographic activity of the deep cervical flexor muscles during performance of the craniocervical flexion test," *Spine* **29**, 2108–2114 (2004).

47. Falla, D., G. Jull, S. O'Leary, and P. Dall'Alba, "Further evaluation of an EMG technique for assessment of the deep cervical flexor muscles," *J. Electromyogr. Kinesiol.* **16**, 621–628 (2004).

48. Falla, D., R. Lindstrøm, L. Rechter, S. Boudreau, and F. Petzke,"Effect of an 8-week exercise program on specificity of neck muscle activity in patients with chronic neck pain," in *Proceedings of the XIXth Congress of the International Society of Electrophysiology & Kinesiology, Brisbane,* 2012.

49. Falla, D., R. Lindstrom, L. Rechter, and D. Farina, "Effect of pain on the modulation in discharge rate of sternocleidomastoid motor units with force direction," *Clin. Neurophysiol.* **121**, 744–753 (2010).

50. Falla, D., S. O'Leary, D. Farina, and G. Jull, "The change in deep cervical flexor activity after training is associated with the degree of pain reduction in patients with chronic neck pain" *Clin. J. Pain* **28**, 628–634 (2012).

51. Falla, D., A. Rainoldi, R. Merletti, and G. Jull, "Myoelectric manifestations of sternocleidomastoid and anterior scalene muscle fatigue in chronic neck pain patients," *Clin. Neurophysiol.* **114**, 488–495 (2003).

52. Falla, D., A. Rainoldi, R. Merletti, and G. Jull, "Spatio-temporal evaluation of neck muscle activation during postural perturbations in healthy subjects," *J. Electromyogr. Kinesiol.* **14**, 463–474 (2004).

53. Farina, D., L. Arendt-Nielsen, R. Merletti, and T. Graven-Nielsen, "Assessment of single motor unit conduction velocity during sustained contractions of the tibialis anterior muscle with advanced spike triggered averaging," *J. Neurosci. Meth.* **115**, 1–12 (2002).

54. Farina, D., E. Fortunato and R. Merletti, "Noninvasive estimation of motor unit conduction velocity distribution using linear electrode arrays," *IEEE Trans. Biomed. Eng.* **47**, 380–388 (2000).

55. Farina, D., M. Fosci and R. Merletti, "Motor unit recruitment strategies investigated by surface EMG variables," *J. Appl. Physiol.* **92**, 235–247 (2002).

56. Farina, D., F. Leclerc, L. Arendt-Nielsen, O. Buttelli, and P. Madeleine, "The change in spatial distribution of upper trapezius muscle activity is correlated to contraction duration," *J. Electromyogr. Kinesiol.* **18**, 16–25 (2008).

57. Farina, D., R. Merletti, and R. M. Enoka, "The extraction of neural strategies from the surface EMG," *J. Appl. Physiol.* **96**, 1486–1495 (2004).

58. Farina, D., M. Pozzo, E. Merlo, A. Bottin, and R. Merletti, "Assessment of average muscle fiber conduction velocity from surface EMG signals during fatiguing dynamic contractions," *IEEE Trans. Biomed. Eng.* **51**, 1383–1393 (2004).

59. Ferreira, M,. P. H. Ferreira, and P. W. Hodges. "Changes in postural activity of the trunk muscles following spinal manipulative therapy," *Man. Ther.* **12**, 240–248 (2007).

60. Flanders, M., and J. Soechting, "Arm muscle activation for static forces in three-dimensional space," *J. Neurophysiol.* **64**, 1818–1837 (1990).

61. Fredin, Y., J. Elert, N. Britschgi, V. Nyberg, A. Vaher and B. Gerdle, "A decreased ability to relax between repetitive muscle contractions in patients with chronic symptoms after whiplash trauma of the neck," *J. Musculoskel. Pain* **5**, 55–70 (1997).

62. Friedli, W. G., M. Hallett and S. R. Simon, "Postural adjustments associated with rapid voluntary arm movements 1. Electromyographic data," *J. Neurol. Neurosurg. Psychiatry* **47**, 611–622 (1984).

63. Gandevia, S. C., Spinal and supraspinal factors in human muscle fatigue, *Physiol. Rev.* **81**, 1725–1789 (2001).

64. Gazzoni, M., D. Farina, and R. Merletti, "Motor unit recruitment during constant low force and long duration muscle contractions investigated by surface electromyography," *Acta Physiol. Pharmacol. Bulg.* **26**, 67–71 (2001).

65. Gogia, P. P., and M. A. Sabbahi, "Electromyographic analysis of neck muscle fatigue in patients with osteoarthritis of the cervical spine," *Spine* **19**, 502–506 (1994).

66. Greig, A., K. Bennell, A. Briggs, and P. Hodges, "Postural taping decreases thoracic kyphosis but does not influence trunk muscle electromyographic activity or balance in women with osteoporosis," *Man. Ther.* **13**, 249–257 (2008).

67. Henry, S., J. Hitt, S. Jones, and J. Bunn, "Decreased limits of stability in response to postural perturbations in subjects with low back pain," *Clin. Biomech. (Bristol, Avon)* **21**, 881–892 (2006).

68. Hodges, P., G. Moseley, A. Gabrielsson, and S. Gandevia, "Experimental muscle pain changes feedforward postural responses of the trunk muscles," *Exp. Brain Res.* **151**, 262–271 (2003).

69. Hodges, P. W. and C. A. Richardson, "Inefficient muscular stabilization of the lumbar spine associated with low back pain. A motor control evaluation of transversus abdominis," *Spine* **21**, 2640–2650 (1996).

70. Hodges, P. W. and C. A. Richardson, "Feedforward contraction of transversus abdominis is not influenced by the direction of arm movement," *Exp. Brain Res.* **114**, 362–370 (1997).

71. Holtermann, A. and K. Roeleveld, "EMG amplitude distribution changes over the upper trapezius muscle are similar in sustained and ramp contractions," *Acta Physiol.* **186**, 159–168 (2006).

72. Holtermann, A., K. Roeleveld, and J. S. Karlsson, "Inhomogeneities in muscle activation reveal motor unit recruitment," *J. Electromyogr. Kinesiol.* **15**, 131–137 (2005).

73. Horak, F. B., P. Esselman, M. E. Anderson, and M. K., "Lynch MK. The effects of movement velocity, mass displaced, and task certainty on associated postural adjustments made by normal and hemiplegic individuals," *J. Neurol. Neurosurg. Psychiatry* **47**, 1020–1028 (1984).

74. Jacobs, J., S. Henry, S. Jones, J. Hitt, and J. Bunn, "A history of low back pain associates with altered electromyographic activation patterns in response to perturbations of standing balance," *J. Neurophysiol.* **106**, 2506–2514 (2011).

75. Johnston, V., G. Jull, T. Souvlis, and N. L. Jimmieson, "Neck movement and muscle activity characteristics in female office workers with neck pain" *Spine* **33**, 555–563 (2008).

76. Jones, S., S. Henry, C. Raasch, J. Hitt, and J. Bunn, "Individuals with non-specific low back pain use a trunk stiffening strategy to maintain upright posture," *J. Electromyogr. Kinesiol.* **22**, 13–20 (2012).

77. Jones, S., J. Hitt, M. DeSarno, and S. Henry, "Individuals with non-specific low back pain in an active episode demonstrate temporally altered torque responses and direction-specific enhanced muscle activity following unexpected balance perturbations," *Exp. Brain Res.* **221**, 413–426 (2012).

78. Jørgensen, M., L. Andersen, N. Kirk, M. Pedersen, K. Søgaard, and A. Holtermann, "Muscle activity during functional coordination training: implications for strength gain and rehabilitation," *J. Strength Cond. Res.* **24**, 1732–1739 (2010).

79. Jull, G., D. Falla, B. Vicenzino, and P. W. Hodges, "The effect of therapeutic exercise on activation of the deep cervical flexor muscles in people with chronic neck pain," *Man. Ther.* **14**, 696–701 (2009).

80. Jull, G., E. Kristjansson, and P. Dall'Alba, "Impairment in the cervical flexors: a comparison of whiplash and insidious onset neck pain patients," *Man. Ther.* **9**, 89–94 (2004).

81. Kadi, F., K. Waling, C. Ahlgren, G. Sundelin, S. Holmner, G. S. Butler-Browne, and L. Thornell, "Pathological mechanisms implicated in localized female trapezius myalgia," *Pain* **78**, 191–196 (1998).

82. Kankaanpaa, M., S. Taimelas, O. Hänninen, and O. Airaksinen, "Back and hip extensor fatigability in chronic low back pain patients and controls," *Arch. Phys. Med. Rehabil.* **79**, 412–417 (1998).

83. Karst, G. M., and G. M. Willet, "Onset timing of electromyographic activity in the vastus medialis oblique and vastus lateralis muscles in subjects with and without patellofemoral pain syndrome," *Phys. Ther.* **75**, 813–823 (1995).

84. Keshner, E. A., D. Campbell, R. T. Katz, and B. W. Peterson, "Neck muscle activation patterns in humans during isometric head stabilization," *Exp. Brain Res.* **75**, 335–344 (1989).

85. Keshner, E. A., M. H. Woollacott, and B. Debu, "Neck, Trunk and Limb Muscle Responses During Postural Perturbations in Humans," *Exp. Brain Res.* **71**, 455–466 (1988).

86. Kleine, B. U., D. Stegeman, D. Mund, and C. Anders, "Influence of motorneuron firing synchronisation on SEMG characteristics in dependence of electrode position," *J. Appl. Physiol.* **91**, 1588–1599 (2001).

87. Kossler, F., F. Lange, G. Caffier, and G. Kuchler G., "External potassium and action potential propagation in rat fast and slow twitch muscles," *Gen. Physiol. Biophys.* **10**, 485–498 (1991).

88. Kramer, C. G., G. Hagg, and B. Kemp, "Real time measurement of muscle fatigue related changes in surface EMG," *Med. Biol. Eng. Comp.* **25**, 627–630 (1987).

89. Krekoukias, G., N. Petty, and L. Cheek, "Comparison of surface electromyographic activity of erector spinae before and after the application of central posteroanterior mobilisation on the lumbar spine," *J. Electromyogr. Kinesiol.* **19**, 39–45 (2009).

90. Kumar, S., Y. Narayan, N. Prasad, A. Shuaib, and Z. A. Siddiqi, "Cervical electromyogram profile differences between patients of neck pain and control," *Spine* **32**, E246–E253 (2007).

91. Lamoth, C., O. Meijer, A. Daffertshofer, P. Wuisman, and P. Beek, "Effects of chronic low back pain on trunk coordination and back muscle activity during walking: changes in motor control," *Eur. Spine J.* **15**, 23–40 (2006).

92. Larsson, R., H. Cai, Q. Zhang, P. A. Oberg, and S. E. Larsson, "Visualization of chronic neck–shoulder pain: Impaired microcirculation in the upper trapezius muscle in chronic cervicobrachial pain," *Occup. Med. (London)* **48**, 189–194 (1998).

93. Larsson, R., P. A. Oberg, and S. E. Larsson, "Changes of trapezius muscle blood flow and electromyography in chronic neck pain due to trapezius myalgia," *Pain* **79**, 45–50 (1999).

94. Latash, M. L., A. S. Aruin, I. Neyman, and J. J. Nicholas, "Anticipatory postural adjustments during self inflicted and predictable perturbations in Parkinson's disease," *J. Neurol. Neurosurg. Psychiatry* **58**, 326–334 (1995).

95. Lee, W. A., T. S. Buchanan, and M. W. Rogers, "Effects of arm acceleration and behavioral conditions on the organization of postural adjustments during arm flexion," *Exp. Brain Res.* **66**, 257–270 (1987).

96. Lindman, R., M. Hagberg, K. Angqvist, K. Soderlund, E. Hultman, and L. Thornell, "Changes in muscle morphology in chronic trapezius myalgia," *Scand. J. Work Environ. Health* **17**, 347–355 (1991).

97. Lindstrom, L., and R. Magnusson, "Interpretation of myoelectric power spectra: a model and its applications," *Proc. IEEE* **65**, 653–662 (1977).

98. Lindstrom, L., R. Magnusson, and R. Petersen, "Muscle fatigue and action potential conduction velocity changes studied with frequency analysis of EMG signals. *Electromyogr. Clin. Neurophysiol.* **10**, 341–356 (1970).

99. Lindstrom, R., J. Schomacher, D. Farina, L. Rechter, and D. Falla, "Association between neck muscle co-activation, pain, and strength in women with neck pain," *Man. Ther.* **16**, 80–86 (2011).

100. Lowery, M. M., C. L. Vaughan, P. J. Nolan, and M. J. O'Malley, "Spectral compression of the electromyographic signal due to decreasing muscle fiber conduction velocity," *IEEE Trans. Rehabil. Eng.* **8**, 353–361 (2000).

101. Ma, C., G. Szeto, T. Yan, S. Wu, C. Lin, and L. Li, "Comparing biofeedback with active exercise and passive treatment for the management of work-related neck and shoulder pain: a randomized controlled trial" *Arch. Phys. Med. Rehabil.* **92**, 849–858 (2011).

102. MacDonald, D., G. Moseley, and P. Hodges, "People with recurrent low back pain respond differently to trunk loading despite remission from symptoms," *Spine (Phila Pa 1976)* **35**, 818–824 (2010).

103. Madeleine, P., F. Leclerc, L. Arendt-Nielsen, P. Ravier, and D. Farina, "Experimental muscle pain changes the spatial distribution of upper trapezius muscle activity during sustained contraction," *Clin. Neurophysiol.* **117**, 2436–2445 (2006).

104. Masuda, K., T. Masuda, T. Sadoyama, M. Inaki, and S. Katsuta, "Changes in surface EMG parameters during static and dynamic fatiguing contractions," *J. Electromyogr. Kinesiol.* **9**, 39–46 (1999).

105. Masuda, T., H. Miyano, and T. Sadoyama, "The position of innervation zones in the biceps brachii investigated by surface electromyography," *IEEE Trans. Biomed. Eng.* **32**, 36–42 (1985).

106. McKenna, M. J., J. Bangsbo, and J. M. Renaud, "Muscle K+, Na+, and Cl disturbances and Na+ K+ pump inactivation: implications for fatigue," *J. Appl. Physiol.* **104**, 288–295 (2008).

107. Merletti, R., D. Farina, and A. Granata, "Non-invasive assessment of motor unit properties with linear electrode arrays," in *Clinical Neurophysiology: From Receptors to Perception*, G. Comi, C. Locking, J. Kimura, and P. M. Rossini, eds., Elsevier, Amsterdam, 1999.

108. Merletti, R., M. Knaflitz, and C. J. De Luca, "Myoelectric manifestations of fatigue in voluntary and electrically elicited contractions," *J. Appl. Physiol.* **69**, 1810–1820 (1990).

109. Merletti, R., and L. Lo Conte, "Advances in processing of surface myoelectric signals: part 1," *Med. Biol. Eng. Comput.* **33**, 362–372 (1995).

110. Merletti, R., A. Rainoldi, and D. Farina, "Surface electromyography for noninvasive characterization of muscle," *Exerc. Sport Sci. Rev.* **29**, 20–25 (2001).

111. Merletti, R., and S. Roy, "Myoelectric and mechanical manifestations of muscle fatigue in voluntary contractions," *J. Orthop. Sports Phys. Ther.* **24**, 342–353 (1996).

112. Merlo, A., D. Farina, and R. Merletti, "A fast and reliable technique for muscle activity detection from surface EMG signals," *IEEE Trans. Biomed. Eng.* **50**, 316–323 (2003).

113. Merlo, E., M. Pozzo, G. Antonutto, R. Merletti, P. E. di Prampero, and D. Farina, "Time–frequency analysis and estimation of muscle fiber conduction velocity from surface EMG signals during esplosive dynamic contractions," *J. Neurosci. Methods* **142**, 267–274 (2005).

114. Milner-Brown, H. S., R. B. Stein, and R. Yemm, "Changes in firing rate of human motor units during linearly changing voluntary contractions," *J. Physiol. (Lond.)* **230**, 371–390 (1973).

115. Moritani, T., M. Muro, and A. Nagata, "Intramuscular and surface electromyogram changes during muscle fatigue," *J. Appl. Physiol.* **60**, 1179–1185 (1986).

116. Moseley, G., M. Nicholas, and P. Hodges, "Does anticipation of back pain predispose to back trouble?," *Brain* **127**, 2339–2347 (2004).

117. Nederhand, M. J., M. J. Ijzerman, H. J. Hermens, C. T. M. Baten, and G. Zilvold, "Cervical muscle dysfunction in the chronic whiplash associated disorder grade II (WAD-II)," *Spine* **25**, 938–1943 (2000).

118. Ng, G., and P. Wong, "Patellar taping affects vastus medialis obliquus activation in subjects with patellofemoral pain before and after quadriceps muscle fatigue," *Clin. Rehabil.* **23**, 705–703 (2009).

119. Nielsen, O. B., and T. Clausen, "The Na+/K(+)-pump protects muscle excitability and contractility during exercise," *Exerc. Sport Sci. Rev.* **28**, 159–164 (2000).

120. Okubo, Y., K. Kaneoka, A. Imai, I. Shiina, M. Tatsumura, S. Izumi and S. Miyakawa, "Electromyographic analysis of transversus abdominis and lumbar multifidus using wire electrodes during lumbar stabilization exercises," *J. Orthop. Sports. Phys. Ther.* **40**, 743–750 (2010).

121. Piper, H., *Elektrophysiologie Menschlicher Muskeln*, Springer Verlag, Berlin, 1912.

122. Pozzo, M., B. Alkner, L. Norrbrand, D. Farina, E. Merlo, and P. Tesch, "Muscle fiber conduction velocity during concentric and eccentric actions on a flywheel exercise device," *Muscle Nerve* **34**, 169–177 (2006).

123. Radebold, A., J. Cholewicki, M. M. Panjabi, and T. C. Patel, "Muscle response pattern to sudden trunk loading in healthy individuals and in patients with chronic low back pain. *Spine* **25**, 947–954 (2000).

124. Reeves, N., J. Cholewicki, and T. Milner, "Muscle reflex classification of low-back pain," *J. Electromyogr. Kinesiol.* **15**, 53–60 (2005).

125. Roy, S. H., C. J. De Luca, and D. A. Casavant, "Lumbar muscle fatigue and chronic lower back pain," *Spine* **14**, 992–1001 (1989).

126. Saunders, S., D. Rath, and P. Hodges, "Postural and respiratory activation of the trunk muscles changes with mode and speed of locomotion," *Gait Posture* **20**, 280–290 (2004).

127. Schomacher, J., S. Boudreau, F. Petzke, and D. Falla, "Localized pressure pain sensitivity is associated with lower activation of the semispinalis cervicis muscle in patients with chronic neck pain," *Clin. J. Pain* **29**, 898–906 (2013).

128. Schomacher, J., D. Farina, R. Lindstroem, and D. Falla, "Chronic trauma-induced neck pain impairs the neural control of the deep semispinalis cervicis muscle," *Clin. Neurophysiol.* **123**, 1403–1408 (2012).

129. Schomacher, J., F. Petzke, and D. Falla, "Localised resistance selectively activates the semispinalis cervicis muscle in patients with neck pain," *Man. Ther.* **17**, 544–548 (2012).

130. Søgaard, K., A. Blangsted, P. Nielsen, L. Hansen, L. Andersen, P. Vedsted, and G. Sjøgaard, "Changed activation, oxygenation, and pain response of chronically painful muscles to repetitive work after training interventions: a randomized controlled trial," *Eur. J. Appl. Physiol.* **112**, 173–181 (2012).

131. Stalberg, E., "Propagation velocity in human muscle fibers in situ," *Acta Physiol. Scand.* **70**, 1–112 (1996).

132. Stegeman, D., J. Blok, H. Hermens, and K. Roeleveld, "Surface EMG models: properties and applications," *J. Electromyogr. Kinesiol.* **10**, 313–326 (2000).

133. Sterling, M., Jull, G., Vicenzino, B., Kenardy, J., and Darnell, R., "Development of motor dysfunction following whiplash injury," *Pain* **103**, 65–73 (2003).

134. Stokes, I., J. Fox, and S. Henry, "Trunk muscular activation patterns and responses to transient force perturbation in persons with self-reported low back pai," *Eur. Spine J.* **15**, 658–667 (2006).

135. Stulen, F. B., and C. J., DeLuca, "Frequency parameters of the myoelectric signal as a measure of muscle conduction velocity," *IEEE Trans. Biomed. Eng.* **28**, 515–523 (1981).

136. Sung, P., A. Lammers, and P. Danial, "Different parts of erector spinae muscle fatigability in subjects with and without low back pain," *Spine J.* **9**, 115–120 (2009).

137. Szeto, G. P., L. M. Straker, and P. B. O'Sullivan, "A comparison of symptomatic and asymptomatic office workers performing monotonous keyboard work. 1. Neck and shoulder muscle recruitment patterns," *Man. Ther.* **10**, 270–280 (2005).

138. Thorn, S., K. Søgaard, L. Kallenberg, L. Sandsjö, G. Sjøgaard, H. Hermens, R. Kadefors, and M. Forsman, "Trapezius muscle rest time during standardised computer work a comparison of female computer users with and without self-reported neck/shoulder complaints," *J. Electromyogr. Kinesiol.* **17**, 420–427 (2007).

139. Tsao, H., and P. W. Hodges, "Persistence of improvements in postural strategies following motor control training in people with recurrent low back pain," *J. Electromyogr. Kinesiol.* **18**, 559–567 (2008).

140. Tucker, K., D. Falla, T. Graven-Nielsen, and D. Farina, "Electromyographic mapping of the erector spinae muscle with varying load and during sustained contraction," *J. Electromyogr. Kinesiol.* **19**, 373–379 (2009).

141. van Bolhuis, B., and C. Gielen, "The relative activation of elbow-flexor muscles in isometric flexion and in flexion/extension movements," *J. Biomech.* **30**, 803–811 (1997).

142. van der Hulst, M., M. Vollenbroek-Hutten, J. Rietman, L. Schaake, K. Groothuis-Oudshoorn, and H. Hermens, "Back muscle activation patterns in chronic low back pain during walking: a "guarding" hypothesis," *Clin. J. Pain* **26**, 30–37 (2010).

143. Vasavada, A. N., B. W. Peterson, and S. L. Delp, "Three-dimensional spatial tuning of neck muscle activation in humans," *Exp. Brain Res.* **147**, 437–448 (2002).

144. Vogt, L., K. Pfeifer, and W. Banzer, "Neuromuscular control of walking with chronic low-back pain," *Man. Ther.* **8**, 21–28 (2003).

145. Wakeling, J. M., V. Von Tscharner, B. M. Nigg, and P. Stergiou, "Muscle activity in the leg is tuned in response to ground reaction forces," *J. Appl. Physiol.* **91**, 307–1317 (2001).

146. Westgaard, R. H., "Muscle activity as a releasing factor for pain in the shoulder and neck," *Cephalalgia* **19**, 251–258 (1999).

147. Woollacott, M. H., C. von Hosten, and B. Rosblad, "Relation between muscle response onset and body segmental movements during postural perturbations in humans," *Exp. Brain Res.* **72**, 593–604 (1988).

148. Zwarts, M. J., and D. F. Stegeman, "Multichannel surface EMG: basic aspects and clinical utility," *Muscle Nerve* **28**, 1–17 (2003).

18

SURFACE EMG BIOFEEDBACK

A. Gallina,[1,2] M. Gazzoni,[1] D. Falla,[3,4] and R. Merletti[1]

[1]*Laboratory for Engineering of the Neuromuscular System, Politecnico di Torino, Italy*
[2]*University of British Columbia—Rehabilitation Science, Vancouver, BC, Canada*
[3]*Institute for Neurorehabilitation Systems, Bernstein Center for Computational Neuroscience, University Medical Center Göttingen, Georg-August University, Göttingen, Germany*
[4]*Pain Clinic, Center for Anesthesiology, Emergency and Intensive Care Medicine, University Hospital Göttingen, Göttingen, Germany*

18.1 THE BEGINNINGS AND PRINCIPLES OF BIOFEEDBACK

The word "biofeedback" was coined in the late 1960s [4] to describe laboratory based, experimental research procedures used to train subjects to alter brain activity, blood pressure, heart rate, and other physiological functions that normally are not under voluntary control. These studies showed that individuals have more control over the so-called "involuntary bodily functions" than once thought possible.

Biofeedback is a technique that measures physiological signals and feeds them (or one of their features) back to the subject in a comprehensible format (usually audio or visual) to help the individual gain awareness and control of physiological processes [21,37,44]. The subject then attempts to alter the feedback signal to modify the physiological response. A biofeedback device can be thought of as a sixth sense which allows the subject to "see" or "hear" physiological functions. Biofeedback can also be described as a "psycho-physiological mirror" providing subjects with a way of monitoring the physiological signals produced by the body and learn from them to self-regulate a targeted pattern of physiologic functioning [42].

Surface Electromyography: Physiology, Engineering, and Applications, First Edition.
Edited by Roberto Merletti and Dario Farina.
© 2016 by The Institute of Electrical and Electronics Engineers, Inc. Published 2016 by John Wiley & Sons, Inc.

Clinical applications of biofeedback include monitoring the electrical activity of muscles (surface electromyography; sEMG), brain, heart, or visceral and vasomotor responses [43]. Regardless of the clinical application, the general procedure of biofeedback has three stages: (1) the measurement of a physiological variable, (2) the conversion of the measurement into some meaningful display or effect, and (3) the feeding back of this information to the subject. The following prerequisites must be met in order to achieve effective biofeedback: (1) It must be possible to identify a measurable response related to the condition under treatment, (2) the response must have variability and a perceptible sensory cue associated with it, (3) the subject must be able to voluntarily control the status of the measured response, (4) the afferent sensory and efferent pathways must be intact, and (5) the patient must be attentive and motivated.

Biofeedback utilizes sensors, generally noninvasive, for detecting mechanical (position or kinematics of human body segments) or electrophysiological (muscle activity, skin temperature, respiration, heart rate, blood pressure, brain electrical activity) signals. Regardless of the type of sensor used, the signals are processed in order to extract the useful information that is displayed back to the subject in a comprehensible and simple form (usually visual or auditory), as indicated in Fig. 18.1a as a concept and in Fig. 18.1b in the case of a sEMG spatial distribution map.

Applied research has demonstrated that biofeedback, alone or in combination with other therapies, is effective for treating a variety of disorders ranging from high blood pressure to tension headache, incontinence, hypertension, and attention disorders.

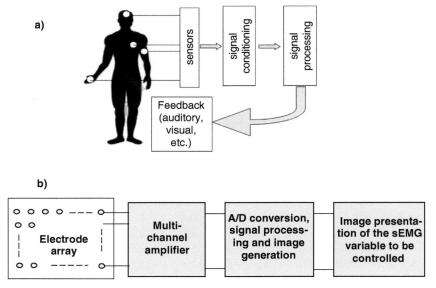

FIGURE 18.1 General flow diagram of a biofeedback system. Biological signals are recorded, using sensors and amplifiers, processed, and transformed into a signal that is fed back to the subject. The audio or visual signal reflects a feature (such as amplitude) or a pattern of the sEMG spatial distribution.

Although it is frequently referred to as a form of treatment, biofeedback should rather be considered as a training technique [19].

This chapter focuses on the most common form of feedback in physical rehabilitation: sEMG biofeedback. The chapter provides an overview of technical considerations and guidelines for the use of sEMG biofeedback and briefly reviews some of its most common clinical applications.

18.2   sEMG BIOFEEDBACK

The principle of sEMG biofeedback is to provide the subject with enhanced information about muscle activity. The most common goals are muscle relaxation, reduction of excessive and/or inappropriate muscle activation, or the learning or relearning of muscle control [10]. The sEMG biofeedback device detects the electrical signals generated by the muscles and translates them into a form that patients can easily understand and try to modify: for example, by activating a light or an audio tone every time the muscle activity exceeds a threshold or by presenting an image of muscle activation that the subject is expected to modify (Fig. 18.1b). The patients can learn to control (or relax) their muscles by trying to modify the system's output (e.g., change an image or slow down the flashing or beeping of an amplitude indicator).

Biofeedback systems are attracting increasing interest from clinicians as more sophisticated equipment is developed that provides precise measurements and attractive displays of physiological signals. Developments such as wireless sensors and electrode sleeves make the task of acquiring data from the individual less intrusive and labor intensive. Presentation of output information in the form of images or as controls of computer games or toys makes the training interesting, challenging or competitive, especially for children. In general, biofeedback is likely to be more effective if it is (i) multimodal, so that more than one sense is involved in the training; (ii) attractive, to facilitate subjects' motivation and attention; and (iii) easy to understand, to be a reliable source of information without overloading the subject [30].

18.3   sEMG-BIOFEEDBACK APPLICATIONS AND CONSIDERATIONS

Many sEMG biofeedback devices and protocols have been described, and several differences and discrepancies between protocols can be observed. The lack of standardization of the application of sEMG biofeedback introduces potential limitations with using these devices. The following sections consider some of the technical considerations when applying and interpreting sEMG biofeedback.

18.3.1   Is Electrode Placement Important?

Almost all sEMG biofeedback devices present the subject with information concerning the sEMG amplitude or muscle activation times. sEMG amplitude and, to some degree, muscle activation or resting time are known to be strongly affected by the location of the electrodes over the muscle and their interelectrode distance (see Chapters 2, 3, and 5). When only two electrodes are used for biofeedback, their size and distance are less critical than in other applications; however, care must be taken to ensure that they are positioned in the same location and at the same distance in different sessions to reduce the variability due to geometric factors (see Chapters 2, 3, and 5). This is particularly relevant when sEMG biofeedback is applied on different days, and subjects are asked to reach (or not to exceed) a certain sEMG amplitude threshold since a slight shift of the electrode position might lead to a large difference in sEMG amplitude with no change in task performance. The importance of electrode position in the estimation of muscle activity, outlined in previous chapters, was highlighted by Holtermann and colleagues [27,28], among many others. In their experiments it was shown that sEMG biofeedback based on the signal detected with pairs of electrodes positioned over different muscle portions may lead to selective activation of regions of the muscle of interest. Hermens and Hutten [26] investigated whether the position of the electrodes on the upper trapezius muscle (both along and across the direction of muscle fibers) has a relevant effect on the estimation of sEMG amplitude and on the estimation of relative rest time (percentage of time the muscle activity is lower than a certain threshold). This work highlighted a large variability of sEMG amplitude with electrode position; which could be partially reduced by normalizing the amplitude values to a reference contraction. Ideally, sEMG biofeedback systems should be designed to allow the sEMG amplitude to be normalized to a reference voluntary contraction to eliminate this limitation. Measurement of the relative rest time proved to be more robust to changes of electrode position; however, the large variability observed for some subjects indicates that this parameter is likely not the most suitable for feedback. The use of electrode grids (Fig. 18.1b) may help to overcome these difficulties at the cost of more sophisticated equipment.

18.3.2   Which sEMG Index Should Be Used?

The most common indices are the sEMG amplitude and the timing of muscle activation (usually expressed as the percentage of time with sEMG amplitude higher than a threshold). sEMG biofeedback can be applied with the aim of training the subject to increase the activation of selected muscles ("up-training") or to relax them ("down-training"). "Up-training" is either adopted when the goal is to train the subject to increase the activation of a target muscle during a task [18,40] or to attempt to exert a higher level of force [49]. "Down-training" is used to reduce the contraction level of overactive muscles due to, for example, incorrect postures [41], spasticity [17], or excessive co-contraction. The effect of "up-training" versus "down-training" has been evaluated in some situations. For instance, Leveau and Rogers [34] investigated which method was more effective to enhance the activity of the vastus medialis relative to the vastus lateralis muscle in healthy volunteers. This study showed that up-training of the vastus medialis was more effective than down-training of vastus lateralis, although other factors might have contributed to these results. The effectiveness of up- versus down-training has also been evaluated in the field of reflex conditioning. Thompson et al. [45] showed that subjects could learn to increase or decrease the intensity of the H-reflex during quiet standing, although "down-training" took longer than "up-training" (5/6 versus 2 sessions).

EMG biofeedback can also be used to inform the subject on the timing of their muscle activity [1] and to promote the occurrence of rest periods in the case where a muscle is continuously active [25]. When muscle relaxation is the goal, an sEMG system with low noise detection and high rejection of power-line interference is required. Chapter 3 focuses on how to achieve these specifications. In many cases, notch filters are included in the conditioning circuits to remove power-line interference. Although these filters are, in general and for other applications, not advisable since they alter the sEMG amplitude and spectrum, many sEMG biofeedback devices adopt this simple solution to reduce interference.

18.3.3 How Should the Information Be Fed Back?

The sEMG biofeedback device can provide the user with different forms of feedback; the most common are audio, visual, vibration, or combined audio/visual. Each modality has advantages and disadvantages: For instance, visual feedback might allow the subjects to view feedback from multiple sources, such as different muscles [34] or different muscle regions [27,28] (Fig. 18.2); however, this form of feedback is problematic if the subject is performing a task which requires them to visually control their actions. Audio feedback can be used during functional activities and in occupational tasks, but it implies sounds that may not be welcome in

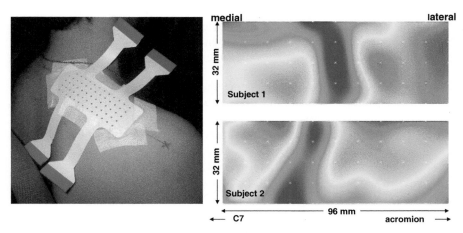

FIGURE 18.2 A 64-electrode array is applied over the upper trapezius muscle. An interpolated map of the sEMG amplitude (average rectified value, ARV, or the root mean square value, RMS) is provided as feedback to the subject. Two maps from two subjects activating the trapezius muscle differently are depicted, showing different activation modalities. The subject could try to activate the muscle in a different way by using the image as a form of biofeedback and try to change the color distribution in the displayed image, therefore learning how to do so.

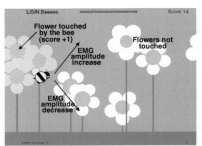 

a) EMG-driven bee touching flowers sliding on the screen.   **b)** EMG-controlled cars on Playmobil tracks.

FIGURE 18.3   Example of two dynamic biofeedback games. **(a)** The sEMG signal is used to control a video game where the subject must modulate muscle contraction to guide the bee to touch the flowers or avoid a danger while the background moves to the left. **(b)** The amplitude of an sEMG signal controls the speed of a slot car. A competition game increases the motivation of the subjects. A physical game, as opposed to a virtual computer game, and the control of a physical object (such as keeping the slot car on the track) may have an impact on the acceptability of the system. (Courtesy of Compagnia di San Paolo, Torino, Italy.)

workplaces. Vibration is an alternative modality which could be used when the feedback system has to be used in every-day situations [6,25], since it is discrete and the vibrators can be hidden under clothing.

Some studies have compared the effectiveness of different feedback modalities for inducing changes in muscle activation. For example, Alexander et al. [2] showed that the activity of the frontalis muscle could be more effectively reduced when subjects were provided with audio rather than visual feedback. However, other studies found no significant differences between these two modalities [7,20,36,47]. Since no feedback modality appears to be superior to another, the best modality can be selected according to the task performed and the subject's preferences. Providing attractive forms of feedback may facilitate learning [30]. Recent developments propose providing images of the sEMG amplitude distribution over a muscle (see Fig. 18.2) or using sEMG signals to control a sound generator, a musical instrument, a game, or a toy (see Fig. 18.3). This motivates the users, especially children.

18.3.4   How Often and When Is It Appropriate to Give Feedback?

A number of studies have investigated the effectiveness of different sEMG biofeedback protocols. For example, studies have compared the difference between providing a subject with "concurrent" feedback as they perform a task versus "terminal" feedback which is given after each trial or after a number of trials. Some studies suggest that concurrent and terminal sEMG biofeedback have comparable effects, for example, for inducing upper trapezius muscle relaxation during a gross motor task [46] or for relaxation of laryngeal muscles during phonation [53].

The rate at which feedback is provided may also be relevant. In one example, subjects were provided with feedback when their upper trapezius muscle was continuously active for more than 20% of a preset time interval. Three epoch lengths were investigated (5, 10, 20 s) and the greatest decrease of both average activity and relative rest time of the upper trapezius muscle were observed when feedback was given every 10 s [48]. Cohen et al. [8] compared continuous and intermittent feedback schedules to facilitate an increase in forearm muscle activity. When tested during a simple task (repetitively squeeze a ball), subjects who received continuous feedback of forearm muscle activity showed greater sEMG amplitude, yet the intermittent feedback schedule led to a greater improvement of the response rate (i.e., the correct pattern of muscle contraction/relaxation). Furthermore, the intermittent schedule was the most resistant to the phenomenon of extinction, which is the progressive loss of the acquired ability after the end of feedback training.

In practice, one may consider to adjust the feedback schedule over time—that is, progressively reduce the amount of feedback given to the subjects. Fading feedback might promote learning by asking the subjects to rely less on feedback once they are able to correctly perform the requested task [32]. Furthermore, the amplitude threshold could be changed during or between training sessions to facilitate the patient to reach progressively higher [16,34,45] or lower [35,45] amplitude values to facilitate improvement [35].

18.4   sEMG BIOFEEDBACK: CLINICAL APPLICATIONS

The following sections summarize the use of sEMG biofeedback for various clinical conditions. Studies reviewing the effectiveness of sEMG biofeedback on disability, function, and pain are included to inform the reader about the utility of this form of intervention.

18.4.1   Stroke

Woodford and Price [51] reviewed whether sEMG biofeedback provides additional benefit when used with standard physiotherapy for the recovery of motor function in stroke patients. Some of the reviewed studies suggested that sEMG biofeedback plus standard physiotherapy produced improvements in motor power, functional recovery, and gait quality when compared to standard physiotherapy alone. However, it was noted that further research is needed since the trials in this area were small and generally not well-designed. The use of sEMG biofeedback for the restoration of functional deficits associated to stroke were also considered by Nelson [38], who reviewed whether sEMG biofeedback was effective for improving movement, dysphagia, and urinary incontinence. Overall it was concluded that sEMG bio-feedback could have a positive impact on stroke rehabilitation, especially for re-training lower limb function. Examples of sEMG biofeedback application for stroke rehabilitation include providing patients with visual feedback of gastrocnemius lateralis and tibialis anterior muscle activity while walking on a treadmill. Twelve training sessions were shown to significantly improve gait function in a group of stroke survivors with mild impairment [1]. Furthermore, Jonsdottir et al. [32] showed that auditory feedback from the lateral gastrocnemius combined with a motor learning approach effectively changed gait variables (peak ankle power at push-off, gait velocity, and stride length) in chronic stroke patients compared to conventional physiotherapy. In this study, patients were asked to increase the amplitude of the sEMG signal during the push-off phase. The feedback schedule and gait characteristics were adjusted according to motor learning paradigms: Patients were trained while walking (task-oriented practice), the amount of feedback given to the subjects decreased across the treatment sessions (fading feedback), and gait parameters were changed within the sessions (variable practice).

EMG biofeedback is also considered as a promising rehabilitation approach for upper limb recovery following stroke [33]. Dogan-Aslan et al. [17] showed that the use of sEMG biofeedback to reduce the activity/spasticity of the wrist flexor muscles led to improved upper limb function after 3 weeks of intervention compared to a conventional intervention. Other sEMG biofeedback applications have been described for patients following stroke including the use of sEMG biofeedback on swallowing for the management of dysphagia [11].

18.4.2 Cerebral Palsy

Several studies have evaluated the role of feedback in the treatment of people with cerebral palsy. Yoo et al. [54] recently described an innovative sEMG-virtual reality biofeedback system which was designed to provide accurate information on triceps and biceps activity during reaching. The use of this combined feedback system led to improved muscle activation in the underactive triceps and decreased activity of the hypertonic biceps in children with cerebral palsy. Dursun et al. [18] investigated whether sEMG biofeedback combined with exercises was more effective than exercises alone on gait rehabilitation. In their study, sEMG biofeedback was provided as visual/auditory cues encouraging the children to increase the activity of their ankle dorsiflexors and to reduce the activity of the triceps surae during selected exercises. The addition of sEMG biofeedback to the exercise program resulted in significant improvements in muscle tone, ankle dorsiflexion range of motion, gait velocity, and stride length—improvements that were maintained at a 3-month follow-up. As a further example, in a study by Bloom et al. [6], sEMG biofeedback was provided to children with cerebral palsy in the form of vibration as the children wore the system for 5–6 hours a day for one month. The goal of the treatment was selected for each patient, but mostly involved increasing the activity of wrist/finger extensors and decreasing of the hyperactivity of finger flexors. The results supported the efficacy of long-term sEMG biofeedback for enhancing functional capacity.

18.4.3 Spinal Cord Injury

EMG biofeedback can be used to facilitate the activation of selected muscles in patients with incomplete spinal cord injury. For instance, Govil and Noohu [22] showed that the addition of sEMG biofeedback when given specifically over gluteus maximus resulted in improvement of sEMG amplitude and various gait parameters (walking velocity, cadence) compared to traditional rehabilitation. Likewise, treatment involving sEMG biofeedback significantly increased voluntary sEMG responses from the rectus femoris muscle in individuals with spinal cord injuries [16]. Moreover, sEMG biofeedback has also been shown to be useful for down-training the excitability of hyperactive spinal stretch reflexes patients with spinal cord injury [50].

18.4.4   Knee Conditions

A recent systematic review [49] identified positive effects of sEMG biofeedback for quadriceps femoris muscle training on patient-oriented outcomes (function and pain) for postsurgical rehabilitation of the knee but not for chronic conditions such as anterior knee pain syndrome or knee osteoarthritis. However, the authors of this review also concluded that there were a limited number of studies, with relatively low quality and large heterogeneity of results. Further randomized controlled trials in this area are warranted.

The effectiveness of sEMG biofeedback after knee surgery was also investigated by Akkaya et al. [3]. In this study, 45 patients that had undergone arthroscopic surgery for partial meniscectomy were randomized into three training groups; all three groups followed the same home exercise program; one group received sEMG biofeedback training in addition to home exercise training, and one group received an electrical stimulation program in addition to training. The most relevant observation was the difference in the average time that patients needed a walking aid: on average, the exercise-only group used aids for 8.3 ± 8.0 days, compared to 4.5 ± 1.5 for the electrical stimulation group and to 1.5 ± 2.5 for the sEMG biofeedback group.

Although the benefit of sEMG biofeedback training for relief of pain in patients with anterior knee pain is questionable [49], the use of biofeedback may facilitate improved coordination between the vastus medialis obliquus and vastus lateralis muscles. For instance, Ng and colleagues [40] showed that the incorporation of sEMG biofeedback into a physiotherapy exercise program facilitated the activation of vastus medialis obliquus muscle relative to the vastus lateralis such that the muscle could be preferentially recruited during daily activities. Several other studies support the benefit of including sEMG biofeedback during exercise for restoring the balance between vastus medialis obliquus and vastus lateralis muscle activation [9,15,31,34,52].

18.4.5   Low Back and Neck Pain

According to a Cochrane review by Henschke et al. [23], there is low evidence that sEMG biofeedback is more effective than placebo or control interventions for relief of low back pain or for improved function. These results were obtained from three studies only, the most recent dated 1996. There is a surprising lack of recent studies investigating the effect of sEMG biofeedback interventions in low back pain.

In contrast, numerous studies have evaluated the effectiveness of sEMG biofeedback in patients with work-related neck pain. Hermens and Hutten [25] described a system in which (i) surface sEMG signals were collected from the upper trapezius bilaterally; (ii) feedback was given to the subject if the muscle was continuously active for a time period longer than a defined value; (iii) the subject experiences the feedback as vibration; and (iv) the device is wearable, so it could be used during every day activities. Using this system four weeks resulted in reduced pain/discomfort with a further reduction of symptoms noted one month after training [25]. In addition, an increase of the relative muscular rest time (i.e., percentage of the time with no activity of the upper trapezius) was identified.

Holtermann et al. [29] reported that five weekly sessions using sEMG biofeedback on upper trapezius muscle activity were sufficient to induce a change in upper trapezius muscle activity (reduced activity, increased frequency of short and long gaps of activity, and increased relative rest time) in a group of computer workers. Changes in muscle activity were observed bilaterally, even if the feedback was given unilaterally. A further randomized controlled trial compared the effectiveness of sEMG biofeedback training to active exercise, passive exercise, and no intervention [35]. In this study, sEMG biofeedback was given for 2 hours a day, twice a week, for 6 weeks. The patients were provided with audio-feedback of their upper trapezius muscle activity and were encouraged to keep the sEMG amplitude below a set threshold, which was changed weekly. The results confirmed that sEMG biofeedback training was more effective than the other interventions for reducing pain and disability, and these improvements were maintained at a six-month follow-up.

18.4.6 Urinary Incontinence

A Cochrane review by Herderschee et al. [24] stated that biofeedback may provide benefit in addition to pelvic floor muscle training in women with urinary incontinence. An example is a study by Dannecker et al. [14] who investigated 430 women, incontinent for an average of 6.7 years. This study concluded that sEMG biofeedback-assisted pelvic floor muscle training is an effective therapy of stress urinary or mixed incontinence and that improvements are maintained at follow-up (3 years on average). Other applications of sEMG to the pelvic floor muscles are discussed in Chapter 14.

18.4.7 Temporomandibular Disorders and Headache

A meta-analysis of sEMG biofeedback treatment of temporomandibular disorders confirmed that sEMG biofeedback treatments were superior to no treatment or psychologic placebo controls [12]. In a further review it was concluded that sEMG biofeedback training of the masticatory muscles combined with adjunctive cognitive-behavioral therapy is an efficacious treatment for temporomandibular disorders [13].

Medium-to-large mean effect sizes for biofeedback in adult migraine and tension-type headache patients have been reported [39]. Significant effects have been observed for headache frequency, perceived self-efficacy, symptoms of anxiety and depression, and medication consumption. Furthermore, reduced muscle tension in pain related areas was observed following sEMG feedback for tension-type headache. Treatment effects remained stable over an average follow-up period of 14 months. Nestoriuc et al. [39] also recently reviewed the efficacy of sEMG biofeedback training for headache. Strong evidence of efficacy and specificity of sEMG biofeedback training as opposed to no treatment and relaxation therapy was reported for tension-type headache. In addition, sEMG biofeedback was shown to be more effective than "wait and see" for migraine headache.

18.5 FUTURE PERSPECTIVES

EMG biofeedback initiated as an invasive technique, usable only in laboratory settings and capable of only providing information about the firing frequency of few motor units [4]. In the following decades, the development of surface electrodes made this technique available for clinical applications, providing individuals with the means to sense and control their gross motor activity during a variety of tasks [5,10,30]. Recent technological advances and discoveries in the field of motor control and learning prompted new solutions for sEMG biofeedback applications, possibly leading to more effective training. These advances include (i) the system used to collect the sEMG signals and to feed the information back to the subject and (ii) the type of exercise schedule which should be used for training.

It was recently suggested that the failure of some sEMG biofeedback training protocols might be related to the fact that information about muscle activity was provided to the subjects as the length of bars or the frequency of beeps, which makes the training uninteresting and scarcely appealing for patients [30]. Thus, among the currently available devices for sEMG biofeedback training, those based on high-density surface sEMG techniques may provide benefits with respect to traditional, bipolar techniques for how the sEMG signal is collected and displayed. The images generated with high-density surface sEMG biofeedback may provide more intuitive feedback possibly improving patients' compliance, interest, and motivation with training (Fig. 18.4). However, the applicability of high-density sEMG as a feedback technique is currently limited by the time needed to set up the detection system. Future developments in this area, such as wearable grids of electrodes and pre-gelled matrices, may facilitate the application. Furthermore, high-density sEMG systems are not usually portable, which implies that the patient cannot take the system home for home-training. Since high-density sEMG relies on a large number of channels, a further disadvantage is that sensory modalities such as sound or vibration are more difficult to use for feeding the information back to the subject; this is a major disadvantage when sEMG biofeedback is used during task-oriented exercises or functional activities which require the patient to observe what they are doing rather than the system display. A possible solution would be to consider a single index that is created based on the information of all channels; alternatively, the relevant information to be transmitted via audio or vibration might be obtained by algorithms that automatically find which channel or which pattern of the grid best represents muscle activity. Training studies investigating whether high-density sEMG biofeedback provides an advantage over sEMG signals collected with traditional bipolar electrodes are under way. Further research is also required to determine the best tasks to perform and the best sEMG biofeedback protocol to apply for a specific clinical condition.

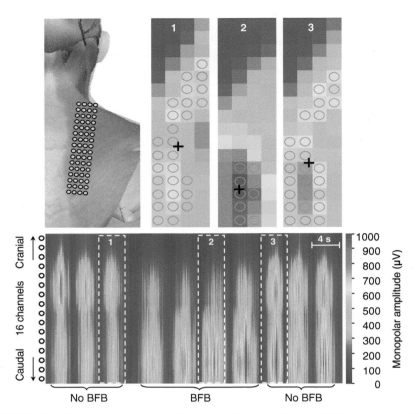

FIGURE 18.4   Top panel: Example of training with a high-density sEMG system. A grid of 64 electrodes (16 rows and 4 columns, with the major axis oriented along the cranial—caudal direction, with electrodes spaced 10 mm apart) was placed on the right trapezius of a healthy subject. The task performed was dynamic shoulder abduction while holding a 1-kg dumbbell. **Bottom panel:** Photograph showing the whole exercise (*X* axis: time, *Y* axis: space along the array, Color scale: EMG monopolar amplitude averaged along the rows). Ten repetitions can easily be identified from the plot. The feedback, delivered as an EMG amplitude image, was shown to the subject only for the movements 4 to 8. In these movements, the subject focused to selectively activate the portion of the muscle situated underneath the caudal portion of the grid. Amplitude distribution is clearly different in these conditions. The maps in the top-right panel show the noninterpolated images corresponding to trials 1, 2, and 3 (bottom panel); the cross is indicating the centroid of each map.

18.6   CONCLUSION

Like all other sEMG-based procedures, sEMG biofeedback must be applied with full awareness of the potential limitations in the detection and interpretation of sEMG signals. Most limitations (choice of electrode size, location, and geometry, see Chapters 2, 3, and 5 for further details) can be overcome with adequate knowledge of sEMG application. Furthermore, other factors, such as the exercise schedule, the characteristic of the patient, and how and when the feedback is provided, are likely to play a major role in the success of sEMG biofeedback training. When applied correctly, sEMG biofeedback can be an effective supplement for the management of numerous clinical conditions.

REFERENCES

1. Aiello, E., D. H. Gates, B. L. Patritti, K. D. Cairns, M. Meister, E. A. Clancy, and P. Bonato,"Visual EMG biofeedback to improve ankle function in hemiparetic gait," in *27th Annual Conference of IEEE Engineering in Medicine and Biology*, pp. 7703–7706, 2005.

2. Alexander, A. B., C. A. French, and N. J. Goodman, "A comparison of auditory and visual feedback in biofeedback assisted muscular relaxation training," *Psychophysiology* **12**, 119–123 (1975).

3. Akkaya, N., F. Ardic, M. Ozgen, S. Akkaya, F. Sahin, and A. Kilic, "Efficacy of electromyographic biofeedback and electrical stimulation following arthroscopic partial meniscectomy: a randomized controlled trial," *Clin. Rehabil.* **26**, 224–236 (2012).

4. Basmajian, J. V., "Control and training of individual motor units," *Science* **141**, 440–441 (1963).

5. Basmajian, J. V., "Biofeedback in rehabilitation: a review of principles and practices," *Arch. Phys. Med. Rehabil.* **62**, 469–475 (1981).

6. Bloom, R., A. Przekop, and T. D. Sanger, "Prolonged electromyogram biofeedback improves upper extremity function in children with cerebral palsy," *J. Child. Neurol.* **25**, 1480–1484 (2010).

7. Chen, W., "Comparison of sensory modes of biofeedback in relaxation training of frontalis muscle," *Perc. Mot. Skills* **53**, 875–80 (1981).

8. Cohen, S. L., J. Richardson, J. Klebez, S. Febbo, and D. Tucker, "EMG biofeedback: the effects of CRF, FR, VR, FI, and VI schedules of reinforcement on the acquisition and extinction of increases in forearm muscle tension," *Appl. Psychophysiol. Biofeedback* **26**, 179–194 (2001).

9. Cowan, S., K. Bennell, K. Crossley, P. W. Hodges, and J. McConnell, "Physical therapy alters recruitment of the vasti in patellofemoral pain syndrome," *Med. Sci. Sports Exerc.* **34**, 1879–85 (2002).

10. Cram, J. R., "Biofeedback applications," in R. Merletti and P. Parker, ed., *Electromyography: Physiology, Engineering And Noninvasive Applications*, IEEE Press and Wiley Interscience, Hoboken, NJ, 2004.

11. Crary, M. A., G. D. C. Mann, M. E. Groher, and E. Helseth, "Functional benefits of dysphagia therapy using adjunctive sEMG biofeedback," *Disphagia* **164**, 160–164 (2004).

12. Crider, A. B., and A. G. Laros, "A meta-analysis of EMG biofeedback treatment of temporomandibular disorders," *J. Orofac. Pain* **13**, 29–37 (1999).

13. Crider, A. B., A. G. Glaros, and R. N. Gevirtz, "Efficacy of biofeedback-based treatments for temporomandibular disorders," *Appl. Psychophysiol. Biofeedback* **30**, 333–345 (2005).

14. Dannecker, C., V. Wolf, R. Raab, H. Hepp, and C. Anthuber, "EMG-biofeedback assisted pelvic floor muscle training is an effective therapy of stress urinary or mixed incontinence: a 7-year experience with 390 patients," *Arch. Gynecol. Obstet.* **273**, 93–97 (2005).

15. Davlin, C. D., W. R. Holcomb, and M. A. Guadagnoli, "The effect of hip position and electromyographic biofeedback training on the vastus medialis oblique: vastus lateralis ratio," *J. Athl. Train.* **34**, 342–346 (1999).

16. De Biase, M. E. M., F. Politti, E. T. Palomari, T. E. P. Barros-Filho, and O. P. De Camargo, "Increased EMG response following electromyographic biofeedback treatment of rectus femoris muscle after spinal cord injury," *Physiotherapy* **97**, 175–179 (2011).

17. Dogan-Aslan, M., G. F. Nakipoglu-Yugan, A. Dogan, I. Karabay, and N. Ozgirgin, "The effect of electromyographic biofeedback treatment in improving upper extremity functioning of patients with hemiplegic stroke," *J. Stroke Cerebrovasc. Dis.* **21**, 187–192 (2012).

18. Dursun, E., N. Dursun, and D. Alican, "Effects of biofeedback treatment on gait in children with cerebral palsy," *Dis. Rehabil.* **26**, 116–120 (2004).

19. Frank, D. L., L. Khorshid, J. F. Kiffer, C. S. Moravec, and M. G. McKee, "Biofeedback in medicine: who, when, why and how?" *Ment. Health Fam. Med.* **7**, 85–91 (2010).

20. Gaudette, M., A. Prins, and J. Kahane, "Comparison of auditory and visual feedback for EMG training," *Percept. Mot. Skills* **56**, 383–386 (1983).

21. Gilbert, C. and Moss, D,. "Biofeedback and biological monitoring," in *Handbook of Mind–Body Medicine for Primary Care*, Sage Publications, Thousand Oaks, CA, p. 110, 2003.

22. Govil, K., and M. M. Noohu, "Effect of EMG biofeedback training of gluteus maximus muscle on gait parameters in incomplete spinal cord injury," *NeuroRehabilitation.* **33**, 147–152 (2013).

23. Henschke, N., R. W. J. G. Ostelo, M. W. Van Tulder, J. W. S. Vlaeyen, S. Morley, W. J. J. Assendelft, and C. J. Main, "Behavioural treatment for chronic low-back pain," *Cochrane Database Syst. Rev.* **20**, (2010).

24. Herderschee, R., E. J. C. Hay-Smith, G. P. Herbison, J. P. Roovers, and M. J. Heineman, "Feedback or biofeedback to augment pelvic floor muscle training for urinary incontinence in women," *Cochrane Database Syst. Rev.* **7**, (2011).

25. Hermens, H. J., and M. M. R. Hutten, "Muscle activation in chronic pain: its treatment using a new approach of myofeedback," *Int. J. Ind. Ergon.* **30**, 325–336 (2002).

26. Hermens, H. J., and M. M. R. Vollenbroek-Hutten, "Effects of electrode dislocation on electromyographic activity and relative rest time: effectiveness of compensation by a normalisation procedure," *Med. Biol. Eng. Comp.* **42**, 502–508 (2004).

27. Holtermann, A., P. J. Mork, L. L. Andersen, H. B. Olsen, and K. Søgaard, "The use of EMG biofeedback for learning of selective activation of intra-muscular parts within the serratus anterior muscle: a novel approach for rehabilitation of scapular muscle imbalance," *J. Electromyogr. Kinesiol.* **20**, 359–365 (2010).

28. Holtermann, A., K. Roeleveld, P. J. Mork, C. Gronlund, J. S. Karlsson, L. L. Andersen, H. B. Olsen, M. K. Zebis, J. Sjogaard, and K. Sogaard, "Selective activation of neuromuscular compartments within the human trapezius muscle," *J. Electromyogr. Kinesiol.* **19**, 896–902 (2009).

29. Holtermann, A., K. Søgaard, H. Christensen, B. Dahl, and K. Blangsted K, "The influence of biofeedback training on trapezius activity and rest during occupational computer work: a randomized controlled trial," *Eur. J. Appl. Physiol.* **104**, 983–989 (2008).

30. Huang, H., S. L. Wolf, and J. He, "Recent developments in biofeedback for neuromotor rehabilitation," *J. Neuroeng. Rehab.* **12**, 1–12 (2006).

31. Ingersoll, C. D., and K. L. Knight, "Patellar location changes following EMG biofeedback or progressive resistive exercises," *Med. Sci. Exerc. Sport Sci.* **23**, 1122–1127 (1991).

32. Jonsdottir, J., D. Cattaneo, M. Recalcati, A. Regola, M. Rabuffetti, M. Ferrarin, and A. Casiraghi, "Task-oriented biofeedback to improve gait in individuals with chronic stroke: motor learning approach," *Neurorehabil. Neural Repair* **24**, 478–485 (2010).

33. Langhorne, P., F. Coupar, and A. Pollock, "Motor recovery after stroke: a systematic review," *Lancet Neurol.* **8**, 741–754 (2009).

34. Leveau, B. F., and C. Rogers, "Selective training of the vastus medialis muscle using EMG biofeedback," *Phys. Ther.* **60**, 1410–1415 (1980).

35. Ma, C., G. P. Szeto, T. Yan, S. Wu, and C. Lin, "Comparing biofeedback with active exercise and passive treatment for the management of work-related neck and shoulder pain: a randomized controlled trial," *Arch. Phys. Med. Rehabil.* **92**, 849–859 (2011).

36. Madeleine, P,. P. Vedsted, A. K. Blangsted, G. Sjogaard, and K. Sogaard, "Effects of electromyographic and mechanomyographic biofeedback on upper trapezius muscle activity during standardized computer work," *Ergonomics* **49**, 921–933 (2006).

37. Moss, D., "Biofeedback," in *Handbook of Complementary and Alternative Therapies in Mental Health*, S. Shannon, ed., Academic Press, San Diego, CA, 2001.

38. Nelson, L. A., "The role of biofeedback in stroke rehabilitation: past and future directions," *Top. Stroke Rehabil.* **14**, 59–66 (2007).

39. Nestoriuc, Y., A. Martin, W. Rief, and F. Andrasik, "Biofeedback treatment for headache disorders: a comprehensive efficacy review," *Appl. Psychophysiol. Biofeedback* **33**, 125–140 (2008).

40. Ng, G. Y. F., A. Q. Zhang, and C. K. Li, "Biofeedback exercise improved the EMG activity ratio of the medial and lateral vasti muscles in subjects with patellofemoral pain syndrome," *J. Electromyogr. Kinesiol.* **18**, 128–133 (2008).

41. Park, S., and W. Yoo, "Effect of EMG-based feedback on posture correction during computer operation," *J. Occup. Health* **54**, 271–277 (2012).

42. Peper, E., R. Harvey, and N. Takabayashi, "Biofeedback: an evidence based approach in clinical practice," *Jap. J. Biofeedback Res.* **36**, 3–10 (2009).

43. Robinson, A. J., *Clinical Electrophysiology: Electrotherapy and Electrophysiologic Testing*, Lippincott Williams & Wilkins, New York, 2007.

44. Schwartz, M. S., and F. Andrasik, *Biofeedback: A Practitioner's Guide*, 3rd ed., Lippincott Williams & Wilkins, New York, 2003.

45. Thompson, A. K., X. Y. Chen, and J. R. Wolpaw, Acquisition of a simple motor skill: task-dependent adaptation plus long-term change in the human soleus H-reflex, *J. Neurosci.* **29**, 5784–5792 (2009).

46. Van Dijk, H., G. E. Voerman, and H. J. Hermens, "The influence of stress and energy level on learning muscle relaxation during gross-motor task performance using electromyographic feedback," *Appl. Psychophysiol. Biofeedback* **31**, 243–252 (2006).

47. Vedsted, P., K. Søgaard, A. K. Blangsted, P. Madeleine, and G. Sjøgaard, "Biofeedback effectiveness to reduce upper limb muscle activity during computer work is muscle specific and time pressure dependent," *J. Electromyogr. Kinesiol.* **21**, 49–58 (2011).

48. Voerman, G. E., L. Sandsjo, M. M. R. Vollenbroek-Hutten, C. G. M. Groothuis-Oudshoorn, and H. J. Hermens, "The influence of different intermittent myofeedback training schedules on learning relaxation of the trapezius muscle while performing a gross-motor task," *Eur. J. Appl. Physiol.* **93**, 57–64 (2004).

49. Wasielewski, N. J., T. M. Parker, and K. M. Kotsko, "Evaluation of electromyographic biofeedback for the quadriceps femoris: a systematic review," *J. Athl. Train.* **46**, 543–554 (2011).

50. Wolf, S. L., and R. L. Segal, "Conditioning of the spinal stretch reflex: implications for rehabilitation," *Phys. Ther.* **70**, 652–656 (1999).

51. Woodford, H. J., and C. Price, "Electromyographic biofeedback for the recovery of motor function after stroke," *Stroke* **38**, 1999–2000 (2007).

52. Yip, S. L., and G. Y. F. Ng, "Biofeedback supplementation to physiotherapy exercise programme for rehabilitation of patellofemoral pain syndrome: a randomized controlled trial," *Clin. Rehabil.* **20**, 1050–1057 (2006).

53. Yiu, E. M., K. Verdolini, and L. P. Chow, "Electromyographic study of motor learning for a voice production task," *J. Speech, Lang. and Hear. Res.* **48**, 1254–1269 (2005).

54. Yoo, J. W., D. R. Lee, Y. J. Sim, J. H. You, and C. J. Kim, "Effects of innovative virtual reality game and EMG biofeedback on neuromotor control in cerebral palsy," *Biomed. Mater. Eng.* **24**, 3613–3618 (2014).

19

EMG IN EXERCISE PHYSIOLOGY AND SPORTS

A. Rainoldi,[1] T. Moritani,[2] and G. Boccia[3]

[1]*Professor of Physiology, Department of Medical Sciences, Motor Science Research Center, SUISM University of Turin, Turin, Italy*
[2]*Professor of Applied Physiology, Graduate School of Human and Environmental Studies, Kyoto University, Sakyo-ku, Kyoto, Japan*
[3]*Motor Science Research Center, SUISM University of Turin, Turin, Italy; and CeRiSM Research Center "Sport, Mountain, and Health," Rovereto (TN), Italy*

19.1 SURFACE EMG FOR STUDYING MUSCLE COORDINATION

Surface EMG is useful in studying many issues of motor control. The issues of co-activation, onset muscle timing, and characterization of exercise will be analyzed in this section under the viewpoint of muscle coordination (see also Chapter 6).

19.1.1 Methodological Issues in Assessing Muscle Coordination

As highlighted in the review of Hug [80], many limitations of sEMG could constrain the appropriateness of EMG in detecting muscle coordination. The limitations of sEMG are intrinsically related to the technique or to signal processing. Intrinsic limitations are that sEMG can be detected only from superficial muscles, provides information related to only a limited volume of the muscle, is dependent on the electrode configuration, can be altered by crosstalk, and is affected by amplitude cancellation.

Surface Electromyography: Physiology, Engineering, and Applications, First Edition.
Edited by Roberto Merletti and Dario Farina.
© 2016 by The Institute of Electrical and Electronics Engineers, Inc. Published 2016 by John Wiley & Sons, Inc.

Filtering EMG. Regarding signal analysis, one drawback revealed by Hug [80] is the lack of standards in the choice of the low-pass filter to extract the linear envelope of sEMG profile (see Chapter 16). A wide range of low-pass filters have been used in literature, from 3 Hz to 40 Hz [64,190]. Since the choice of the cutoff frequency and of the filter order lead to different EMG pattern, this issue is of considerable importance. A hypothesis is that there should be an "optimal" order and low-pass cutoff frequency for each application (e.g., biofeedback, gait analysis, control of devices).

To increase the signal-to-noise ratio [24], a number of trials or cycles of each task should be averaged. Because of differences in the task durations, time normalization is commonly used to convert time axis in a percentage axis [82,190] (e.g., gait cycle, pedaling cycle, or other tasks).

EMG Normalization. Since comparing sEMG amplitude between different muscles does not provide any information about the degree of activation, normalization of EMG amplitude is recommended. In this case, sEMG activity is expressed as a percentage of a previously recorded reference value, that is, MVC [7,48]. Since muscle activation is dependent on joint angle [156], some studies using MVC normalization methods reported an activity level above 100% in maximal cycling [73] and during a pitch in baseball [89]. To avoid lengthy procedures related to MVC assessment and to face the problem that sEMG amplitude during MVC without training could be 20/40% lower than that obtained after training [75], a number of studies normalized the EMG pattern with respect to the peak obtained in a specific movement [168]. However, this method (named peak dynamic method) does not provide information about the degree of muscle activation. In summary, at this time, there is no agreement for the best normalization procedure [25]; therefore, precise comparison about the degree of muscle activity during a specific task cannot yet be performed.

Assessing Timing Activation. Two additional issues concern the interpretation of coordination regarding the timing of muscle activation. The first is the threshold value chosen to consider a muscle as "active." There are more than ten methods to determine muscle onset: usually a percentage of the peak EMG (for example, 20% of MVC value) or one to five standard deviation above the mean of baseline activity have been used [180]. Another method has been proposed by Merlo et al. [122] and is not based on threshold crossing but on the identification of motor unit action potentials with the use of the continuous wavelet transform. When repetitive or cyclic movements are studied, muscle activation timing is obtained by time normalization and averaging many consecutive cycles. This EMG profile generally depicts the evolution of the amplitude throughout the stride or crank cycle, and its duration is expressed as percentage of total duration of the complete cycle.

Another issue is the so called electromechanical delay, which is the time lag between muscle activation and muscle force production. As reported [148], it ranges from 30 to 100 ms depending on mechanical properties of both tendon and aponeurosis. Hence intersubject timing variability should be taken into account in the interpretation of sEMG onset.

19.1.2   Co-activation

Role of Co-activation.   The simplest pattern of activation among a group of muscles is the relation between agonist and antagonist muscles across a joint. Since the biomechanical definition of agonistic and antagonistic muscle does not necessarily coincide with the corresponding anatomical definitions [158], agonist and antagonist muscles are herein respectively defined as the muscle producing a moment coincident with, or opposite to, the direction of the joint moment [5].

Co-activation is defined as antagonist muscle activity occurring during voluntary agonist contraction [157]. The modulation of co-activation is a strategy used by CNS to achieve opposite objectives: maximal force output and stabilization to ensure joint integrity. Although co-activation may impair the full activation of agonist muscle by reciprocal inhibition [186], it can (i) assist ligaments in maintaining joint stability under heavy loads [10] and (ii) provide breaking mechanisms during high-velocity movements [109]. The effect of co-activation is the stiffening of the joint, when a precision or dangerous motor task is required.

Surface EMG Variables to Assess Co-activation.   Co-activation level could be expressed as a percentage of a reference EMG values recorded in different conditions: during MVC as agonist in isometric contraction; slow concentric contraction like 15°/s [10]; slow eccentric contraction [1]; the maximum value obtained during a movement. It is worth noting that the eccentric normalization is a proper method since any antagonist activation is an eccentric contraction, a condition related to higher torque and lower EMG activation [93]. Gracies [63] proposed a co-contraction index defined as the ratio of the RMS of a muscle when acting as antagonist to the intended effort to the RMS of the same muscle when acting as an agonist to the opposite effort.

Motor Learning and Co-activation.   Co-activation has a protective effect, but could be seen as a counterproductive action because to maximize the torque expressed by a joint it necessary to minimize the activation of the antagonist muscles. As expected, reduced activation of antagonist muscles has been related to the force gains obtained after a strength training period. For instance, in Carolan and Cafarelli [27] a 20% decrease of hamstring co-activation during the early period of training were associated to a 33% increase of net knee extensor torque; that is, the increase in torque exertion depends both on the greater capacity of quadriceps to generate force and on the deactivation of antagonist hamstring muscles.

Reduced level of muscle co-activation is also related to the achievement of motor skill [11,12]. In order to obtain a more economical coordination strategy [16], a progressive inhibition of unnecessary muscular activity (e.g., co-activation) is obtained during the course of specific task training. For instance, lower co-activation was found in isometric or isokinetic movements across the elbow in skilled tennis player [12,13] and across the knee in high jumpers [4].

McGuire et al. [118] referred to maximal isometric contractions as a motor skill, not only for the acquisition of muscle strength but also for the variability of force output. In their study, participants performed three training sessions consisting in 10 rapid maximal contractions of the wrist flexor muscles. Force variability was defined as both the variance of the torque–time curves and the stability of the torque at maximum. The peak-to-peak amplitude of the V-wave was measured to assess changes in neural drive in agonist muscle (see Chapter 12 for details). The primary mechanism to increase maximum strength was a reduction in co-activation because the peak-to-peak amplitude of the V-wave for the agonist remained unchanged with training. The authors argued that during the early stages of training, participants learned to manage "minimally sufficient" levels of co-activation, allowing an increased expression of agonist muscle strength while at the same time providing enough joint stiffness to reduce the variability of force production.

Task Differences.   The role of velocity in co-activation is still a topic of discussion: Some studies found a positive relation between angular velocity and antagonist co-activation [66,93]; others did not find any influence [13,79,149]. Only few works investigated the role of co-activation in ballistic sport specific actions. Sbriccoli et al. [173] compared two actions, a "constrained" (isokinetic) vs a "free" (karate front kick) ones, in two groups of elite and amateur karateka. They showed that elite karateka used a more effective tuning of agonist–antagonist muscles. Elite karateka adopted lower co-activation in isokinetic movements and higher co-activation in the front kick with respect to amateurs. Indeed when the movement is safe, such as that performed using an isokinetic machine, they decrease the antagonist activation to increase the net torque expressed; when the co-activation of flexor muscles provides a stronger braking action, which is essential for a correct technique execution, they increase the antagonist activation.

The level of co-activation across a joint is different, depending on which muscle acts as agonist or antagonist. As shown in Bazzucchi et al. [13], when triceps brachii acts as an antagonist of elbow flexion, it shows an activation that is 16% greater than shown by biceps brachii when it is an antagonist of elbow extension. Since the authors clearly discussed both confounding factors such normalization procedure and relative subcutaneous tissue thickness, it is possible to conclude that the observed differences in co-activation are probably due to the different involvement of those muscles in daily activity.

Age Effect.   The loss of strength in elderly people, beyond the histological and neuromuscular issues at the level of agonist muscles, could be caused by an ineffective tuning of antagonist muscles. Indeed some studies report an increase in co-activation in elderly compared to young people at the level of elbow and knee joint both in men and women [72,87,94,113] and at the level of ankle in postural balance task [144].

19.1.3   Onset Timing

Many other strategies of the neuromuscular control are aimed to protect joints from potential dangerous situation. Critical is the timing of onset muscle activation prior (i.e., preparatory activity) or posterior (i.e., latency) to a sudden mechanical event. In general the variable adopted to address these issues is the time difference between a mechanical event occurring at a joint and the muscle onset aimed to protect the joint. As previously mentioned, the method used to define the sEMG threshold of muscle activation is pivotal to detect the muscle onset.

Preparatory Activity.   The absorption of impacts resulting from the contact with a landing surface during running, jumping, and landing has been widely addressed in the literature. This topic has an important clinical relevance to appropriately plan and control the absorption of impacts which might injury the muscle–skeletal system [172]. The so-called preparatory, or pre-landing, muscle activation is a neuro-muscular strategy occurring before the impact during the downward flight of landing. Lower limb muscles are activated in order to provide the appropriate levels of stiffness of the joints to smoothly absorb the impact of landing. This mechanism is involved partly to counteract latency period due to electromechanical delay [22] and partly to prepare force buildup before toe down in landing [61]. An imbalanced or ineffective neuromuscular recruitment pattern during landing or pivoting maneuvers may lead to increase in injury risk [18]. The preparatory activity has been shown to modulate both EMG amplitude and onset timing on the basis of the landing height and control strategies.

One of the most common injuries in sports is the rupture of the anterior cruciate ligaments (ACL). This injury occurs when high dynamic loading of knee joint are not adequately absorbed by the muscle activation and the passive ligaments structure are subjected to excessive torque [19,110]. In particular, anterior directed shear of the tibia, possibly responsible for ACL injury, should be counteracted not only by the ACL but also by appropriated activation of knee flexor muscles [47]. The timing of non-contact ACL injury ranges between 17 to 50 ms after initial ground contact [105]; within this time lapse, feedback and reactive motor mechanism are too slow to occur. Thus, substantial neural pre-activation of knee flexor muscles just before ground contact seems to be essential [191] during fast dangerous movements such as landing and side cutting. This is consistent with the findings that a reduced pre-activation of knee flexor muscles associated with an elevated activation of knee extensors (which can lead to an augmented anterior shear of the tibia) are two main risk factors in ACL injuries [191].

Latency or Delayed Onset.   Besides preparatory activities, the stretch reflex activation could also be seen as a protective joint mechanism. The short latency component of stretch reflex is a mechanism responsible for rapid muscle recruitment elicited by mechanical muscle stretch. The stretch reflex of muscles involved in ankle control could represent a neuromuscular protection mechanism to avoid ankle sprain. This is an important issue since ankle sprain is one of the most common injuries in athlete lower limbs.

The reaction time of evertor muscles to simulated ankle sprain has been advocated to characterize sensorimotor deficit in the so-called "functional ankle instability." The reaction time has been measured as the time lapse between platform tilt and sEMG

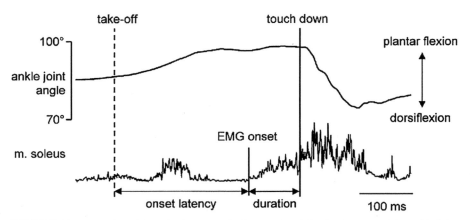

FIGURE 19.1   EMG activity of soleus muscle (**bottom**) and the time course of the ankle joint angle (**top**) during a drop jump from 0.2-m height. Reproduced from Santello [172] with permission.

onset of muscles involved in ankle control, such as peroneus longus, peroneus brevis, tibialis anterior, and extensor digitorum (see Fig. 19.1). Several tilt platforms were used to simulate ankle sprain while remaining within an injury-free range of motion. Some had only one tilt plate, allowing one predetermined limb tilted, and other have two movable plates so that the subject is unaware which side will tilt. Most platforms expose the limb only to supination while others combining plantar flexion and supination, reflecting better the mechanics of joint injury [103]. Using these methods several authors have reported imbalances showing 10–20 ms of delay in the onset latency in subject with functional ankle instability [91,102,103,128]. Thus measuring evertors reaction time in sudden inversion movements has been proposed as a reliable procedure to assess sensorimotor imbalances (Fig. 19.2) [15]. However, this issue is still under debate since recent meta-analysis did not find unequivocal impairment in subjects with functional ankle instability [142]. Considering the neuromuscular basis of imbalanced muscle latency, it could be assumed that training program can influence the onset timing of muscles involved in ankle sprain. Balance and strength training has been adopted to prevent ankle sprain, but contradictory results on muscle latency were found. While some authors reported neuromuscular training as an effective program to reduce muscle latency [175], others did not [45].

19.2   USE OF sEMG TO CHARACTERIZE TRAINING EXERCISE

sEMG could be used as a useful tool to (a) precisely characterize strength and rehabilitation exercise, (b) know the reciprocal activation of each muscle involved in an exercise, and (c) determine which exercise could be suitable for a specific training aim. Moreover, sEMG could help sport scientists and practitioners to choose the most effective configuration of physical exercises to a specific target. Indeed each exercise could be performed in various forms [182] by modifying the device or parameters

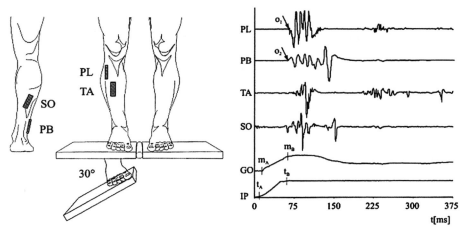

FIGURE 19.2 **Left panel:** Schematic view of the tilting platform. The subject had to stand on the tilting part with more than 95% of body weight on it. When the EMG showed baseline activity, the trapdoor was released (PL, peroneus longus; PB, peroneus brevis; TA, tibialis anterior; SO, soleus). **Right panel:** Representative graph of a single measurement. It shows that rear foot inversion lasts longer than tilting movement. Whether inversion is limited actively by peroneal contraction alone or by passive ligamentous complex cannot be distinguished (in this study, the inversion angle (GO) was not analyzed. Reproduced from Benesch et al. [15] with permission.

such as overload, body position, range of motion of the movement, speed, and grip. Each form corresponds to a different type of training stress and to a different physiological adaptation.

19.2.1 Preferential Activation

The hypothesis that one synergic muscle could be preferentially activated with respect to another one is an interesting topic of discussion. In particular, the relative contribution of synergistic muscle heads belonging to the same muscle group has been widely addressed (see Chapter 5 and the issue of load sharing). Muscle groups such as quadriceps, gastrocnemius, triceps brachii, and biceps femoris has been investigated. For instance, a debate exists around whether in the quadriceps muscle the vastus medialis obliquus (VMO) can be preferentially activated with respect to vastus lateralis (VL) muscle during knee extension exercises.

An imbalance in the activation of VMO and VL may be an etiological factor in the development of patella–femoral pain syndrome [34,116], and VMO atrophy contributes to patella instability [62,150]. To reduce such an imbalance, VMO has been proposed to be strengthened through exercises involving VMO more than other quadriceps muscle [62], and sEMG has been used to verify the relative activation of VMO and VL [178]. To perform this analysis, sEMG amplitude of each muscle head has been assessed during MVC of knee extension and used as reference for successive analysis. The VMO/VL sEMG amplitude ratio during MVC has been taken as reference. Then, each successive exercise is characterized by its own VMO/VL ratio, where the larger the ratio the greater the activation of VMO.

As sometimes occurs, practice habits are not based on solid methodological studies. Contrary to the widely held belief that training quadriceps at 30° knee flexion mostly activates VMO, recent work of Spairani et al. [179] failed to find a difference in VMO/VL activation by changing the knee angle from 30° to 90°. Moreover, some authors suggested that tibial external rotation [189] and hip adduction [76] enhanced VMO activity while performing a knee extension task. Other exercise alterations such as foot pronation or supination, ankle dorsiflexion, or plantar-flexion [108] have been used in the hope to increase VMO intervention. Irish et al. [85] showed that closed kinetic chain exercise (such as squat and lunge) implied a greater VMO/VL ratio than open kinetic chain exercises (such as pure knee extension). However, in the review of Smith et al. [176], only three studies (with some methodological limitations) out of 20 referred preferential VMO activation by altering lower limb joint orientation or co-activation. Because of these findings, in presence of patella-femoral disorder, it has been recommended to focus on general quadriceps strengthening opposite to specific VMO training.

19.2.2 Multi-articular Muscles

SEMG has been widely used to study muscle coordination during locomotion such as walking, running, and cycling. SEMG is often used to highlight the role of each muscle along the locomotion cycles. Typically up to 12 muscles of hip and inferior limbs are sampled. The pattern of muscle activation during locomotion can be analyzed in terms of activity level and/or activation timing, [88] as earlier discussed.

Pedaling is a standardized movement and could be a useful framework to extract information about the role of mono-articular and bi-articular muscles. Ericson [50] showed that 120-W cycling workload induces higher EMG activity in mono-articular than in bi-articular muscles. Mono-articular muscles such as VL and soleus showed 44% and 32% amplitude of their respective sEMG at MVC, while bi-articular rectus femoris and gastrocnemius lateralis showed 22% and 18%, respectively. In addition, mono- and bi-articular muscles show different timing activations. Indeed, has been reported [169] that both VMO and VL muscles, being mono-articular knee extensors, exhibited a rapid onset and cessation with relatively constant activity among subjects during the down-stroke phase. Conversely the rectus femoris, being bi-articular muscle, demonstrated a more gradual rise and decline. Moreover, Ryan and Gregor [168] noted that the mono-articular muscles (gluteus maximum, VL, VM, tibialis anterior and soleus) play a relatively invariant role as primary power producers. Conversely the bi-articular muscles (biceps femoris, semitentidosus, semimembranosus, rectus femoris, gastrocnemius medialis, and gastrocnemius medialis) behave differently and with greater variability [81,82]. According to the theory proposed by Ingen Schenau et al. [84] and largely reported in the literature following their study, bi-articular muscles appear to be primarily active in the transfer of energy between joints at critical times in the pedaling cycle and in the control of the direction of force production on the pedal.

19.2.3   Characterizing Strength Training Exercises

sEMG can be also used to characterize muscle activation patterns in strength training exercises. Since each exercise recruits muscles in a specific way, it is possible to focus the activity of a particular muscle and select the right exercise variation. Tomasoni et al. [185] recruited strength-trained athletes to test different forms of bench press exercise (barbell bench press, dumbbell bench press, inclined dumbell bench press). They recorded sEMG from prime mover muscles (pectoralis major, long and lateral head of triceps brachii, anterior deltoid, and serratus anterior) and used the sEMG normalization technique to compare muscles. Athletes exercised at 70% of one repetition maximum, which is the maximum amount of weight that one can lift in a single repetition for a given exercise. The most important finding was that long and lateral heads of triceps brachii were more active (i.e., greater sEMG amplitude relative to their sEMG maximum) in the exercises performed with barbells than in those performed with two dumbbells, while pectoralis major showed comparable activation in the two conditions. It seems that, when passing from the use of dumbells to barbells, the elbow extensors increase their contribution in the upper limbs push.

The same methodology has been used in the literature to quantify the contribution of muscles involved in the exercises not as prime movers but as stabilizers and neutralizer. For instance, Freeman et al. [54] compared different forms of push-ups introducing asymmetric hand placement and labile support surfaces. They measured not only the activations of prime movers upper limbs muscles but also abdominal and lumbar muscles, which act as trunk stabilizers. They noted that asymmetric handstand and ballistic push-up exercise evoked the highest level of muscle activation in the abdominal and back extensor musculature. This is an important issue since the load at the level of lumbar spine in free weight exercises should be considered. Indeed, it can both elicit back pain in some patients or it can be suggested as a therapeutic or prevention intervention in some other cases.

19.2.4   Links Between Coordination and Fatigue in Isometric Task

Muscle fatigue has effects in the coordination and pattern of load sharing among muscles. Tamaki et al. [184] reported the activity of synergistic muscles of triceps surae (soleus, medial and lateral gastrocnemius) during prolonged low-level isometric contraction. During the 210-min time course of each isometric task, alternate activity among synergist muscles of the triceps surae was observed, with some muscles becoming more active while others becoming inactive or less active. The activities seemed to occur complementary: When the lateral gastrocnemius sEMG amplitude increased, the medial gastrocnemius became inactive. It is suggested that alternate activities might facilitate the maintenance of the ankle plantar flexion tasks for as long as 210 minutes. Indeed the interval between occurrences of alternate activity tended to be shorter and more frequent in the second than in the first half of the exercise periods.

Kouzaki and Shinohara [104] confirmed this hypothesis assessing the alternating activity of quadriceps muscles (VM and VL versus rectus femoris) during prolonged isometric knee extension. They found a negative correlation between the frequency of

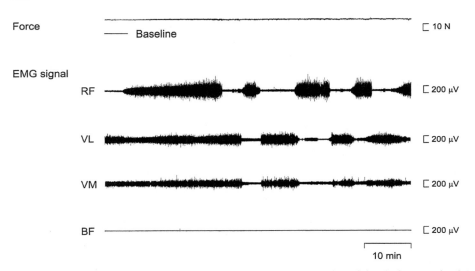

FIGURE 19.3 Representative data demonstrating alternate muscle activity during sustained knee extension at 2.5% of MVC force. Knee extension force, sEMG of rectus femoris (RF), vastus lateralis (VL), vastus medialis (VM), and biceps femoris long head (BF) are shown. Reproduced from Kouzaki and Shinohara [104] with permission.

alternating activity and mechanical fatigue; that is, the subjects who alternated more frequently the activity of a muscle pair showed lower mechanical fatigue after the fatiguing task (see Fig. 19.3).

19.2.5 Links Between Coordination and Fatigue in Dynamic Task

The occurrence of fatigue induces significant alterations in muscle coordination in all muscles of the kinetic chain involved in the task. For instance, during a fatiguing cycling task [46], quadriceps sEMG amplitude remained constant, whereas hip extensor muscles, gluteus, and biceps femoris increased their sEMG activity by 29% and 15%, respectively. The authors interpreted the increase of activity in gluteus maximus and biceps femoris as an instinctive coordination strategy compensating for potential fatigue and loss of force of the knee extensors (i.e., VL and VM) with a higher moment of the hip extensors. The question of benefits of these adaptations is open to discussion.

Alternating or shifting muscle activity during the time course of a fatiguing task seems to be an optimal strategy to face fatigue. So et al. [177] compared kinematic and EMG data during a 6-minute test in five groups of rowers, from young slow rowers to Olympic athletes. Olympic rowers used an alternating muscle strategy despite the fact that they showed the lower mechanical fluctuation in intensity. Indeed they involved all muscle groups in the beginning of the task, with particular emphasis of back muscles and then switching the emphasis between the quadriceps and the back in the middle of the test, lowering the activity of back muscles and increasing that of quadriceps. Such shifts may occur without the awareness of an athlete, as an instinctive CNS strategy to cope with fatigue.

In the reported examples the changes of muscle activation pattern during the course of fatigue was unintentional. If alternating and switching are strategies to counteract fatigue, voluntary intervention could indeed provide better control of such mechanisms. For that purpose, direct sEMG biofeedback variables would be useful for improving the activity modulation of muscles pattern.

19.3   TRAINING-INDUCED MUSCLE STRENGTH GAIN: NEURAL FACTORS VERSUS HYPERTROPHY

A motor unit (MU) consists of a motoneuron in the spinal cord and the muscle fibers it innervates [26] (see Fig. 19.4). The number of MUs per muscle in humans may range from about 100 for a small hand muscle to 1000 or more for large limb muscles [74]. It has also been shown that different MUs vary greatly in force generating capacity, that is, a 100-fold or more difference in twitch force [59,181]. In voluntary contractions, force is modulated by a combination of motor unit recruitment and changes in motor unit activation frequency (rate coding) [106,126]. The greater the number of motor units recruited and their discharge frequency, the greater the force will be. During motor unit recruitment the muscle force, when activated at any constant discharge

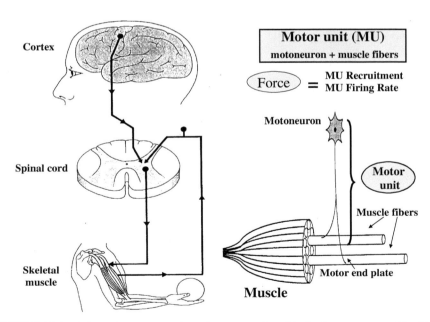

FIGURE 19.4   Schematic representation of a motor unit and its basic components. Modified from Sale [171].

frequency, is approximately 2–5 kg/cm^2 and in general is relatively independent of species, gender, age, and training status [3].

The electrical activity in a muscle is determined by the number of motor units recruited and their mean discharge frequency of excitation, that is, the same factors that determine muscle force [20,131,138]. Thus, direct proportionality between EMG and force might be expected (see also Chapters 10 and 13). Under certain experimental conditions, this proportionality can be well demonstrated by recording the smoothed rectified or integrated EMG (iEMG)[1] [41,125,129,131], and reproducibility of EMG recordings are remarkably high—for example the test–retest correlation ranging from 0.97 to 0.99 [97,129,131]. However, the change in the surface EMG should not automatically be attributed to changes in either motor unit recruitment or excitation frequencies as the EMG signal amplitude is further influenced by the individual muscle fiber potential, degree of motor unit discharge synchronization, and fatigue [20,21,137,141]. Nonetheless, carefully controlled studies have successfully employed surface EMG recording techniques and demonstrated the usefulness of iEMG as a measure of muscle activation level under a variety of experimental conditions [66,67,69,71,96,129,130,136,170,171].

The above short summary suggests that muscle strength can be modulated by motor unit activity which in turn is under the influence of central motor drive [40]. It is a common observation that repeated testing of the strength of skeletal muscles results in increasing test scores in the absence of measurable muscle hypertrophy [33,41]. Such increasing test scores are typically seen in daily or even weekly retesting at the inception of a muscle strength training regimen. In some cases, several weeks of intensive weight training resulted in significant improvement in strength without a measurable change in girth [41,100]. It has also been shown that when only one limb is trained, the paired untrained limb improves significantly in subsequent retests of strength but without evidence of hypertrophy [33,83,129,130].

In an earlier study of Rasch and Morehouse [164], it was demonstrated that strength gains from a six week training in tests when muscles were employed in a familiar way, but little or no gain in strength was observed when unfamiliar test procedures were employed. These data suggest that the higher scores in strength tests resulting from the training programs reflected largely the acquisition of skill and training-induced alterations in antagonist muscle activity, that is, enhanced reciprocal inhibition that contributes to greater net force production, reduced energy expenditure, and more efficient coordination [167].

All of the above findings support the importance of "neural factors," which, although not yet well-defined, certainly contribute to the display of maximal muscle force which we call strength. On the other hand, a strong relationship has been demonstrated both between absolute strength and the cross-sectional area of the

---

[1]The term integrated EMG (iEMG) was widely used in the past as a gross measure of muscle activation. It is actually incorrect since it is equal to zero if EMG is not rectified and is always positive and monotonically increasing if EMG is rectified. It was (sometimes until now!) wrongly used instead of the average rectified value (ARV). We decided to herein maintain it only for a matter of historical references. The authors used such a term only when referring and quoting previous papers in which iEMG was used. Such a term should actually be avoided.

muscle [166] and between strength gain and increase in muscle girth or cross-sectional area [83]. It is quite clear, therefore, that human voluntary strength is determined not only by the quantity (muscle cross-sectional area) and quality (muscle fiber types) of the involved muscle mass, but also by the extent to which the muscle mass has been activated (neural factors) (see Moritani [133,134] for reviews).

A reasonable hypothesis for describing the time course of strength gain with respect to its two major determinants is that suggested by De Lorme and Watkins [38], who postulated that: "The initial increase in strength on progressive resistance exercise occurs at a rate far greater than can be accounted for by morphological changes within the muscle. These initial rapid increments in strength noted in normal and disuse-atrophied muscles are, no doubt, due to motor learning. It is impossible to say how much of the strength increase is due to morphological changes within the muscle or to motor learning."

We now have available electromyographic instrumentation and methodology which makes it possible to separate muscle activation level (motor learning) from hypertrophic effects (morphological changes) as described by deVries [41] and Moritani and deVries [129,130]. Figure 19.5 shows a schema for evaluation of percent contributions of neural factors and hypertrophy to the gain of strength output. If strength gain is brought about by "neural factors" such as learning to disinhibit and/or to increase muscle excitation level, then we should expect to see increases in maximal neural activation (iEMG) without any change in force per fiber or motor units innervated as shown in Fig. 19.5a. On the other hand, if strength gain were

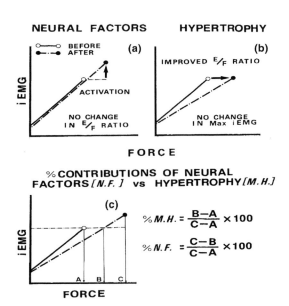

FIGURE 19.5 Schema for evaluation of percent contributions of neural factors and hypertrophy to the gain of strength during the course of muscle training based on Moritani and deVries. Reproduced from Mitchell et al. [128] with permission.

entirely attributable to muscle hypertrophy, then we should expect the results shown in Fig. 19.5b. Here the force per fiber (or per unit activation) is increased by virtue of the hypertrophy, but there is no change in maximal activation (iEMG). Figure 19.5c shows our method for evaluation of the percent contributions of the two components when both factors may be operative in the time course of muscle strength training.

Figure 19.6 illustrates the time course of strength gain with respect to the calculated percent contributions of neural factors and hypertrophy, calculated with the equation in Figure 19.5, during the course of 8-week strength training of the arm flexors of young college students. The results clearly demonstrated that the neural factors played a major role in strength development at early stages of strength gain and then hypertrophic factors gradually dominated over the neural factors for the young subjects in the contribution to the further strength gain (see Moritani and deVries [129,130] for more details). The strength gain seen for the untrained contralateral arm flexors provided further support for the concept of cross education.

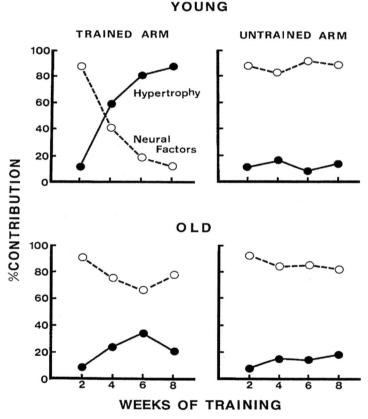

FIGURE 19.6 The time course of strength gain showing the percent contributions of neural factors and hypertrophy in the trained and contralateral untrained arms of young (**above**) and old (**below**) subjects. Reproduced from Moritani and deVries [130] with permission.

It is reasonable to assume that the nature of this cross education effect may entirely rest on the neural factors presumably acting at various levels of the nervous system which could result in increasing the maximal level of muscle activation.

A subsequent study [130] with older men (mean age of 70 years) demonstrated that the old subjects achieved the strength gain by virtue of neural factors as indicated by the increases in the maximal iEMG in the absence of hypertrophy. It is suggested that the training-induced increase in the maximal level of muscle activation (neural factors) through greater motor unit discharge frequency and/or motor unit recruitment may be the only mechanism by which the aged subjects increase their strength in the absence of any significant evidence of hypertrophy. These results are entirely consistent with those reported by Komi et al. [100], who demonstrated that changes in iEMG and force take place almost in parallel during the course of 12 weeks of training.

Häkkinen et al. [70] demonstrated that the subject trained with high-intensity loads of combined concentric and eccentric contractions showed an accelerated increase in force together with the parallel increase in iEMG during the first 8 weeks of training while showing minor muscle fiber hypertrophy. Greater muscle hypertrophy of both slow- and fast-twitch fibers was observed during the last 8 weeks of training that resulted in further strength gain with no significant change in the iEMG. Subsequent studies [35,37,67–71,86,101,145] confirmed these observations and provided evidence for the concept that in strength training the increase in voluntary neural drive accounts for the larger proportion of the initial strength increment and thereafter both neural adaptation and hypertrophy take place for further increase in strength, with hypertrophy becoming the dominant factor [70,129]. Seynnes et al. [174] have recently investigated the early skeletal muscle hypertrophy and architectural changes in response to a 35-day high-intensity resistance training (RT) program. It was clearly demonstrated that changes in muscle size are detectable after only 3 weeks of RT and that remodeling of muscle architecture precedes gains in muscle cross-sectional area. Muscle hypertrophy seems to contribute to strength gains earlier than previously reported.

Interestingly, there has been some evidence suggesting that strength development can be achieved by involuntary contractions initiated by electrical stimulation [51]. However, these experiments resulted in a considerably smaller strength gain than the values found in normal voluntary training. Since the motor pathways are probably minimally involved in electrical training, it seems likely that a training stimulus resides in the muscle tissue itself and hence the hypertrophic factor is the principal constituent for strength development. Subsequent studies [35,117] have indicated that the muscle training using electrically evoked contractions (80 maximal isometric tetani for 10 s) produced no increase in maximal voluntary strength, suggesting that neural drive has to be present in the training in order to produce large increases in maximal voluntary strength.

Strength training studies are typically carried out for a period of 5–20 weeks and have shown that the early increases in the voluntary strength are associated mainly with neural adaptation while hypertrophy begins to occur at the latter stage of training. Serious athletes, however, train over a period of many months or years. Häkkinen et al. [69] have studied the effects of strength training for 24 weeks with intensities ranging between 70% and 120% of maximal voluntary force. The increase in strength correlated with significant increase in the neural activation (iEMG) of the leg extensor muscles during the most intensive training months along with significant enlargement of fast-twitch fiber area. During subsequent detraining, a great decrease in the maximal strength was correlated with the decrease in maximal iEMG of the leg extensors. It was suggested that selective training-induced hypertrophy could contribute strength development but muscle hypertrophy may have some limitations during long-term strength training, especially in highly trained subjects. This suggestion has recently been confirmed by a one-year training study indicating the limited potential for strength development in elite strength athletes [71].

On the other hand, there has been evidence indicating that lifelong high-intensity physical activity could potentially mitigate the loss of motor units associated with aging well into the seventh decade of life [155]. We have recently reported that the winner of an international contest to find the world's fastest drummer (WFD) can perform repetitive wrist tapping movements with one hand using a handheld drumstick at 10 Hz, much faster than the maximum tapping frequency of 5–7 Hz in the general population [58]. The WFD showed more rapid sEMG amplitude rise, earlier decline of sEMG activity, and more stable muscle activation time than the non-drummers (NDs) and ordinary drummers (ODs). In addition, there was a significant correlation between the EMG rise rate and the duration of drum training in the group of drummers (i.e., ODs and WFD). Our subsequent spike shape analysis revealed that the WFD had exceptional motor unit activity such as higher motor unit discharge rate, more motor unit recruitment, and/or higher motor unit synchronization to achieve extraordinary fast 10-Hz drumming performance [58]. Interestingly, Claflin et al. [32] have investigated the effects of movement velocity during resistance training on the size and contractile properties of individual muscle fibers from human VL muscles of young (20–30 years old) men and women and older (65–80 years old) men and women. In each group, one-half of the subjects underwent a traditional progressive resistance training (PRT) protocol that involved shortening contractions at low velocities against high loads, while the other half performed a modified PRT protocol that involved contractions at 3.5 times higher velocity against reduced loads. Contrary to their hypothesis, the velocity at which the PRT was performed did not affect the fiber-level outcomes substantially. They concluded that, compared with low-velocity PRT, resistance training performed at velocities up to 3.5 times higher against reduced loads is equally effective for eliciting an adaptive response in type 2 fibers from human skeletal muscle.

19.4 INVESTIGATION OF MUSCLE DAMAGE BY MEANS OF SURFACE EMG

Nearly everyone has experienced delayed-onset muscle soreness (DOMS) at some time; many have suffered from this common ailment on numerous occasions. DOMS is characterized by stiffness, tenderness, and pain during active movements and weakness of the affected musculature. A number of investigators have demonstrated that the eccentric component of dynamic work plays a critical role in determining the occurrence and severity of exercise-induced muscle soreness [17,55,146]. It has been also demonstrated that type II fibers are predominantly affected by this type of muscular contraction [55,92].

It is well established that eccentric (lengthening) muscle action requires less oxygen and lower amount of ATP than concentric muscle action [36]. Both surface [98] and intramuscular EMG studies [142] have demonstrated that motor unit recruitment patterns are qualitatively similar in both types of contractions, but for a given MU the force at which motor unit recruitment occurs is greater in eccentric muscle action than in either isometric or concentric (shortening) muscle actions. Based on these findings and the results of EMG studies cited earlier, it is most likely that DOMS associated with eccentric component of dynamic exercise might be in part due to high mechanical forces produced by a relatively small number of active MUs which may in turn result in some degree of disturbance in structural proteins in muscle fibers, particularly those of high recruitment threshold MUs.

Despite the fact that DOMS is a well-known phenomenon in sports as well as working life, the exact pathophysiological mechanisms underlying it are still not well understood. According to Armstrong [6], a number of hypotheses may exist to explain the etiology and cellular mechanisms of DOMS. The following model may be proposed: (1) High tension, particularly associated with eccentric muscle action in the contractile and elastic system of the muscle, results in structural damage; (2) muscle cell membrane damage leads to disruption of Ca^{2+} homeostasis in the injured muscle fibers resulting in necrosis that peaks about 2 days post-exercise; and (3) products of macrophage activity and intracellular contents accumulate in the interstitium, which stimulate free nerve endings of group IV sensory neurons in the muscles leading to the sensation of DOMS.

Earlier EMG work by deVries [42,44] demonstrated that symptomatic soreness and tenderness seemed to parallel reduced EMG amplitude. DeVries has thus proposed the spasm theory: DOMS is caused by tonic, localized spasm of motor units as a result of a vicious cycle in which the activity induced ischemia in turn leads to further pain and reflex activity. Later workers have been unable to demonstrate any EMG activity in resting painful muscles [2]. Berry et al. [17] have shown that muscle soreness only occurred in muscles that had contracted eccentrically and did not occur at the time of greatest myoelectric signal changes. Although elevated EMG activity was accompanied with eccentric muscle action, there seems to exist dissociation in the time course of these two parameters [17,136,146]. The data of Newham et al. [146] have also demonstrated that eccentric muscle action has a long lasting effect on the muscle's ability to generate force after exercise. When the quadriceps muscle was stimulated at low frequencies, it was not able to develop the same force as it had under similar conditions before eccentric exercise. On the other hand, high-frequency stimulations elicited similar force before and after such exercise. The underlying mechanism of this so-called "low-frequency fatigue" has been postulated to be impaired excitation-contraction coupling [49] due to reduced release of calcium or possibly because of impaired transmission in the transverse tubular system, as a result of muscle damage in the period of ischemic activity.

19.4.1   Acute Effects of Static Stretching on Muscle Soreness

We conducted a series of studies to determine the physiological effects of static muscle stretching upon DOMS which was induced experimentally by heel raise (10 rep, 10 sets) with a 70% MVC equivalent weight attached on a universal shoulder press equipment or step test [17,136]. Electrophysiological parameters—for example, maximal mass action potential (M-wave), H-wave, and H/M ratio for determination of alpha motoneuron excitability—were measured during standing position (control), 24 h post experimental fatigue, and immediate post static stretching. Changes in the standing EMG signal up to 48 h post experimental fatigue were subjected to frequency power spectral analysis in order to determine the degree of muscle fatigue and resting action potential amplitude.

Surface EMG power spectral analyses revealed that (1) the experimentally induced DOMS was associated with significantly higher resting action potential amplitude and lower mean power frequency, suggesting the existence of some degree of muscle spasm and a possible synchronization of tonic motoneurons [17,136], and (2) static muscle stretching (three sets of 20-s duration) showed immediate and quite noticeable effects of restoring these electrophysiological parameters back towards the control level. Results on alpha motoneuron excitability indicated that there was very little change, if any, in the maximal amplitude of the H waves for the control leg while experimental leg post H wave was markedly reduced by static stretching. The mean relative reduction in the H/M ratio from pre- to post-test for the control and experimental legs was 0.63% ($p > 0.05$) and 21.5% ($p < 0.01$), respectively. These results are entirely consistent with earlier studies [43,44] and further suggest that the inverse myotatic reflex which originates in the Golgi tendon organs (GTO) may be the basis for the relief of DOMS by static stretching (see Fig. 19.7). Since H reflex

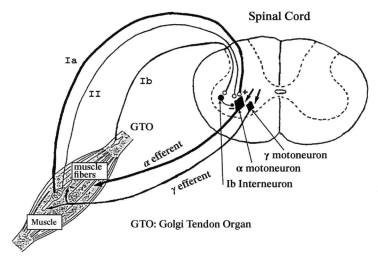

FIGURE 19.7   A simplified schematic representation of basic neural components involved in stretch reflex and Golgi tendon organ Ib inhibition.

involves tonic motor units [119], it is likely that the Ib afferent inhibitory effects from GTO could be mediated through the tonic MUs, thus reducing the evoked H wave amplitude.

We further investigated the physiological effects of static stretching upon DOMS in conjunction with the spinal alpha motoneuron pool excitability and peripheral muscle blood flow in seven healthy male subjects. All subjects performed heel raises (30 rep, 5 sets) with 20-kg load 24 hours prior to testing. Electrophysiological measurements included the Hoffman reflex amplitude (H amplitude) as a measure of spinal alpha motoneuron pool excitability. The directly evoked muscle action potential (M-wave) remained constant for each subject throughout the experiments. Blood flow measurement was performed by near-infrared spectroscopy (NIRS) with venous occlusion technique. In the experimental condition (EXP), those measurements were obtained before and after static muscle stretching (35 s, 3 sets) under experimentally induced muscle soreness. During the control condition (CON), the same measurements were made before and after standing rest for a period of 4 min. The order of the experimental treatments (EXP or CON) was chosen at random.

Figure 19.8 represents a typical set of H-reflex data obtained 24 h after experimentally induced muscle soreness prior to muscle stretching and immediately after muscle stretching. The data clearly indicated that, for the same elicited M-wave, H-reflex amplitude was considerably reduced after muscle stretching. Group data demonstrated that the static stretching brought about a statistically significant reduction in the H/M ratio (23.5%, $p < 0.01$) of the EXP conditions while no such changes were observed in CON trials. These changes were accompanied by nearly 78.5% increase ($p < 0.01$) in blood flow after stretching the gastrocnemius muscle with the experimentally induced soreness. These finding was entirely consistent with earlier studies, suggesting that the inverse myotatic reflex (Ib inhibition) may be the basis for the relief of muscle soreness by static stretching. The increase in blood flow after stretching found in the present study suggested that static stretching could bring about a relief of spasm, which could have caused local muscles ischemia and pain (see Fig. 19.9). Our data strongly suggest that static stretching plays a significant role in

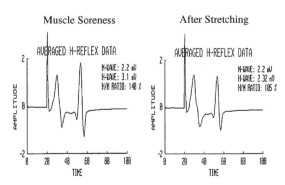

FIGURE 19.8 A typical set of H-reflex data during the experimentally induced muscle soreness and after static muscle stretching.

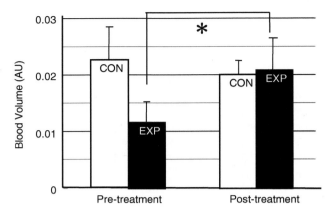

FIGURE 19.9 Blood flow (volume, arbitrary units) changes during the experimentally induced muscle soreness and after static muscle stretching.

relief of DOMS by reducing spinal motoneuron pool excitability and enhancing muscle blood flow.

19.4.2 Fusimotor Sensitivity After Prolonged Stretch-Shortening Cycle Exercise

Evidences have been presented that both short- and long-duration fatiguing exercises lead to deterioration in neuromuscular performance (exhaustive exercise and intensive effort [147], as well as long-lasting low-intensity effort [8]. The underlying mechanisms which mediate modifications of reflex activity after repeated stretch-shortening cycle (SSC) exercise remain an open question. To elucidate which factor is more influential on the stretch reflex reduction after repeated SSC, T-reflex was employed in this study. T-reflex, which is elicited by tendon tap and shares reflex pathway (e.g., alpha-motoneuron pool) with H-reflex except for spindle-mediated fusimotor component, can help to factor out modification of spindle and fusimotor activity when compared to H-reflex. We therefore performed comparison of T-reflex and H-reflex of the triceps surae before, immediately after, 2 h after, and 24 h after two hours of exhaustive running in terms of EMG activity and impact force on the tendon. Five consecutive EMG responses of T-reflex and H-reflex were averaged and analyzed for peak-to-peak amplitude. Results revealed that immediately after the running, T- and H-wave amplitudes were significantly depressed while maximal M-wave remained constant. On the other hand, 2 h after the running, H-reflex amplitudes showed clear-cut rising ($p < 0.001$) and, by contrast, the T-reflex amplitude did not show such a significant elevation. All the EMG amplitudes returned to the pre-exercise level in 24 h.

The impact force on the Achilles tendon (coefficient of rebound force) showed a reduction immediately after the running ($p < 0.05$) and recovered in 24 h. The difference between H- and T-reflex amplitudes 2 h after the exhaustive running might suggest that the sensitivity of fusimotor activity was reduced by 2 h of running. Furthermore, the reduced impact force might reflect deteriorated stiffness regulation of muscle–tendon complex. This may also suggest the degradation of spindle activity. These results support the hypothesis that stretch reflex reduction might be attributed to disfacilitation of alpha motoneuron pool caused by degradation of spindle-mediated fusimotor support and/or fatigue of the muscle spindle itself due to the possible depletion of intrafusal muscle glycogen [8,9].

19.5 RELATIONSHIPS BETWEEN EMG FEATURES AND MUSCLE FIBER FEATURES

Fiber composition is usually investigated by biopsy and histochemical analysis. Johnson et al. [90], analyzed 36 muscles during an autopsy study on six male cadavers (aged 17–30 years) and provided the percentage of type I and type II fibers found in each muscle. This is probably the most cited paper about this issue, but other authors focused their efforts on fewer muscles and a greater number of subjects producing more reliable data [56,60,151,183,192].

The information obtained from bioptical specimens are actually not representative of the muscle as a whole, thus the need of repeated sampling decreases the subjects' compliance supporting the validation of alternative noninvasive methods of fiber type estimate. Hence the main issue is to assess if it is possible to extract, from a wider portion of muscle and using superficial electrodes, information related to the histological properties of human muscles.

A number of physiological parameters were considered in the past to be related to muscle fiber types and motoneurons for their noninvasive assessment. The amplitude estimators (ARV, RMS) and the power spectrum estimators (MNF, MDF) of the recorded surface EMG signals (see Chapters 4, 5, and 10), and the muscle fiber conduction velocity (CV, see Chapter 5) were shown to be related to the pH decrease due to the increment of metabolites produced during a fatiguing contraction.

The "size principle" described by Henneman and Mendell [74] was first proposed based upon results from cat motoneurons, strong evidence has been presented that in muscle contraction there is a specific sequence of recruitment in order of increasing motoneurons and motor unit size [39,106,126,138]. Earlier studies have demonstrated in humans that, for a muscle group with mainly type I slow-twitch fibers (adductor pollicis), rate coding plays a prominent role in force modulation [126,138]. On the contrary, in a muscle group composed of both type I and II fast-twitch fibers, motor unit recruitment seems to be the major mechanism for generating extra force above 40% to 50% of MVC [39,106,138].

Similarly, muscle fatigability during a sustained isometric contraction, as reflected by the progressive recruitment of new motor units and by the increase of iEMG in time, is also dependent upon muscle fiber type composition (see Fig. 19.10). Earlier EMG studies indicated that ARV of the surface EMG increased progressively as a function of time during sustained muscular contraction with a constant force output [43,135]. The work of deVries [43] and Viitasalo and Komi [187] suggested that EMG fatigue curves (iEMG versus time) could provide a measure of motor unit fatigability. Furthermore, Komi and Tesch [99] demonstrated that human muscles

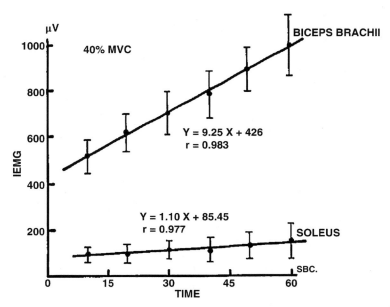

FIGURE 19.10   The mean iEMG data as a function of time at 40% of maximal voluntary contraction (MVC) for the biceps brachii and soleus muscles. Reproduced from Moritani [134] with permission.

characterized by a predominance of fast-twitch fibers showed a greater susceptibility to fatigue, and this was reflected by a sharp decline in force output as well as by a pronounced decrease in the mean power frequency of EMG power spectrum. These data suggested a possibility that muscle fiber features could be well represented by EMG signal characteristics during contractions at varying force levels and muscle fatigue. Figures 19.10 and 19.11 represent a typical set of iEMG and power spectra obtained from the biceps brachii and soleus muscles during sustained isometric contractions at 40% MVC, respectively.

Earlier studies regarding changes of EMG power spectral parameters, such as mean or median spectral frequency (MNF, MDF) and the level of muscle contractions, are somewhat contradictory. For example, a series of studies by Petrofsky and Lind [153,154] showed no systematic relationship between tension levels and MNF for hand grip muscles. Hagberg and Ericson [65] demonstrated in the elbow flexors that MNF increased with contraction strength at low contraction levels but became independent of contraction level above 25–30% MVC, whereas Muro et al. [143] and Broman et al. [23] demonstrated almost linear increases in MNF with force of contraction up to near MVC levels. These different results might be at least in part due to differences in the muscle groups studied, electrode size, and interelectrode distance which could act as various low-pass filters, muscle fiber types [138,141], and underlying motor unit firing statistics and action potential conduction velocity [23].

Figure 19.12 shows EMG signal features demonstrating the influence of electrode size and interelectrode distance upon the amplitude and frequency components (see

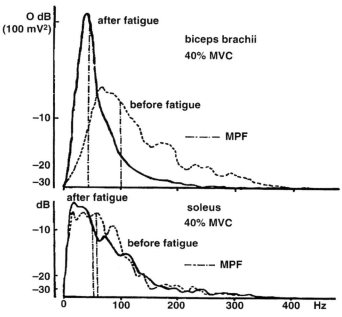

FIGURE 19.11   A typical set of sEMG frequency power spectral changes for the biceps brachii and soleus muscles before and after muscle fatigue at 40% of MVC. Reproduced from Moritani [134] with permission.

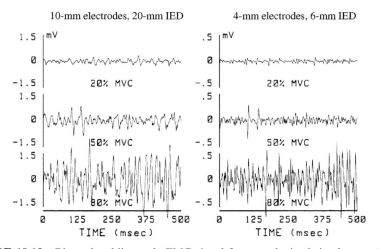

FIGURE 19.12   Biceps brachii muscle EMG signal features obtained simultaneously from large (10-mm electrode diameter, 20-mm interelectrode distance) and small (4-mm electrode diameter, 6-mm interelectrode distance) bipolar silver/silver chloride electrodes.

(a)   FREQUENCY POWER SPECTRA

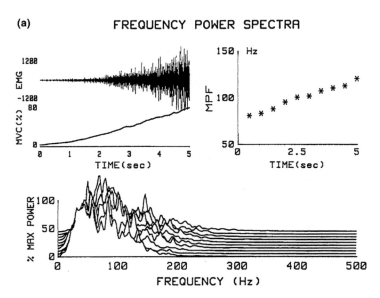

FIGURE 19.13 A typical set of computer outputs showing raw sEMG, force, normalized power spectra and mean power frequency (MPF) plots during ramp force output for a normal subject **(a)** and a highly trained power lifter **(b)**. Reproduced from Moritani et al. [137] with permission.

also Chapter 2). These data were obtained from the biceps brachii muscle contracting at three levels of MVC while recording the EMG signals simultaneously from large (10-mm electrode diameter, 20-mm interelectrode distance) and small (4-mm electrode diameter, 6-mm interelectrode distance) bipolar silver/silver chloride electrodes. Note the marked differences in both amplitude and frequency features of the signals (see also Chapters 2 and 3). Our data demonstrated that surface EMG with larger electrodes and wider interelectrode distance revealed no systematic increases in MNFs, but showed progressive increases in EMG amplitude. The power spectral data obtained from the same muscle with small electrodes and narrow interelectrode distance demonstrated a large and significant increase in MNFs during the ramp contraction up to 80% of MVC [132] (see Fig. 19.13a). For comparison, the results obtained from one of the United States representatives for the 1984 World Power Lifting championship are shown in Fig. 19.13b. These results are entirely consistent with those reported by Petrofsky and Lind [153,154].

In the last decade, however, it was shown that, even if conduction velocity, amplitude, and MNF of sEMG are somehow related to the type of recruited motor units, a large number of additional factors blur the phenomenon. A more extensive discussion of these factors is provided in Chapter 10. Amplitude and spectral features are actually related to the number and type of recruited motor units, whose relative position and depth within the muscle cannot be assessed, and no correction for the filtering effect of the volume conductor thickness can be adopted at this time. Thus the filtering effect due to the tissue between each active motor unit and the recording

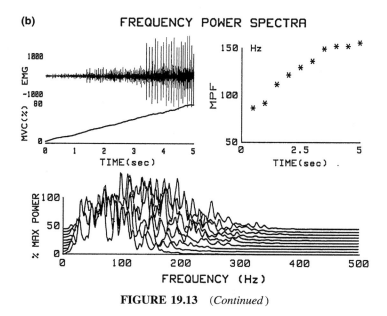

(b)   FREQUENCY POWER SPECTRA

FIGURE 19.13   (*Continued*)

site plays a role which can be accounted for only in particularly controlled experimental conditions. Other factors are discussed in Chapter 10.

In the same way the frequency content of the surface EMG signal is somehow related to the recruited motor unit pool and it was demonstrated, both in animal model [108] and in humans [14,60,99,115], that its time course during fatiguing contractions shows a steeper decrease if the muscle is characterized by more fast than slow fibers. Nevertheless, even in this case, a recent debate described a number of confounding factors (depth of motor unit within the volume conductor, properties of the volume conductor layers, pinnation angle, motor unit synchronization, detection system geometry, among others) which strongly blur the relationship between sEMG properties and muscle fiber type constituency [53]. For these reasons (even if with particular caution), muscle fiber conduction velocity (CV) seems the most suitable candidate to relate, under a physiological framework, the modifications in EMG signals with both the motor unit pool histochemical characteristics and cross-section fiber size [127].

Hopf et al. [78] electrically evoked single twitches in human biceps brachii muscle estimating both the contraction times (defined as the time from the onset of the deflection to the peak) and the muscle fiber conduction velocity (using invasive technique). A negative correlation between contraction times of the elicited twitches and muscle fiber conduction velocity was found (see Fig. 19.14). In the same direction, Sadoyama et al. [169] showed a strong correlation between different fiber type relative area in biopsies and conduction velocities estimated during voluntary contractions from two different groups of athletes (sprinters vs distance runners) (see Fig. 19.15). Identical findings were observed by Kupa et al. [107] and in the work of

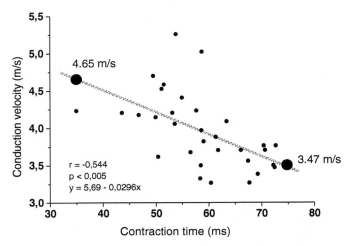

FIGURE 19.14 Diagram showing the first experimental evidence of correlation between muscle fiber contraction times and conduction velocity in electrically evoked contractions. Modified from Hopf et al. [78].

Rainoldi et al. [161] in which CV rate of changes during fatiguing contractions were found to be significantly different between sprinters and long-distance runners, matching the expected fiber type composition. A further confirmation of such an approach was provided in the work of Rainoldi et al. [160], where surface EMG signals were recorded from the vastus medialis longus, vastus medialis obliquus, and VL muscles during isometric knee extension contractions; differences in CV initial

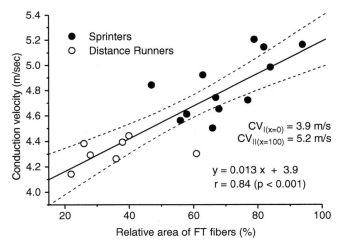

FIGURE 19.15 Correlation between relative areas of fast twitch fibers and conduction velocities estimated during voluntary contractions from two different groups of athletes (sprinters versus distance runners) which can reasonably be considered at the edges of the possible phenotype distribution. Modified from Sadoyama et al. [169].

values among the three muscles were found in agreement with histological evidences provided in the literature.

As known, oxygen distribution in exercising skeletal muscle is regulated by oxygen transport in the blood vessels, as well as by oxygen diffusion and consumption in the tissue [112], hence the amount of oxygen supply is one of the pivotal factor affecting contraction and causing fatigue [77]. In their work, Casale et al. [29] concluded that acute exposure to hypobaric hypoxia did not significantly affect the muscle-fiber membrane properties (no peripheral effect) but impacted on motor-unit control properties (central control strategies for adaptation). Hence, the lack of oxygen induced by high altitude induced central effects resulting in the recruitment of more MUs oxygen-independent than those used at sea level to maintain the requested force task. This finding demonstrated that it is possible to assess, also in clinical settings, if variations in sEMG manifestations of fatigue are central or peripheral adaptations by means of two different contraction modalities, namely voluntary or electrically induced [30]. To reach the same goal—that is, to distinguish between peripheral and central effects of fatigue—a different approach was proposed by Mesin et al. [124] in which a bidimensional vector based on CV estimate and sEMG signal fractal dimension provided selective sensitivity to peripheral or central fatigue.

Moreover, estimates of CV initial values and of CV rates of change (that is, myoelectric manifestations of fatigue) reflect differences, with respect to a control group, due to conditioning [28,95,123,161,163] and pathologies [30,52,111,120,152, 162], aging [28,121] as evidence of changes/alterations in fiber types.

A recent finding further confirmed the role of oxygen in modulating fatigue and motor unit recruitment in a group of patients affected by chronic obstructive pulmonary disease (COPD). Such a pathology is characterized by persistently poor airflow [31] and by a number of side effects. Among others, lower limb muscles of COPD patients showed a significant redistribution of fiber type ratios characterized by a reduction in the proportion of type I fibers with a global shifting towards type II fiber type [114,188]. In a recent protocol [165] based on a prolonged isometric contraction of quadriceps at 70% of MVC, COPD and healthy age-matched subjects were compared in terms of myoelectric manifestations of fatigue (Fig. 19.16). The greater proportion of type II muscle fibers in COPD led to a greater rate of EMG fatigue measured as a greater decrease of CV over time.

A specific protocol was recently proposed as a noninvasive technique to distinguish between two extremely different phenotypes highlighting the effect of oxygen availability in endurance and power trained athletes. As described by Rainoldi and Gazzoni [159], intermittent (3 s of contraction and 1 s of rest in between contractions) and continuous contractions (no rest and same total workload) were proposed to a group of endurance and a group of power trained athletes. Findings showed that while no differences were observed in power-trained athletes, passing from intermittent to continuous contractions, fatigue (estimated by the normalized slope of CV) increased by 200% in the endurance group due to the lack of availability of oxygen.

All these findings seem to confirm that, in carefully controlled experimental conditions, it is possible to correlate modifications of sEMG variables with different muscular phenotypes or muscular adaptation processes. Such an approach seems now

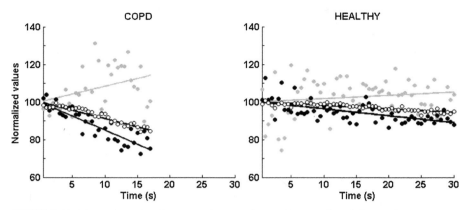

FIGURE 19.16   Fatigue plot diagrams of two subjects: COPD patient (**left panel**) and healthy adult (**right panel**). Time courses of ARV (*gray circle*), MNF (*black circle*), and CV (*white circle*) are represented for each epoch (0.5-s length) of the isometric contraction at 70% of MVC. Each variable is normalized with respect to its initial value, and the slope of the regression line represents an index of myoelectric fatigue: The greater the slope of sEMG variables, the greater the manifestations of fatigue. Modified from Rinaldo et al. [165].

finally ready to be used as a further tool for neuromuscular adaptations monitoring since it is noninvasive and repeatable.

REFERENCES

1. Aagaard, P., E. B. Simonsen, J. L. Andersen, S. P. Magnusson, F. Bojsen-Moller, and P. Dyhre-Poulsen, "Antagonist muscle coactivation during isokinetic knee extension," *Scand. J. Med. Sci. Sports* **10**, 58–67 (2000).

2. Abraham, W. M., "Factors in delayed muscle soreness," *Med. Sci. Sports Exerc.* **9**, 11–26 (1977).

3. Always, S. E., J. Stray-Gundersen, W. H. Grumbt, and W. J. Gonyea, "Muscle cross sectional area and torque in resistance-trained subjects," *Eur. J. Appl. Physiol.* **60**, 86–90 (1990).

4. Amiridis I. G., A. Martin, B. Morlon, L. Martin, G. Cometti, M. Pousson, and J. van Hoecke, "Co-activation and tension-regulating phenomena during isokinetic knee extension in sedentary and highly skilled humans," *Eur. J. Appl. Physiol. Occup. Physiol.* **73**, 149–156 (1996).

5. Andrews J. G., and J. G. Hay, "Biomechanical considerations in the modeling of muscle function," *Acta. Morphol. Neerl. Scand.* **21**, 199–223 (1983).

6. Armstrong R. B., "Mechanisms of exercise-induced delayed onset muscular soreness: a brief review," *Med. Sci. Sports Exerc.* **16**, 529–538 (1984).

7. Arsenault A. B., D. A. Winter, R. G. Marteniuk, and K. C. Hayes, "How many strides are required for the analysis of electromyographic data in gait?," *Scand. J. Rehabil. Med.* **18**, 133–135 (1986).

8. Avela J., H. Kyrolainen, and P. V. Komi, "Altered reflex sensitivity after repeated and prolonged passive muscle stretching," *J. Appl. Physiol.* **86**, 1283–1291 (1999).

9. Avela J., H. Kyrolainen, and P. V. Komi, "Neuromuscular changes after long-lasting mechanically and electrically elicited fatigue," *Eur. J. Appl. Physiol.* **85**, 317–325 (2001).

10. Baratta R., M. Solomonow, B. H. Zhou, D. Letson, R. Chuinard, and R. D'Ambrosia, "Muscular coactivation. The role of the antagonist musculature in maintaining knee stability," *Am. J. Sports Med.* **16**, 113–122 (1988).

11. Basmajian J. V., "Motor learning and control: a working hypothesis," *Arch. Phys. Med. Rehabil.* **58**, 38–41 (1977).

12. Bazzucchi I., M. E. Riccio, and F. Felici, "Tennis players show a lower coactivation of the elbow antagonist muscles during isokinetic exercises," *J. Electromyogr. Kinesiol.* **18**, 752–759 (2008).

13. Bazzucchi I., P. Sbriccoli, G. Marzattinocci, and F. Felici, "Coactivation of the elbow antagonist muscles is not affected by the speed of movement in isokinetic exercise," *Muscle Nerve* **33**, 191–199 (2006).

14. Beck, T. W., T. Housh, A. C. Fry, J. T. Cramer, J. Weir, B. Schilling, M. Falvo, and C. Moore, "MMG-EMG cross spectrum and muscle fiber type," *Int. J. Sports Med.* **30**, 538–544 (2009).

15. Benesch, S., W. Putz, D. Rosenbaum, and H. Becker, "Reliability of peroneal reaction time measurements," *Clin. Biomech.* **15**, 21–28 (2000).

16. Bernstein, N. A., *The Coordination and Regulation of Movements*, Pergamon Press, 1967.

17. Berry, C. B., T. Moritani, and H. Tolson, "Electrical activity and soreness in muscles after exercise," *Am. J. Phys. Med. Rehab.* **69**, 60–66 (1990).

18. Besier, T. F., D. G. Lloyd, T. R. Ackland, and J. L. Cochrane, "Anticipatory effects on knee joint loading during running and cutting maneuvers," *Med. Sci. Sports Exerc.* **33**, 1176–1181 (2001).

19. Beynnon, B. D., and B. C. Fleming, "Anterior cruciate ligament strain in-vivo: a review of previous work," *J. Biomech.* **31**, 519–522 (1998).

20. Bigland-Ritchie, B., "EMG/force relations and fatigue of human voluntary contractions," *Exerc. Sports Sci. Rev.* **9**, 75–117 (1981).

21. Bigland-Ritchie, B., Jones, D. A., and Woods, J. J., "Excitation frequency and muscle fatigue: electrical responses during human voluntary and stimulated contractions," *Exp. Neurol.* **64**, 414–427 (1979).

22. Bigland-Ritchie, B., R. Johansson, O. C. Lippold, and J. J. Woods, "Contractile speed and EMG changes during fatigue of sustained maximal voluntary contractions," *J. Neurophysiol.* **50**, 313–324 (1983).

23. Broman, H., G. Bilotto, and C. J. De Luca, "Myoelectric signal conduction velocity and spectral parameters: influence of force and time," *J. Appl. Physiol.* **58**, 1428–1437 (1985).

24. Bruce, E. N., M. D. Goldman, and J. Mead, "A digital computer technique for analyzing respiratory muscle EMG's," *J. Appl. Physiol.* **43**, 551–556 (1977).

25. Burden, A., and R. Bartlett, "Normalisation of EMG amplitude: an evaluation and comparison of old and new methods," *Med. Eng. Phys.* **21**, 247–257 (1999).

26. Burke, R. E., "Motor units: anatomy, physiology and functional organization," in *Handbook of Physiology* **2**, American Physiological Society, pp. 345–422, 1981.

27. Carolan, B., and E. Cafarelli, "Adaptations in coactivation after isometric resistance training," *J. Appl. Physiol.* **73**, 911–917 (1992).

28. Casale, R., A. Rainoldi, J. Nilsson, and P. Bellotti, "Can continuous physical training counteract aging effect on myoelectric fatigue? A surface electromyography study application," *Arch. Phys. Med. Rehabil.* **84**, 513–517 (2003).

29. Casale, R., D. Farina, R. Merletti, and Rainoldi A., "Myoelectric manifestations of fatigue during exposure to hypobaric hypoxia for 12 days," *Muscle Nerve* **30**, 618–625 (2004).

30. Casale, R., P. Sarzi-Puttini, F. Atzeni, M. Gazzoni, D. Buskila, and A. Rainoldi, "Central motor control failure in fibromyalgia: a surface electromyography study," *BMC Musculoskelet. Disord.* **10**, 78 (2009).

31. Celli, B. R., W. MacNee, and A. E. T. Force, "Standards for the diagnosis and treatment of patients with COPD: a summary of the ATS/ERS position paper," *Eur. Respir. J.* **23**, 932–946 (2004).

32. Claflin, D. R., L. M. Larkin, P. S. Cederna, J. F. Horowitz, N. B. Alexander, N. M. Cole, A. T. Galecki, S. Chen, L. V. Nyquist, B. M. Carlson, J. A. Faulkner, and J. A. Ashton-Miller, "Effects of high- and low-velocity resistance training on the contractile properties of skeletal muscle fibers from young and older humans," *J. Appl. Physiol.* **111**, 1021–1030 (2011).

33. Coleman, E. A., "Effect of unilateral isometric and isotonic contractions on the strength of the contralateral limb," *Res. Q.* **40**, 490–495 (1969).

34. Cowan S. M., K. L. Bennell, P. W. Hodges, K. M. Crossley, and J. McConnell, "Delayed onset of electromyographic activity of vastus medialis obliquus relative to vastus lateralis in subjects with patellofemoral pain syndrome," *Arch. Phys. Med. Rehabil.* **82**, 183–189 (2001).

35. Davies, C. T., P. Dooley, M. J. N. McDonagh, and M. White, "Adaptation of mechanical properties of muscle to high force training in man," *J. Physiol.* **365**, 277–284 (1985).

36. Davies, C. T. M., and C. Barnes, "Negative (Eccentric) work. II. Physiological responses to walking uphill and downhill on a motor-driven treadmill," *Ergonomics* **15**, 121–131 (1972).

37. Davies, J., D. F. Parker, O. M. Rutherford, and D. A. Jones, "Changes in strength and cross sectional area of the elbow flexors as a result of isometric strength training," *Eur. J. Appl. Physiol.* **57**, 667–670 (1988).

38. De Lorme, T. L., and A. L. Watkins, *Progressive Resistance Exercise*, Appleton Century Inc., New York, pp. 262–273, 1951.

39. De Luca, C. J., R. S. LeFever, M. P. McCue, and A. P. Xenakis, "Behaviour of human motor units in different muscles during linearly varying contractions," *J. Physiol.* **329**, 113–128 (1982).

40. De Luca, C. J., and Z. Erim, "Common drive of motor units in regulation of muscle force," *Trends Neurosci.* **17**, 299–305 (1994).

41. deVries, H. A., "Efficiency of electrical activity as a measure of the functional state of muscle tissue," *Am. J. Phys. Med.* **47**, 10–22 (1968a).

42. deVries, H. A., "Electromyographic observations of the effects of static stretching upon muscular distress," *Res. Q.* **32**, 468–479 (1961b).

43. deVries, H. A., "Method for evaluation of muscle fatigue and endurance from electromyographic fatigue curve," *Am. J. Phys. Med.* **47**, 125–135 (1968b).

44. deVries, H. A., "Prevention of muscular distress after exercise," *Res. Q.* **32**, 177–185 (1961a).

45. Dias, A., P. Pezarat-Correia, J. Esteves, and O. Fernandes, "The influence of a balance training program on the electromyographic latency of the ankle musculature in subjects with no history of ankle injury," *Phys. Ther. Sport* **12**, 87–92 (2011).

46. Dorel, S., J. M. Drouet, A. Couturier, Y. Champoux, and F. Hug, "Changes of pedaling technique and muscle coordination during an exhaustive exercise," *Med. Sci. Sports Exerc.* **41**, 1277–1286 (2009).

47. Draganich, L. F., and J. W. Vahey, "An in vitro study of anterior cruciate ligament strain induced by quadriceps and hamstrings forces," *J. Orthop. Res.* **8**, 57–63 (1990).

48. Dubo, H. I., M. Peat, D. A. Winter, A. O. Quanbury, D. A. Hobson, T. Steinke, and G. Reimer, "Electromyographic temporal analysis of gait: normal human locomotion," *Arch. Phys. Med. Rehabil.* **57**, 15–20 (1976).

49. Edwards, R. H. T., "Human muscle function and fatigue," in *Human Muscle Fatigue: Physiological Mechanisms*, R. Porter and J. Whelan, eds., Pitman, London, pp. 1–18, 1981.

50. Ericson, M., "On the biomechanics of cycling. A study of joint and muscle load during exercise on the bicycle ergometer," *Scand. J. Rehabil. Med. Suppl.* **16**, 1–43 (1986).

51. Eriksson, E., T. Häggmark, K. H. Kiessling, and J. Karlsson, "Effect of electrical stimulation on human skeletal muscle," *J. Sports Med.* **2**, 18–22 (1981).

52. Falla, D., A. Rainoldi, R. Merletti, and G. Jull, "Myoelectric manifestations of sterno-cleidomastoid and anterior scalene muscle fatigue in chronic neck pain patients," *Clin. Neurophysiol.* **114**, 488–495 (2003).

53. Farina, D., "Counterpoint: spectral properties of the surface EMG do not provide information about motor unit recruitment and muscle fiber type," *J. Appl. Physiol.* **105**, 1673–1674 (2008).

54. Freeman, S., A. Karpowicz, J. Gray, and S. McGill, "Quantifying muscle patterns and spine load during various forms of the push-up," *Med. Sci. Sports Exerc.* **38**, 570–577 (2006).

55. Friden, J., M. Sjostrom, and B. Ekblom, "Myofibrillar damage following intense eccentric exercise in man," *Int. J. Sports Med.* **4**, 170–176 (1983).

56. Froese E. A., and M. E. Houston, "Torque–velocity characteristics and muscle fiber type in human vastus lateralis," *J. Appl. Physiol.* **59**, 309–314 (1985).

57. Fujii, S., and T. Moritani, "Rise rate and timing variability of surface electromyographic activity during rhythmic drumming movements in the world's fastest drummer," *J. Electromyogr. Kinesiol.* **22**, 60–66 (2012).

58. Fujii, S., and T. Moritani, "Spike shape analysis of surface electromyographic activity in wrist flexor and extensor muscles of the world's fastest drummer," *Neurosci. Lett.* **514**, 185–188 (2013).

59. Garnett, R. A. F., M. J. O'Donovan, J. A. Stephens, and A. Taylar, "Motor unit organization of human medial gastrocnemius," *J. Physiol.* **287**, 33–43 (1979).

60. Gerdle B., K. Henriksson-Larsen, R. Lorentzon, and M. L. Wretling, "Dependence of the mean power frequency of the electromyogram on muscle force and fibre type," *Acta Physiol. Scand.* **142**, 457–465 (1991).

61. Gollhofer, A., and H. Kyrolainen, "Neuromuscular control of the human leg extensor muscles in jump exercises under various stretch-load conditions," *Int. J. Sports Med.* **12**, 34–40 (1991).

62. Grabiner M. D., T. J. Koh, and L. F. Draganich, "Neuromechanics of the patellofemoral joint," *Med. Sci. Sports Exerc.* **26**, 10–21 (1994).

63. Gracies, J. M. "Pathophysiology of spastic paresis. II: emergence of muscle overactivity," *Muscle Nerve* **31**, 552–571 (2005).

64. Guidetti, L., G. Rivellini, and F. Figura, "EMG patterns during running: intra- and interindividual variability," *J. Electromyogr. Kinesiol.* **6**, 37–48 (1996).

65. Hagberg, M., and B. E. Ericson, "Myoelectric power spectrum dependence on muscular contraction level of elbow flexors," *Eur. J. Appl. Physiol.* **48**, 147–156 (1982).

66. Hagood, S., M. Solomonow, R. Baratta, B. H. Zhou, and R. D'Ambrosia, "The effect of joint velocity on the contribution of the antagonist musculature to knee stiffness and laxity," *Am. J. Sports Med.* **18**, 182–187 (1990).

67. Häkkinen, K., and Komi P. V., "Effect of explosive type strength training on electromyographic and force production characteristics of leg extensor muscles during concentric and various stretch shortening cycle exercises," *Scand. J. Sports Sci.* **7**, 65–76 (1985).

68. Häkkinen, K., and P. V. Komi, "Electromyographic changes during strength training and detraining," *Med. Sci. Sport. Exerc.* **15**, 455–460 (1983).

69. Häkkinen, K., M. Alen, and P. V. Komi, "Changes in isometric force– and relaxation–time, electromyographic and muscle fiber characteristics of human muscle during strength training and detraining," *Acta Physiol. Scand.* **125**, 573–585 (1985).

70. Häkkinen, K., P. V. Komi, and P. Tesch, "Effect of combined concentric and eccentric strength training and detraining on force–time, muscle fiber and metabolic characteristics of leg extensor muscles," *Scand. J. Sports Sci.* **3**, 50–58 (1981).

71. Häkkinen, K., P. V. Komi, M. Alen, and H. Kauhanen, "EMG, muscle fiber and force production characteristics during 1 year training period in elite weight-lifters," *Eur. J. Appl. Physiol.* **56**, 419–427 (1987).

72. Häkkinen, K., R. U. Newton, S. E. Gordon, et al., "Changes in muscle morphology, electromyographic activity, and force production characteristics during progressive strength training in young and older men," *J. Gerontol. A Biol. Sci. Med. Sci.* **53**, 415–423 (1998).

73. Hautier, C. A., L. M. Arsac, K. Deghdegh, J. Souquet, A. Belli, and J. R. Lacour, "Influence of fatigue on EMG/force ratio and cocontraction in cycling," *Med. Sci. Sports Exerc.* **32**, 839–843 (2000).

74. Henneman, E., and L. M. Mendell, "Functional organization of motoneuron pool and its inputs," *Handbook of Physiol.* **2**, 423–507 (1981).

75. Hermens, H. J., B. Freriks, C. Disselhorst-Klug, and G. Rau, "Development of recommendations for SEMG sensors and sensor placement procedures," *J. Electromyogr. Kinesiol.* **10**, 361–374 (2000).

76. Hodges P. W., and C. A. Richardson, "The influence of isometric hip adduction on quadriceps femoris activity," *Scand. J. Rehabil. Med.* **25**, 57–62 (1993).

77. Hogan, M. C., R. S. Richardson, and S. S. Kurdak, "Initial fall in skeletal muscle force development during ischemia is related to oxygen availability," *J. Appl. Physiol.* **77**, 2380–2384 (1994).

78. Hopf, H. C., R. L. Herbort, M. Gnass, H. Günther, and K. Lowitzsch, "Fast and slow contraction times associated with fast and slow spike conduction of skeletal muscle fibers in normal subject and in spastic hemiparesis," *Z. Neurol.* **206**, 193–202 (1974).

79. Hubley-Kozey, C., and E. M. Earl, "Coactivation of the ankle musculature during maximal isokinetic dorsiflexion at different angular velocities," *Eur. J. Appl. Physiol.* **82**, 289–296 (2000).

80. Hug, F., "Can muscle coordination be precisely studied by surface electromyography?," *J. Electromyogr. Kinesiol.* **21**, 1–12 (2011).

81. Hug, F., D. Bendahan, Y. Le Fur, P. J. Cozzone, and L. Grelot, "Heterogeneity of muscle recruitment pattern during pedaling in professional road cyclists: a magnetic resonance imaging and electromyography study," *Eur. J. Appl. Physiol.* **92**, 334–342 (2004).

82. Hug, F., N. A. Turpin, A. Guevel, and S. Dorel, "Is interindividual variability of EMG patterns in trained cyclists related to different muscle synergies?," *J. Appl. Physiol.* **108**, 1727–1736 (2010).

83. Ikai, M., and T. Fukunaga, "A study on training effect on strength per unit cross-sectional area of muscle by means of ultrasonic measurements," *Int. Z. Angew. Physiol.* **28**, 173–180 (1970).

84. Ingen Schenau, G. J., P. J. Boots, G. de Groot, R. J. Snackers, and van W. W. Woensel, "The constrained control of force and position in multi-joint movements," *Neuroscience* **46**, 197–207 (1992).

85. Irish, S. E., A. J. Millward, J. Wride, B. M. Haas, and G. L. Shum, "The effect of closed-kinetic chain exercises and open-kinetic chain exercise on the muscle activity of vastus medialis oblique and vastus lateralis," *J. Strength Cond. Res.* **24**, 1256–1262 (2010).

86. Ishida K., T. Moritani, and K. Itoh, "Changes in voluntary and electrically induced contractions during strength training and detraining," *Eur. J. Appl. Physiol.* **60**, 244–248 (1990).

87. Izquierdo, M,. J. Ibañez, E. Gorostiaga, M. Garrues, A. Zúñiga, A. Antón, J. L. Larrión, K. Häkkinen, "Maximal strength and power characteristics in isometric and dynamic actions of the upper and lower extremities in middle-aged and older men," *Acta. Physiol. Scand.* **167**, 57–68 (1999).

88. Jacobs, R., M. F. Bobbert, and G. J. van Ingen Schenau, "Function of mono- and biarticular muscles in running," *Med. Sci. Sports Exerc.* **25**, 1163–1173 (1993).

89. Jobe, F. W., D. R. Moynes, J. E. Tibone, and J. Perry, "An EMG analysis of the shoulder in pitching. A second report," *Am. J. Sports Med.* **12**, 218–220 (1984).

90. Johnson, M. A., J. Polgar, D. Weightman, and D. Appleton, "Data on the distribution of fiber types in thirty-six human muscles: an autopsy study," *J. Neurol. Sci.* **18**, 111–129, (1973).

91. Johnson, M. B., and C. L. Johnson, "Electromyographic response of peroneal muscles in surgical and nonsurgical injured ankles during sudden inversion," *J. Orthop. Sports Phys. Ther.* **18**, 497–501 (1993).

92. Jones D. A., D. J. Newham, J. M. Round, and E. J. Tolfree, "Experimental human muscle damage: morphological changes in relation to other indices of damage," *J. Physiol.* **375**, 435–448 (1986).

93. Kellis, E., and V. Baltzopoulos, "Muscle activation differences between eccentric and concentric isokinetic exercise," *Med. Sci. Sports Exerc.* **30**, 1616–1623 (1998).

94. Klein, C. S., C. L. Rice, and G. D. Marsh, "Normalized force, activation, and coactivation in the arm muscles of young and old men," *J. Appl. Physiol.* **91**, 1341–1349 (2001).

95. Kohn, T. A., B. Essén-Gustavsson, and K. H. Myburgh, "Specific muscle adaptations in type II fibers after high-intensity interval training of well-trained runners," *Scand. J. Med. Sci. Sports* **21**, 765–772 (2011).

96. Komi, P. V., and E. R. Buskirk, "Effect of eccentric and concentric muscle conditioning on tension and electrical activity of human muscle," *Ergonomics* **15**, 417–434 (1972).

97. Komi, P. V., and E. R. Buskirk, "Reproducibility of electromyographic measurements with inserted wire electrodes and surface electrodes," *Electromyography* **4**, 357–367 (1970).

98. Komi, P. V., and J. T. Viitasalo, "Changes in motor unit activity and metabolism in human skeletal muscle during and after repeated eccentric and concentric contractions," *Acta Physiol. Scand.* **100**, 246–256 (1977).

99. Komi, P. V., and P. Tesch, "EMG frequency spectrum, muscle structure, and fatigue during dynamic contractions in man," *Eur. J. Appl. Physiol. Occup. Physiol.* **42**, 41–50 (1979).

100. Komi, P. V., J. T. Viitasalo, R. Rauramaa, and V. Vihko, "Effect of isometric strength training on mechanical, electrical and metabolic aspects of muscle function," *Eur. J. Appl. Physiol.* **40**, 45–55 (1978).

101. Komi, P. V., "Training of muscle strength and power: Interaction of neuromotoric, hypertrophic and mechanical factors," *Int. J. Sports Med.* **7**, 10–15 (1986).

102. Konradsen, L., G. Peura, B. Beynnon, and P. Renstrom, "Ankle eversion torque response to sudden ankle inversion Torque response in unbraced, braced, and pre-activated situations," *J. Orthop. Res.* **23**, 315–321 (2005).

103. Konradsen, L., M. Voigt and C. Hojsgaard "Ankle inversion injuries. The role of the dynamic defense mechanism," *Am. J. Sports Med.* **25**, 54–58 (1997).

104. Kouzaki, M., and M. Shinohara, "The frequency of alternate muscle activity is associated with the attenuation in muscle fatigue," *J. Appl. Physiol.* **101**, 715–720 (2006).

105. Krosshaug, T., A. Nakamae, B. P. Boden, et al., "Mechanisms of anterior cruciate ligament injury in basketball: video analysis of 39 cases," *Am. J. Sports Med.* **35**, 359–367 (2007).

106. Kukulka, C. G., and H. P. Clamann, "Comparison of the recruitment and discharge properties of motor units in human brachial biceps and adductor pollicis during isometric contractions," *Brain Res.* **219**, 45–55 (1981).

107. Kupa, E. J., S. H. Roy, S. C. Kandarian, and C. J. De Luca, "Effects of muscle fiber type and size on EMG median frequency and conduction velocity," *J. Appl. Physiol.* **79**, 23–32 (1995).

108. Lam, P. L., and G. Y. Ng, "Activation of the quadriceps muscle during semisquatting with different hip and knee positions in patients with anterior knee pain," *Am. J. Phys. Med. Rehabil.* **80**, 804–808.

109. Lestienne, F., "Effects of inertial load and velocity on the braking process of voluntary limb movements," *Exp. Brain. Res.* **35**, 407–418 (1979).

110. Li, G., T. W. Rudy, M. Sakane, A. Kanamori, C. B. Ma, and S. L. Woo, "The importance of quadriceps and hamstring muscle loading on knee kinematics and in-situ forces in the ACL," *J. Biomech.* **32**, 395–400 (1999).

111. Linssen, W. H., D. F. Stegeman, E. M. Joosten, R. A. Binkhorst, M. J. Merks, H. J. ter Laak, and S. L. Notermans, "Fatigue in type I fiber predominance: a muscle force and surface EMG study on the relative role of type I and type II muscle fibers," *Muscle Nerve* **14**, 829–837 (1991).

112. Liu, G., F. Mac Gabhann, and A. S. Popel, "Effects of fiber type and size on the heterogeneity of oxygen distribution in exercising skeletal muscle," *PLoS ONE* **7** (2012).

113. Macaluso, A., M. A. Nimmo, J. E. Foster, M. Cockburn, N. C. McMillan, and G. De Vito "Contractile muscle volume and agonist–antagonist coactivation account for differences in torque between young and older women," *Muscle Nerve* **25**, 858–863 (2002).

114. Maltais, F., M. J. Sullivan, P. LeBlanc, B. D. Duscha, F. H. Schachat, C. Simard, J. M. Blank, and J. Jobin, "Altered expression of myosin heavy chain in the vastus lateralis muscle in patients with COPD," *Eur. Respir. J.* **13**, 850–854 (1999).

115. Mannion, A. F., G. A. Dumas, J. M. Stevenson and R. G. Cooper, "The influence of muscle fiber size and type distribution on electromyographic measures of back muscle fatigability," *Spine* **23**, 576–584 (1998).

116. McClinton, S., G. Donatell, J. Weir, and B. Heiderscheit, "Influence of step height on quadriceps onset timing and activation during stair ascent in individuals with patellofemoral pain syndrome," *J. Orthop. Sports Phys. Ther.* **37**, 239–244 (2007).

117. McDonagh, M. J. N., and C. T. M. Davies, "Adaptive response of mammalian skeletal muscle to exercise with high loads," *Eur. J. Appl. Physiol.* **52**, 139–155 (1984).

118. McGuire, J., L. A. Green, and D. A. Gabriel, "Task complexity and maximal isometric strength gains through motor learning," *Physiol. Rep.* **2**, 12218 (2014).

119. McIlwain, J. S., and K. C. Hayes, "Dynamic properties of human motor units in the Hoffmann—reflex and M response," *Am. J. Phys. Med.* **56**, 122–135 (1977).

120. Melchiorri, G., and A. Rainoldi, "Mechanical and myoelectric manifestations of fatigue in subjects with anorexia nervosa," *J. Electromyogr. Kinesiol.* **18**, 291–297 (2008).

121. Merletti, R., D. Farina, M. Gazzoni, and M. P. Schieroni, "Effect of age on muscle functions investigated with surface electromyography," *Muscle Nerve* **25**, 65–76 (2002).

122. Merlo, A., D. Farina, and R. Merletti. "A fast and reliable technique for muscle activity detection from surface EMG signals," *IEEE Trans. Biomed. Eng.* **50**, 316–323 (2003).

123. Mero, A. A., J. J. Hulmi, H. Salmijärvi, M. Katajavuori, M. Haverinen, J. Holviala, T. Ridanpää, K. Häkkinen, V. Kovanen, J. P. Ahtiainen, and H. Selänne, "Resistance training induced increase in muscle fiber size in young and older men," *Eur. J. Appl. Physiol.* **113**, 641–650 (2013).

124. Mesin, L., C. Cescon, M. Gazzoni, R. Merletti, and A. Rainoldi, "A bi-dimensional index for the selective assessment of myoelectric manifestations of peripheral and central muscle fatigue," *J. Electromyogr. Kinesiol.* **19**, 851–863 (2009).

125. Milner-Brown, H. S., and R. B. Stein, "The relation between the surface electromyogram and muscular force," *J. Physiol.* **246**, 549–569 (1975).

126. Milner-Brown, H. S., R. B. Stein, and R. Yemm, "Changes in firing rate of human motor units during linearly changing voluntary contractions," *J. Physiol.* **230**, 371–390 (1973).

127. Minetto, M. A., A. Botter, O. Bottinelli, D. Miotti, R. Bottinelli, and G. D'Antona. "Variability in muscle adaptation to electrical stimulation," *Int. J. Sports Med.* **34**, 544–553 (2013).

128. Mitchell, A., R. Dyson, T. Hale, and C. Abraham, "Biomechanics of ankle instability. Part 1: Reaction time to simulated ankle sprain," *Med. Sci. Sports Exerc.* **40** 1515–1521 (2008).

129. Moritani, T., and H. A. deVries, "Neural factors versus hypertrophy in the time course of muscle strength gain," *Am. J. Phys. Med.* **58**, 115–130 (1979).

130. Moritani, T., and H. A. deVries, "Potential for gross muscle hypertrophy in older men," *J. Gerontol.* **35**, 672–682 (1980).

131. Moritani, T., and H. A. deVries, "Reexamination of the relationship between the surface integrated electromyogram (IEMG) and force of isometric contraction," *Am. J. Phys. Med.* **57**, 263–277 (1978).

132. Moritani, T., and M. Muro, "Motor unit activity and surface electromyogram power spectrum during increasing force of contraction," *Eur. J. Appl. Physiol.* **56**, 260–265 (1987).

133. Moritani, T., "Neuromuscular adaptations during the acquisition of muscle strength, power and motor tasks," *J. Biomech.* **26**, 95–107 (1993).

134. Moritani, T., "Time course of adaptations during strength and power training," in *Strength and Power in Sport*, P. V. Komi, ed., Blackwell Scientific, Boston, pp. 266–278, 1992.

135. Moritani, T., A. Nagata, and M. Muro, "Electromyographic manifestations of muscular fatigue," *Med. Sci. Sports Exerc.* **14**, 198–202 (1982).

136. Moritani, T., K. Ishida, and S. Taguchi, "Physiological effects of stretching upon DOMS: electrophysiological analyses," *Descente Sports Sci.* **8**, 212–220 (1987).

137. Moritani, T., M. Muro, and A. Nagata, "Intramuscular and surface electromyogram changes during muscle fatigue," *J. Appl. Physiol.* **60**, 1179–1185 (1986b).

138. Moritani, T., M. Muro, A. Kijima, and M. J. Berry, "Intramuscular spike analysis during ramp force output and muscle fatigue," *Electromyogr. Clin. Neurophysiol.* **26**, 147–160 (1986a).

139. Moritani, T., M. Muro, K. Ishida, and S. Taguchi, "Electromyographic analyses of the effects of muscle power training," *J. Sports Med. Sci.* **1**, 23–32 (1987).

140. Moritani, T., S. Muramatsu, and M. Muro, "Activity of motor units during concentric and eccentric contractions," *Am. J. Phys. Med.* **66**, 338–435 (1988).

141. Moritani, T., F. D. Gaffney, T. Charmichael, and J. Hargis, "Interrelationships among muscle fiber types, electromyogram, and blood pressure during fatiguing isometric contraction" in *Biomechanics IX-A*, D. A. Winter, ed., Human Kinetics Publishers, Champaign, IL, 287–292 (1985).

142. Munn, J., S. J. Sullivan, and A. G. Schneiders, "Evidence of sensorimotor deficits in functional ankle instability: a systematic review with meta-analysis," *J. Sci. Med. Sport* **13**, 2–12 (2010).

143. Muro, M., A. Nagata, and T. Moritani, "Analysis of myoelectric signals during dynamic and isometric contractions," in H. Matsui, and K. Kobayashi eds., *Biomechanics VIII-A*, Human Kinetics Publishers, Champaign, IL, pp. 432–439 (1983).

144. Nagai, K., M. Yamada, K. Uemura, Y. Yamada, N. Ichihashi, and T. Tsuboyama, "Differences in muscle coactivation during postural control between healthy older and young adults," *Arch. Gerontol. Geriatr.* **53**, 338–343 (2011).

145. Narici, M. V., G. S. Roi, L. Landoni, A. E. Minetti, and P. Cerretelli, "Changes in force, cross-sectional area and neural activation during strength training and detraining of the human quadriceps," *Eur. J. Appl. Physiol.* **59**, 310–319 (1989).

146. Newham, D. J., K. R. Mills, B. M. Quigley, and R. H. T. Edwards, "Pain and fatigue after concentric and eccentric muscle contractions," *Clin. Sci.* **64**, 55–62 (1983).

147. Nicol, C., P. V. Komi, T. Horita, and H. Kyrolainen, "Reduced stretch-reflex sensitivity after exhausting stretch-shortening cycle exercise," *Eur. J. Appl. Physiol.* **72**, 401–409 (1996).

148. Norman, R. W., and P. V. Komi, "Electromechanical delay in skeletal muscle under normal movement conditions," *Acta Physiol. Scand.* **106**, 241–248 (1979).

149. Osternig, L. R., J. Hamill, D. M. Corcos, and J. Lander, "Electromyographic patterns accompanying isokinetic exercise under varying speed and sequencing conditions," *Am. J. Phys. Med.* **63**, 289–297, (1984).

150. Panagiotopoulos, E., P. Strzelczyk, M. Herrmann, and G. Scuderi, "Cadaveric study on static medial patellar stabilizers: the dynamizing role of the vastus medialis obliquus on medial patellofemoral ligament," *Knee Surg. Sports Traumatol. Arthrosc.* **14**, 7–12 (2006).

151. Paoli, A., Q. F. Pacelli, P. Cancellara, L. Toniolo, T. Moro, M. Canato, D. Miotti, and C. Reggiani, "Myosin isoforms and contractile properties of single fibers of human latissimus dorsi muscle," *Biomed. Res. Int.* (2013).

152. Pedrinelli, R., L. Marino, G. Dell'Omo, G. Siciliano, and B. Rossi, "Altered surface myoelectric signals in peripheral vascular disease: correlations with muscle fiber composition," *Muscle Nerve* **21**, 201–210 (1998).

153. Petrofsky, J. S., and A. R. Lind, "Frequency analysis of the surface electromyogram during sustained isometric contractions," *Eur. J. Appl. Physiol.* **43**, 173–182 (1980a).

154. Petrofsky, J. S., and A. R. Lind, "The influence of temperature on the amplitude and frequency components of the EMG during brief and sustained isometric contractions," *Eur. J. Appl. Physiol.* **44**, 189–200 (1980b).

155. Power, G. A., B. H. Dalton, D. G. Behm, A. A Vandervoort, T. J. Doherty, and C. L. Rice, "Motor unit number estimates in Masters runners: use it or lose it?," *Med. Sci. Sports Exerc.* **42**, 1644–1650 (2010).

156. Prasartwuth, O., T. J. Allen, J. E. Butler, S. C. Gandevia, and J. L. Taylor. "Length-dependent changes in voluntary activation, maximum voluntary torque and twitch responses after eccentric damage in humans," *J. Physiol.* **571**, 243–252 (2006).

157. Psek, J. A., and E. Cafarelli, "Behavior of coactive muscles during fatigue," *J. Appl. Physiol.* **74**, 170–175 (1993).

158. Raikova, R. T., "Some mechanical considerations on muscle coordination," *Motor Control* **4**, 89–96 (2000).

159. Rainoldi, A., and M. Gazzoni, "Strength and conditioning: biological principles and practical applications," *Neuromuscular Physiology*, M. Cardinale, R. Newton, and K. Nosaka, eds., 2011, 17–27, Wiley-Blackwell, Chichester, West Sussex, UK.

160. Rainoldi, A., D. Falla, R. Mellor, K. Bennell, and P. Hodges, "Myoelectric manifestations of fatigue in vastus lateralis, medialis obliquus and medialis longus muscles," *J. Electromyogr. Kinesiol.* **18**, 1032–1037 (2008a).

161. Rainoldi, A., Gazzoni M., and G. Melchiorri, "Differences in myoelectric manifestations of fatigue in sprinters and long distance runners," *Physiol. Meas.* **29**, 331–340 (2008c).

162. Rainoldi, A., M. Gazzoni, and R. Casale, "Surface EMG signal alterations in carpal tunnel syndrome: a pilot study," *Eur. J. Appl. Physiol.* **103**, 233–242 (2008b).

163. Rainoldi, A., M. Gazzoni, R. Merletti, and M. A. Minetto, "Mechanical and EMG responses of the vastus lateralis and changes in biochemical variables to isokinetic exercise in endurance and power athletes," *J Sports Sci.* **26**, 321–331 (2008d).

164. Rasch, P. J., and L. E. Morehouse, "Effect of static and dynamic exercise on muscular strength and hypertrophy," *J. Appl. Physiol.* **11**, 29–34 (1957).

165. Rinaldo, N., G. Coratella, G. Boccia, D. Dardanello, A. Rainoldi, and F. Schena, "Different myoelectric manifestations of fatigue in ambulatory muscles of COPD patients

vs. healthy male subjects," *Proceedings of European Sport Medicine congress of EFSMA September 25–28*, 2013, Strasbourg, France.

166. Rodahl, K., and S. M. Horvath, *Muscle as a Tissue*, Academic Medicine, McGraw-Hill, New York, 1962.

167. Rutherford, O. M., and D. A. Jones, "The role of learning and coordination in strength training," *Eur. J. Appl. Physiol.* **55**, 100–105 (1986).

168. Ryan, M. M., and R. J. Gregor, "EMG profiles of lower extremity muscles during cycling at constant workload and cadence," *J. Electromyogr. Kinesiol.* **2**, 69–80 (1992).

169. Sadoyama, T., T. Masuda, H. Miyata, and S. Katsuta, "Fiber conduction velocity and fiber composition in human vastus lateralis," *Eur. J. Appl. Physiol.* **57**, 767–771 (1988).

170. Sale, D. G., "Neural adaptation to resistance training," *Med. Sci. Sports Exerc.* **20**, S135–S145 (1988).

171. Sale, D. G., "Neural adaptation to strength training," in *Strength and Power in Sport*, P. V. Komi, ed., Blackwell Scientific, Boston, pp. 249–265, 1991.

172. Santello, M., "Review of motor control mechanisms underlying impact absorption from falls," *Gait Posture* **21**, 85–94 (2005).

173. Sbriccoli, P., V. Camomilla, A. Di Mario, F. Quinzi, F. Figura, and F. Felici, "Neuromuscular control adaptations in elite athletes: the case of top level karateka," *Eur. J. Appl. Physiol.* **108**, 1269–1280 (2009).

174. Seynnes, E. R., M. de Boer, and M. V. Narici, "Early skeletal muscle hypertrophy and architectural changes in response to high-intensity resistance training.," *J. Appl. Physiol.* **102**, 368–373 (2007).

175. Sheth, P., B. Yu, E. R. Laskowski, and K. N. An, "Ankle disk training influences reaction times of selected muscles in a simulated ankle sprain," *Am. J. Sports Med.* **25**, 538–543 (1997).

176. Smith, T. O., D. Bowyer, J. Dixon, R. Stephenson, R. Chester, and S. T. Donell, "Can vastus medialis oblique be preferentially activated? A systematic review of electromyographic studies," *Physiother. Theory Pract.* **25**, 69–98 (2009).

177. So, R. C., M. A. Tse, and S. C. Wong, "Application of surface electromyography in assessing muscle recruitment patterns in a six-minute continuous rowing effort," *J. Strength Cond. Res.* **21**, 724–730 (2007).

178. Souza, D. R., and M. T. Gross, "Comparison of vastus medialis obliquus: vastus lateralis muscle integrated electromyographic ratios between healthy subjects and patients with patellofemoral pain," *Phys. Ther.* **71**, 310–316; discussion 317–320 (1991).

179. Spairani, L., M. Barbero, Cescon, C., et al., "An electromyographic study of the vastii muscles during open and closed kinetic chain submaximal isometric exercises," *Int. J. Sports Phys. Ther.* **7**, 617–626 (2012).

180. Staude, G. H., "Precise onset detection of human motor responses using a whitening filter and the log-likelihood-ratio test," *IEEE Trans. Biomed. Eng.* **48**, 1292–1305 (2001).

181. Stephens, J. A., and T. P. Usherwood, "The mechanical properties of human motor units with special reference to their fatigability and recruitment threshold," *Brain Res.* **125**, 91–97 (1977).

182. Stone, M., Plisk S., and Collins D., "Training principles: evaluation of modes and methods of resistance training—a coaching perspective," *Sports Biomech.* **1**, 79–103 (2002).

183. Suter, E., W. Herzog, J. Sokolosky, J. P. Wiley, and B. R. Macintosh, "Muscle fiber type distribution as estimated by Cybex testing and by muscle biopsy," *Med. Sci. Sports Exerc.* **25**, 363–370, (1993).

184. Tamaki, H., K. Kitada, T. Akamine, F. Murata, T. Sakou, and H. Kurata, "Alternate activity in the synergistic muscles during prolonged low-level contractions," *J. Appl. Physiol.* **84**, 1943–1951 (1998).

185. Tomasoni, E., M. Romanazzi, G. Boccia, and A. Rainoldi, "sEMG assessment of upper limb muscles during dynamical contractions in different instability conditions," *Sport Sci. Health* **8**, 56–57 (2012).

186. Tyler, A. E., and R. S. Hutton, "Was Sherrington right about co-contractions?," *Brain Res.* **370**, 171–175 (1986).

187. Viitasalo, J. T., and P. V. Komi, "Effects of fatigue on isometric force– and relaxation–time characteristics in human muscle," *Acta Physiol. Scand.* **111**, 87–95 (1981).

188. Whittom, F., J. Jobin, P. M. Simard, et al., "Histochemical and morphological characteristics of the vastus lateralis muscle in patients with chronic obstructive pulmonary disease," *Med. Sci. Sports Exerc.* **30**, 1467–1474 (1998).

189. Willis, F. B., E. J. Burkhardt, J. E. Walker, M. A. Johnson, and T. D. Spears, "Preferential vastus medialis oblique activation achieved as a treatment for knee disorders," *J. Strength Cond. Res.* **19**, 286–291 (2005).

190. Winter, D. A., and H. J. Yack, "EMG profiles during normal human walking: stride-to-stride and inter-subject variability," *Electroencephalogr. Clin. Neurophysiol.* **67**, 402–411 (1987).

191. Zebis, M. K., L. L. Andersen, J. Bencke, M. Kjaer, and P. Aagaard, "Identification of athletes at future risk of anterior cruciate ligament ruptures by neuromuscular screening," *Am. J. Sports Med.* **37**, 1967–1973 (2009).

192. Zijdewind, I., and D. Kernell, "Fatigue associated EMG behavior of the first dorsal interosseous and adductor pollicis muscles in different groups of subjects," *Muscle Nerve* **17**, 1044–1054 (1994).

20

SURFACE ELECTROMYOGRAPHY FOR MAN–MACHINE INTERFACING IN REHABILITATION TECHNOLOGIES

D. Farina and M. Sartori

Department of Neurorehabilitation Engineering, Bernstein Focus Neurotechnology Göttingen, Bernstein Center for Computational Neuroscience, University Medical Center Göttingen, Georg-August University, Göttingen, Germany

20.1 INTRODUCTION

Neurorehabilitation systems are used to replace, restore, or neuromodulate impaired motor functions. For this purpose, an interface with the patient's nervous system needs to be established. This interfacing may occur at various levels along the neuromuscular pathways. It can be established directly at the brain, at the spinal cord, at the nerve, or at the muscle levels. Among these possibilities, muscle interfacing, by means of electromyography (EMG) signals, is currently the only viable solution for clinical applications on a large scale [17,34].

In applications for man–machine interfacing, muscles are used as biological amplifiers of efferent nerve activity because of the one-by-one association between action potentials traveling along the axons of motor neurons and the electrical activity generated in the innervated muscle fibers [16,17,27]. With respect to direct nerve interfacing, muscle recordings do not necessarily need invasive techniques and provide greater signal to noise ratios. Therefore, the use of EMG can be seen as a general neural interface providing information on the activity of the motor neurons innervating the target muscle [19].

Surface Electromyography: Physiology, Engineering, and Applications, First Edition.
Edited by Roberto Merletti and Dario Farina.
© 2016 by The Institute of Electrical and Electronics Engineers, Inc. Published 2016 by John Wiley & Sons, Inc.

The concept of muscle signals as biologically amplified nerve signals has been maximally exploited in the surgical intervention known as targeted muscle reinnervation (TMR) [27]. This procedure is usually applied to patients with highly proximal amputations. It consists in redirecting residual nerves that would normally innervate muscles of missing limbs and degrees of freedoms (DOFs), to accessory intact muscles. In this context, nerves that used to innervate distal muscles such as those in the amputated hand or wrist are redirected into proximal, intact muscles such as the pectoralis muscle. Once the reinnervation is complete, the newly innervated muscles are used as source of EMG signals that now also reflect the activity of the nerves previously controlling the phantom limb and DOFs. In this context, the EMG activity recorded from the reinnervated muscles is not used for the functional mechanical action of the underlying muscle actuator. Rather, muscles are solely used for their ability of amplifying the weak neural signals conveyed by the nerves. Figure 20.1 shows this concept in a patient 10 months after TMR intervention and three years after shoulder disarticulation. Figure 20.1 shows high-density EMG activity using color maps. In this case, it is possible to observe distinctive EMG patterns across the reinnervated muscles recurring as a function of the activated DOF in the phantom limb. This suggests that TMR intervention was successful in distributing nerve signals spatially across large shoulder and trunk muscles so that function-specific EMG patterns could be recorded using dedicated arrays.

This chapter describes the use of surface EMG for establishing intuitive man–machine interfaces for rehabilitation settings and technologies. The EMG is primarily used as source of neural information, as described above, to provide control signals for external devices. Section 20.2 describes the general concepts underlying the extraction of control signals from surface EMG recordings. Sections 20.3–20.5 describe examples of neurorehabilitation technologies based on surface EMG. Since the field is very broad, the applications outlined in this chapter are a limited set of representative examples that aim at exemplifying the main concepts.

20.2 EXTRACTION OF CONTROL SIGNALS FROM THE SURFACE EMG

The surface EMG signal can be used to trigger or continuously control external assistive devices including powered orthoses and prostheses [19]. EMG-to-device control can be achieved with two main EMG-processing approaches [51]. The first is a data-driven approach. It is based on extracting information from the available EMG data without extrapolating or reconstructing information on the associated neurophysiological and musculoskeletal processes. This class of methods is based on machine learning techniques. The idea is to associate a desired control command or an observed action with the underlying EMG patterns [18,33,35]. The second is a model-driven approach. It makes use of the surface EMG as the input to subject-specific, physically correct models of the musculoskeletal system. These are used to reconstruct all the neurophysiological and musculoskeletal transformations from the EMG onset to body function. The forward computation of joint moments

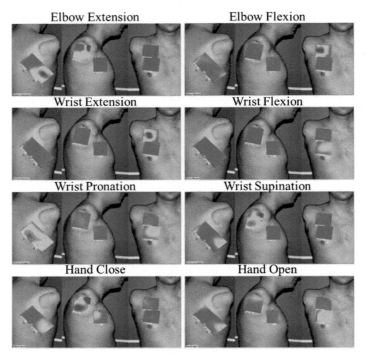

FIGURE 20.1 Multichannel surface EMG signals have been recorded from a patient who underwent TMR. The color maps represent the EMG amplitude during eight tasks (see labels in each panel) attempted by the subject with his phantom limb. The locations of the color maps correspond to the location of the EMG electrode grids (8 × 8 electrodes, 10-mm interelectrode distance) used for signal recordings. The EMG represented amplitude is the root mean square value, which has been normalized, for each task, with respect to the maximum across all grids (red colors are high values and blue colors are low values in the color maps). Values of root mean square have been interpolated in space by a factor 8 for better graphical representation. The activity represents the degree of activation of the nerves previously innervating the phantom limb and now redirected to innervate trunk muscles. The muscles from which the recordings are obtained are exclusively used to amplify and make available the nerve activity. Reproduced from Amsuess et al. [3] with permission.

from EMG recordings and joint kinematics is an example of these approaches (i.e., also see Chapter 9) [20,31,42,43,45,48].

20.2.1 Data-Driven: Machine Learning

From a surface EMG recording (single or multichannel), it is possible to extract features of various complexities that characterize the signal in the time or frequency (or joint) domain (see Chapters 4 and 5). These features are not necessarily related to the physical and physiological processes underlying the generation of the EMG signal. However, they are useful to characterize signal properties descriptive and distinctive of a specific task or generated force. For example, the description of the EMG from its power spectrum does not imply that the EMG is generated as the sum of pure sinusoids, yet it is useful for describing specific signal patterns.

A widely used approach to process the EMG for man–machine interfacing is signal classification or clustering. According to this scheme, it is possible to identify distinctive and consistent EMG patterns for an executed or attempted motor task. It is therefore possible to associate a task (class) with the EMG recording in each time interval. The association can be performed either by learning the mapping between EMG features and classes from a (training) set of labeled samples (supervised classification) or without any training set (clustering). Examples of common classifiers used for EMG are neural networks, support vector machines, linear discriminant analysis, and *k*-nearest neighbourhoods whereas classic features are EMG amplitude, number of zero crossings, and waveform length [38]. A classic example of EMG classification in neurotechnologies is the control of active prostheses (i.e., see Section 20.3).

When the EMG is mapped into continuous signals, instead of a discrete number of classes, the approach is called regression. Regression allows proportionally controlling a device based on time-varying features such as EMG amplitude. As for EMG classification, the regression into continuous signals can be learned based on labeled examples in a training set or be unsupervised. Linear regression is a simple example of supervised method. In linear regression, the mapping function is linear and the unknown coefficients are estimated from the data available in the training set. Many other approaches, in addition to linear regression, have been proposed in the specialized literature for mapping EMG into kinematics and force (e.g., Jiang and co-workers [23,24]).

It is also possible to extract continuous control signals from the EMG without a training set with labeled samples. For example, multichannel EMG signals can be factorized into a set of low-dimensional signals, used for control, and a set of coefficients. This factorization usually requires some constraints, such as positivity of the extracted low-dimensional signals. In this context, the non-negative matrix factorization (NMF) approach has been commonly used [29]. This algorithm is also commonly applied for analyzing a multi-muscle EMG recording into motor modules and activation coefficients (i.e., also see Section 20.4).

20.2.2 Model-Driven: EMG-Driven Musculoskeletal Modeling

EMG signals can also be used as neural input to models of the human musculoskeletal system [8,31,42,43,45,47]. These are referred to as EMG-driven models and are used to compute instantaneous modulations of a number of internal EMG-dependent variables of physiological nature. Importantly, these variables would not be accessible by the sole analysis of EMG data or by the isolated analysis of human movement. Therefore, the combination of EMG with subject-specific musculoskeletal modeling allows establishing a deeper interface with the subject than traditional EMG processing and movement analysis techniques. Predicted variables may include muscle force [5,45], joint moment [8,31,45], joint stiffness [39,44], and bone-to-bone compression [21]. These variables can be applied to generate direct EMG-to-device control commands in man–machine interfacing scenarios [7,10,20,36]. EMG-driven modeling is described in greater detail in Chapter 9. A typical example is the projection of multi-muscle EMG signals to the joint moment level for the proportional control of powered exoskeletons [10,20,46]. In this application, the exoskeleton provides assistive joint moments proportionally to the user's strength. The user's strength is quantified online via EMG-dependent estimates of muscle force and joint moment (i.e., see Section 20.4 for more details) [20]. The exoskeleton controller continually compensates for the discrepancy between model-based estimates of the user's strength and desired target joint strength via the injection of assistive joint moment. This approach has the benefit of keeping the user actively involved in the human–machine control loop. Furthermore, it results in control strategies that account for the evolving user's motor ability, which may vary throughout a rehabilitation therapy. In a similar way, neuromusculoskeletal models can be used to drive or synthetize human-like viscoelastic properties in prosthetic limbs [15,32]. Contrary to data-driven approaches that do not make assumptions on the relation between EMG and biomechanical variables, model-based approaches are based on explicitly modeling these associations and transformations, often on a subject-specific basis, which requires subject-specific model scaling and calibration procedures [10,20,43,45,47,48].

20.3 FUNCTION REPLACEMENT: ACTIVE PROSTHESES

In the following paragraphs of this section, we will outline representative examples of neurotechnologies that make use of EMG signals for device control purposes. We will provide examples in the three main areas of neurorehabilitation: replacement, restoration, and neuromodulation. In these examples, either data-driven or model-driven approaches are used for processing the EMG and for generating control signals to external devices.

Replacement implies substituting anatomically or functionally missing body parts with robotic devices (i.e., prostheses). In the following paragraphs of this section, we will refer mainly to upper limb prostheses that require direct control by the user.

State-of-the-art commercial prostheses for upper limb function are based on very simple algorithms for EMG processing. A common scheme is shown in Fig. 20.2, which consists in the proportional control of one or two DOFs sequentially. According to this approach, the EMG signals recorded from two muscle sites are used to proportionally (i.e., according to signal amplitude) control one DOF at a time. For example, the user may flex the prosthesis wrist by contracting one muscle and extend the wrist by contracting a second muscle. The control is simple but extremely robust, and robustness is a prerequisite for these clinical applications. The same scheme is sometimes used to control more than one DOF, by switching mode of operation with co-contraction. When the user co-contracts the two targeted muscles, the system switches into the proportional control of another DOF (e.g., wrist rotation) and returns to the previous DOF following another co-contraction. The resultant control scheme is a state machine with two states (i.e., the selected DOFs). It has been shown that extension of this concept to more than two DOFs is not practically

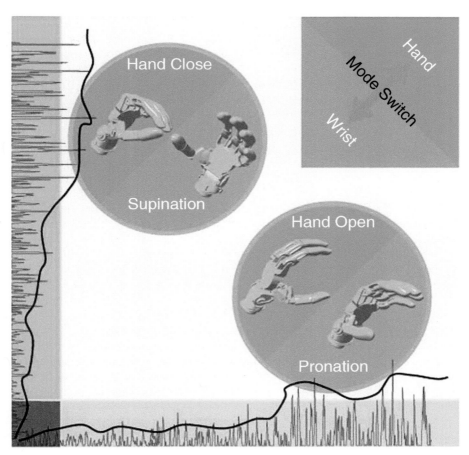

FIGURE 20.2 Schematic diagram of a conventional myoelectric control approach implemented in several commercial prostheses. Two channels of surface EMG (shown schematically in the horizontal and vertical axes) are used to control one DOF at a time. The EMGs are detected from two remnant muscles above the amputation. For example, in transradial amputees, one channel may be obtained from the flexors of forearm and the other channel from the extensors of the forearm. At a given time, the two channels control either wrist flexion and extension or hand open and close. A threshold for each channel is selected, and when the threshold is exceeded, the corresponding DOF is activated. When both thresholds are exceeded concurrently, the controller mode is switched to the other DoF. Despite its .limited functionality, this approach constitutes a very robust way for prosthesis control. Reproduced from Amsuess et al. [2] with permission.

feasible [2]. Although robust, this control method (and similar methods developed from this concept) is unintuitive and provides limited re-gain in functionality. For this reason, the abandonment rates of this type of prostheses are very high, with peaks of 40% to 50% [6].

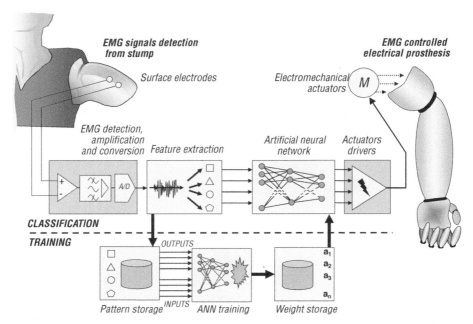

FIGURE 20.3   Block diagram of a classic myocontrol system for active prosthesis. From the surface EMG signals detected from muscles above the amputation and under the control of the user, a feature set is extracted and then classified to define the commands to be delivered to the actuators of the prosthesis. In this scheme, an artificial neural network (ANN) is shown as classifier, as a representative example. Reproduced from Pozzo et al. [41] with permission.

In order to increase the number of functions that can be controlled by EMG signals, various methods for pattern recognition of multichannel EMG recordings have been developed over the past decades. These methods assume that there is an association between the activity of the remnant muscles above the amputation and the intended movement of the phantom limb. A classic scheme for pattern-recognition-based control is shown in Fig. 20.3. This approach extracts features from the EMG data, reduces their dimensionality (if needed), and stratifies the features into a finite set of classes each one of these referring to a specific motor tasks. With respect to the proportional direct control of industrial systems, this method allows increasing substantially the number of controlled DOFs. For example, it has been shown that multichannel EMG signals detected at the forearm can be classified with high accuracy for more than 10 hand tasks (i.e., classes) [30]. This often requires a sample of data for which the labeling is known (supervised pattern recognition).

The high performance achieved by classification of EMG signals is, however, often achieved in ideal laboratory conditions. That is, in offline scenarios and not during the daily use of an assistive device. When transferred to more challenging conditions, pattern recognition of the EMG has substantially lower performance levels than what is observed in laboratory settings. This impedes the direct translation of these methods to large clinical numbers [24]. For example, it is now well known that donning and doffing of the prosthesis correspond to a change in the relative location of the electrodes (that are mounted inside the socket) with respect to the muscles. Therefore, the myocontrol performance worsens substantially when donning/doffing the prosthesis [3]. A similar effect has been observed for changing arm position. If the myocontrol algorithm is trained in one arm posture, performance worsens for different arm postures [47]. Deterioration of performance is also due to a change in the strategy of the user in activating the control muscles [3].

Beside the nonideal daily-life conditions that impact EMG classification, this approach has also limitations intrinsic in the underlying control scheme. Specifically, the control is sequential; that is, the user cannot select more than one DOF motion at a time. Moreover, contrary to clinical/commercial devices, the pattern-recognition-based control operates according to on–off bases and is in general not proportional (although proportionality can be added in such scheme). These intrinsic characteristics of pattern-recognition-based control make it distant from the natural control of the human limbs that is based on the simultaneous and proportional activation of multiple DOFs [25].

Alternative to classifying or clustering the EMG into a finite number of tasks/motions, it is also possible to establish a continuous mapping between the EMG signal (or signal features) and kinematic variables that define the prosthesis position in time. This approach is based on regression analysis and has been outlined in the previous sections. Various regression methods have been compared for myocontrol of prostheses [23]. An interesting recent method merges a classification scheme with regression by separating the control of the hand from the control of the wrist. Specifically, the wrist DOFs are controlled with a regression approach while the grasping type is classified. With this approach, users could control a modern multiple DOF prosthesis (i.e., the Michelangelo Hand by OttoBock) in an intuitive manner, without the use of co-contraction switches. Figure 20.4 shows an example of control sequences using this method.

When developing algorithms for myocontrol of prostheses, it is important to emphasize the relevance of the validation approach. For several decades, academic work in the field has been mainly limited to tests on able-bodied subjects, in laboratory conditions, and on fixed posture, often reporting results only for offline analyses. It has been recently recognized, however, that these validation approaches are not representative of the real use of the prostheses, so that translation of these results to clinical applications is not possible. This problem may have contributed to a

FIGURE 20.4 Exemplary sequence of a prosthesis user performing simultaneous wrist flexion and supination followed by a pinch grip to grasp a mug. The control algorithm uses regression for wrist control and classification for the selection of grasp type. Reproduced from Sebastian Asmuess and Dario Farina [2,3] with permission.

gap between the academic achievements in the field and the industrial exploitation of these achievements in the last three decades [24]. As recently highlighted [1], myocontrol algorithms need to be validated in real users (amputees), wearing real prostheses (with customized sockets), during activities that resemble in all respects those of daily living. Although complicated by the need of many expertise and facilities, this validation approach is the only one that provides a direct information on the benefit of a new algorithm with respect to the state of the art.

20.4 FUNCTION RESTORATION: ORTHOTICS

Restoring musculoskeletal function implies re-establishing neuromuscular pathways that are impaired. Alternatively, mechanical forces can be injected into an individual's impaired musculoskeletal system to support actuation and movement of limbs. The use of EMG-driven musculoskeletal modeling is highly relevant to these applications. In this context, the ability of predicting EMG-dependent muscle and joint dynamics is crucial for enabling intuitive control of wearable assistive devices, such as powered orthoses [10,20,44,45,47].

The EMG-driven modeling methodology described in Chapter 9 can be used to estimate the user's strength during dynamic motor tasks such as locomotion [31,45]. A viable way to estimate strength-related variables is to convert multi-muscle EMG signals into single-joint or multi-joint torque estimates [9,20,42–47]. If the user's strength can be estimated on a real-time and context-dependent basis (i.e., performed motor task), then control strategies can be established that provide only the minimally required level of physical support [10,20]. Figure 20.5 shows how EMG signals recorded from 16 muscle groups can be used to predict joint moments simultaneously produced about six DOFs in the lower extremity during a variety of biomechanically different locomotion tasks. Figure 20.5 also shows that joint moments can be equally well predicted using a five-dimensional set of excitation primitives, or XPs [43]. These are five Gaussian curves that are parameterized as a function of the gait cycle. They recruit a total 34 musculotendon units (MTUs) apportioned into seven groups of synergistic muscles [43]. In this context, a five-dimensional representation of the 16-dimensional EMG signals can be used to drive 34 MTUs in a lower extremity model and predict the resulting musculoskeletal forces during locomotion. This approach allows drastically reducing the number of needed EMG channels in experimental scenarios, thus increasing the robustness of the methodology, which would otherwise require 16 recording sites [43]. In this context, gait segmentation algorithms can be used to index in real-time the locomotion gait cycle and recruit the virtual MTUs in the musculoskeletal model based on the five XPs [43].

In this context, exoskeleton controllers can be implemented to operate at the joint torque level in proportion to the predicted user's effort. The user's effort could be quantified using EMG-driven or XP-driven modeling. An example of this is outlined in Fig. 20.6A, which illustrates a powered orthosis of the knee joint controlled by an EMG-driven model of the single knee flexion–extension (KneeFE) DOF [20]. Figure 20.6B depicts the EMG sensors used to record the muscle activity. In this

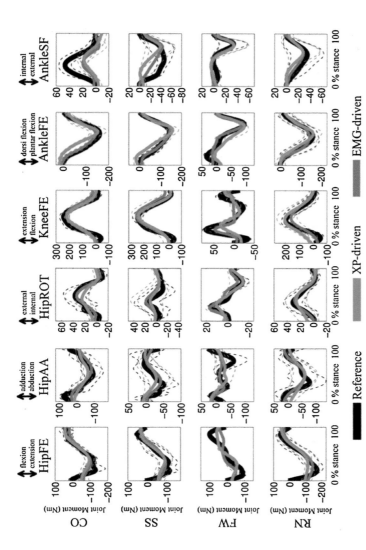

FIGURE 20.5 Experimental joint moments as well as those predicted using the excitation primitive (XP)-driven and electromyography (EMG)-driven musculoskeletal models, respectively. The ensemble average (*filled lines*) and standard deviations (*dotted lines*) curves are depicted for the predicted (i.e., XP-driven and EMG-driven) and experimental (i.e., Reference) joint moments about six degrees of freedom (DOFs) including: hip flexion–extension (HipFE), hip adduction–abduction (HipAA), hip internal–external rotation (HipROT), knee flexion–extension (KneeFE), ankle plantar–dorsi flexion (AnkleFE), and ankle subtalar–flexion (AnkleSF). Results are shown for four motor tasks including: fast walking (FW), running (RN), side-stepping (SS), and cross-over (CO) cutting maneuvers. The reported data are from the stance phase with 0% being heel-strike and 100% toe-off events. Data are from two healthy men (age 28 and 26 years, height 183 and 167 cm, mass 67 and 73 kg). Data are averaged across eight repetitions of each motor task. Figure as originally published in Sartori et al. [43].

particular application, only five thigh muscles were used to determine the subject's effort including vastus medialis, vastus lateralis, rectus femoris, lateral hamstrings, and medial hamstrings. Figure 20.6C depicts the powered orthosis control structure in greater details. In this case, the EMG-driven model uses KneeFE angles and EMG signals to estimate the force the user's muscles can produce at a given instant of time as well as the resulting KneeFE moments. When the subject's muscle force predicted from EMGs are not sufficient to produce enough joint moment, they can be conveniently amplified by a predefined support ratio to produce the user's target moment—that is, the KneeFE moment that the subject's muscles should produce to properly execute the desired movement at the given instant of time. In parallel to this pathway, the current net joint moment produced by the human–machine system is evaluated. This represents the total joint moment contribution produced by the user wearing the powered orthosis and therefore accounts for (a) the motor support given by the orthosis, (b) the effort produced by the user, and (c) the effect of the external forces, such as gravity. The orthosis controller is then fed with the difference between the user's target moment and the current human–machine net moment. This difference defines the amount of extra joint moment the orthosis actuation system should add, during the next time frame, to obtain the target moment at the knee joint. One cycle of this loop is executed within the muscle's electromechanical delay (EMD), which is estimated to be around 10 ms in the lower extremity muscles [37]. The EMD represents the time delay between the EMG onset and the muscle force onset. Setting the real-time deadline to match the physiological EMD allows synchronizing the subject's actual muscle contraction with the device actuation enabling the device to become a natural extension of the subject's neuromuscular skeletal system [10,20].

Figure 20.7 shows results from a stair climbing experiment. In this, the KneeFE moment curves and joint trajectories are shown. The powered orthosis contribution gradually increases with the support ratio. With a ratio of 0.5, the orthosis support results in a decreased user's effort with no substantial changes in the resulting knee join trajectories with respect to the case during which no support was provided (i.e., support ratio = 0.0). For a ratio of 1.0, the joint moment produced by the subject was equally balanced by that contributed by the powered orthosis. However, the joint angle trajectories showed several high-frequency oscillations, that is, at $t \approx 3.0$ s and $t \approx 7.5$ s. These observed oscillations and instabilities result from baseline muscle activity being amplified by the high support ratio. In these gait phases, the instrumented leg is not in direct contact with the ground. Therefore, small variations in the feed-forward KneeFE torque lead to large knee joint accelerations. The same variations have less effect during stance phases with foot–ground contact such as push-up (Fig. 20.7).

It is worth noting that human–machine interfaces that estimate the user's strength solely by detecting the onset of measured external forces (i.e., joint moment, angle, or human–machine contact force) cannot produce any support until the subject's has actually moved and produced detectable external forces. In this scenario, the benefit of having a powered orthosis control system based on an EMG-driven model is that the user's effort can be estimated even if the subject cannot move to begin with. The EMG signals can be recorded and evaluated even if the resulting muscle forces are not

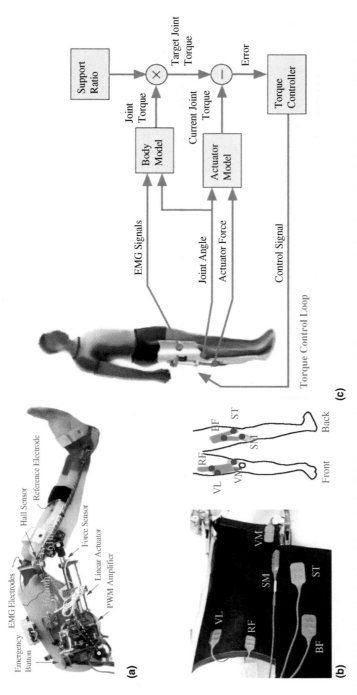

FIGURE 20.6 (**A**) The knee joint orthosis. This is powered by an electric linear actuator attached on the thigh and shank segments. (**B**) Attachment of electrodes to the soft tissue of the orthosis for the rectus femoris (RF), vastus medialis (VM), vastus lateralis (VL), semimembranosus (SM), semitendinosus (ST), and biceps femoris (BF) (*left*). Approximate sensor placement of the powered orthosis system. The body model (i.e., EMG-driven model of the knee joint) computes the knee joint flexion–extension moment (Joint Torque) that the operator's muscles are producing by evaluation of the EMG signals. This torque is multiplied by a support ratio and passed to the torque control loop that controls the actuator to produce the desired torque. Reproduced from Fleischer et al. [20] with permission.

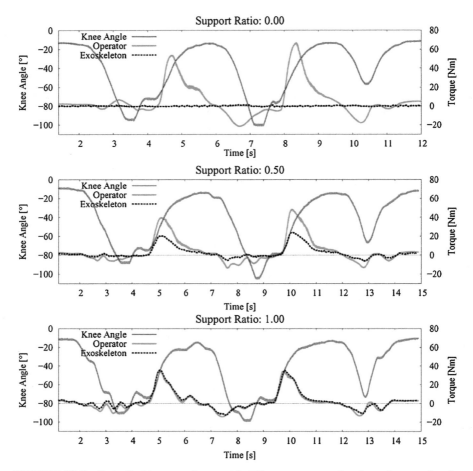

FIGURE 20.7   Stair climbing experiment with different support ratios. It can be seen that the user can reduce the knee joint torque contribution as the powered orthosis support increases. Refer to the text for details on the interpretation. Reproduced from Fleischer et al. [20] with permission.

sufficient to produce detectable movement by external sensors. An additional benefit provided by EMG-driven modeling is that it allows implementing control strategies in which the subject is supported as much as it is needed. That is, in order to receive support, the subject is motivated to activate their muscles and produce EMG signals in their muscles, thus being actively involved in the human–machine control loop. This is crucial to stimulate neural plasticity and obtain a more effective rehabilitation process [22].

The ability of accurately predicting the force produced by the muscles crossing a given joint is also crucial for the determination of how humans modulate the joint stiffness according to the demand of the locomotion task and terrain [39,44,48]. Joint stiffness modulation happens subconsciously through activation-dependent changes in intrinsic MTU properties [16,28,39,44]. Individuals wearing assistive devices are often challenged when locomoting across different terrains. This results from the inability of current wearable orthotic technology to modulate joint viscoelasticity physiologically and properly adapt to the mechanical demand underlying the performed motor task or terrain. With the modern actuation technology, the mechanisms underlying the physiological stiffness modulation could be effectively integrated in powered orthotic control systems [26,39].

In this scenario, EMG-driven modeling is crucial to enable such an application because it allows muscle co-activation and co-contraction, which are commonly observed in dynamic locomotion tasks and predominantly contribute to modulate joint viscoelasticity. In this case, an EMG-driven model can be used to estimate MTU force, which can be then converted into MTU stiffness as well as in net joint stiffness [44].

Figure 20.8 (right-hand graph) shows the force–length–velocity–activation space for the soleus muscle fibers created from a subject-specific EMG-driven model for a given value of fiber velocity. Given instantaneous muscle fiber states (i.e., muscle activation, fiber length, and contraction velocity), it is possible to compute the directional derivative of fiber force along the fiber length axis. This reflects the stiffness of the muscle fibers [28,44]. Figure 20.8 (left-hand graph) also shows the instantaneous net stiffness reconstructed about the ankle plantar-dorsi flexion (AnkleFE) DOF from the individual muscle fiber stiffness and series elastic tendon stiffness from seven MTUs, including soleus, gastrocnemius medialis, gastrocnemius lateralis, peroneus longus, brevis, tertius, and tibialis anterior.

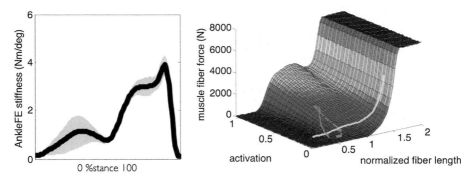

FIGURE 20.8 (**Right**) Instantaneous force–length–activation surface (i.e., 3D colored surface) of the soleus muscle for a given value of fiber contraction velocity, with projected fiber force-length work loop (i.e., red curve) relative to one single walking stance phase. The yellow curve is relative to a given surface cross section for a given level of muscle activation level. Given the muscle fiber states (i.e., activation, length, and contraction velocity) fiber stiffness is computed as the directional derivative of the force surface along the fiber length axis [44]. (**Left**) EMG-driven model-based estimation of ankle plantar-dorsi flexion (AnkleFE) stiffness averaged across six locomotion trials performed by one healthy male subject (age: 27 years, height: 1.82 m, weight: 67 kg). The thick black line represents mean stiffness, whereas the shaded area represents ± one standard deviation. In this, the stance phase is intended with 0% corresponding to the heel-strike event and 100% corresponding to the toe-off event [44].

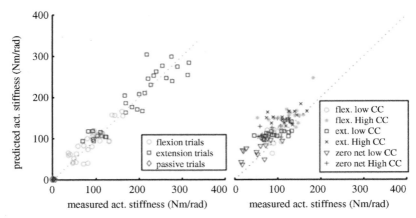

FIGURE 20.9   Active isometric stiffness prediction accuracy based on EMG for conditions without (left) and with (right) co-contraction. The dashed line in both figures represents the unity line. Each data point represents a trial. Reproduced from Pfeifer et al. [39] with permission.

Figure 20.9 shows EMG-based predictions of knee joint stiffness being in agreement with experimental joint stiffness recorded in conditions both with and without co-contraction (i.e., see Pfeifer et al. [39]). The EMG-based predictions were in general slightly higher than the experimental values. Nevertheless, results demonstrate the ability of EMG-driven modeling to predict joint stiffness in the intact human. Data are reported for both active and passive isometric flexion and extension trials executed by five healthy male subjects (age: 27–31 years, height: 175–193 cm, weight: 71–89 kg). Experimental joint stiffness data were obtained using conventional perturbation techniques [39]. EMG signals were recorded from seven muscles including: vastus medialis, vastus lateralis, rectus femoris, lateral hamstrings, medial hamstrings, gastrocnemius medialis, and gastrocnemius lateralis.

20.5   NEUROMODULATION: EMG-DRIVEN ELECTRICAL STIMULATION AND REHABILITATION ROBOTICS

Neuromodulation consists in modifying neural circuitries on the basis of the plasticity of the central nervous system. Surface EMG can be used in neuromodulation devices as a signal indicating the motor command. For example, in functional electrical therapy (FET), the muscles are activated by electrical stimulation of the innervating nerves or motor points and the stimulation is triggered by the presence of residual EMG activity. This triggering closes the loop between the motor intention and the consequent afferent volley that is enhanced by the electrical stimuli. The reestablished closed loop not only supports a function by the muscle activation but also induces cortical plasticity [4] beneficial for motor recovery [40].

A specific application of EMG-triggered electrical stimulation is the suppression of pathological tremor. In this case, the tremor phase and magnitude is estimated from

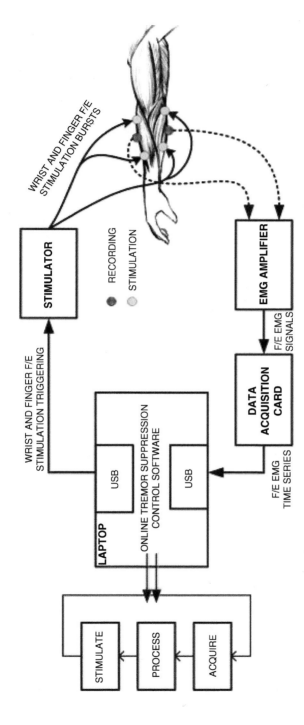

FIGURE 20.10 Components of an EMG-driven system for pathological tremor suppression by peripheral stimulation: EMG amplifier, USB data acquisition card, standard laptop with custom-made control software, and programmable stimulator. The EMG from the wrist flexor and extensor muscles is recorded and processed to estimate the tremorogenic activation according to Dosen et al. [14]. The electrical stimulation is delivered to the same muscles, out of phase with the tremorogenic bursts and below the motor threshold (afferent stimulation). Reproduced from Dosen et al. [14] with permission.

the analysis of the EMG by separating tremor and voluntary muscle activity [12]. The stimulation is then delivered out of phase with respect to the extracted tremor oscillations, which decreases the spinal gain in tremor amplification [14]. Figure 20.10 shows a block diagram of this EMG-driven system for neuromodulation.

A similar concept such as EMG-triggered stimulation is applied in rehabilitation robotics [11,13,51,52,54]. In this case, the patient's limbs are mobilized by the use of robots. The movement is assisted if the patient's EMGs have certain characteristics, such as being above a certain amplitude threshold. Current state-of-the-art methodologies set the robot support level based on simple EMG features such as the root mean squared value. In this context, the ability of predicting physiological variables such as muscle and joint dynamics (as in Chapter 9) can result in advanced robot-based therapies that can target the recovery of specific musculoskeletal features.

20.6 CONCLUSIONS

Man–machine interfaces for neurorehabilitation consist in connecting the patient's nervous system with external assistive devices. These devices may replace, restore, or neuromodulate impaired motor functions. The interfacing can be established at various levels of the central nervous system, but viable clinical approaches on a large scale are only currently possible by using muscle recordings. These reflect the neural activity sent to the muscles from the spinal cord output circuitries. Therefore, muscles can be considered, in the context of this chapter, as biological amplifiers of nerve activity. This chapter provided representative examples of prosthetics, orthotics, and electrical stimulation systems based on EMG signals used for extracting a variety of control commands for assistive devices of upper and lower extremities. The information conveyed by the EMG can be extracted and processed based either on the available data or on models built from our anatomical and physiological knowledge of EMG generation and musculoskeletal force production. The ability of characterizing how EMG features evolve in a context-dependent manner (i.e., as a function of muscle fatigue, training, impairment) will open up new avenues for establishing robust interfaces between the patient and the machine.

ACKNOWLEDGMENT

Support for this work was provided by the European Research Council (ERC) via the ERC Advanced Grant DEMOVE (267888).

REFERENCES

1. Amsuess, S, Gobel P, Graimann B, Farina D. "A Multi-Class Proportional Myocontrol Algorithm for Upper Limb Prosthesis Control: Validation in Real-Life Scenarios on Amputees," *IEEE Trans Neural Syst Rehabil Eng,* **23**, 825–836 (2014)

2. Amsuess, S, Vujaklija I, Gobel P, Roche A, Graimann B, Aszmann O, Farina D. "Context-Dependent Upper Limb Prosthesis Control for Natural and Robust Use,". *IEEE Trans Neural Syst Rehabil Eng,* In Press, 2015.

3. Amsuess, S., L. P. Paredes, N. Rudigkeit, B. Graimann, M. J. Herrmann, and D. Farina, "Long term stability of surface EMG pattern classification for prosthetic control," in *Proceedings of the Annual International Conference of the IEEE Engineering in Medicine and Biology Society, EMBS,* pp. 3622–3625, 2013.

4. Barsi, G. I., D. B. Popovic, I. M. Tarkka, T. Sinkjær, and M. J. Grey, "Cortical excitability changes following grasping exercise augmented with electrical stimulation," *Exp. Brain Res.* **191**, 57–66 (2008).

5. Besier, T. F., M. Fredericson, G. E. Gold, G. S. Beaupré, and S. L. Delp, "Knee muscle forces during walking and running in patellofemoral pain patients and pain-free controls," *J. Biomech.* **42**, 898–905 (2009).

6. Biddiss, E. A., and T. T. Chau, "Upper limb prosthesis use and abandonment, a survey of the last 25 years," *Prosthet. Orthot. Int.* **31**, 236–257 (2007).

7. Blana, D., J. G. Hincapie, E. K. Chadwick, and R. F. Kirsch, "A musculoskeletal model of the upper extremity for use in the development of neuroprosthetic systems," *J. Biomech.* **41**, 1714–1721 (2008).

8. Buchanan, T. S., D. G. Lloyd, K. Manal, and T. F. Besier, "Neuromusculoskeletal modeling, estimation of muscle forces and joint moments and movements from measurements of neural command," *J. Appl. Biomech.* **20**, 367–395 (2004).

9. Cavallaro, E., J. Rosen, J. C. Perry, S. Burns, and B. Hannaford, "Hill-based model as a myoprocessor for a neural controlled powered exoskeleton arm—parameters optimization," in *Proceedings of the 2005 IEEE International Conference on Robotics and Automation, ICRA,* pp. 4514–4519, 2005.

10. Cavallaro, E. E., J. Rosen, J. C. Perry, and S. Burns, "Real-time myoprocessors for a neural controlled powered exoskeleton arm," *IEEE Trans. Biomed. Eng.* **53**, 2387–2396 (2006).

11. Cesqui, B., P. Tropea, S. Micera, and H. I. Krebs, "EMG-based pattern recognition approach in post stroke robot-aided rehabilitation, a feasibility study," *J. Neuroeng. Rehabil.* **10**, 75 (2013).

12. Dideriksen, J. L., F. Gianfelici, L. Z. P. Maneski, and D. Farina, "EMG-based characterization of pathological tremor using the iterated Hilbert transform," *IEEE Trans. Biomed. Eng.* **58**, 2911–2921 (2011).

13. Dipietro, L., M. Ferraro, J. J. Palazzolo, H. I. Krebs, B. T. Volpe, and N. Hogan, "Customized interactive robotic treatment for stroke, EMG-triggered therapy," *IEEE Trans. Neural Syst. Rehabil. Eng.* **13**, 325–334 (2005).

14. Dosen, S., S. Muceli, J. Dideriksen, J. Romero, E. Rocon, J. Pons, and D. Farina, "Online tremor suppression using electromyography and low level electrical stimulation," *IEEE Trans. Neural Syst. Rehabil. Eng.* **12**, 1–11 (2014).

15. Eilenberg, M. F., H. Geyer, and H. Herr, "Control of a powered ankle–foot prosthesis based on a neuromuscular model," *IEEE Trans. Neural Syst. Rehabil. Eng.* **8**, 164–173 (2010).

16. Enoka, R. M., *Neuromechanics of Human Movement*, 4th ed., Human Kinetics Publishers, Champaign, IL, pp. 1–560 2008.

17. Farina, D., and O. Aszmann, "Bionic Limbs: Clinical Reality and Academic Promises," *Sci. Transl. Med.* **6,** 257ps12 (2014).

18. Farina, D., N. Jiang, H. Rehbaum, A. Holobar, B. Graimann, H. Dietl, and O. C. Aszmann, "The extraction of neural information from the surface EMG for the control of upper-limb prostheses, emerging avenues and challenges," *IEEE Trans. Neural Syst. Rehabil. Eng.* **22**, 797–809 (2014).

19. Farina, D., and F. Negro, "Accessing the neural drive to muscle and translation to neurorehabilitation technologies," *IEEE Rev. Biomed. Eng.* **5**, 3–14 (2012).

20. Fleischer, C., and G. Hommel, "A human–exoskeleton interface utilizing electromyography," *IEEE Trans. Robot.* **24**, 872–882 (2008).

21. Gerus, P., M. Sartori, T. F. Besier, B. J. Fregly, S. L. Delp, S. A. Banks, M. G. Pandy, D. D. D'Lima, and D. G. Lloyd, "Subject-specific knee joint geometry improves predictions of medial tibiofemoral contact forces," *J. Biomech.* **46**, 2778–2786 (2013).

22. Gordon, K. E., and D. P. Ferris, "Learning to walk with a robotic ankle exoskeleton," *J. Biomech.* **40**, 2636–2644 (2007).

23. Hahne, J. M., F. Bießmann, N. Jiang, H. Rehbaum, S. Member, D. Farina, F. C. Meinecke, K. R. Muller, and L. C. Parra, "Linear and nonlinear regression techniques for simultaneous and proportional myoelectric control," *IEEE Trans. Neural Syst. Rehabil. Eng.* **22**, 269–279 (2014).

24. Jiang, N., S. Dosen, K. R. Muller, and D. Farina, "Myoelectric control of artificial limbs, is there a need to change focus?" [In the spotlight], *IEEE Signal Process. Mag.* **29**, 150–152 (2012).

25. Jiang, N., K. B. Englehart, and P. A. Parker, "Extracting simultaneous and proportional neural control information for multiple-dof prostheses from the surface electromyographic signal," *IEEE Trans. Biomed. Eng.* **56**, 1070–1080 (2009).

26. Karavas, N., A. Ajoudani, N. Tsagarakis, J. Saglia, A. Bicchi, and D. Caldwell, "Tele-impedance based stiffness and motion augmentation for a knee exoskeleton device," in *Proceedings of 2013 IEEE International Conference on Robotics and Automation (ICRA)*, pp. 2186–2192, 2013.

27. Kuiken, T. A., L. A. Miller, R. D. Lipschutz, B. A. Lock, K. Stubblefield, P. D. Marasco, P. Zhou, and G. A. Dumanian, "Targeted reinnervation for enhanced prosthetic arm function in a woman with a proximal amputation, a case study," *Lancet* **369**, 371–380 (2007).

28. Latash, M. L., and V. M. Zatsiorsky, "Joint stiffness, myth or reality?," *Hum. Mov. Sci.* **12**, 653–692 (1993).

29. Lee, D. D., and H. S. Seung, "Learning the parts of objects by non-negative matrix factorization," *Nature* **401**, 788–791 (1999).

30. Liu, L., P. Liu, E. A. Clancy, E. Scheme, and K. B. Englehart, "Electromyogram whitening for improved classification accuracy in upper limb prosthesis control," *IEEE Trans. Neural Syst. Rehabil. Eng.* **21**, 767–774 (2013).

31. Lloyd, D. G., and T. F. Besier, "An EMG-driven musculoskeletal model to estimate muscle forces and knee joint moments in vivo," *J. Biomech.* **36**, 765–776 (2003).

32. Markowitz, J., P. Krishnaswamy, M. F. Eilenberg, K. Endo, C. Barnhart, and H. Herr, "Speed adaptation in a powered transtibial prosthesis controlled with a neuromuscular model," *Philos. Trans. R. Soc. Lond. B. Biol. Sci.* **366**, 1621–1631 (2011).

33. Matrone, G. C., C. Cipriani, M. C. Carrozza, and G. Magenes, "Real-time myoelectric control of a multifingered hand prosthesis using principal components analysis," *J. Neuroeng. Rehabil.* **9**, 40 (2012).

34. Merletti, R., A. Rainoldi, and D. Farina, "Surface electromyography for noninvasive characterization of muscle," *Exerc. Sport Sci. Rev.* **29**, 20–25 (2001).

35. Muceli, S., N. Jiang, and D. Farina, "Extracting signals robust to electrode number and shift for online simultaneous and proportional myoelectric control by factorization algorithms," *IEEE Trans. Neural Syst. Rehabil. Eng.* **3**, 623–633 (2013).

36. Nakamura, Y., K. Yamane, Y. Fujita, and I. Suzuki, "Somatosensory computation for man–machine interface from motion-capture data and musculoskeletal human model," *IEEE Trans. Robot.* **21**, 58–66 (2005).

37. Nordez, A., T. Gallot, S. Catheline, A. Guével, C. Cornu, and F. Hug, "Electromechanical delay revisited using very high frame rate ultrasound," *J. Appl. Physiol.* **106**, 1970–1975 (2009).

38. Parker, P., K. Englehart, and B. Hudgins, "Myoelectric signal processing for control of powered limb prostheses," *J. Electromyogr. Kinesiol.* **16**, 541–548 (2006).

39. Pfeifer, S., H. Vallery, M. Hardegger, R. Riener, and E. J. Perreault, "Model-based estimation of knee stiffness," *IEEE Trans. Biomed. Eng.* **59**, 2604–2612 (2012).

40. Popović, D. B., and T. Sinkjær, "Neuromodulation of lower limb monoparesis, functional electrical therapy of walking," *Acta Neurochir. Suppl.* **94**, 387–393 (2007).

41. Pozzo M, Farina D, and Merletti R, "Electromyography: detection, processing and applications," in Handbook of biomedical technology and devices, Moore JE and Zouridakis G, Eds. CRC Press, 2003.

42. Sartori M, Farina D, Lloyd DG. "Hybrid neuromusculoskeletal modeling to best track joint moments using a balance between muscle excitations derived from electromyograms and optimization". *J Biomech.* **47**, 3613–3621 (2014).

43. Sartori M, Gizzi L, Lloyd DG, Farina D. "A musculoskeletal model of human locomotion driven by a low dimensional set of impulsive excitation primitives". *Front Comput Neurosci* **7**, 79 (2013).

44. Sartori M, Maculan M, Claudio P, Reggiani M, Farina D. "Modeling and simulating the neuromuscular mechanisms regulating ankle and knee joint stiffness during human locomotion". *Journal of Neurophysiology,* **DOI**:10.1152/jn.00989.2014 (2015).

45. Sartori M, Reggiani M, Farina D, Lloyd DG. "EMG-driven forward-dynamic estimation of muscle force and joint moment about multiple degrees of freedom in the human lower extremity". *PLoS One,* **7**, 1–11 (2012).

46. Sartori M, Reggiani M, Lloyd DG. "A neuromusculoskeletal model of the human lower limb: Towards EMG-driven actuation of multiple joints in powered orthoses". *In Proceedings of the 2011 IEEE International Conference on Rehabilitation Robotics (ICORR),* 2011, pp. 709–14.

47. Sartori M, Reggiani M, Pagello E, Lloyd DG. "Modeling the human knee for assistive technologies". *IEEE Trans Biomed Eng.* **59**, 2642–9 (2012).

48. Sartori M, Reggiani M, van den Bogert AJ, Lloyd D. "Estimation of musculotendon kinematics in large musculoskeletal models using multidimensional B-splines". *Journal of Biomechanics.* **45**, 595–601 (2012).

49. Scheme E, Fougner A, Stavdahl, Chan ADC, Englehart K. "Examining the adverse effects of limb position on pattern recognition based myoelectric control". *In Proceedings of the 2010 Annual International Conference of the IEEE Engineering in Medicine and Biology Society, EMBC.* 2010, pp. 6337–40.

50. Shamaei K, Sawicki GS, Dollar AM. "Estimation of Quasi-Stiffness and Propulsive Work of the Human Ankle in the Stance Phase of Walking". *PLoS ONE,* **8**:e59935 (2013).

51. Song R, Tong K, Hu X, Zhou W. "Myoelectrically controlled wrist robot for stroke rehabilitation". *J Neuroeng Rehabil.* **10**, 52 (2013).

52. Stein J, Narendran K, McBean J, Krebs K, Hughes R. "Electromyography-controlled exoskeletal upper-limb-powered orthosis for exercise training after stroke". *Am J Phys Med Rehabil.* **86**(4), 255–61 (2007).

53. Valero-Cuevas FJ, Hoffmann H, Kurse MU, Kutch JJ, Theodorou EA. Computational Models for Neuromuscular Function. *IEEE Rev Biomed Eng.* **2**, 110–35 (2009).

54. Vivian M, Tagliapietra L, Reggiani M, Farina D, Sartori M. "Design of a subject-specific EMG model for rehabilitation movement". *In: Jensen W, Andersen OK, Akay M, eds. Replace, Repair, Restore, Relieve – Bridging Clinical and Engineering Solutions in Neurorehabilitation.* Berlin: Springer International Publishing; 2014, pp. 813–22.

INDEX

Surface Electromyography: Physiology, Engineering, and Applications, First Edition.
Edited by Roberto Merletti and Dario Farina.
© 2016 by The Institute of Electrical and Electronics Engineers, Inc. Published 2016 by John Wiley & Sons, Inc.

IEEE Press Series in Biomedical Engineering

The focus of our series is to introduce current and emerging technologies to biomedical and electrical engineering practitioners, researchers, and students. This series seeks to foster interdisciplinary biomedical engineering education to satisfy the needs of the industrial and academic areas. This requires an innovative approach that overcomes the difficulties associated with the traditional textbooks and edited collections.

Series Editor: Metin Akay, University of Houston, Houston, Texas